Geology of Siliciclastic Shelf Seas

Geological Society Special Publications
Series Editor A. J. FLEET

GEOLOGICAL SOCIETY SPECIAL PUBLICATION NO. 117

Geology of Siliciclastic Shelf Seas

EDITED BY

M. DE BATIST & P. JACOBS
Renard Centre of Marine Geology,
University of Gent, Belgium

1996

Published by

The Geological Society

London

THE GEOLOGICAL SOCIETY

The Society was founded in 1807 as The Geological Society of London and is the oldest geological society in the world. It received its Royal Charter in 1825 for the purpose of 'investigating the mineral structure of the Earth'. The Society is Britain's national society for geology with a membership of around 8000. It has countrywide coverage and approximately 1000 members reside overseas. The Society is responsible for all aspects of the geological sciences including professional matters. The Society has its own publishing house, which produces the Society's international journals, books and maps, and which acts as the European distributor for publications of the American Association of Petroleum Geologists, SEPM and the Geological Society of America.

Fellowship is open to those holding a recognized honours degree in geology or cognate subject and who have at least two years' relevant postgraduate experience, or who have not less than six years' relevant experience in geology or a cognate subject. A Fellow who has not less than five years' relevant postgraduate experience in the practice of geology may apply for validation and, subject to approval, may be able to use the designatory letters C. Geol. (Chartered Geologist).

Further information about the Society is available from the Membership Manager, The Geological Society, Burlington House, Piccadilly, London W1V 0JU, UK. The Society is a Registered Charity, No. 210161.

Published by The Geological Society from:
The Geological Society Publishing House
Unit 7, Brassmill Enterprise Centre
Brassmill Lane
Bath BA1 3JN
UK
(*Orders*: Tel. 01225 445046
Fax 01225 442836)

First published 1996

The publishers make no representation, express or implied, with regard to the accuracy of the information contained in this book and cannot accept any legal responsibility for any errors or omissions that may be made.

British Library Cataloguing in Publication Data
A catalogue record for this book is available from the British Library.

ISBN 1-897799-67-5
ISSN 0305-8719

Typeset by Aarontype Limited,
Easton, Bristol, UK

Printed by The Alden Press,
Osney Mead, Oxford, UK

Distributors

USA
AAPG Bookstore
PO Box 979
Tulsa
OK 74101-0979
USA
(*Orders*: Tel. (918) 584-2555
Fax (918) 560-2652)

Australia
Australian Mineral Foundation
63 Conyngham Street
Glenside
South Australia 5065
Australia
(*Orders*: Tel. (08) 379-0444
Fax (08) 379-4634)

India
Affiliated East-West Press PVT Ltd
G-1/16 Ansari Road
New Delhi 110 002
India
(*Orders:* Tel. (11) 327-9113
Fax (11) 326-0538)

Japan
Kanda Book Trading Co.
Tanikawa Building
3-2 Kanda Surugadai
Chiyoda-Ku
Tokyo 101
Japan
(*Orders*: Tel. (03) 3255-3497
Fax (03) 3255-3495)

Contents

New techniques in continental shelf research

Preface

Continental shelves separate the continents from the world's oceans, and if humans are to further exploit or make sensible use, or eventually inhabit, the Earth's oceans and seas, the continental shelves are the places to start from. A good knowledge of the structural and stratigraphical geology of these shelves will help in the prediction of *loci* of potential exploitable resources (gravel and sand accumulations, shelf sand bodies and incised valley fills as modern analogues for hydrocarbon reservoirs, marine placers...). It will also allow prediction of the short-, medium- and possibly even long-term behaviour and stability of the shelves, and applications related to offshore construction, harbour or recreational development projects. Better insight into the sediment and morphodynamical processes acting on shelves in relation to meteo-oceanographical factors will allow prediction of coastal evolution and the fates of major transportation routes and of fishing areas, and the definition of major sediment transport pathways. Problems related to geotechnical and environmental issues have become increasingly relevant for the continental shelf environment and have initiated new pulses of technological development.

In this volume, we have aimed to present a selection of some of the recent research activities and developments in the field of continental shelf geology. Most papers are European, and often reflect cooperative research work carried out in the framework of the EC-financed Marine Science and Technology Program (MAST), but there are also contributions from the US, Canada and South America.

The chapters in this volume are organised around four major themes: 1, stratigraphy and sedimentary geology of siliciclastic shelves; 2, modern siliciclastic shelves; 3, nearshore and coastal environments; 4, new techniques in continental shelf research.

Stratigraphy and sedimentary geology of siliciclastic shelves

This section addresses the reconstruction from the geological record of sedimentary and geological processes which affect continental shelves.

The first three papers in this section essentially discuss the same concept of the complex facies and sequence architecture in epicontinental basins – in a so-called ramp-type setting – taking examples from the North Sea Basin. **Michelsen & Danielsen** use multi-channel seismic and well-log data to establish a sequence stratigraphical model for the Oligocene deposits in the Danish North Sea, and clearly illustrate the resolution-related limitations of the seismic tools, the usefulness of the well-log data and the characteristic facies pattern within this type of deposits. The strata post-dating the mid-Miocene in the Danish North Sea are discussed by **Konradi**, who uses biostratigraphical arguments to reconstruct the evolution of the sedimentary environment through time. **Jacobs & De Batist** integrate high-resolution seismic and core data to develop a sequence stratigraphical and geometrical model for the Eocene deposits in the southernmost North Sea Basin, thereby highlighting some characteristic facies patterns occurring in these ramp-type shelf deposits.

Mellere & Steel present a detailed analysis of sedimentary facies associations within the Mid-Jurassic Bearreraig Sandstone Formation (Scotland), which they use to reconstruct the tectonically complex palaeogeographic setting of this shallow-marine environment. The sedimentary facies of Late Cretaceous to Early Tertiary deposits from Patagonia (Argentina) are analysed by **Spalletti**, and used to reconstruct the evolution of the depositional environment.

Modern siliciclastic shelves: architecture, sea level, tectonics and sediment supply

This section addresses the architecture and recent evolution of siliciclastic shelves as related to changes in sea level, tectonic evolution and sediment supply. A wide range of continental shelf environments, both tectonically active and passive and with different depositional regimes, are considered in this section.

Anderson *et al.* present a compilation of a large amount of high-resolution seismic and core data from the East Texas shelf to illustrate the relative role of sediment supply, rapid changes in sea level, shelf gradient and tectonics, and of autocyclic phenomena in controlling the overall packaging of facies into systems tracts for the Pleistocene to Recent evolution of the area. **Ercilla & Alonso** and **Hernandez-Molina *et al.*** present integrated continental shelf and margin studies of the Spanish Mediterranean realm, highlighting the influence of tectonic style of the margin, of the typical rapid changes in sea level during the Quaternary, and of sediment supply on the evolution of the considered margin segments. **Correggiari *et al.*** discuss the Late Quaternary evolution of the epicontinental Adriatic Sea and how large bedforms can be used to deduce changes in the oceanographical regime of the area in relation to changes in sea level. Two papers focus on the evolution of continental shelves and margins in high-latitude settings. **Bart & Anderson** clearly point out the complexity of the stratigraphical record of glacial activity on the Antarctic Peninsula continental shelf, and how this record should be investigated and interpreted for reconstructing glacial history. **Sejrup *et al.*** investigate the sediment fluxes – as well as the processes accounting for these fluxes and their timing – which pass from the Norwegian continental shelf, through fjords and the Norwegian Channel, towards the North Sea Fan. **Lericolais *et al.*** revises the tectonic and depositional setting of the Hurd Deep in the English Channel, using a variety of advanced geophysical investigation techniques.

Nearshore and coastal environments

This section deals mostly with sediment transport agents and processes and on coastal evolution studies in various settings. **Dominguez** discusses the relative importance of the processes shaping the depositional system off the São Francisco river in Brasil and uses his findings to comment on the classical delta classification systems. **Barrie & Conway** address the evolution of a nearshore macrotidal sand transport system in Canada and discuss in this context the importance of tectonic uplift. **Cleary *et al.*** present a well-documented study of the influence of the inherited geological framework, upon the evolution of shoreface system off the coast of North Carolina. A detailed study of the sediment transport processes affecting one of the beaches of the island of Norderney (Germany) is presented by **Eitner *et al.***

New techniques in continental shelf research

In this section, a number of new approaches and techniques are presented whose development has been triggered by the increasing interest in continental shelf studies. **De Meijer *et al.*** present the promising results of radiometrical techniques for the assessment of the selective transport of heavy and light minerals in coastal sands. Results of an integrated geophysical, geotechnical and geological study for determining the physical parameters of shallow-marine sediments from the Baltic Sea are presented by **Missiaen *et al.*** **Davis** reviews the role of various geophysical techniques for offshore site investigations, and comments on the advances and developments that these techniques have undergone in the past years.

Acknowledgements: This volume emanated from a three-day conference hosted by the Renard Centre of Marine Geology of the University of Gent on 24–26 May 1994. Exactly a hundred scientists from various European countries, from the US and from South America participated in this conference at which 47 talks and 28 posters were presented. Some of the presenters were invited to contribute to this special publication and we solicited further contributions to extend the coverage of the volume. Neither the conference nor this special publication would have been possible without the financial, logistic and moral support of the Geological Society of London and its Publishing House (D. Ogden, A. Hills), the Belgian National Fund for Scientific Research, GEOLAB, ASLK, the Instituut voor Zee Wetenschappelijk Onderzoek (IZWO) and the Management Unit of the Mathematical Model of the North Sea and the Scheldt Estuary (MUMM). We also wish to thank the following referees who devoted part of their valuable time to reviewing the manuscripts: J. R. L. Allen, J. B. Anderson, D. Ardus, S. Berné, G. Boillot, M. Collins, R. W. Dalrymple, A. Davis, R. A. Davis, G. De Moor, J. F. Donoghue, B. W. Flemming, P. Gayes, J. P. Henriet, E. C. Kosters, J. Luternauer, A. Maldonado, T. F. Moslow, M. Paul, G. Postma, H. Roberts, D. Rubin, R. Steel, E. Steurbaut, M. Stoker, J. P. M. Syvitski, J. Terwindt, B. Tessier, F. Trincardi, P. R. Vail, N. Vandenberghe, J. Verbeek, T. Vorren, R. G. Walker, J. Wehmiller and J. Wells.

M. De Batist is senior research assistant of the Belgian National Fund for Scientific Research.

M. De Batist and P. Jacobs
January 1996

Sequence and systems tract interpretation of the epicontinental Oligocene deposits in the Danish North Sea

O. MICHELSEN & M. DANIELSEN

Department of Earth Sciences, Aarhus University, C.F. Møllers Allé, DK-8000 Århus C, Denmark

Abstract: A sequence stratigraphical scheme has been established for the siliciclastic Cenozoic deposits in the southeastern North Sea Basin. The present paper addresses the problem of recognizing systems tracts by means of logs, using the Oligocene sequences as an example. The Oligocene sequences are characterized by an overall prograding seismic reflection pattern, and systems tracts cannot be interpreted from seismic sections alone.

The sequences are dominated by marine clay deposits, and abrupt changes in lithological facies are rarely seen. The gamma-ray log trends indicate retrograding, aggrading and prograding stacking patterns.

The lowstand deposits are identified as prograding sedimentary bodies. Coarse-grained sharp-based lowstand deposits are found basinwards of the depositional shoreline break of the preceding sequence, and further basinwards they are more fine-grained. Fan deposits are not recognized. A thin interval of transgressive deposits including upward fining clayey sediments is found at the top of the lowstand deposits and in the landward direction at the top of the older highstand deposits. The maximum flooding surface is identified by a high gamma-ray peak. The overlying highstand deposits thicken in the landward direction, and are here characterized by an upward coarsening trend. They thin in the basinward direction and become a condensed clay interval.

Deposition took place during a tectonically quiet period in an epicontinental basin with a gently southwestward dipping sea floor. The lowstand deposits comprise forced regressive deposits and prograding deposits, and a deep ramp model is, therefore, suggested for these deposits.

A depositional sequence is defined as a relatively conformable genetically related succession of strata bounded by unconformities or their correlative conformities (e.g. Posamentier *et al.* 1988; Van Wagoner *et al.* 1988, 1990). A sequence is interpreted as deposited in the period between two sea-level falls, and the bounding surfaces of the sequence are defined as unconformities created by subaerial exposure of the shelf during the sea-level falls. The boundary is thus associated with an abrupt change in the depositional environment, from a relatively deep-water environment below the boundary to shallower-water or non-marine conditions above. A sequence is divided into three systems tracts: the lowstand (or shelf margin), transgressive and highstand systems tracts. The smallest unit in the sequence is the parasequence, which is a natural succession of beds showing an upward shallowing trend. The parasequences are separated by flooding surfaces. A number of parasequences can be grouped into a parasequence set with either a prograding, aggrading, or retrograding trend.

In addition to the transgressive and highstand systems tracts, Hunt & Tucker (1992) suggest the forced regressive wedge systems tract and the lowstand prograding wedge systems tract, which correspond to the above-mentioned lowstand systems tract. The sequence boundary is located between these two systems tracts, at the top of the forced regressive wedge (Hunt & Tucker 1992; Helland-Hansen & Gjelberg 1994).

The present paper addresses the problem of recognizing sequence stratigraphical surfaces and systems tracts by means of conventional seismic sections and petrophysical logs from deep wells. In the marine siliciclastic Cenozoic deposits in the southeastern part of the epicontinental North Sea Basin, the systems tracts cannot be recognized on seismic sections, possibly because they are below seismic resolution. Variations in the log patterns indicate changes in the stacking pattern of the marine sediments. A subdivision into systems tracts based on logs, primarily the gamma-ray log, is therefore the theme of the present paper. The systems tracts are identified on the gamma-ray log as intervals with upward decreasing, constant or upwards increasing gamma-ray values. These log trends are interpreted as an upward increase, constant or upward decrease in grain-size, corresponding to prograding, aggrading, or retrograding systems, respectively. Parasequences are below seismic resolution, but may be identified by the gamma-ray log trends.

From De Batist, M. & Jacobs, P. (eds), 1996, *Geology of Siliciclastic Shelf Seas*, Geological Society Special Publication No. 117, pp. 1–13.

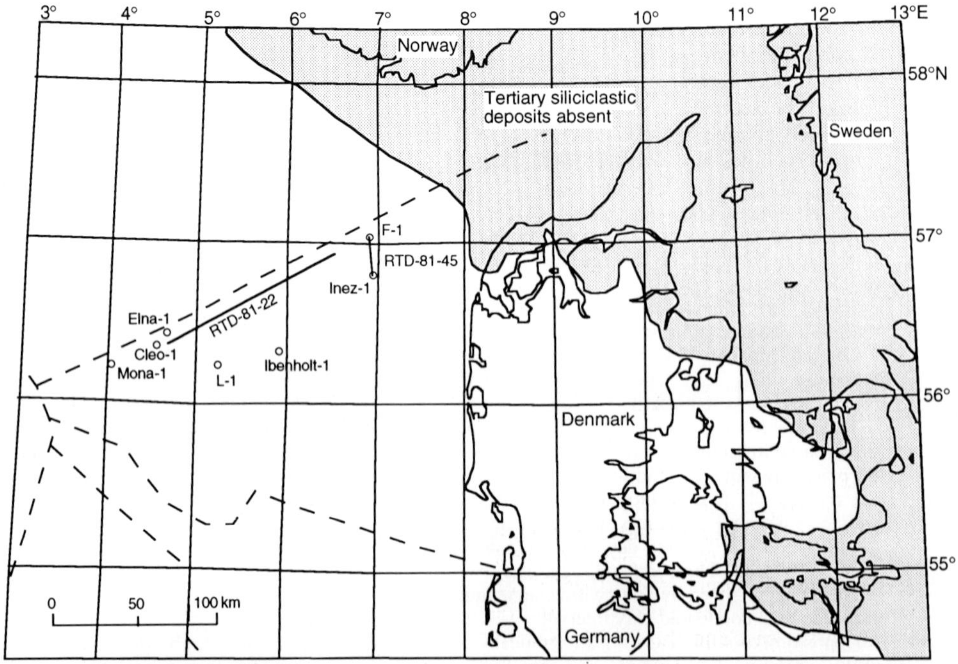

Fig. 1. Map of the southeastern North Sea area showing the present distribution of the siliciclastic Tertiary deposits. The study area and the locations of wells and seismic sections discussed in this paper are indicated.

An integrated stratigraphical study of the Cenozoic deposits in the southeastern North Sea has been carried out earlier, comprising analyses of 20 000 km of seismic sections, petrophysical logs from 76 wells, well samples, and biostratigraphical studies (foraminifera, dinoflagellates and calcareous nannofossils) on samples from 14 wells (Michelsen *et al.* 1996). The study area comprised the Danish North Sea sector and the adjacent Norwegian, German, and Dutch sectors. The study area of the present paper is restricted to the northern part of the Danish North Sea (Fig. 1).

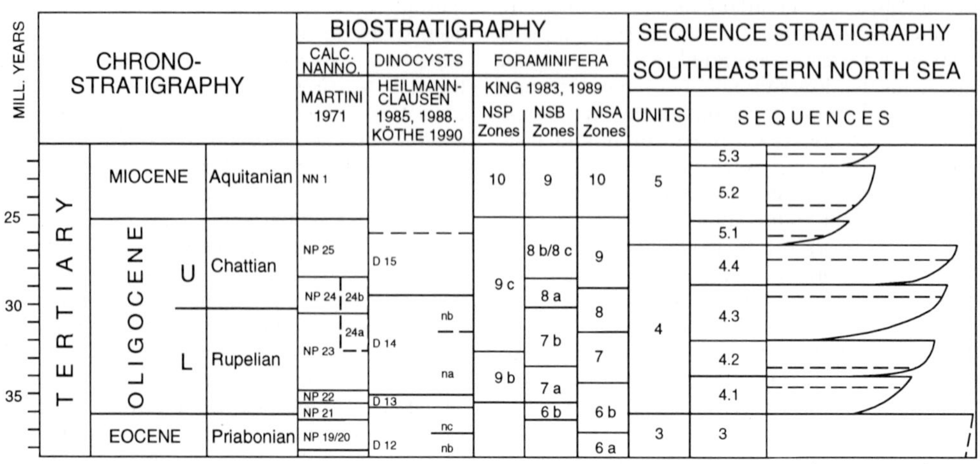

Fig. 2. Stratigraphical scheme for the Oligocene sequences in the southeastern North Sea (modified from Michelsen *et al.* 1996, Fig. 24). The applied biostratigraphical zonation is shown.

Twenty-one sequences were defined within the respective depocentres. In some cases, it was difficult to correlate them across larger areas, because the sequences locally are beyond seismic resolution. The sequences were, therefore, grouped into seven major sequence stratigraphical units. Unit 4 includes most of the Oligocene deposits (Fig. 2). The boundaries of this unit are easily recognized on seismic sections in the southeastern North Sea. Unit 4 is subdivided into four sequences, 4.1 to 4.4, on the basis of seismic sections and petrophysical logs from the depocentre of unit 4.

Oligocene deposits

Basinal setting

During the Cenozoic, the North Sea region constituted a large epicontinental basin with a north–south axis above the older Central Trough structures (Nielsen *et al.* 1986). The basin was flanked by the positive areas of Scandinavia to the east and the British Isles to the west. The Cenozoic deposits in the central part of the North Sea Basin reach a thickness of more than 3000 m representing most of the erathem. Clay and silty clay, deposited in a sublittoral to upper bathyal environment, constitute the major part of the succession in the North Sea.

The Danish North Sea was located in the central part of the basin during the Late Palaeocene and Eocene, and a hemipelagic sedimentation dominated. The thickness of the Upper Palaeocene is here less than 50 m often 15–20 m thick. The seismic reflection pattern is characterized by strong concordant reflections. The log motif of the deposits has an equal appearance in most of the wells, indicating only minor lateral variations in lithology.

The Upper Palaeocene and lowermost Eocene succession comprises (from below) marl, a non-calcareous, dark grey, silty clay, a smectite-rich, greenish and reddish clay, a finely laminated, silty clay with a high content of organic matter, and (uppermost) the volcanic ash series. The overlying Eocene deposits consist of fine clay, being reddish in the lower part and greenish in the upper part. The thickness is less than 50 m in the major part of the Danish area. A continuous seismic reflection marks the top of the unit in the Danish sector. These Upper Palaeocene and Eocene deposits constitute the distal and condensed part of the sequences having their depocentres north and west of the Danish North Sea.

The wide distribution and rather uniform thicknesses of the Upper Palaeocene and Eocene deposits in the southeastern North Sea indicate a basin with an even and gently dipping sea floor. The Cenozoic deposits are largely unfaulted. Minor disturbances are observed along the pre-Cenozoic fault zones in the Central Trough area. Small-scale faulting and episodes of inversion are referred to differential subsidence and inversion along these older fault trends (Clausen & Korstgård 1993*a*, *b*). The depocentres of the Oligocene sequences are located east of the Central Trough area, and the sequences extend into this area as condensed deposits. The major part of the Oligocene sediments were deposited in a tectonically quiet ramp-like basin, and the onset of the late Cenozoic uplift of the Scandinavian region (Doré 1992) may have caused a south- and westward dipping sea floor in the eastern part of the basin.

Distribution and lithology

The base of the Oligocene deposits marks the beginning of a new period of deposition in the southeastern North Sea. The deposits show a clear change in lithology from fine-grained, clay-dominated, distal Palaeocene and Eocene deposits below the lower boundary to more proximal silty clay deposits above. The higher gamma-ray values as compared to those of the underlying Eocene deposits are apparently associated with a higher content of illite and mica (Danielsen 1989). This change may be related to a different source area of the sediments. The sediment transport changed from an eastward direction during the Middle–Late Eocene to south-southwestward directions during the Oligocene. The maximum thickness, approx. 900 m is found at the border between the Norwegian and the Danish sectors (Fig. 3). The change in lithology and in sediment transport direction was probably controlled by the start of the late Cenozoic uplift of the Scandinavian region.

Olive grey to brownish grey, silty clays are dominant, and are characterized by a higher content of silt, mica, illite, and kaolinite than in the underlying Upper Palaeocene and Eocene deposits (Danielsen 1989). The depositional environment was open marine with generally well-oxidized bottom water, ranging from a nearshore upper sublittoral environment in the proximal part of the unit to lower sublittoral/upper bathyal distally (Michelsen *et al.* 1996). Thick sections of sand have been drilled in the northeastern part of the Danish sector. These

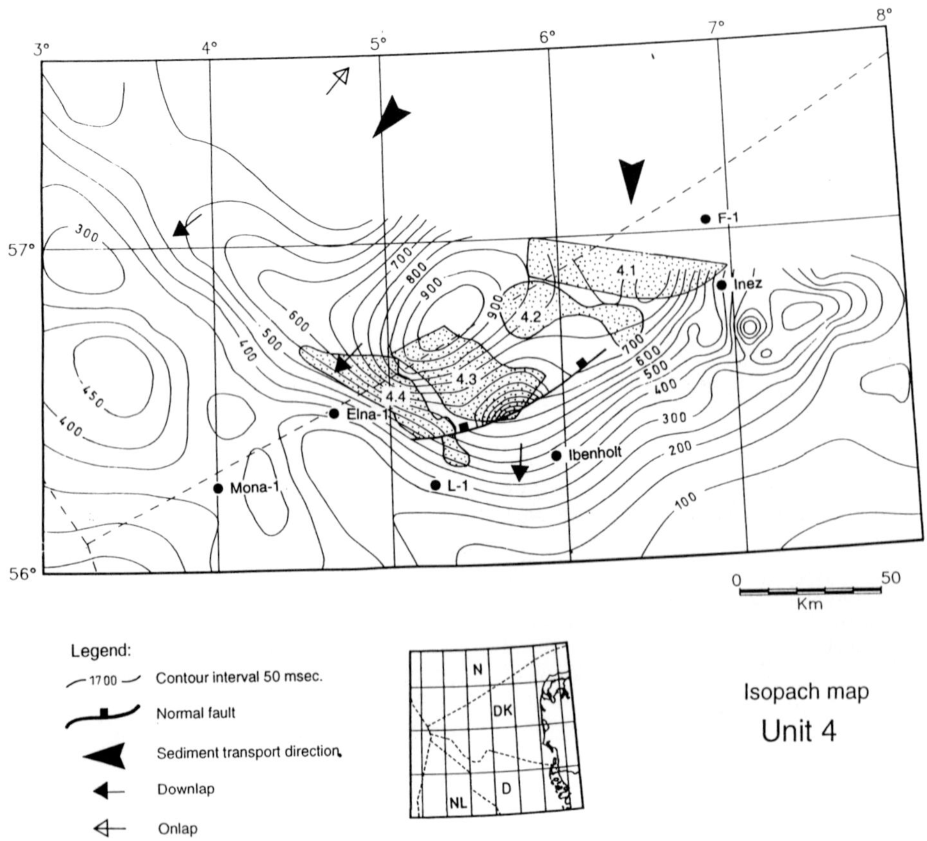

Fig. 3. Isopach map of the major sequence stratigraphical unit 4 (modified from Michelsen *et al.* 1996, Fig. 14). The locations of the depocentres of sequences 4.1–4.4 and the wells discussed in this paper are indicated.

sand bodies are interpreted as nearshore upper sublittoral sediments, deposited during a southward migration of the shoreline in the epicontinental Tertiary North Sea Basin.

Sequence stratigraphical surfaces

The Oligocene sequence stratigraphical unit 4 is characterized by an overall prograding seismic reflection pattern, which is a marked contrast to the concordant reflection pattern of the underlying Upper Palaeocene and Eocene deposits (Fig. 4). A significant basinward shift in onlap is seen above the upper unit boundary, representing the initial part of a new major prograding unit.

Sequence boundaries and maximum flooding surfaces within unit 4 have been interpreted on the basis of seismic sections and petrophysical logs. The sequence boundaries were identified mainly in the sequence depocentres by seismic onlaps and downlaps (Fig. 4). In the northern part of the depocentres toplaps and truncation features are observed. The seismic sequence boundary was tied to a log interval, determined from uncertainties caused by the seismic resolution and the velocity calculation. A more precise location of the boundary within this log interval was identified, using the gamma-ray log and, occasionally, the sonic log (see also Van Wagoner *et al.* 1990).

A narrow stratigraphical interval with high gamma-ray values was found to correlate with an internal seismic downlap surface. The interval is characterized by low to upward decreasing sonic velocities or occasionally associated with a high-velocity sonic peak. We have interpreted this interval as a maximum flooding surface or condensed section (*sensu* Loutit *et al.* 1988;

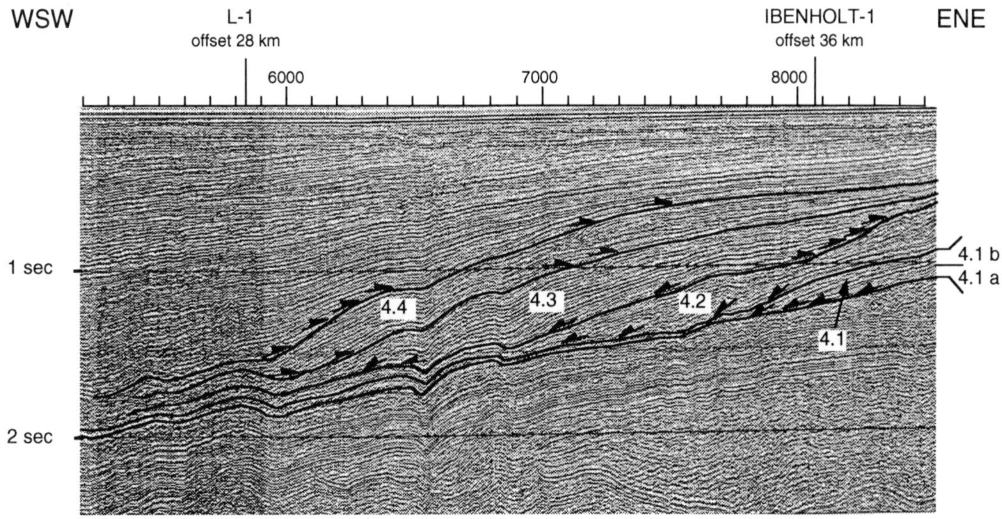

Fig. 4. Sequences 4.1–4.4 shown on the seismic section RTD81–22. The sequence boundaries are characterized by onlaps and downlaps. Note the overall progradational pattern of the sequences. For locations see Fig. 1.

Galloway 1989). The gamma-ray peaks are sometimes associated with sediments rich in glauconite or organic matter.

Systems tract interpretation

A systems tract interpretation is hampered because the sequence boundaries and the maximum flooding surface are the only sequence stratigraphical surfaces identifiable on the seismic sections. The maximum flooding surface occurs at the boundary between the transgressive and the highstand systems tracts (Posamentier *et al.* 1988; Van Wagoner *et al.* 1988). This surface is often situated close to the upper boundary of our sequences in wells located basinwards of the sequence depocentre. Here, the relatively thick interval between the lower sequence boundary and the maximum flooding surface most likely includes both the lowstand systems tract and the transgressive systems tract, but diagnostic features are not clearly identifiable. The boundary between the lowstand and the transgressive systems tracts has not been distinguished on the seismic sections.

The sediments within the study area are generally fine-grained, and in most cases it is difficult to relate the facies variations on the basis of cuttings samples to systems tracts. A systems tract interpretation based on log patterns is, therefore, the theme of the following chapter. It must be emphasized that correlation of internal sequence stratigraphical surfaces between the different well sections is performed with little chronostratigraphical control. The stratigraphical resolution found by the biozones of the Oligocene succession mainly equals that of the sequences (Fig. 2).

Log patterns of the Oligocene sequences

Overview

Each of the four Oligocene sequences is subdivided (on seismics) into two parts, separated by the internal downlap surface (the maximum flooding surface), whereas the log pattern often indicates a tri-partition. Lateral and vertical variations of the log pattern occur within the sequence, but the differences are often subtle and difficult to interpret in terms of changes in the depositional environment. Our interpretation of the depositional environment is based on differences in log trends, and supported by the sparse information from cuttings samples and from biostratigraphical analyses. Upward decreasing gamma-ray values are interpreted to reflect upward coarsening sediments, and an upward increasing gamma-ray trend as reflecting upward fining sediments. Generally, an upward decreasing gamma-ray trend is expected to represent the highstand and lowstand systems tracts, and the upward increasing trend the transgressive systems tract.

Highstand and lowstand systems tracts are both characterized by basinward prograding parasequences (e.g. Van Wagoner *et al.* 1988). To distinguish between these two systems tracts, it is, among other things, important to examine whether a thick prograding deposit is located landwards or basinwards of the depositional shoreline break of the preceding sequence (Armentrout *et al.* 1993). We have therefore arranged our log-profiles in Figs 5, 7, 8 and 9 in accordance with the palaeotopography interpreted from the seismic sections, assuming an even and gently southwestward dipping sea floor. The mapping of the sequence geometry presented in Michelsen *et al.* (1996) shows that the oldest sequence (4.1) is located closest to the Fennoscandian Shield and 4.2, 4.3 and 4.4 are gradually displaced towards the centre of the basin (see also Fig. 3).

Sequence 4.1

The depocentre for sequence 4.1, as identified on seismic sections and petrophysical logs, is located in the northeastern part of the Danish North Sea (Fig. 3). A complex sequence (Figs 5 and 6) is seen in the F-1 and Inez-1 wells. Sequence 4.1 is here interpreted to include two sequences, 4.1a and 4.1b. Only sequence 4.1b will be discussed in the present paper, since the few log data of sequence 4.1a do not contribute significantly to the discussion of the systems tracts.

The base of sequence 4.1b is located at the base of a blocky gamma-ray pattern with low values in Inez-1 (Fig. 5). The log pattern shows the presence of several upward coarsening intervals (parasequences), which are stacked into two parasequence sets: a lower set with a blocky pattern of constant, rather low gamma-ray values and an upper upward coarsening set. Cuttings samples reveal a coarse-grained marine quartz sand as the dominating lithology. The maximum flooding surface is identified above the blocky log pattern in the Inez-1 well, which is located basinwards of the sequence 4.1a depocentre (Fig. 6). The two parasequence sets are, therefore, interpreted as lowstand deposits. The lower set may have been deposited during forced regressive conditions, and the upper set as lowstand prograding deposits during the initial rise of the relative sea level. A significantly high sonic velocity peak occurs below the boundary between the two lowstand units.

The upward increasing gamma-ray trend between the blocky interval and the maximum flooding surface is interpreted as a transgressive deposit, and the decreasing trend above the maximum flooding surface as highstand deposits.

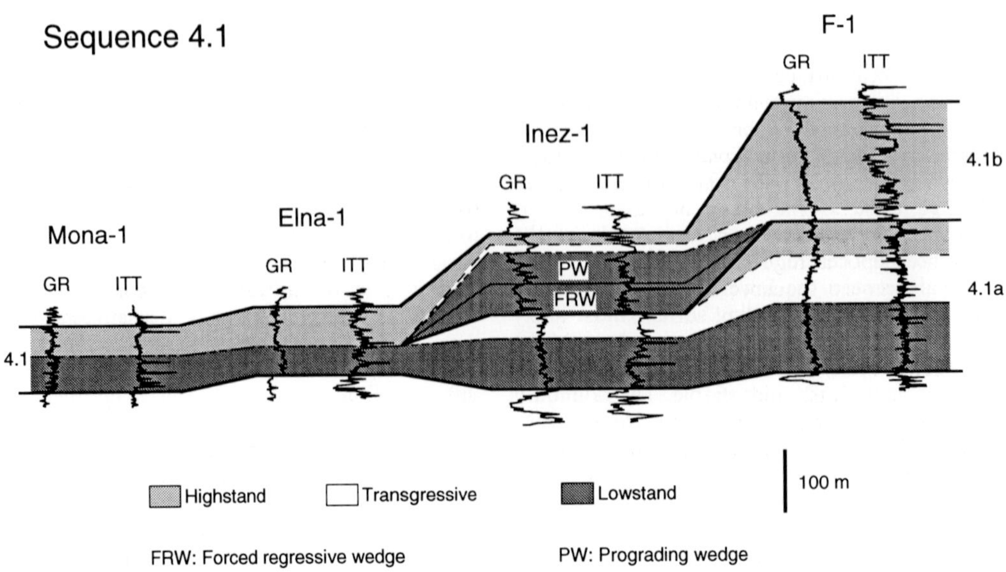

Fig. 5. Systems tract interpretation of sequence 4.1 (comprising sequences 4.1a and 4.1b) outlined by log profiles of the F-1, Inez-1, Elna-1 and Mona-1 wells. The lowstand systems tract (in Inez-1) is subdivided into forced regressive deposits and lowstand prograding deposits. For locations see Fig. 1.

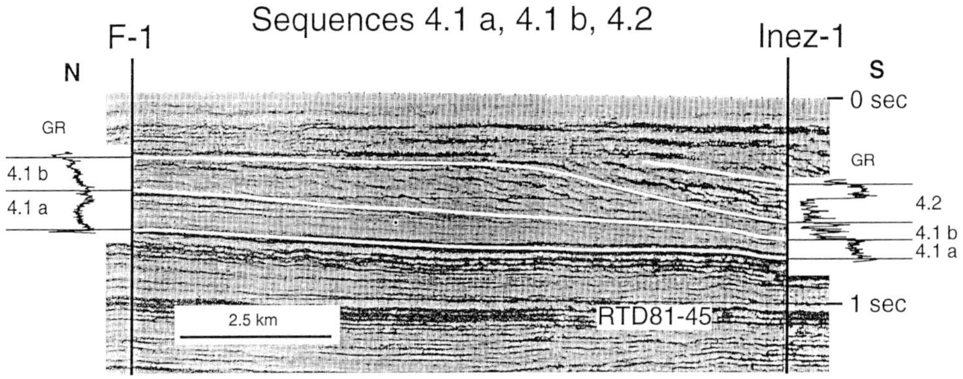

Fig. 6. Systems tract interpretation of sequence 4.1 (comprising sequences 4.1a and 4.1b) and partly sequence 4.2 outlined by log profiles of the F-1 and Inez-1 wells, correlated with the seismic section RTD81–45. For locations see Fig. 1.

Sequence 4.1b is identified in the F-1 well on the basis of a seismic correlation of the boundaries defined in Inez-1 (Fig. 6). The maximum flooding surface is indicated by a gamma-ray high and a high sonic velocity peak and by the seismic downlap surface lowermost in the sequence. A seismic correlation of the maximum flooding surface between the two wells is, however, questionable, owing to a slightly chaotic reflection pattern near Inez-1. Neither is it possible to distinguish the top lowstand surface on the seismic section near Inez-1. The thick interval above the maximum flooding surface in F-1 comprises a number of upward decreasing gamma-ray trends, which may be interpreted as three prograding parasequence sets. This log interval correlates with a prograding seismic reflection pattern (Fig. 6). Toplaps are clearly present on the seismic section, and the upward decreasing gamma-ray trend is thus interpreted as highstand deposits.

The complex sequence 4.1 can be traced towards the central part of the basin (Fig. 4). In this distal part of the sequence, the gamma-ray values indicate the presence of more fine-grained sediments, and the log pattern only comprises a lower unit with constant to slightly upward increasing gamma-ray values and an upper unit with constant and slightly lower values. Correspondingly upward increasing to decreasing sonic velocities are seen, e.g. in the Cleo-1 and Mona-1 wells (Fig. 5). The deposits here probably represent a condensed interval. A consistent interpretation, including the lateral extension of the depositional systems seen in the proximal part of the sequence, is not possible.

Sequence 4.2

The depocentre of sequence 4.2 is located basinwards of the sequence 4.1 depocentre (Fig. 3). Sequence 4.2 can be traced from the depocentre and basinwards into the central part of the basin. Integrated interpretation of the log and seismic reflection patterns allows identification of the maximum flooding surface, which proximally is located low in the sequence, e.g. the F-1 well, and distally high in the sequence, e.g. the Mona-1 well (Fig. 7).

The interval interpreted as lowstand deposits is represented by a significant lateral variation of the log pattern. A thick interval with a blocky log pattern with low gamma-ray values is present in the Inez-1 well, which is positioned basinwards of the depositional shoreline break of sequence 4.1b (Figs 6 and 7). The interval is dominated by coarse-grained marine quartz sand. The thin interval with a corresponding log pattern lowermost in F-1 is here suggested as transgressive deposits (Fig. 7), but the deposits may alternatively constitute a part of the lowstand systems tract.

The distal part of sequence 4.2 is characterized by clayey sediments, and the vertical variation of the gamma-ray log pattern in the interval below the maximum flooding surface is very subtle. The interval suggested as lowstand deposits includes the major part of the sequence. It is characterized by constant to slightly upward decreasing gamma-ray values (Mona-1 in Fig. 7). The sonic curve has a serrated appearance, with some high-velocity peaks. It consists of two cycles

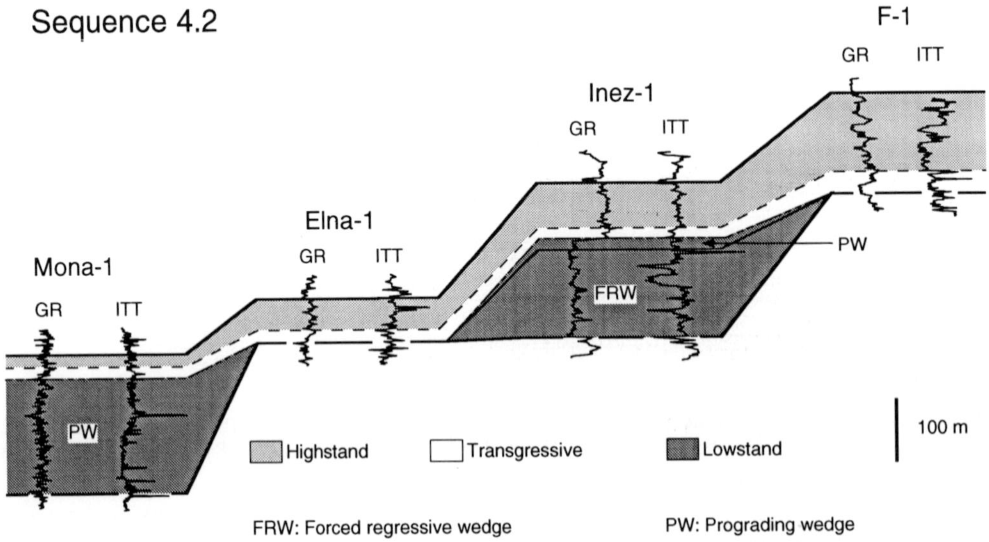

Fig. 7. Systems tract interpretation of sequence 4.2 outlined by log profiles of the F-1, Inez-1, Elna-1 and Mona-1 wells. The lowstand systems tract in Inez-1 is subdivided into forced regressive deposits and lowstand prograding deposits. The lowstand systems tract in Mona-1 is identified as lowstand prograding deposits. For locations see Fig. 1.

with upward decreasing to upward increasing sonic velocities, separated by a significantly high sonic velocity peak.

The genetic and stratigraphical relationships between these fine-grained distal lowstand deposits and the coarse-grained proximal lowstand deposits (in Inez-1) are not clear to us. Both must be interpreted as prograding units. The distal deposits (Mona-1) may be mainly lowstand prograding deposits, and the proximal deposits (Inez-1) may belong to a forced regressive wedge. The upward fining interval uppermost in the proximal deposits may, however, be interpreted as a part of the lowstand prograding wedge (see Fig. 7). A high sonic velocity peak occurs at the base of this interval. A corresponding distinction between the two types of lowstand deposits is not obvious within the fine-grained distal deposits. They may rather be regarded as one unit of lowstand prograding deposits, though two cycles of sonic velocities occur in Mona-1.

The interval interpreted as transgressive deposits is thin in all well sections, and mostly identified by a distinct upward increase in gamma-ray values (Fig. 7).

Proximally compared to the depocentre, a pronounced upward decreasing gamma-ray trend is seen above the maximum flooding surface, e.g. in F-1 (Fig. 7). The interval is interpreted as representing highstand deposits. The interval thins markedly in the basinward direction, and constant to slightly upward decreasing gamma-ray values and constant to upward increasing sonic velocities are seen basinwards of the depositional shoreline break of sequence 4.1b.

Sequences 4.3 and 4.4

The log patterns of sequences 4.3 and 4.4 show only minor lateral and vertical variations. Blocky log motifs such as those described above are not observed in any of the wells penetrating these two sequences (Figs 8 and 9).

The depocentres of sequences 4.3 and 4.4 are located basinwards of the sequence 4.2 depocentre (Fig. 3). Sequence 4.3 can be traced to the central part of the basin, and sequence 4.4 only to a position east of Mona-1. Sequence 4.4 is not identified in the Mona-1, and a biostratigraphical hiatus is recognized at this level in the well (Michelsen *et al.* 1996). The maximum flooding surfaces of both sequences are primarily identified by high gamma-ray values and relatively low sonic velocities (Figs 8 and 9). The maximum flooding surfaces correlate with internal downlap surfaces seen on a few seismic sections.

In both sequences, the surface between the lowstand deposits and the transgressive deposits is suggested at the transition from an interval with almost constant gamma-ray readings below the

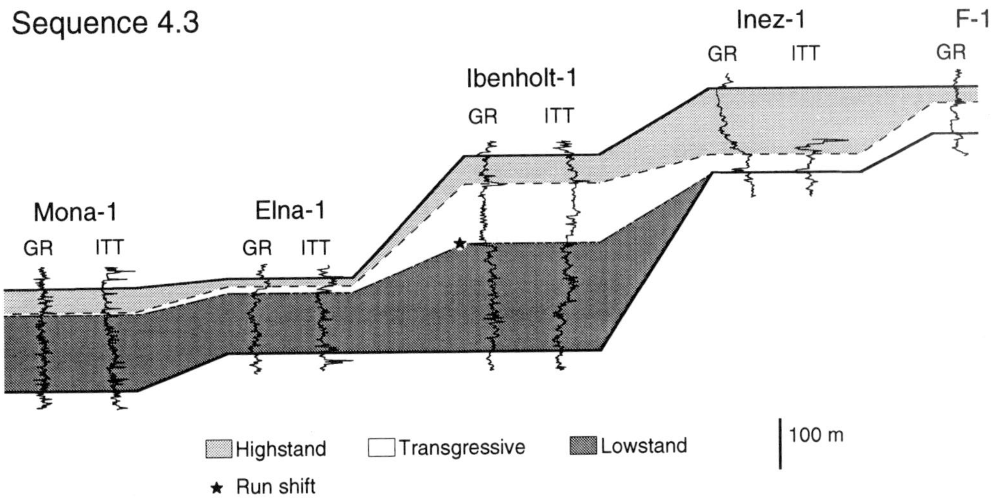

Fig. 8. Systems tract interpretation of sequence 4.3 outlined by log profiles of the F-1, Inez-1, Ibenholt-1, Elna-1 and Mona-1 wells. The lowstand systems tract is identified as lowstand prograding deposits. For locations see Fig. 1.

surface to an upward increasing gamma-ray trend above. The suggested lowstand deposits are rather fine-grained and characterized by slightly upward decreasing to constant gamma-ray values (Figs 8 and 9). A thin interval with coarse-grained sediments occurs in the basal part of the lowstand systems tract in sequence 4.3 (Elna-1, see Fig. 8) and in sequence 4.4 (Elna-1, L-1 and Ibenholt-1, see Fig. 9). The sonic velocities show an overall trend from low to upward decreasing velocities in the lower part, and upward increasing to slightly decreasing velocities in the upper part. The depocentre of the lowstand deposits on both sequences is located basinwards of the depositional shoreline break of the preceding sequence. These fine-grained lowstand deposits probably correspond to the distal lowstand deposits in sequence 4.2, and may represent the lowstand prograding wedge. The apparent absence of coarse-grained deposits in sequences 4.3 and 4.4 is probably due to the fact that no wells are located close to the depositional shoreline break of sequence 4.2. There is a significant distance between the sequence 4.2 depocentre and the

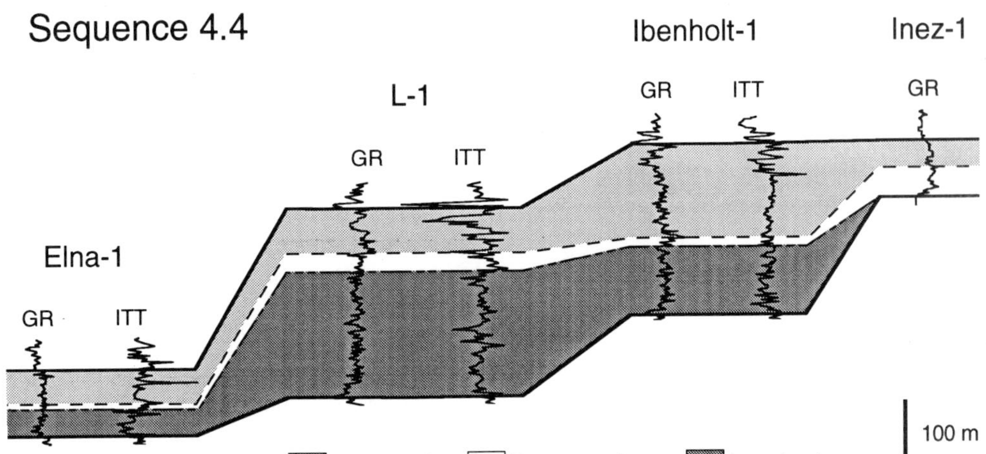

Fig. 9. Systems tract interpretation of sequence 4.4 outlined by log profiles of the Inez-1, Ibenholt-1, L-1 and Elna-1 wells. The lowstand systems tract is identified as lowstand prograding deposits. For locations see Fig. 1.

Ibenholt-1 and L-1 wells, and the two wells comprise the most proximal part of the thick lowstand deposits of sequences 4.3 and 4.4 (Fig. 3). It must be noted that Inez-1 (with the blocky log pattern interpreted as forced regressive deposits in sequence 4.2) is located on the flank of sequence 4.1 depocentre (Fig. 3).

A rather thin interval with upward increasing gamma-ray values is interpreted as transgressive deposits in both sequences. The thick transgressive interval of sequence 4.3 indicated in Ibenholt-1 is questionable. Interpretation problems may be caused by the fact that the sequence here spans the transition between two log runs.

The intervals between the maximum flooding surface and the upper sequence boundary are interpreted as highstand deposits (Figs 8 and 9). Upward decreasing gamma-ray values are seen in wells located on the top of the depocentre of the preceding sequence, e.g. in sequence 4.3 of Inez-1. Basinwards, constant to slightly upward decreasing gamma-ray values are measured. The sonic velocities generally vary from constant to upward decreasing trends. The maximum thickness of the highstand deposits is located landwards of the depositional shoreline break (sequence 4.3), and at the depositional shoreline break (sequence 4.4). The highstand deposits thin markedly in the basinward direction.

Discussion and conclusion

Sequence boundaries and maximum flooding surfaces of the Oligocene sequences have been interpreted on the basis of seismic sections and petrophysical logs. The seismic reflection pattern generally suggests aggrading and prograding features (Fig. 4), and only the sequence boundaries and the maximum flooding surfaces can be recognized within the sequences.

A systems tract interpretation cannot be established on the basis of the seismic sections alone. The upper part of the sequences, above the maximum flooding surface, probably represents the highstand systems tract, and the lower part the lowstand and transgressive systems tracts. The bounding surface between the two latter systems tracts cannot be identified on the seismic sections. A subdivision of the sequences has to be based on log interpretation, and the three log trends described in the previous section can probably be referred to systems tracts.

Lowstand systems tract

Two different gamma-ray trends are recognized in the data set (Figs 5, 7 and 10).

(1) A blocky log pattern with low gamma-ray values is present basinwards of the depositional shoreline break of the preceding sequence.

(2) In the distal part of the sequence, constant to slightly upward decreasing gamma-ray values characterize the lowstand deposits. The interval transit time curve in this distal position often has a serrated appearance and is characterized by several high-velocity peaks, compared to the overlying part of the sequence.

Both types of lowstand deposits are interpreted as prograding deposits. We assume that the coarse-grained deposits (the blocky log pattern) reflect a forced regressive deposition in an epicontinental basin with a gently dipping sea floor. Lowstand deposits associated with the process of forced regression in a ramp margin setting are characterized by sharp-based shoreface sediments in the proximal part of the sequence and more gradational deposits further basinwards (Posamentier *et al.* 1992).

The intervals of blocky log pattern with constantly low gamma-ray values in the basal part of sequences 4.1b and 4.2 (both in the Inez-1 well) are positioned basinwards of and topographically below the depositional shoreline break of the preceding highstand deposits (Figs 5, 6 and 7). The log features of these lowstand deposits show a slightly prograding pattern, and the log intervals are sharp-based at the lower sequence boundary. These coarse-grained, marine sediments are, therefore, interpreted as deposited during a sea-level fall, forcing a basinward migration of the shoreface environment. The upper part of the lowstand systems tract of sequences 4.1b and 4.2 (in Inez-1) shows a different gamma-ray pattern. A significantly coarsening upward trend occur in 4.1b and a slightly fining upward trend in 4.2, which are suggested to represent the proximal part of lowstand prograding wedges (*sensu* Hunt & Tucker 1992).

The lowstand deposits represented by constant to slightly upward decreasing gamma-ray trends occur further basinwards (e.g. in sequence 4.2, Mona-1 (Fig. 7)), and may be regarded as the distal and more fine-grained part of the lowstand prograding wedge. Seismic downlaps are generally present on the lower sequence boundary in this distal part of the sequence. The fine-grained lowstand deposits described from sequences 4.3 and 4.4 probably correspond to those seen in the distal part of sequence 4.2 and, therefore, may also represent distal lowstand prograding deposits. The apparent absence of coarse-grained lowstand deposits in sequences

4.3 and 4.4 may be due to the fact that none of the wells penetrate the proximal part of the lowstand systems tracts.

These fine-grained lowstand deposits are here regarded as one unit of lowstand prograding deposits, though two cycles of sonic velocities are observed (in Mona-1). The two cycles are separated by a high sonic velocity peak, which is comparable with the peaks separating the forced regressive deposits from the prograding wedges in Inez-1 (sequence 4.1b and 4.2). This sonic feature may reflect an erosional surface. However, this is not confirmed by lithology studies of cuttings samples, so the presence of the peak may not yet be used to identify the boundary between the two lowstand wedges. The presence of more coarse-grained sediments low in the lowstand systems tract is also indicated by the gamma-ray values locally, e.g. in Elna-1 (sequence 4.3) and L-1 (sequence 4.4). Forced regressive deposits may thus be present, but they cannot be consistently identified and delineated on the basis of the available data.

Sequence boundaries are most easily located at the base of the lowstand systems tract of our sequences (in accordance with Posamentier *et al.* 1988), and mainly identified by seismic onlaps, downlaps and truncation features. The forced regressive and lowstand prograding wedges have not been identified on seismic sections, and are not determined unequivocally in all well sections. It is, therefore, difficult to locate the sequence boundary between these two systems tracts, as suggested by Hunt & Tucker (1992) and Helland-Hansen & Gjelberg (1994).

The chronostratigraphical position of the fine-grained lowstand deposits in relation to the coarse-grained deposits is uncertain. A general problem is that the wells are too widely spaced to determine precisely the lateral variation in lithology, and that the stratigraphical relationships of the depositional systems within each sequence are difficult to unravel on the basis of the available biostratigraphy.

Transgressive systems tract

The interval between the interpreted lowstand deposits and the maximum flooding surface is characterized by upward increasing gamma-ray values, and is interpreted to represent transgressive deposits (Fig. 10).

The transgressive systems tract includes sediments deposited landwards of the depositional shoreline break during relative sea-level rise, and has typically retrogradational character (Posamentier *et al.* 1988). This systems tract tends to be a thin, mud-dominated interval, and the coeval basinal facies is sediment starved, forming part of the condensed section (Armentrout *et al.* 1993).

The interval with upward increasing gamma-ray values interpreted as transgressive deposits is always present in our sequence. The interval seems to increase in thickness in the landward direction. The thin section of clay basinwards of the depositional shoreline break may be interpreted as a condensed interval.

Highstand systems tract

The interval between the maximum flooding surface and the upper sequence boundary is characterized by upward decreasing or constant gamma-ray values (Fig. 10). The interval transit time log often shows a cyclic character with upward decreasing to upward increasing sonic velocities. These upward coarsening deposits represent a prograding unit.

The highstand systems tract is expected to be widely distributed landwards of the depositional shoreline break and dominated by prograding parasequence sets (Posamentier *et al.* 1988). According to Armentrout *et al.* (1993), this systems tract includes log patterns of aggrading and prograding, upward coarsening parasequences.

The thick interval characterized by upward decreasing gamma-ray values in our sequences 4.1b and 4.2 (both in the F-1 well) is located at the top of the underlying sequence depocentres (Figs 5 and 7). The seismic features of this interval in sequence 4.1b show a prograding character (Fig. 6), and the upward coarsening interval is interpreted as the highstand systems tract. It is obvious that the coarse-grained highstand sediments only occur uppermost in the thick proximal part of the systems tract, which is characterized by seismic toplaps. The interval with highstand deposits thins markedly in the basinward direction, and is characterized by a constant gamma-ray trend with higher values (Fig. 7). The thin distal part of the systems tract is dominated by clay and may be interpreted as a condensed interval.

It is somewhat problematic to refer these sequences to an established sequence stratigraphical model. The deposits are characterized by marine, clay-dominated sediments, and abrupt changes in lithological facies are rarely seen. The sediments were deposited in an epicontinental basin. The initial uplift of the Scandinavian region probably created a gently southwestward dipping sea floor. The seismic reflection pattern for the sequence shows an overall prograding

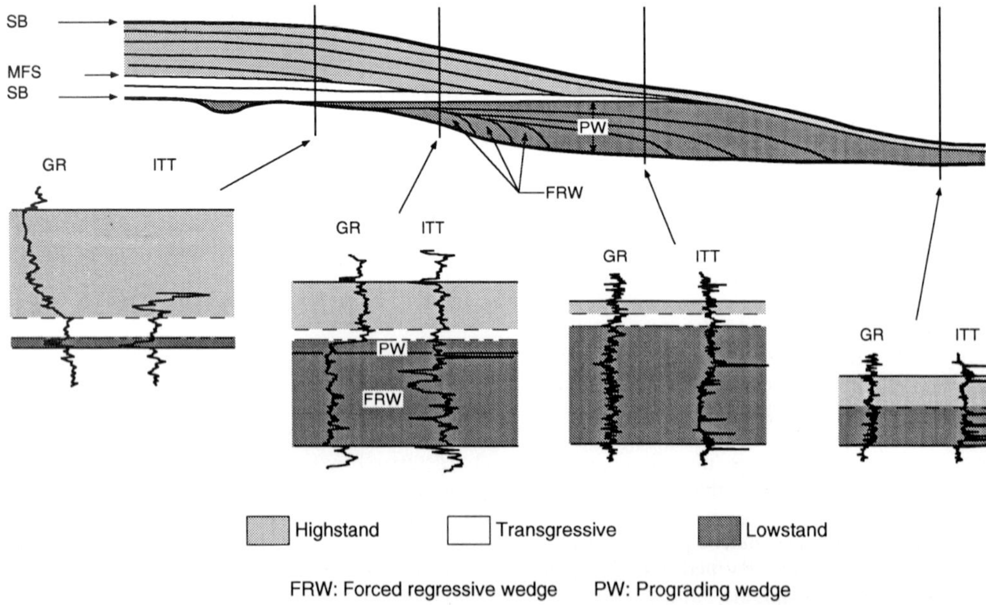

Fig. 10. Principles of interpretation of sequence stratigraphical surfaces in log profiles illustrated by a sequence model of a deep ramp setting (modified from Vail *et al.* 1991) and log from the studied North Sea wells. Wells with a thick highstand systems tract characterized by upward decreasing gamma-ray values are located landwards of the depositional shoreline break of the preceding sequence. The highstand interval thins markedly in the basinward direction, and a constant to slightly upward decreasing gamma-ray trend is seen. Wells including a thick lowstand systems tract characterized by a blocky log pattern with low gamma-ray values (representing the forced regressive wedge and the prograding wedge) are located basinwards of the depositional shoreline break. Wells with an overall constant to slightly upward decreasing log trend and higher gamma-ray values (representing the prograding wedge) are located further basinwards.

character. The gamma-ray log trends indicate retrograding, aggrading and prograding stacking patterns.

The deep ramp model (Vail *et al.* 1991) may be applied (or modified) for the Oligocene deposits in the eastern North Sea Basin, as illustrated in Fig. 10 by the characteristic lateral variation in the log trends of the interpreted systems tracts.

Based on log trends, we have identified the lowstand deposits as prograding sedimentary bodies, which were deposited basinwards of the depositional shoreline break of the preceding highstand systems tract and at a topographic level below the break. Coarse-grained, sharp-based lowstand deposits are found basinwards of the depositional shoreline break, comprising forced regressive deposits succeeded by lowstand prograding deposits. Further basinwards more fine-grained lowstand prograding deposits constitute the lowstand systems tract (Fig. 10). Lowstand fan deposits are not observed on seismic sections, and diagnostic gamma-ray log features such as those described by Armentrout *et al.* (1993) are not present.

A thin interval of transgressive deposits including upward fining clayey sediments is found at the tops of the lowstand deposits, and in the landward direction at the top of the older highstand deposits. The transgressive deposits seem to thicken in the landward direction, and the top of these deposits are marked by the maximum flooding surface, indicated by a high gamma-ray peak. The overlying highstand deposits thicken in the landward direction, where they have a characteristic upward coarsening trend (Fig. 10). They thin in the basinward direction and become more clayey, probably representing a condensed interval.

The paper was written while Mette Danielsen was in receipt of financial support from the Danish Energy Agency (EFP-programme). Lars Henrik Nielsen (Geological Survey of Denmark) has read the manuscript and suggested improvements. Ole Rønø Clausen has contributed with valuable computer-preparations of the illustrations; and Lissie Jans made the drawings. Ronald J. Steel and Peter Vail reviewed this manuscript and suggested valuable improvements.

References

ARMENTROUT, J-M., MALECEK, S. J., FEARN, L. B. *et al.* 1993. Log-motif analysis of Paleogene depositional systems tract, Central and Northern North Sea: defined by sequence stratigraphic analysis. *In*: PARKER, J. R. (ed.) *Petroleum Geology of Northwest Europe: Proceedings of the 4th Conference*. The Geological Society, London, 45–57.

CLAUSEN, O. R. & KORSTGÅRD, J.A . 1993*a*. Small-scale faulting as an indicator of deformation mechanism in the Tertiary sediments of the northern Danish Central Trough. *Journal of Structural Geology*, **15**(11), 1343–1357.

—— & —— 1993*b*. Tertiary tectonic evolution along the Arne-Elin Trend in the Danish Central Trough. *Terra Nova*, **5**, 233–243.

DANIELSEN, M. 1989. *En Sedimentologisk Undersøgelse af Tertiære Sedimenter i de Danske Nordsøboringer Lulu-1 og Inez-1*. PhD Thesis, Aarhus University.

DORÉ, A. G. 1992. The base Tertiary surface of southern Norway and the northern North Sea. *Norsk Geologisk Tidsskrift*, **72**, 259–265.

GALLOWAY, W. E. 1989. Genetic stratigraphic sequences in basin analysis, I: Architecture and genesis of flooding-surface bounded depositional units. *AAPG Bulletin*, **73**(2), 125–142.

HELLAND-HANSEN, W. & GJELBERG, J. G. 1994. Conceptual basis and variability in sequence stratigraphy: a different perspective. *Sedimentary Geology*, **92**, 31–52.

HUNT, D. & TUCKER, M. E. 1992. Stranded parasequences and the forced regressive wedge systems tract: deposition during base-level fall. *Sedimentary Geology*, **81**, 1–9.

LOUTIT, T. S., HARDENBOL, J., VAIL, P. R. & BAUM, G. R. 1988. Condensed sections: The key to age dating and correlation of continental margin sequences. *In*: WILGUS, C. K., HASTINGS, B. S., KENDALL, C. G. ST. C., POSAMENTIER, H. W., ROSS, C. A. & VAN WAGONER, J. C. (eds) *Sea-level Change – An Integrated Approach*. Society of Economic Paleontologists and Mineralogists, Special Publication, **42**, 71–109.

MICHELSEN, O., THOMSEN, E., DANIELSEN, M., HEILMANN-CLAUSEN, C., JORDT, H. & LAURSEN, G. V. 1996. Cenozoic sequence stratigraphy in the eastern North Sea. *In*: DE GRACIANSKY, P. CH., HARDENBOL, J., JACQUIN, T., VAIL, P. R. & FARLEY, M. B. (eds) *Mesozoic–Cenozoic Sequence Stratigraphy of Western European Basins, 2*. Society of Economic Paleontologists and Mineralogists, Special Publication.

NIELSEN, O. B., SØRENSEN, S., THIEDE, J. & SKARBØ, O. 1986. Cenozoic differential subsidence of North Sea. *AAPG Bulletin*, **70**(3), 276–298.

POSAMENTIER, H. W., ALLEN, G. P., JAMES, D. P. & TESSON, M. 1992. Forced regressions in a sequence stratigraphic framework: concepts, examples, and exploration significance. *AAPG Bulletin*, **76**(11), 1687–1709

——, JERVEY, M. T. & VAIL, P. R. 1988. Eustatic controls on clastic deposition, I. Conceptual framework. *In*: WILGUS, C. K., HASTINGS, B. S., KENDALL, C. G. ST. C., POSAMENTIER, H. W., ROSS, C. A. & VAN WAGONER, J. C. (eds) *Sea-level Change – An Integrated Approach*. Society of Economic Paleontologists and Mineralogists, Special Publication, **42**, 109–124.

VAIL, P. R., AUDEMARD, F., BOWMAN, S. A., EISNER, P. N. & PEREZ-CRUZ, G. 1991. The stratigraphic signatures of tectonics, eustasy and sedimentology – an overview. *In*: EINSELE, G., RICKEN, W. & SEILACHER, A. (eds) *Cycles and Events in Stratigraphy*. Springer, Berlin, 617–659.

VAN WAGONER, J. C., MITCHUM, R. M., CAMPION, K. M. & RAHMANIAN, V. D. 1990. *Siliciclastic Sequence Stratigraphy in Well Logs, Cores and Outcrops: Concepts for High-resolution Correlation of Time and Facies*. American Association of Petroleum Geologists, Methods in Exploration, **7**.

——, POSAMENTIER, H. W., MITCHUM, R. M., VAIL, P. R., SARG, J. F., LOUTIT, T. S. & HARDENBOL, J. 1988. An overview of the fundamentals of sequence stratigraphy and key definitions. *In*: WILGUS, C. K., HASTINGS, B. S., KENDALL, C. G. ST. C., POSAMENTIER, H. W., ROSS, C. A. & VAN WAGONER, J. C. (eds) *Sea-level Change – An Integrated Approach*. Society of Economic Paleontologists and Mineralogists, Special Publication, **42**, 39–45.

Foraminiferal biostratigraphy of the post-mid-Miocene in the Danish Central Trough, North Sea

P. B. KONRADI

Geological Survey of Denmark and Greenland, Thoravej 8, DK-2400 Copenhagen, Denmark

Abstract: The foraminiferal content of ditch cutting samples from three exploration wells in the Danish North Sea has been investigated in the sections above the mid-Miocene marker. The wells are Cleo-1, situated at the edge of the Central Trough, and Kim-1 and M-10 situated in the central part of the trough, in the northern and the southern part, respectively. Based on analysis of the benthic foraminiferal faunas, the strata in the boreholes are subdivided in accordance with the standard North Sea zonation. The interpretations of the faunal assemblages indicate the oldest strata to be deposited in an open marine, outer shelf environment. Up-hole through the wells, the content of planktonic species diminishes gradually; the benthic assemblages indicate shallowing water depth reaching an inner shelf to littoral environment. This is interpreted to reflect the filling of the North Sea Basin through the Middle to Late Miocene when sedimentation mainly occurred in the north. In Early Pliocene the sedimentation centre had shifted to the south whereas in Late Pliocene a hiatus is found in the south and northeast and sedimentation only took place in the northwest in the basin centre. The faunas in the Pleistocene deposits indicate an inner shelf to littoral environment, with periods of reduced salinity and non-marine sedimentation.

The following is part of an ongoing research project on the foraminiferal faunas of the interval above the mid-Miocene marker in the Danish sector of the North Sea. The mid-Miocene marker is more or less coincident with a geological event, which is expressed in several ways. In seismic investigations in the North Sea, it is seen as a marker horizon or prominent unconformity (Cameron *et al.* 1993). It is also registered in log sequence analyses, usually as two distinct gamma ray peaks (Kristoffersen & Bang 1982). In the microfossil assemblages, the event is seen as a change in the faunas between Zones NSB 11 and NSB 12 (King 1989, Fig. 9.12), and in the equivalent change between the Zones B 8 and B 9 of IGCP 124 Working Group (1988, Fig. 79). It is especially evidenced by the occurrences of *Uvigerina* species (von Daniels 1986) and of *Bolboforma* species (Spiegler & von Daniels 1991).

Material and methods

The present study is concerned with three exploration wells: Cleo-1, situated on the north-eastern edge of the Danish Central Trough, and Kim-1 and M-10, situated in the central part of the Central Trough, to the north and to the south respectively (Fig. 1). The geological setting of the Danish Central Trough is given by Michelsen *et al.* (1995). The study is based on ditch cutting samples stored at the archives of the Geological Survey of Denmark. The samples were washed on 0.1 mm and 0.063 mm screens. From the residue on the 0.1 mm screen a minimum of 300 foraminifera were picked, if possible, and counted. In samples with abundant inorganic material, the foraminifera were concentrated by means of a heavy liquid with a specific gravity of 1.8 g/cm^3, and the residues were checked for remaining foraminifera.

Owing to the method by which ditch cutting samples are generated at well site, they only allow the first downhole occurrence of species to be used in biostratigraphical interpretations

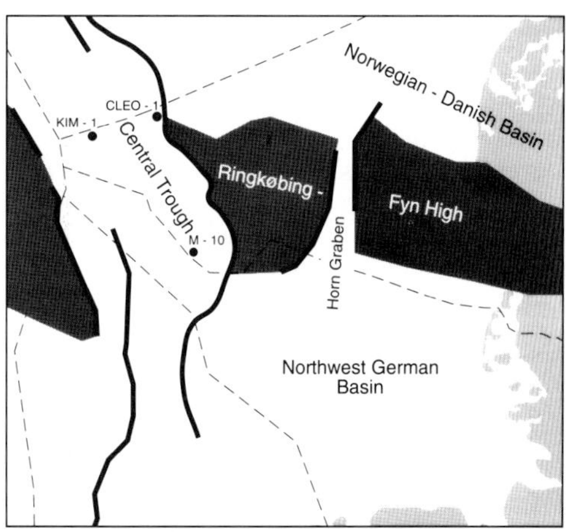

Fig. 1. Positions of the wells Cleo-1, Kim-1 and M-10 in the Danish North Sea in relation to the Central Trough.

From De Batist, M. & Jacobs, P. (eds), 1996, *Geology of Siliciclastic Shelf Seas*, Geological Society Special Publication No. 117, pp. 15–22.

(King 1983). For this reason the zones will be described from the top in descending stratigraphical order. Therefore, 'tops' of specific foraminiferal species together with *Bolboforma* spp. are used to divide the intervals into zones in accordance with the NSB zonation of King (1989, and partly 1983, see below). In the following, the terms 'marker species' and 'substitute marker' refer to the species which define the zones or subzones and which are given in the range charts of King (1983, Fig. 3; 1989, Figs 9.12 and 9.13).

King (1989) revised Zones NSB 16 and NSB 17 of his 1983 paper and introduced a new Zone NSB 16x for the North Sea north of 57° N. As the investigated wells are situated south of 57° N, Zones NSB 16 and NSB 17 of King (1983) are applied, and consequently also Subzone NSB 15b of 1983. The Pliocene–Pleistocene boundary is placed in accordance with Thompson *et al.* (1992). Ecological interpretation of the faunas is based on Murray (1991). The investigated strata are equivalent to the major sequence stratigraphical unit 7 of Michelsen *et al.* (1996).

Biostratigraphy

The investigations will be described by reference to the analysis of well M-10. The analyses of wells Cleo-1 and Kim-1 have been published by Konradi (1995).

Well M-10

Well M-10 is situated at 55°28′ N and 5°05′ E in the Salt Dome Province in the southern part of the Danish Central Trough area. The present water depth of the site is 43 m.

The samples were taken at 30 ft (9 m) or (below 3320 ft) at 20 ft (6 m) intervals. A selected number of these samples were investigated. A total of 125 species were identified. A summary of the foraminiferal investigations is given in Fig. 2 as percentage distributions of selected species, with information about the number of species, faunal dominance and the diversity (Walton 1964).

The correlation of the investigated strata to the NSB zonation is based on the following.

Zone NSB 17. Zone NSB 17 is identified only in the uppermost sample investigated from 430 ft (131 m). It has a very low yield and may even represent a reworked assemblage. It is dominated by the marker species *Elphidium excavatum* (Terquem).

In the samples from 520 to 910 ft (158 to 277 m), no calcareous microfossils are found, and the interval is regarded as being non-marine, probably fluvial, deposited in a glacial period when the general sea level was lowered.

Subzone NSB 16b. Subzone NSB 16b is recognized in the interval from 1000 to 1510 ft (305 to 460 m), based on the first downhole occurrence of the marker species *Elphidiella hannai* (Cushman and Grant). The assemblages are dominated by *Elphidium excavatum*. The common occurrence of *Nonion orbiculare* (Brady) in the upper part indicates shallow waters. The faunas reveal increasing sedimentation depth downhole. The diversity is low and the dominance high. The fauna is indicative of a littoral and cold environment.

Subzone NSB 16a. Subzone NSB 16a is represented in the two samples from 1570 ft (479 m) and 1630 ft (497 m), based on the occurrence of the marker species *Elphidium oregonense* (Cushman and Grant). The fauna further includes *Elphidium excavatum*, *Elphidiella hannai* and *Bulimina marginata* (d'Orbigny). In these samples the dominance has decreased and the diversity increased slightly compared to the subzone above. The fauna indicates an outer littoral to inner shelf environment. Subzones NSB 16a and 16b and Zone NSB 17 are of Pleistocene age (King 1983).

Zone NSB 15. Zone NSB 15 is not found in this well and a hiatus probably exists at this level.

Subzone NSB 14b. Subzone NSB 14b is identified in the interval from 1630 ft (497 m) to 1960 ft (597 m), characterized by the first downhole occurrence of the marker species *Monspeliensina pseudotepida* (van Voorthuysen). The fauna is dominated by *Elphidium excavatum* and *Elphidiella hannai* and common species are *Cassidulina reniforme* (Nørvang) and *Buccella* spp. The following species have their first downhole appearances in this subzone: *Bulimina aculeata* (d'Orbigny), *Cassidulina laevigata* (d'Orbigny), *Cassidulina carinata* (Silvestri) and *Sigmoilopsis schlumbergeri* (Silvestri). In this subzone, the abundance of species and diversity increase slightly downhole whereas the dominance slightly decreases. The middle part of the interval has a low yield. The environment is interpreted as inner shelf.

Subzone NSB 14a. Subzone NSB 14a is identified in the interval from 1990 ft (607 m) to 2800 ft

(853 m) and is characterized by the first downhole occurrences of the substitute markers *Florilus boueanus* (d'Orbigny) and *Cassidulina pliocarinata* (van Voorthuysen) and of *Nonion affine* (Reuss). Above 2190 ft (668 m), the fauna is dominated by *Elphidium excavatum*, *Monspeliensina* pseudotepida and *Bulimina aculeata* and below by *Cassidulina carinata*, *Nonion affine*

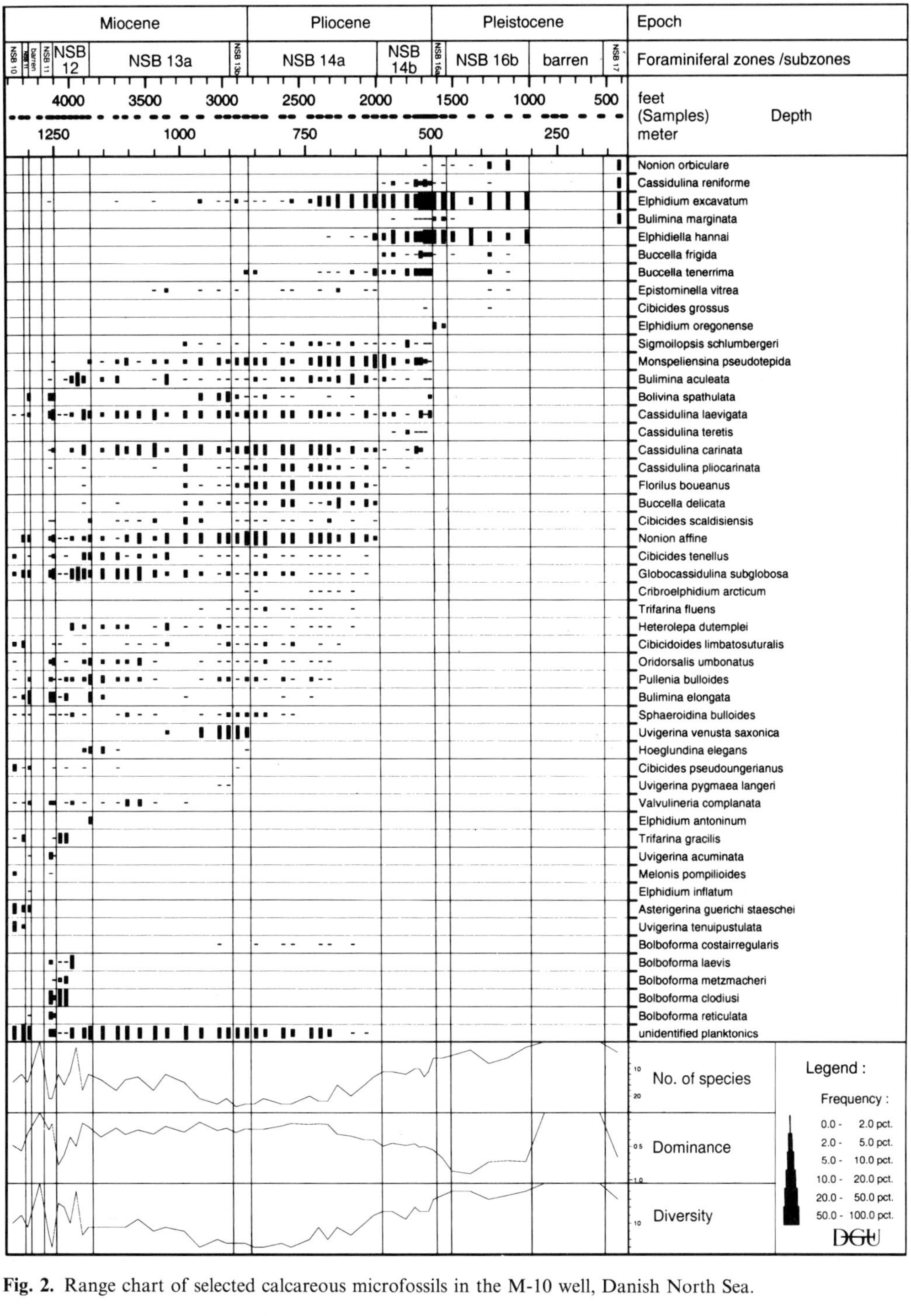

Fig. 2. Range chart of selected calcareous microfossils in the M-10 well, Danish North Sea.

and planktonic specimens. Additional common species in the subzone include *Cassidulina laevigata* and *Buccella delicata* (Voloshinova). *Bolboforma costairregularis* (Toering and van Voorthuysen) is recorded sporadically in this subzone. Above 2190 ft (668 m) the fauna indicates an inner to middle shelf environment. Below this depth the number of species and the diversity rise abruptly and the first downhole occurrences of *Oridorsalis umbonatus* (Reuss) and *Pullenia bulloides* (d'Orbigny) are recorded, indicating a change to a middle to outer shelf environment in the lower part of this zone.

Subzone NSB 13b. Subzone NSB 13b is identified in the samples from 2860 ft (867 m) to 2920 ft (890 m) by the first downhole occurrence of the marker species *Uvigerina venusta saxonica* (von Daniels and Spiegler). Other common species are *Monspeliensina pseudotepida*, *Cassidulina laevigata*, *C. carinata*, *Florilus boueanus*, *Nonion affine*, *Sphaeroidina bulloides* (d'Orbigny) and planktonic species. The diversity and the abundance of species are high and the fauna represents an open marine, middle to outer shelf environment. The top of the Miocene is placed at the top of this subzone (King 1989).

Subzone NSB 13a. Subzone NSB 13a is recognized in the interval from 2920 ft (890 m) to 3800 ft (1158 m) based on the first downhole occurrence of the marker species *Uvigerina pygmaea langeri* (von Daniels and Spiegler). The fauna in the upper part is dominated by *Uvigerina venusta saxonica* and *Bolivina spathulata* (Williamson), whereas the lower part is dominated by *Cassidulina carinata* and *Globocassidulina subglobosa* (Brady). Common species include *Monspeliensina pseudotepida*, *Bulimina aculeata*, *Cassidulina laevigata*, *Nonion affine* and *Cibicides* spp. In the lower part *Oridorsalis umbonatus* and *Pullenia bulloides* are accessory species. Planktonic specimens constitute an important part of the fauna. The samples from 3260 to 3380 ft (994 to 1030 m), as well as the lowermost sample, have low yields. The faunas in the samples from the lower part of the subzone are badly preserved. The number of species and the diversity gradually decrease downhole in the upper part of the subzone. The faunas indicate an open marine, outer shelf environment.

Zone NSB 12. Zone NSB 12 is identified in the interval from 3860 to 4080 ft (1177 to 1244 m) based on the occurrence of the marker species *Elphidium antoninum* (d'Orbigny) in the top sample. Common species are *Bulimina aculeata*, *B. elongata* (d'Orbigny), *Cassidulina carinata*, *C. laevigata*, *Nonion affine*, *Globocassidulina subglobosa*, *Cibicides* spp. and planktonic specimens. In the lowermost two samples from 4040 and 4080 ft (1231 and 1244 m), *Trifarina gracilis* (Reuss) is very common and an influx of abundant *Bolboforma clodiusi* (von Daniels and Spiegler) is seen. In this zone, the samples have low to very low yield (sample at 3960 ft (1207 m)) and specimens of foraminifera are poorly preserved (except for the sample at 4040 ft (1231 m)). Variation in the preservation quality is probably the reason for the fluctuations of the number of species, the diversity and the dominance. The fauna indicates a middle to outer shelf environment.

Zone NSB 11. Zone NSB 11 is represented by the samples from 4100 to 4280 ft (1250 to 1305 m) based on the first downhole occurrence of the substitute markers *Uvigerina acuminata* (Hosius) and *Melonis pompilioides* (Fichtel and Moll). The marker species *Asterigerina guerichi staeschei* (ten Dam and Reinhold) and *Elphidium inflatum* (Reuss) are found in the sample from 4280 ft (1305 m). The fauna is dominated by *Bulimina elongata* and common species are *Cassidulina laevigata*, *Nonion affine*, *Globocassidulina subglobosa*, *Bolivina spathulata*, *Oridorsalis umbonatus* and planktonic specimens. The first downhole occurrence of *Bolboforma reticulata* (von Daniels and Spiegler) is recorded from the top of this zone. The sample from 4200 ft (1280 m) contains no calcareous microfossils. The diversity is generally high and the dominance low. The assemblages indicate a middle to outer shelf environment.

Zone NSB 10. Zone NSB 10 is identified in the two lowermost investigated samples from 4320 and 4380 ft (1317 and 1335 m) based on the first downhole occurrence of the marker species *Uvigerina tenuipustulata* (van Voorthuysen). Common species are *Asterigerina guerichi staeschei*, *Trifarina gracilis*, *Melonis pompilioides* and *Cibicides* spp. Planktonic specimens dominate the fauna, which indicates an open marine, middle shelf environment.

Comparison between the wells

The composition of faunal assemblages in the investigated wells is clearly related to the depositional environmental settings especially water depth (Fig. 3).

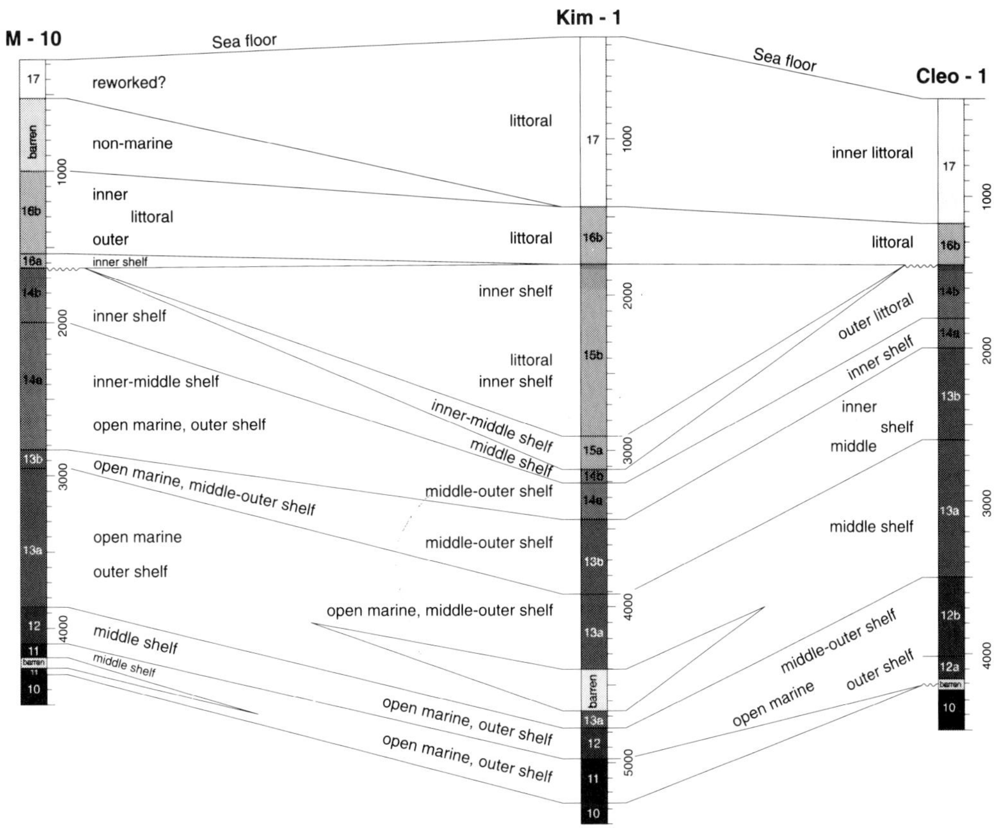

Fig. 3. Comparison of the wells Cleo-1, Kim-1 and M-10 and the environmental interpretation of the faunal assemblages. The distances between the boreholes are not to scale.

In all three wells, Zone NSB 10 is identified in the lowermost investigated samples.

Zone NSB 11 is identified in M-10 and in Kim-1 with thicknesses of 180 ft (55m) and 220 ft (67 m) respectively. This zone is not recognized in Cleo-1 but a 60 ft (18 m) barren interval is found between Zones NSB 10 and NSB 12. A hiatus is proposed at this site. The faunas of Zone NSB 11 in Kim-1 indicate an open marine, outer shelf environment and at M-10 a middle to outer shelf environment. At the time of deposition of the sediments of this zone the Kim-1 site was positioned further offshore in the basin in comparison to M-10.

In Cleo-1, Zone NSB 12 has a thickness of 570 ft (174 m) and can be subdivided into two subzones, NSB 12a and NSB 12b, whereas in Kim-1 and M-10, Zone NSB 12 cannot be subdivided and has thicknesses of 150 ft (46 m) and 220 ft (67 m) respectively. An impoverished fauna is found in Cleo-1 at 4020 ft (1225 m), the lowermost part of Subzone NSB 12b, and in M-10 at 3960 ft (1207 m). These levels are considered to be correlatable. An equivalent interval is not identified in Kim-1. The existence of Elphidiidae in Cleo-1 indicates a shallower environment than at the Kim-1 site. In M-10, *Elphidium antoninum* is only found in the sample from the top of the interval, and Zone NSB 12 in M-10 is therefore thought to have been deposited in an intermediate position in comparison to Cleo-1 and Kim-1. A seismic section between these two wells shows that the major sequence stratigraphical unit is made up of several lens-shaped sequences building out into the basin from the northeast to the southwest. At the time of deposition of sediments representing Zone NSB 12, the major sedimentation took place around the Cleo-1 site, whereas the Kim-1 site was situated further out in the basin. The faunal assemblages in Kim-1 indicate an open marine, outer shelf environment; in M-10 the environment was outer to middle shelf and in Cleo-1 there is a shift from open marine, outer shelf in

Subzone NSB 12a to middle shelf in the upper part of Subzone NSB 12b, also indicating the filling up of the basin.

Zone NSB 13 in the wells is subdivided into Subzones 13a and 13b. Subzone NSB 13a in Cleo-1 has a thickness of 840 ft (256 m), in Kim-1 790 ft (240 m) and in M-10 850 ft (259 m). In this period sedimentation apparently took place equally at all three sites. In Kim-1, an interval with an impoverished fauna or without any fauna is found in the lower half of Subzone NSB 13a. In M-10, an interval with a low yield is seen from 3260 to 3380 ft (994 to 1030 m). These two intervals are thought to correlate. An equivalent interval is not identified in Cleo-1. Subzone NSB 13b in Cleo-1 has a thickness of 570 ft (174 m), in Kim-1 it is 420 ft (128 m) and in M-10 only 90 ft (27 m). In this period sedimentation took place at the Cleo-1 site and the Kim-1 site, whereas the M-10 site was situated outside the deposition centre. The faunas in the zone indicate shallower water depth at the Cleo-1 site and a change from middle shelf to inner shelf. At the Kim-1 and M-10 sites, the faunas reflect open marine, outer shelf to middle to outer shelf environments. The Miocene–Pliocene boundary is placed in the Cleo-1 well at 2190 ft (668 m) depth and in the Kim-1 well at 3480 ft (1061 m) depth, at the first downhole occurrence of *Valvulineria complanata* (=*V. mexicana grammensis*) in the upper part of Subzone NSB 13b (King 1989, Fig. 9.13). In the M-10 well the boundary is placed at the top of the comparatively thin Subzone NSB 13b at 2830 ft (863 m).

The thickness of Zone NSB 14 in the Cleo-1 well is 450 ft (137 m), in the Kim-1 well 210 ft (64 m) and in the M-10 well 1170 ft (357 m). In all three wells the zone can be divided into Subzones 14a and 14b. At the time of deposition of sediments representing this zone, the depocentre had obviously changed to the southerly position of the M-10 site. This is also reflected in the fauna. In M-10, the lower part of Subzone NSB 14a seems to be deposited in an outer shelf environment. This changes up-hole into a middle to inner shelf environment in the upper part of Subzone NSB 14a and further to an inner shelf environment in Subzone NSB 14b. An equivalent, but less pronounced change is registered in Kim-1 and Cleo-1. At the latter site, a conspicuous change is seen in the fauna at 1800 ft (549 m) between the two subzones from an inner shelf environment to a more littoral facies. This indicates a possible hiatus at this level. A comparable faunal change is not seen in the M-10 or Kim-1 wells. In the latter well, the fauna indicates a middle shelf environment.

At the Cleo-1 site and the M-10 site, Zone NSB 15 is not identified. This indicates a hiatus here. Zone NSB 15 in the Kim-1 well has a thickness of 1320 ft (402 m). This is a revision of the 720 ft (219 m) stated by Konradi (1995) as the top of Zone NSB 15 in Kim-1 is now defined by the first downhole occurrence of *Cibicides grossus* at 1810 ft (552 m) (in accordance with King 1983). The zone is subdivided into Subzones NSB 15a and NSB 15b. In the latter subzone, the faunas suggest a fluctuation in sea level from inner shelf to littoral and back to inner shelf. An abrupt faunal change between the two subzones at 2910 ft (887 m) indicates a possible hiatus at this level.

The Pliocene–Pleistocene boundary is placed between Zones NSB 15 and NSB 16 (King 1983). In the Kim-1 well this boundary is located at 1810 ft (552 m). In Cleo-1 (1440 ft) and M-10 (1630 ft) it is coincident with the hiatus between Zones NSB 14 and NSB 16. This also shows that in this period the Kim-1 site was situated within the deepest part of the basin, as also seen in seismic profiles.

Zone NSB 16 can be divided into Subzones NSB 16a and NSB 16b. The former subzone is only identified in the M-10 well, with a thickness of 90 ft (27 m), characterized by the species *Elphidium oregonense*. This species is considered to be typical of outer littoral facies (van Voorthuysen 1952). Subzone NSB 16a is therefore regarded as indicating shallow water facies, as also stated by King (1989). At the Kim-1 site, this subzone is not identified, probably because water depth was too great as this site was situated in the centre of the basin, as seen in the seismic profiles. Sedimentation is thought to have been continuous here at the Pliocene–Pleistocene boundary. Subzone NSB 16a has not been recorded at Cleo-1. Either it was not deposited or it was later eroded. Subzone NSB 16b is 210 ft (64 m) thick in Cleo-1, deposited in a littoral facies and its top is situated at 1170 ft (357 m). In Kim-1, the subzone is 360 ft (110 m) thick and has its top at 1140 ft (439 m). The faunal assemblages here are interpreted to indicate slightly shallowing water depth up-hole in a littoral facies. In M-10, the subzone is 510 ft (155 m) thick and the assemblages also indicate shallowing water depth. Here, its top is at 1000 ft (305 m) and it is overlain by a 390 ft (119 m) interval with no calcareous microfossils and which is considered to be of fluvial origin.

Zone NSB 17 is represented by sediment thicknesses of 660 ft (192 m) in the Cleo-1 well and 900 ft (274 m) in the Kim-1 well. At both sites, the faunas indicate a littoral and cold environment, but in Cleo-1 they are impoverished, showing a more extreme environment perhaps due to its proximity to the coast. In the M-10 well, only the uppermost sample at 430 ft (131 m) is assigned to this zone.

Conclusion

Based on investigations of the foraminiferal faunal assemblages, the post-mid-Miocene deposits from the exploration wells Cleo-1, Kim-1 and M-10 can be subdivided according to the NSB zonation of King (1983, 1989). The sediments are of Middle to Late Miocene, Pliocene and Early to Middle Pleistocene age. In Cleo-1 and Kim-1 the Miocene–Pliocene boundary is placed in the upper part of Subzone NSB 13b, whereas in M-10 it is placed at the top of the subzone. The Pliocene–Pleistocene boundary is placed at the base of Subzone NSB 16a in M-10, or at the base of Subzone NSB 16b in Cleo-1 and Kim-1, where Subzone NSB 16a is not found.

At the Cleo-1 site, the sediments were deposited at a shallower water depth than at the Kim-1 and the M-10 sites, which were situated further out in the basin. Moreover, the main sedimentation took place earlier at Cleo-1, in late Middle and Late Miocene, than at Kim-1 and M-10, as the sediments were building out into the basin from the northeast. In the Early Pliocene, the main sedimentation centre was situated to the south at the M-10 site. In the Late Pliocene, the sedimentation centre shifted north and sediments are found only at the Kim-1 site. A hiatus corresponding to that time interval is found at the Cleo-1 and M-10 wells.

Notable hiatuses are identified in Cleo-1 in the Middle Miocene and at the Pliocene–Pleistocene boundary. The latter hiatus is also seen in M-10. This hiatus is probably mirrored in Kim-1 as a fluctuation in water depth in the Late Pliocene. Hiatuses are also suggested in the faunal assemblages in Cleo-1 between Subzones NSB 14a and 14b and at Kim-1 between Subzones NSB 15a and 15b, in both cases indicating a sudden drop in sedimentation depth. Whether this is caused by a period of non-deposition or by erosion cannot be determined.

As a whole, the foraminiferal assemblages in the three boreholes Kim-1, M-10 and Cleo-1 evidence the gradual shoaling of the North Sea due to filling of the basin after the mid-Miocene event. Contrary to what would have been expected from sequence stratigraphical models, no sudden changes in the faunas reflecting fluctuating sea levels are identified in the investigated faunal assemblages until the Late Pliocene.

This study is part of an EFP-92 Programme, partly financed by the Danish Energy Agency, Grant no. 1313/92–0003.

References

CAMERON, T. D. J., BULAT, J. & MESDAG, C. S. 1993. High resolution seismic profile through a Late Cenozoic delta complex in the southern North Sea. *Marine and Petroleum Geology*, **10**, 591–599.

IGCP 124 WORKING GROUP 1988. Benthic foraminifera: the description and the interregional zonation (B zones). *In*: VINKEN, R. (ed.) *The Northwest European Tertiary Basin, Results of the International Geological Correlation Programme Project No. 124*. Geologisches Jahrbuch A, **100**, 145–151.

KING, C. 1983. *Cainozoic Micropalaeontological Biostratigraphy of the North Sea*. Institute of Geological Sciences, Report **82/7**.

——1989. Cenozoic of the North Sea. *In*: JENKINS, D. G. & MURRAY, J. W. (eds) *Stratigraphical Atlas of Fossil Foraminifera*. Ellis Horwood, Chichester. 418–489.

KONRADI, P. B. 1995. Foraminiferal biostratigraphy of the post mid-Miocene in two boreholes in the Danish North Sea. *Danmarks Geologiske Undersøgelse, Serie C*, **12**, 101–112.

KRISTOFFERSEN, F. N. & BANG, I. 1982. Cenozoic excl. Danian limestone. *In*: MICHELSEN, O. (ed.) *Geology of the Danish Central Graben*. Danmarks Geologiske Undersøgelse, Serie B, **8**, 62–71.

MICHELSEN, O., DANIELSEN, M., HEILMANN-CLAUSEN, C., JORDT, H., LAURSEN, G. V. & THOMSEN, E. 1995. Occurrence of major sequence stratigraphic boundaries in relation to the basin development in Cenozoic deposits of the southeastern North Sea. *In*: STEEL R. J., FELT, V. L., JOHANNESSON, E. P. & MATHIEU, C. (eds) *Sequence Stratigraphy on the Northwest European Margin*. Norwegian Petroleum Society (NPF), Special Publication, **5**, 415–427.

——, THOMSEN, E., DANIELSEN, M., HEILMANN-CLAUSEN, C., JORDT, H. & LAURSEN, G. V. 1996. Cenozoic sequence stratigraphy in the eastern North Sea. *In*: DE GRACIANSKY, P. CH., HARDDENBOL, J., JACQUIN, T., VAIL, P. R. & FARLEY, M. B. (eds) *Mesozoic–Cenozoic Sequence Stratigraphy of European Basins, 2*. Society of Economic Paleontologists and Mineralogists, Special Publication.

MURRAY, J. W. 1991. *Ecology and Palecology of Benthic Foraminifera*. Longman, London.

SPIEGLER, D. & VON DANIELS, C. H. 1991. A stratigraphic and taxonomic atlas of *Bolboforma* (Protophytes, *Incertae sedis*, Tertiary). *Journal of Foraminiferal Research*, **21**, 126–158.

THOMPSON, R., CAMERON, T. D. J., SCHWARTZ, C., JENSEN, K. A., MAENHAUT VAN LEMBERGE, V. & SHA, L. P. 1992. The magnetic properties of Quaternary and Tertiary sediments in the southern North Sea. *Journal of Quaternary Science*, **7**, 319–334.

VAN VOORTHUYSEN, J. H. 1952. *Elphidium oregonense* Cushman and Grant, A possible marker for the Amstelian (Lower Pleistocene) in North America and northwestern Europe. *Contributions from the Cushman Foundation for Foraminiferal Research*, **3**, 22–23.

VON DANIELS, C. H. 1986. *Uvigerina* in the NW European Neogene. *Utrecht Micropaleontological Bulletins*, **35**, 67–119.

WALTON, W. R. 1964. Recent foraminiferal ecology and paleoecology. *In*: IMBRIE, J. & NEWELL, N. D. (eds) *Approaches to Paleoecology*. Wiley, New York, 151–237.

Sequence stratigraphy and architecture on a ramp-type continental shelf: the Belgian Palaeogene

P. JACOBS & M. DE BATIST

Renard Centre of Marine Geology, University of Gent, Krijgslaan 281/S8, B-9000 Gent, Belgium

Abstract: In Palaeogene times, the 'Southern Bight' of the North Sea functioned as an intracratonic, shallow-marine, siliciclastic basin and accumulated a few hundred metres of gently dipping sediment packages. A fine-scale seismic-stratigraphical model for the Palaeogene was formulated on the basis of a dense, high-resolution reflection seismic grid. In total 13 major seismic-stratigraphical units were defined, based on geometry and seismic facies characteristics. The seismic stratigraphy has been complemented with the results of four cored wells near the Belgian coast, containing a nearly continuous, 200 m thick sediment succession of Eocene age. Facies analyses of these cores suggest that part of these sediments were deposited on a muddy shelf and part in a delta environment. Evidence from relevant onshore outcrops has been used to complete the geological history of the Palaeogene, with special emphasis on the Eocene. A sedimentation model for the Eocene is presented, and relative sea-level changes, regional tectonic events and changes in sediment input are discussed. Genetic interpretation of the various lithological units and the large-scale architecture of the ramp-type margin enable evaluation of sequence-stratigraphical concepts, initially defined for a typical shelf-slope–basin section along an Atlantic-type continental margin.

The concepts of sequence stratigraphy (Vail *et al.* 1977; Posamentier *et al.* 1988; Posamentier & Vail 1988; Van Wagoner *et al.* 1987, 1988) have initiated a tremendous 'revival' in stratigraphical research in the past decade, as they proved – or claimed – to be able to explain stratal geometries and facies distributions in an easy, logical way. They were originally developed for 'typical' Atlantic-type passive margin settings, characterized by clearly defined shelf, slope and basin-floor provinces, by moderate regional subsidence and by continuous sediment supply, and exposed to Mesozoic-type changes in relative sea level. Attempts to apply these concepts to various sedimentary basins around the world have shown that some of the variables that were kept simple in the original model (subsidence, sediment supply, autocyclic shifts of depocentre, tectonics, basin morphology, etc.) may exert a stronger than anticipated influence on the stratigraphical architecture.

In this study we use a dense grid of high-resolution reflection seismic profiles, offshore cores and nearby outcrop observations to establish the sequence stratigraphy and architecture of the Palaeogene – and the Eocene in more detail – in the northwestern part of the Belgian Basin (see also Vandenberghe *et al.* 1996), and to illustrate how the particular characteristics of this basin may impede 'blind' application of the 'simple' sequence-stratigraphical concepts.

Geological setting

The 'Belgian Basin' (Fig. 1), a bight-like extension of the southernmost North Sea Basin, can be classified as an intracratonic basin in a ramp-type margin shelf setting. The basin developed on top of the London–Brabant Massif, a relatively stable continental block of Palaeozoic age that was not flooded before Late Cretaceous times and continued to shelter the area from strong subsidence throughout the Tertiary. The Cenozoic stratigraphical record consists almost completely of siliciclastic marine to marginal marine sediment series (Ziegler 1982). Throughout the Palaeogene, a shallow shelf environment persisted and the area was periodically flooded during periods of high relative sea level. Water depths during these highstand periods probably never exceeded 100 m as demonstrated by sedimentological and micropalaeontological studies of comparable deposits in the UK southern North Sea sectors (Cameron *et al.* 1992). During Thanetian and Ypresian times the shallow sea extended westwards, well into the English channel. The rising Weald–Artois High started to form a barrier closing the connection to the English Channel from Lutetian times onwards (Cameron *et al.* 1992) and possibly even earlier (Dupuis *et al.* 1984). The Neogene was a period of sediment starvation as the depocentre shifted even further northward into the main North Sea Basin (Balson 1989; Cameron *et al.* 1989). In

From De Batist, M. & Jacobs, P. (eds), 1996, *Geology of Siliciclastic Shelf Seas*,
Geological Society Special Publication No. 117, pp. 23–48.

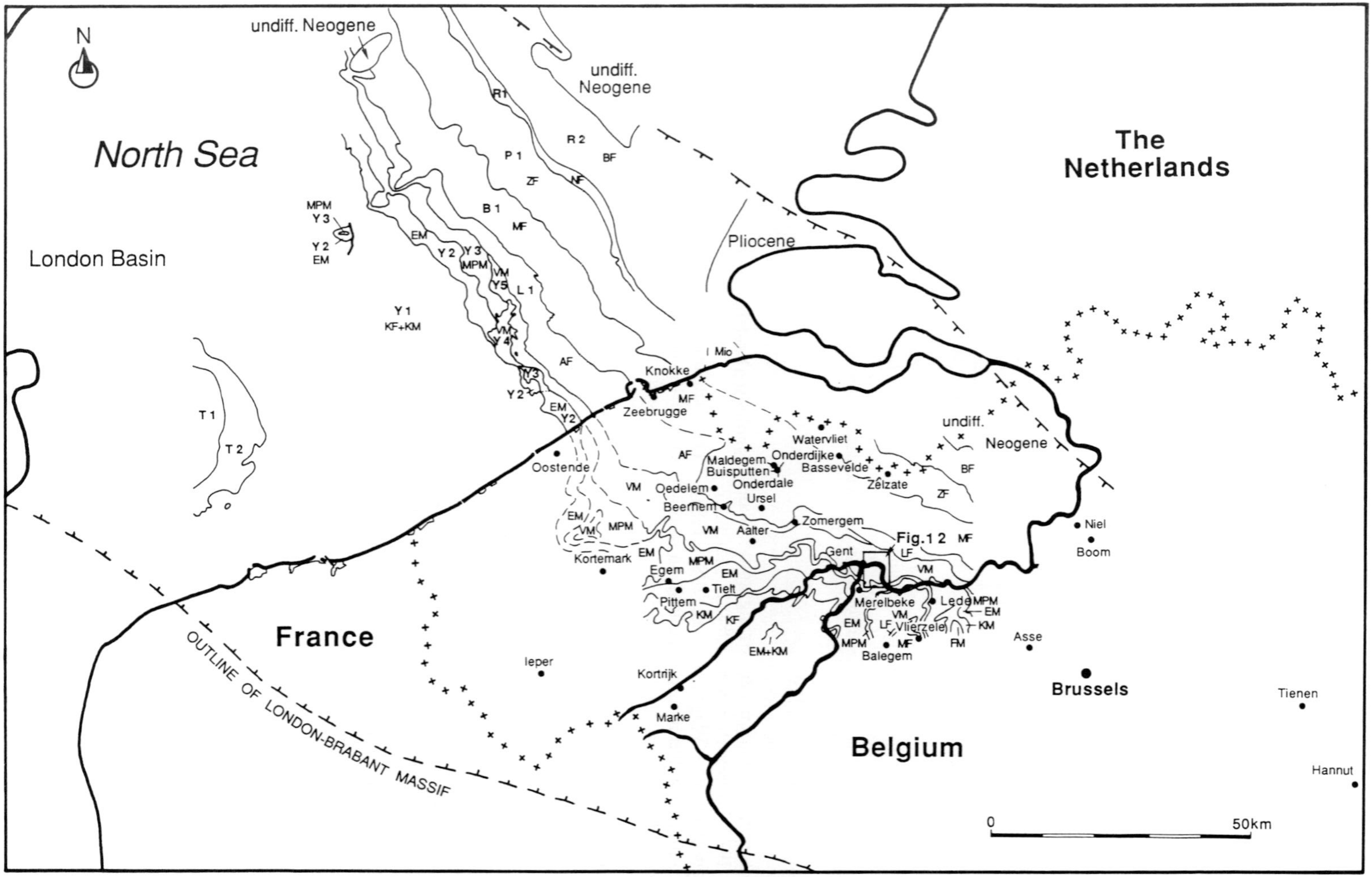

Fig. 1. Location of the study area in the southern North Sea Basin, and simplified geological map of the Belgian Basin showing the subcrop pattern of the Palaeogene offshore seismic-stratigraphical units and the onshore lithostratigraphical units beneath the Quaternary cover. All localities and type localities of lithostratigraphical units used in the text and the location of study area of Fig. 12 are indicated. KF: Kortrijk Formation; KM: Kortemark Member; EM: Egem Member; MPM: Merelbeke and Pittem Members; VM: Vlierzele Member; AF: Aalter Formation; MF: Maldegem Formation.

Quaternary times, the area emerged repetitively in response to glacio-eustatic sea-level falls. The Holocene flooded shelf has remained essentially sediment starved.

In the Belgian Basin, the Palaeogene strata onlap the Late Cretaceous chalk and dip less than 0.5% to the NNE. Offshore (Fig. 1), they crop out locally on the sea bed between the discontinuous sediment cover of the Quaternary 'Flemish Banks'. Onshore in northern Belgium (Fig. 1), numerous well-known outcrops of the parallel WNW–ESE-oriented strata exemplify the classic Lower Cenozoic geology of the Belgian Basin.

Available data and stratigraphical framework

Offshore seismic stratigraphy

A high-resolution reflection seismic grid with a total length of about 16 000 km has been acquired in the Belgian sector of the continental shelf and adjacent parts of the Dutch, French and UK sectors (between 51°–52° N and 2°–3.5° E) with a variety of different seismic tools (Fig. 2). Detailed interpretation following Mitchum *et al.* (1977), allowed De Batist (1989) and De Batist & Henriet (1995) to identify 13 seismic-stratigraphical units and a number of subunits within the Palaeogene succession. They have been labelled with a character-digit symbol, indicating their most probable chronostratigraphical position: T1 and T2 (Thanetian), Y1 to Y5 (Ypresian), L1 and L2 (Lutetian), B1 (Bartonian), P1 (Priabonian) and R1 and R2 (Rupelian). The main seismic-stratigraphical characteristics of these units are listed in Table 1 (from De Batist & Henriet 1995), and are illustrated on a synoptic seismic and schematic type section, constructed as a composite or 'collage' of several seismogram sections acquired with comparable source signatures (Fig. 3).

Unit boundaries are surfaces of consistent reflector termination. Downlap is frequently observed on the basal surfaces, whereas coastal onlap occurs only sporadically. Erosional truncation and valley incisions are common features at the top of the units, but the seismic data do not always provide sufficient arguments to characterize all of them as unconformities *sensu* Van Wagoner *et al.* (1988), i.e. surfaces of subaerial exposure and erosion and their correlative submarine surfaces of erosion. Most of the units have a pronounced sheet-like shape, with planar dipping boundaries at their base and top, and show only minor thickness variations. Each unit is also characterized by a distinct seismic facies and/or by typical facies variations, indicative for the depositional environment and its evolution.

The subcrop pattern of these units at the base of the Quaternary cover, where present, is shown on Fig. 1. The stratal relationships and geometries are illustrated by means of a number of interpreted line-drawings of seismic sections through the Belgian Basin (Fig. 4).

Offshore lithostratigraphy

Four shallow cored boreholes were drilled in front of the Belgian coast, through the Quaternary drift into the Tertiary substratum: the GR1, SWB, SEWB and VR1 wells (Fig. 2). These boreholes provide the lithological and micropalaeontological data required to complement the geometrical information obtained from interpretation of the extensive seismic data base. The boreholes cut through a composite, 200 m thick, marine sediment series of Eocene age, roughly forming a SW–NE/S–N dip section from Oostende to north of Zeebrugge, and have been described in detail by Jacobs & Sevens (1993*a*), and Jacobs (1995*b*). Grain-size and sedimentary facies analyses, completed with sediment-genetic interpretations, were performed on these cores, which allowed the complete section to be correlated lithostratigraphically with equivalent sediment series onshore (Fig. 5). Biostratigraphy was established from samples from all four wells, and compared to the biostratigraphy encountered in nearby onshore wells 22W-276 and 11E-138 (King 1990; Steurbaut 1990). Calcareous microfossil conservation was poor because of secondary oxidation and reworking, but palynomorphs (Fig. 5) provided valuable indications on age (following reference zonations of Costa & Manum (1988), and Powell (1988)), and diatoms on palaeobathymetry and depositional environment.

Onshore lithostratigraphy

The Palaeogene stratigraphy of onshore northwestern Belgium was established by Rutot (1882, 1883), Mourlon (1888), Vandenbroeck (1893) and Leriche (1912, 1922), and also by Gulinck (1965, 1969*a*; *b*). In more recent years, detailed outcrop and borehole studies added to the knowledge of the classic Palaeogene stratigraphy of Belgium: the stratigraphy of the Lower Eocene was revised by Steurbaut & Nolf (1986), the transitional layers between the

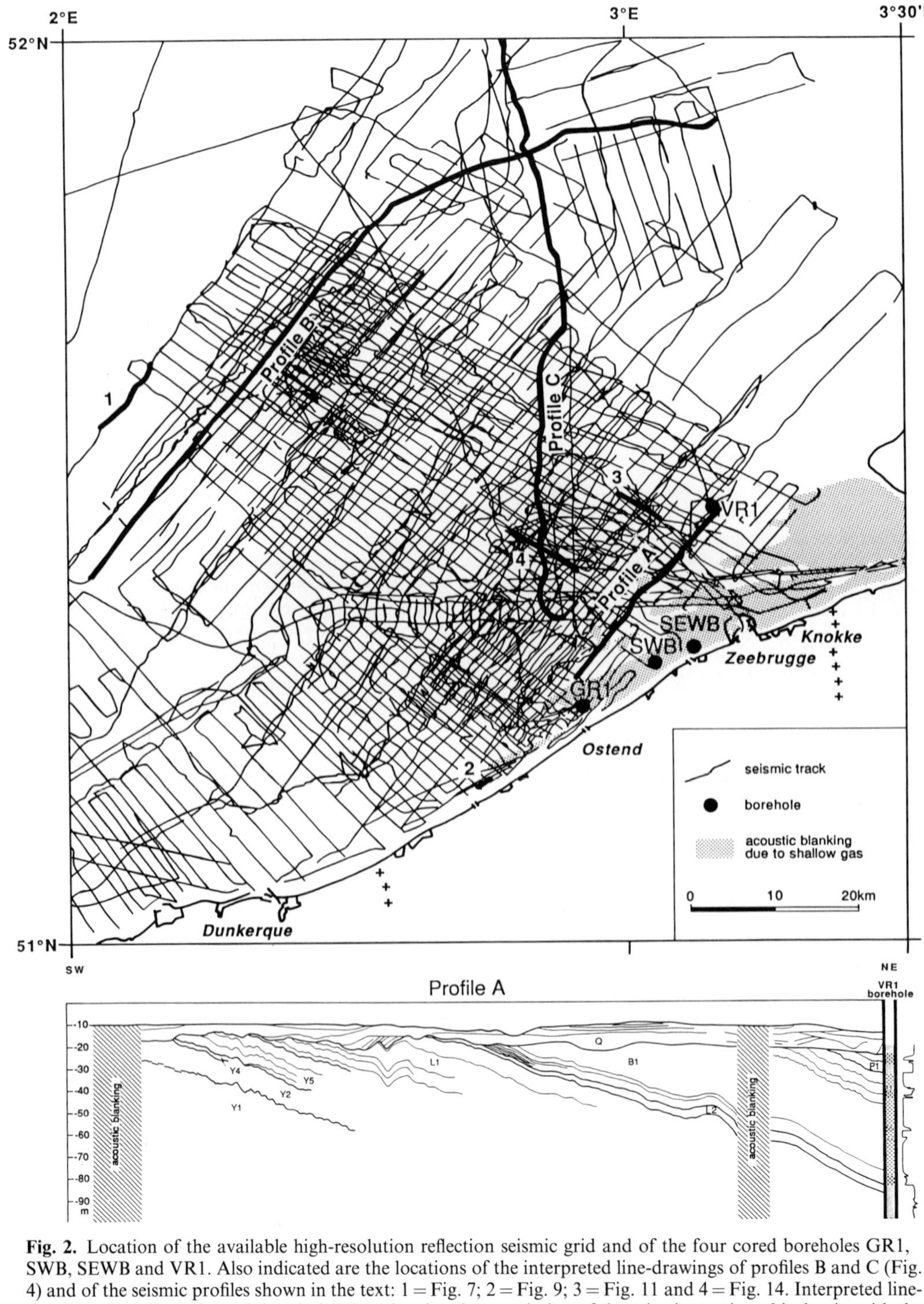

Fig. 2. Location of the available high-resolution reflection seismic grid and of the four cored boreholes GR1, SWB, SEWB and VR1. Also indicated are the locations of the interpreted line-drawings of profiles B and C (Fig. 4) and of the seismic profiles shown in the text: 1 = Fig. 7; 2 = Fig. 9; 3 = Fig. 11 and 4 = Fig. 14. Interpreted line-drawing of profile A through borehole VR1, showing the correlation of the seismic-stratigraphical units with the litho-units encountered in the core.

Table 1. *Seismic-stratigraphical characteristics of the Palaeogene in the offshore Belgian Basin (from De Batist & Henriet 1995)*

Unit	Nature of base	Nature of top	Seismic facies	Geometry
R2	Conformity	Erosional truncation channel incision	Regular pattern of continuous, parallel, high- to medium-amplitude reflectors	Planar dipping base; strongly incised & channelized top Thickness: 0–60 m ($v = 1650\,m\,s^{-1}$)
R1	Discrete downlap	Discrete truncation	Discontinuous, subparallel to wavy reflectors with variable amplitude	Planar dipping base; planar dipping top Thickness: ±35 m ($v = 1700\,m\,s^{-1}$)
P1	Discrete onlap	Discrete truncation	Vertical succession of two seismic facies units: (2) homogeneous pattern of continuous, parallel, low- to middle-amplitude reflectors (1) continuous, parallel, draping reflectors of variable amplitude	Planar dipping base; planar dipping top; divergent to NE Thickness: 40–90 m ($v = 1700\,m\,s^{-1}$)
B1	Conformity	Conformity	Vertical succession of seven seismic facies units: (7) reflection-free with a low-amplitude, discontinuous, draping reflector (6) medium-amplitude, draping reflectors (5) reflection-free (4) convex mounds of medium-amplitude, prograding, hummocky reflectors (3) reflection-free (2) subparallel reflectors with shingled reflector on top (1) regular set of continuous, parallel, high-frequent reflectors	Planar dipping base; Planar dipping top; Thickening to N and NE Thickness: 45–60 m ($v = 1580\,m\,s^{-1}$)
L2	?	?	?	Local distribution, very thin
L1	Conformity	Tectonically influenced truncation	Vertical succession of two seismic facies units (2) discontinuous, parallel to subparallel reflectors of variable amplitude; towards the top 2 or 3 discontinuous, subparallel, very high-amplitude reflectors (1) 2 continuous, high-amplitude parallel reflectors, in the S and 3rd discontinuous, low-amplitude reflector	Planar dipping base: Planar dipping top: Thickness: 25–30 m ($v = 1700\,m\,s^{-1}$)
Y5	Downlap	Truncation and toplap	Three seismic facies units of local areal extent (3) low-amplitude, parallel reflectors (2) parallel-oblique clinoforms (1) reflection-free	Planar dipping base; planar dipping top; Wedge shape, pinching out towards N Thickness: 0–17 m ($v = 1700\,m\,s^{-1}$)
Y4	Downlap	Truncation	Three SE prograding subunits with sigmoidal to parallel-oblique clinoforms of variable amplitude	Linear channelized base; planar dipping top; channel-fill shape (10 km wide, N70° E orientation)
Y3	Discrete downlap	Truncation	Low-amplitude, discontinuous, parallel reflectors or parallel-oblique clinoforms	Planar dipping base; planar dipping top, locally channelized Thickness: 0–25 m ($v = 1600\,m\,s^{-1}$)

Table 1. *Continued*

Unit	Nature of base	Nature of top	Seismic facies	Geometry
Y2	Downlap	Truncation	Reflection-free or very low-amplitude, parallel-oblique, prograding clinoforms	Planar dipping base; planar dipping top Thickness: ±30 m ($v = 1750\,m\,s^{-1}$)
Y1	Conformity	Discrete truncation	Low-amplitude, discontinuous, parallel reflectors, affected by intraformational deformations: top part: undisplaced, faulted blocks with alternately tilting and downwarping bedding terminations central part: major tilted blocks, convolute structures with broad synclines and cusp anticlines, diapirs lower part: block-faulting, tilted and bent blocks, randomly dipping fault planes	Planar dipping base; planar dipping top; thickening to NE Thickness: 150–180 m ($v = 1620\,m\,s^{-1}$)
T2	Downlap	Truncation	Oblique or shingled clinoforms and low-amplitude, discontinuous, subparallel or hummocky reflectors; incised channels at base and at some metres above base	Planar dipping base, locally channelized: planar dipping top Thickness: 15–20 m ($v = 1800\,m\,s^{-1}$)
T1	Onlap	Truncation	Few parallel reflectors of variable amplitude, separated by reflection-free intervals	Irregular base (erosional surface): planar dipping top Thickness: 15–30 m ($v = 1800\,m\,s^{-1}$)

Eocene and the Oligocene by Jacobs (1975, 1978) and the Oligocene by Vandenberghe (1978) and by Vandenberghe & Van Echelpoel (1987). The present knowledge of the Palaeogene lithostratigraphy (Fig. 6) was summarized by Maréchal & Laga (1988) with contributions from various authors.

Palaeocene

In the following sections, the geometry and stratigraphy of the Palaeogene in the Belgian Basin based on offshore and onshore seismic, core and outcrop data are discussed. A summary of the seismic data and lithostratigraphy is given in Fig. 6.

Thanetian

Offshore, the oldest stratigraphical unit of Palaeogene age is the shallow-marine Hannut Formation, equivalent to seismic-stratigraphical Unit T1 of De Batist & Henriet (1995). It is separated from the underlying Late Cretaceous chalk by a pronounced regional onlap surface. Seismic profiles show this erosional surface to be smooth and nearly parallel to the reflectors in the underlying chalk in most of the Belgian Basin. In the extreme west, however, in the prolongation of a major SW–NE-trending structural lineament (the 'North Hinder Deformation Zone' of De Batist 1989), Thanetian palaeohighs delineated by onlap terminations correspond to anticlinal structures (Fig. 7) trending N70° E to N95° E. Unit T1's seismic facies consists of few parallel reflectors of variable amplitude, separated by reflection-free intervals. These are interpreted as the alternations between clay layers and glauconitic fine sands that characterize the Hannut Formation.

It is overlain by the Tienen Formation (seismic-stratigraphical Unit T2 of De Batist & Henriet (1995)), which is known to consist onshore of a variety of continental, fluvial and lagoonal facies (Leriche 1928; De Geyter 1981; Laga & Vandenberghe 1990). On some offshore seismic sections distinct channel cut-and-fill structures can be observed at the base of the unit (Fig. 8; see also De Batist & Henriet (1995, Fig. 9)) and sometimes also at some metres above the base, thus locally defining two erosional surfaces. A tectonic origin for these

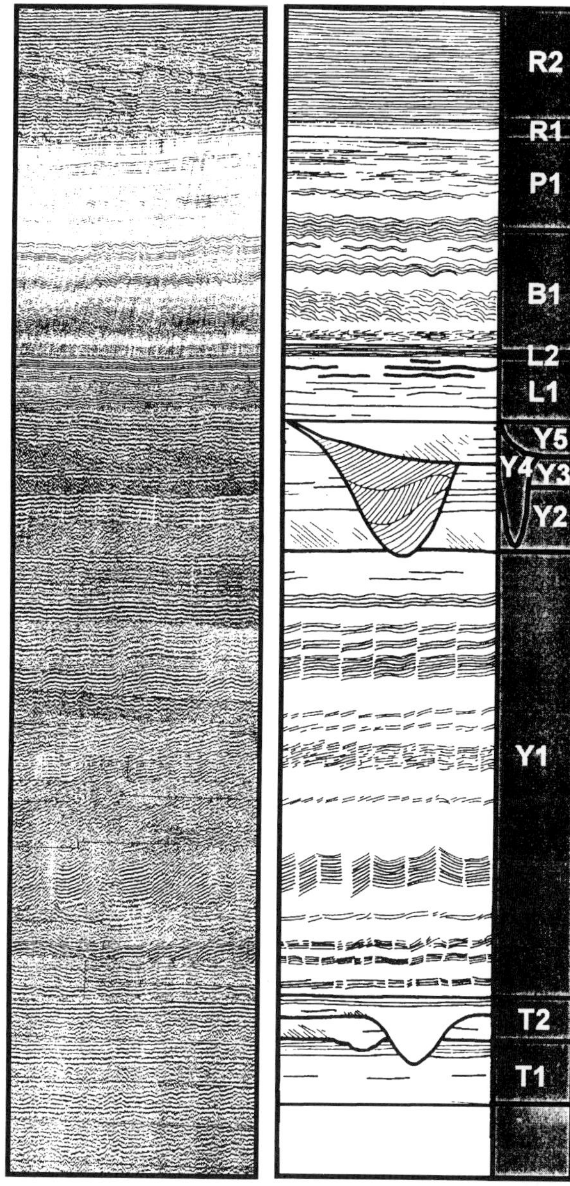

Fig. 3. Synoptic seismic type section, constructed as a composite or 'collage' of several seismogram sections acquired with comparable source signatures, and a synoptic line-drawing illustrating the most important seismic-stratigraphical characteristics of the Palaeogene in the offshore Belgian Basin.

incisions, possibly related to a first phase of updoming of the Weald–Artois High southwest of the study area, is very likely. Unit T2's seismic facies is characterized by oblique or shingled clinoforms and low-amplitude, discontinuous, subparallel or hummocky reflectors. All seismic characteristics suggest that the Tienen Formation is present offshore in its fluvial facies.

Eocene

Ypresian

The oldest stratigraphical unit of Eocene age in the Belgian Basin consists of the 'Ieper clay', which includes the Kortrijk Formation and the Kortemark Member of the Tielt Formation. It is

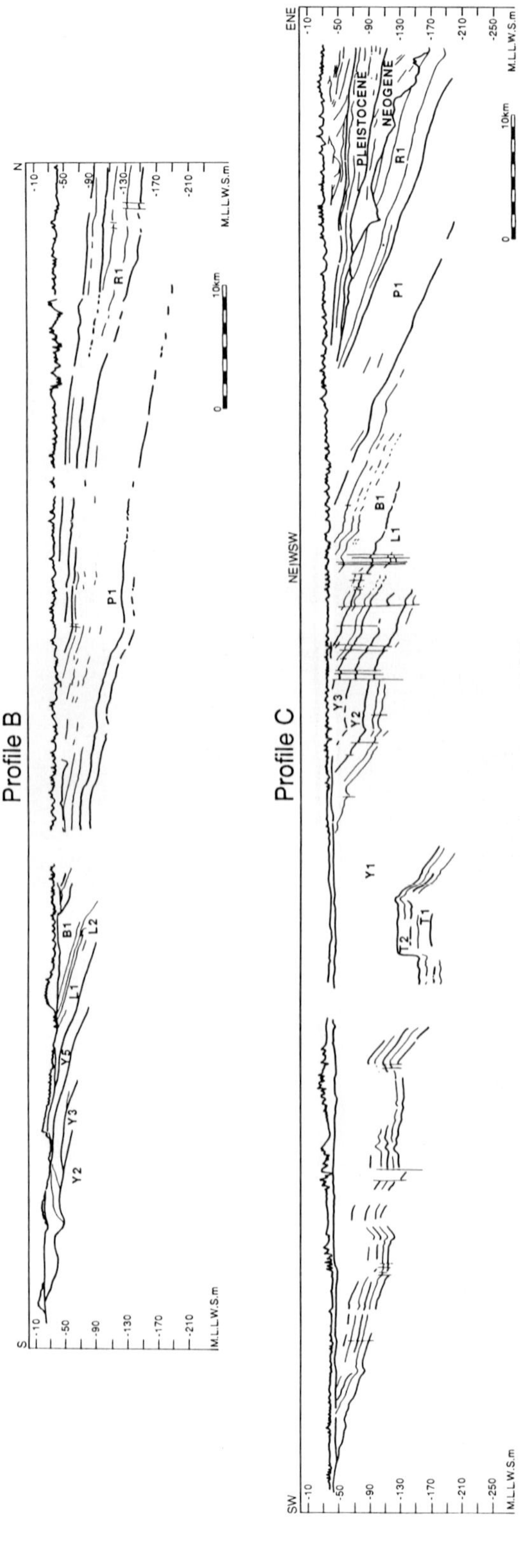

Fig. 4. Interpreted line-drawings of selected seismic sparker profiles through the offshore Belgian Basin illustrating some of the stratal relationships and geometries. Note the limited depth information (max. 100 m), which is due to the type of seismic source and receiver used during acquisition and to the occurrence of a strong sea-floor multiple. See Fig. 2 for location.

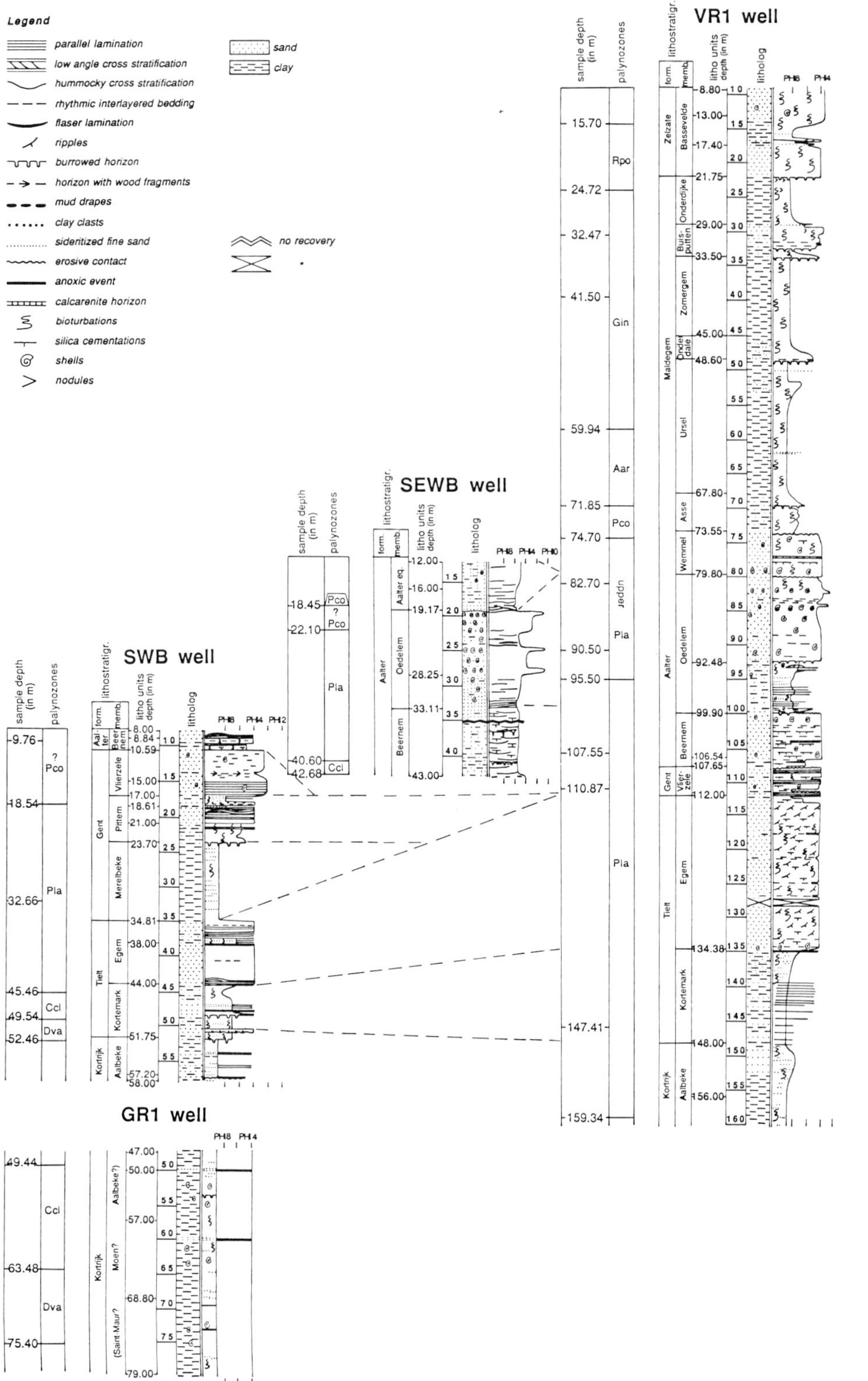

Fig. 5. Composite litholog of the offshore boreholes GR1, SWB, SEWB and VR1, with correlation of identified litho-units. See Fig. 2 for location.

equivalent to seismic-stratigraphical Unit Y1 of De Batist & Henriet (1995) and largely correlates with the London Clay Formation of southern England. Owing to its thickness of more than 150 m its outcrop dominates the geological map, both offshore and onshore. Seismic-stratigraphical Unit Y1 is characterized by a homogeneous seismic facies of low-amplitude, discontinuous,

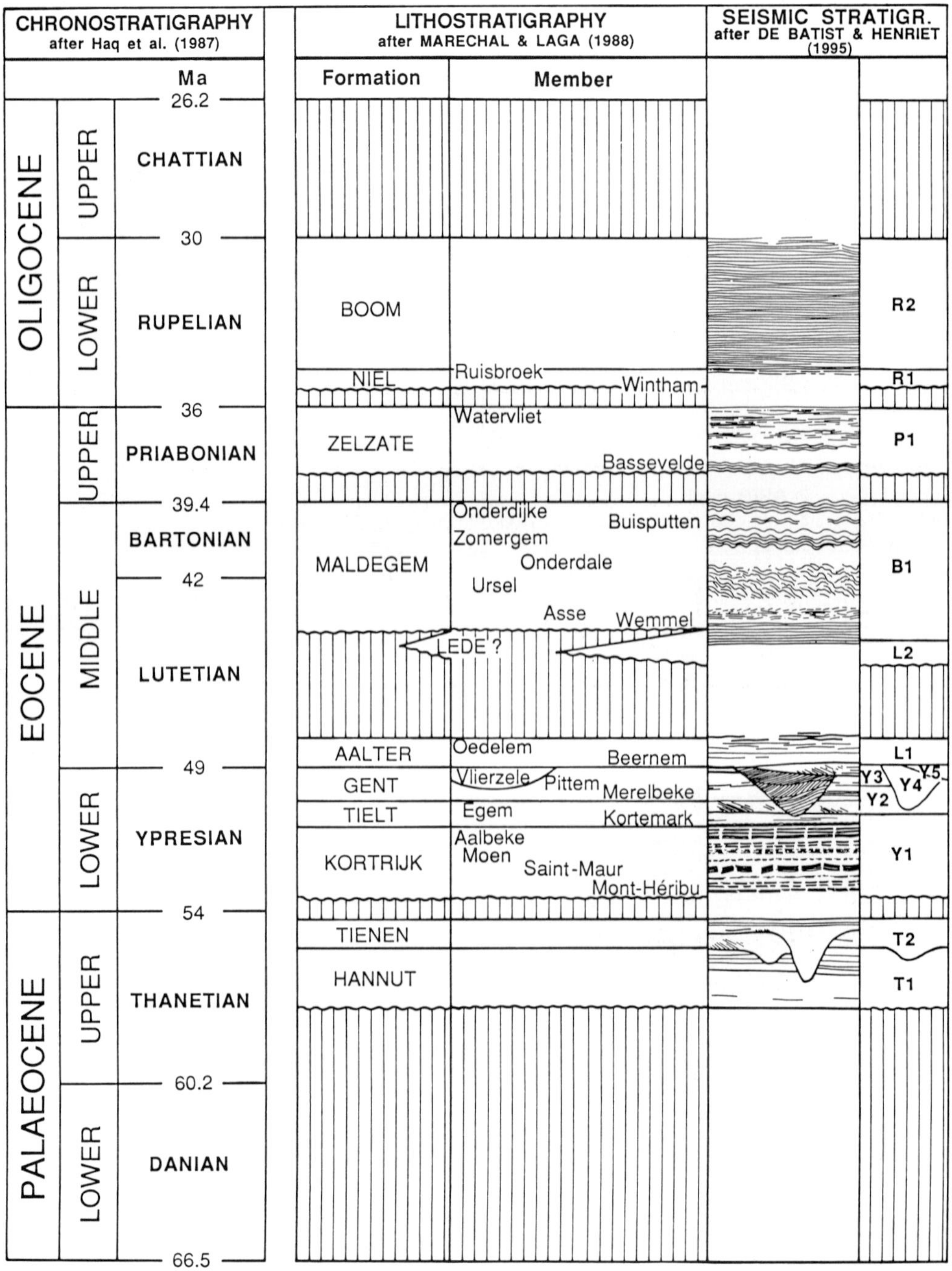

Fig. 6. Synoptic chrono-litho-seismic-stratigraphical correlation diagram for the offshore and onshore Belgian Basin.

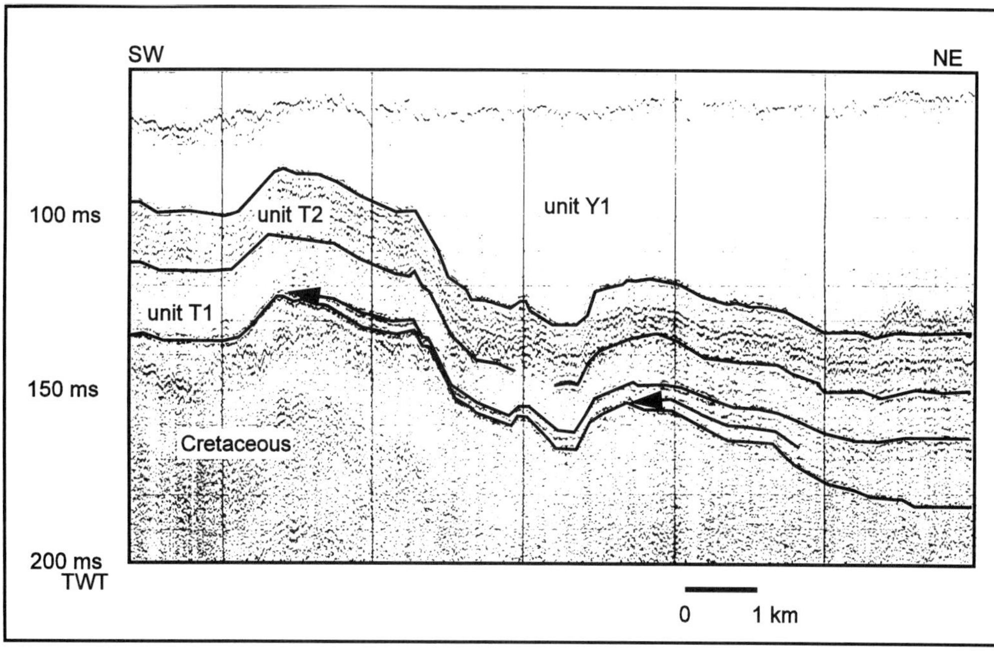

Fig. 7. Interpreted sparker profile showing the characteristic basal onlap on seismic-stratigraphical Unit T1 (Hannut Formation) against the structurally controlled palaeomorphology at the top of the Cretaceous chalk. See Fig. 2 for location.

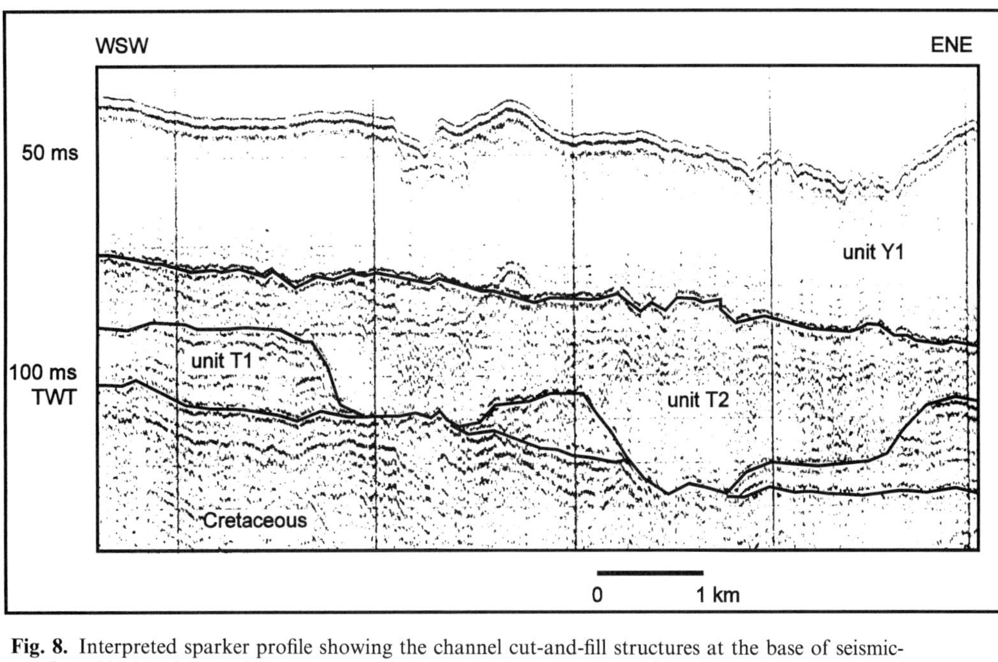

Fig. 8. Interpreted sparker profile showing the channel cut-and-fill structures at the base of seismic-stratigraphical Unit T1 (Tienen Formation). Location is at about 51°47′ N–01°46′ E, just outside Fig. 2.

parallel reflectors, suggesting a relatively long period of stable, low-energy depositional environments. In cores, the massive green-grey clay that makes up most of the Kortrijk Formation is interpreted to be deposited in a mud-shelf environment below storm-wave base, characterized by very low sediment accumulation rates of about $1.5\,cm\,ka^{-1}$ (Jacobs & Sevens 1988).

In the 'homogeneous' clay section of the GR1 well, submicroscopical clay-particle orientations indicate transgressive and regressive trends (Van Bavinchove, pers. comm.). During a transgressive phase the depositional environment deepens, causing a migration of the anaerobic environment towards the coast. This results in a progressive transition from an aerobic over a dysaerobic towards an anaerobic environment, accompanying a vertical change in clay texture from a disturbed (through bioturbation) and random orientation of the clay particles, over an undisturbed but preferential orientation towards a laminated and preferential orientation. The colour of the clay gradually becomes darker. This process reverses during a regression (O'Brien & Slatt 1990).

A dark, organic-rich clay layer in the GR1 well marks an anoxic event representing a condensed section. This is also documented by the submicroscopical texture that reveals the clay particles to be very well parallel-oriented, which is due to very low sedimentation rates and absence of bioturbation with reduced detrital sediment supply but high organic material deposition (Van Bavinchove, pers. comm.). Organic-rich laminae present in the GR1 well are most probably related to peak discharge of rivers. Bioturbated horizons indicate short periods of non-deposition or sediment reworking (hiatus).

In the upper part of the 'Ieper clay' section in the VR1 well, stratigraphically equivalent with the Kortemark Member, centimetric, fining upward, yellowish, silty sand laminae with sharp lower and upper boundaries point to increasing storm influence. They consist almost entirely of early-diagenetic authigenic botryoidal and framboidal siderite in a poor clay matrix (Van Bavinchove, pers. comm.), formed shortly after burial and before compaction, just beneath the zone of bacterial sulphate reduction. The

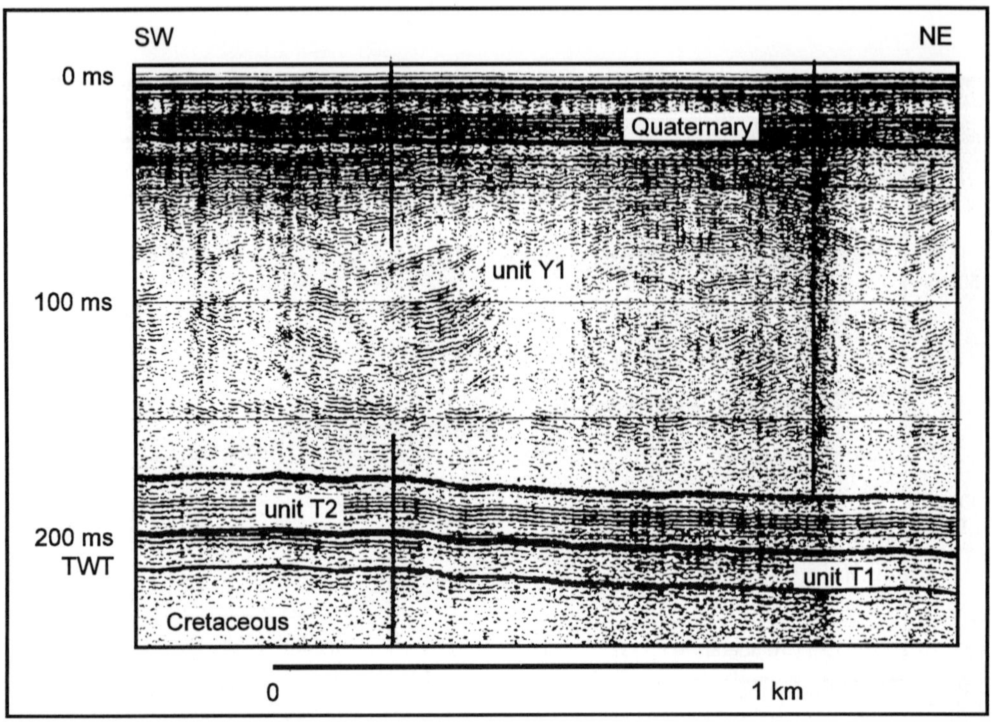

Fig. 9. Interpreted sparker profile off the Belgian coast showing the disturbed seismic facies of seismic-stratigraphical Unit Y1 (Kortrijk Formation) due to the presence of intraformational 'sediment tectonic' deformations. See Fig. 2 for location.

occurrence of slightly higher-amplitude reflectors at the top of Unit Y1 could be the seismic expression of this.

Palaeowater depths of the 'Ieper clay', based on palynomorph Dva to Ccl zones, range from 20–40 m at the base to 20 m at the top, while diatoms indicate that the depositional environment varies from oxygen-depleted central shelf through inner shelf with less salinity to coastal and inner shelf.

In the Kortrijk Formation section present in the GR1 and SWB wells, a stacked pattern can be discerned in a total of four fining upward clayey litho-units of overall transgressive nature. They are capped by one coarsening upward regressive litho-unit, consisting of sandy clay with burrows at the top, and by a sharp-based bioturbated silty to fine-sandy clay unit with centimetric silty sand lenses (Jacobs *et al.* 1990; Jacobs & Sevens 1993*a*; Jacobs 1995*b*). These litho-units cannot be identified on the seismic sections of Unit Y1, because the internal reflection pattern is strongly disturbed by a wide range of post-depositional, compaction-related intraformational 'sediment tectonic' deformations (Fig. 9), including faulting, clay diapirism, etc. (Henriet *et al.* 1988, 1991). Some of these structures can also be observed in onshore clay quarries (Verschuren 1992).

The 'Ieper clay' is overlain by the Egem Member of the Tielt Formation. This Egem Member is equivalent to the seismic-stratigraphical Unit Y2 of De Batist & Henriet (1995). The boundary between Y1 and Y2 is a laterally continuous, very high-amplitude reflector (see Fig. 3) throughout the Belgian Basin (De Batist *et al.* 1989). Based on its seismic characteristics it is interpreted as a sequence boundary overlain by downlapping lowstand deposits with reflection-free or very low-amplitude, parallel-oblique, prograding clinoforms. In the offshore boreholes, the Egem Member consists of sharp-based, glauconitic, bioturbated fine sands with low-angle cross-bedding and fine laminations, ripples and flaser lamination (mud drapes) indicative of a deltaic origin with a southern sediment supply (Jacobs *et al.* 1990; Jacobs & Sevens 1993*a*; Jacobs 1995*a*, *b*). Figure 10 shows some aspects of the sedimentary facies of the Egem Member in its type locality. Here, a 20 cm thick silica-cemented clayey sand layer with shell ghosts occurring at the top of the Egem sands, displays at its surface polygonal desiccation cracks, filled in with clay and iron coatings. This surface is interpreted to

Fig. 10. Flat-bottomed erosion gullies of deltaic origin in the Egem Member (seismic-stratigraphical Unit Y2) visible in the Egem quarry.

indicate subaerial exposure (top lowstand deposits), as it is covered with abundant and coarse authigenic glauconite, probably representing a lateral time equivalent of the (transgressive) Merelbeke clay, that is absent in the Egem outcrop.

Seismic-stratigraphical Unit Y3 of De Batist & Henriet (1995) corresponds to the Merelbeke Member (at the base) and the Pittem Members (at the top), which are both part of the Gent Formation. Low-amplitude, discontinuous, parallel reflectors or parallel-oblique clinoforms characterize this Unit Y3. Facies and palaeobathymetry analyses performed on the offshore cores indicate that the sedimentary environment deepens up to a nearshore mudshelf (15 m palaeowater depth tending towards 20 to 30 m). In the SWB well, the strongly bioturbated transgressive Merelbeke silty clay with thin sand laminae is heavily burrowed at the top and truncated, suggesting delta-complex influence (Jacobs & Sevens 1993*a*; Jacobs 1995*b*). It is overlain by the bioturbated, sandy clay Pittem Member with mud drapes, rhythmic interlayered bedding, shell clasts and local silica cementations. The Pittem Member is interpreted to be deposited in a tidal environment during relative sea-level highstand, although palaeowater depths of palynomorph Pla zone seem to increase (reworking?), but salinity decreases and continental influx becomes prominent.

Seismic-stratigraphical Unit Y4 of De Batist & Henriet (1995) is confined to an erosional depression (channel or basin?), trending N70° E, some 20 km N off Oostende. It is locally more than 10 km wide and deeply incised in the underlying strata. On seismic sections, Unit Y4 displays three infilling stages by subunits composed of sigmoidal to parallel-oblique clinoforms of variable amplitude, prograding from NW to SE (Fig. 11). This seismic unit most likely correlates with a basal part of the Vlierzele Member, which is part of the Gent Formation. This interpretation is supported by comparable erosional features at the base of this Vlierzele Member onshore, as observed by Birchall (pers. comm.) (Fig. 12), who interpreted them as caused by valley incision in response to falling relative sea level.

Seismic-stratigraphical Unit Y5 (De Batist & Henriet 1995), separated from Unit Y4 by a sharp erosion surface, rapidly pinches out in the offshore direction (Fig. 11). A number of different facies subunits of local areal extent can be identified within this unit. Some consist of low-amplitude parallel reflections; others of eastward prograding parallel-oblique clinoforms; some are reflection-free. Unit Y5 probably correlates with the cross-bedded sands of the Vlierzele Member, which have been described onshore in several outcrops and that have been interpreted as a tidal ridge system (Houthuys 1990).

In the offshore borehole SWB, the Vlierzele Member is present in its typical facies, as green glauconitic fine sands with low-angle parallel lamination and belonging to palynomorph Pla or ?Pco zone. Wave-influenced sand shoals characterized by thin brown sand laminae and a brown clayey matrix in the top zone are indicative of an intertidal depositional environment transitioning into a supratidal environment. Organic debris and wood fragments are probably remnants of reworked soils and

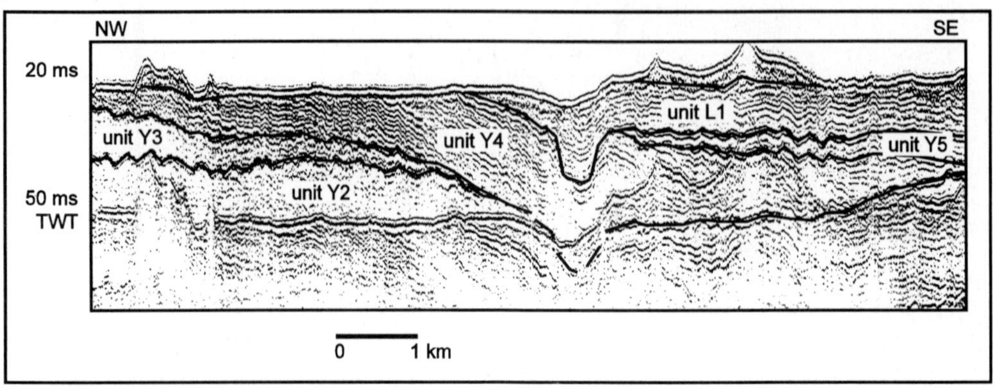

Fig. 11. Interpreted sparker profile showing the truncation associated with the erosional bases of seismic-stratigraphical Units Y4 and Y5 (Vlierzele Formation) and the progradational seismic facies of these units. Note that the shape of the erosional basin or channel of Unit Y4 is obliterated on this profile by a synclinal fold affecting these strata. See Fig. 2 for location.

(a)

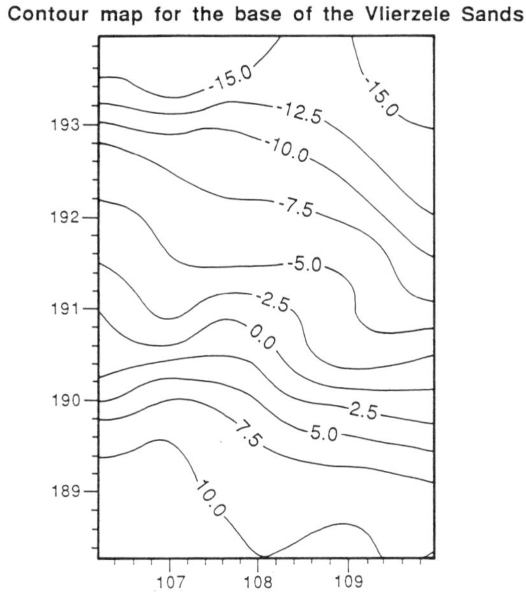

(b)

Fig. 12. (**a**) Isohypses of the erosional base of the Vlierzele Member suggest local valley incision in response to falling relative sea level (after R. Birchall, pers. comm.). For location of study area, see Fig. 1. Lambert coordinates are indicated. (**b**) In the Balegem quarry, valley infilling is documented by green glauconitic fine sands with low-angle parallel lamination of the Vlierzele sands Member litho-unit (seismic-stratigraphical Unit Y4).

vegetation in the vicinity, suggesting emersion. The time-equivalent facies in the VR1 well is more clayey and deposited on the central (to outer?) shelf with an oxygenated bottom under 40 m palaeowater depth. The three coarsening upward cycles suggest transition from a lagoonal environment towards a minor mouthbar or crevasse splay deposition in a deltaic (delta plain) environment. At present, it remains unclear which seismic-stratigraphical unit (Y4 or Y5?) is represented as the Vlierzele Member in these wells. Due to its very restricted occurrence, it is likely that Unit Y4 is not encountered in either of the wells, and that both lithofacies are to be attributed to Unit Y5.

The erosion surface at the base of Unit Y4 is clearly a sequence boundary, while the channel infill and the overlying Unit Y5 could represent the lowstand and the transgressive (parallel facies) to highstand (prograding facies) deposits respectively, separated by a ravinement surface. Conversely, the erosion surface at the base of Unit Y5 could also be interpreted as a separate sequence boundary. In any case, the seismic data seem to suggest that the offshore Vlierzele Member is rather more complex than previously believed, and may contain a basal facies that is not known onshore.

Lutetian

The Vlierzele Member is overlain by the Aalter Formation, largely correlative with seismic-stratigraphical Unit L1 (De Batist & Henriet 1995) and consisting of a variety of very shallow marine deposits with great vertical and lateral facies diversity, reflecting strongly varying depositional environments. In the SEWB well the succession starts with the Beernem Member, consisting of slightly coarsening upward grey-green bioturbated glauconitic clayey fine sands with local silica cementations, shell fragments (and large specimens of *Cardita planicosta* bivalves in the VR1 well). Vertical burrows of *Callianassa* extend from erosion surfaces and penetrate into parallel-laminated fine sands. Deposition took place in subtidal gullies and mixed intertidal flats with high-energy conditions caused by repeated channel incision and infill with lateral migration high on the shelf. Palaeowater depths decrease from 40 m (central to ?outer shelf, with oxygenated bottom) to 30 and 20 m (inner shelf).

Flaser lamination in the SEWB well, obliterated by bioturbation and parallel lamination, points to tidal influence during the deposition of the grey-green glauconitic fine sands to clayey sands of the overlying, highly fossiliferous Oedelem Member, which is characterized by an irregular but sharp lower boundary with clay clasts and burrows backfilled with sand. Coquinas exclusively composed of *Turritella* gastropods form barriers (Fig. 13), which in a later stage are reworked in a lagoonal facies as indicated by the deposition of sandy clays with reworked shells (Jacobs & Sevens 1993*a*; Jacobs 1995*b*). Slow sedimentation in an inner (to ?central) shelf environment with oxygenated bottom under 30 to 20 m palaeowater depth

Fig. 13. In the SEWB cored borehole, the Oedelem Member litho-unit (seismic-stratigraphical Unit L1) contains coquinas composed of *Turritella* gastropods, that form barriers protecting a lagoon open to the sea.

evolves towards sedimentation on an inner and coastal shelf with decreasing salinity, under 20 to 10 m palaeowater depth.

The seismic facies succession within Unit L1 is in agreement with these types of deposits: (1) two continuous, high-amplitude parallel reflectors and a third discontinuous, low-amplitude reflector at the base; (2) discontinuous, parallel to subparallel reflectors of variable amplitude. A number of discontinuous, subparallel, very high-amplitude reflectors at the top of L1 correlate with calcareous sandstone beds that have been reported in the top section of the Oedelem Member in the Zeebrugge area (Depret 1983). They are indicative of the upward shallowing of the Aalter Formation (palynomorph Pco zone), the upper part of which has been interpreted as a stacked pattern of highstand systems tract (HST) parasequences.

Off Zeebrugge, the strata are offset by normal faults and deformed by asymmetrical folds. Here, the characteristic strong reflectors of the upper part of Unit L1 are truncated at the crests of the folds (Fig. 14). Two folding phases can be recognized from the seismic data (De Batist 1989). As the main folding phase clearly took place after the Bartonian, the observed truncation pattern indicates an initial deformation phase shortly before deposition of the overlying seismic-stratigraphical Unit L2 (De Batist & Henriet 1995). The upper boundary of Unit L1 therefore represents a tectonically enhanced unconformity. This seismically well-defined unconformity coincides with a considerable hiatus in the coastal area onshore, where the Brussel and Lede Formations of central-northern Belgium are generally absent. Seismic-stratigraphical Unit L2, overlying this unconformity, is a very thin and locally distributed, patchy deposit, which is most likely not present in the VR1 well. The seismic characteristics of Unit L2 suggest that it may represent a transgressive lag deposit, overlying a ravinement surface that coincides with the sequence boundary. Unit L2's exact nature and stratigraphical position remains uncertain, although it is most probably stratigraphically equivalent to the Lede Formation. Jacobs & Sevens (1993*b*) interpret this Lede Formation in onshore outcrops as a shallow marine, transgressive deposit, consisting of slightly fining upward, yellowish-grey, fine sands with numerous *Nummulites variolarius* and a sharp, erosive lower boundary with weathered *Cardita planicosta* specimens, silicified *Nummulites laevigatus*, well-rounded but corroded Brussel Formation sandstone pebbles, and ray and shark teeth. Apparent stratification is absent, but coarse-grained sand layers represent storm deposits, and (normally three) calcarenite horizons show great lateral continuity, and their gradual

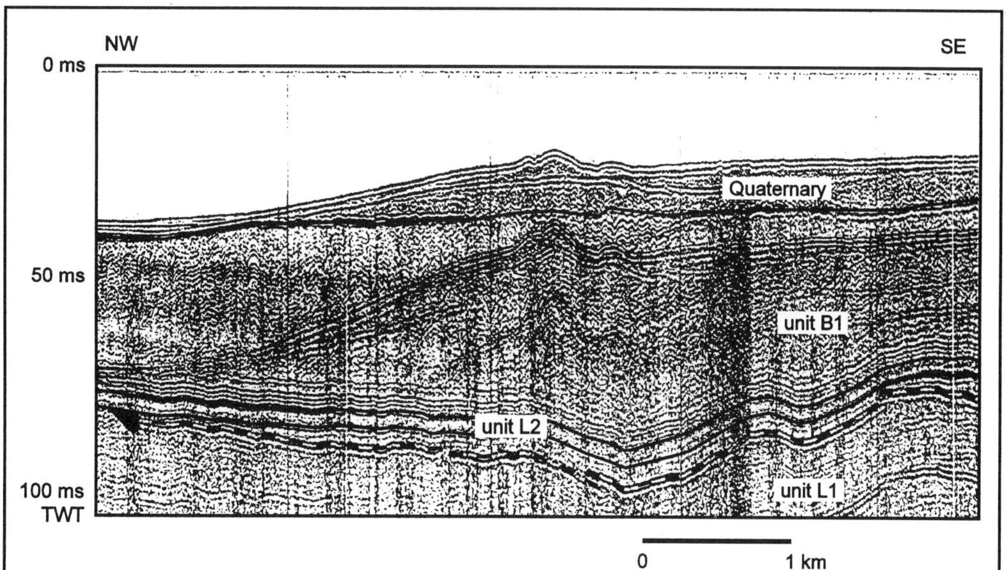

Fig. 14. Interpreted sparker profile showing the tectonically enhanced unconformity at the top of seismic-stratigraphical Unit L1 (Aalter Formation), the thin overlying Unit L2, and Unit B1 (Maldegem Formation) affected by folding. See Fig. 2 for location.

transition can sometimes be traced into coarse-grained sand layers that acted as local aquifers. The calcarenite build-up, characterized by a dense and well-cemented internal zone with shell ghosts sandwiched between two crumbly and poorly cemented outer rims with friable shells, suggests progressive formation by calcite precipitation, as meteoric water is pumped through local aquifers in these shallow marine deposits during the ensuing relative sea-level fluctuations (Tucker 1993).

Lutetian–Bartonian

Onshore, the Lede Formation is overlain by the Maldegem Formation, which is also encountered in the VR1 well offshore. The formation is characterized by a regular succession of seven distinct litho-units showing typical alternations of sands and clays: the Wemmel sands Member at the base, the Asse clay Member, the Ursel clay Member, the Onderdale sands Member, the Zomergem clay Member, the Buisputten sands Member and the Onderdijke clay Member. The Maldegem Formation, with a total thickness of about 60 m correlates well with seismic-stratigraphical Unit B1 of De Batist & Henriet (1995). On seismic profiles throughout the Belgian Basin, Unit B1 is characterized by a very distinctive and laterally very continuous succession of seven seismic facies units (Fig. 15): from bottom to top (1) a regular set of continuous, parallel, high-frequency reflectors, (2) subparallel reflectors with shingled reflector on top, (3) a reflection-free interval,

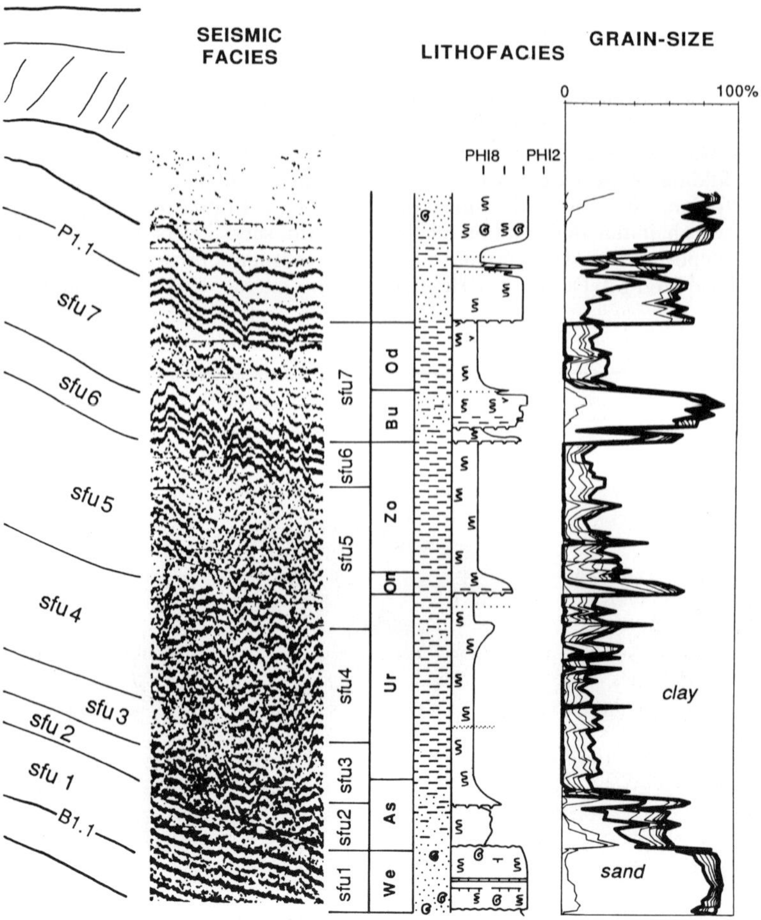

Fig. 15. The very distinctive and laterally continuous succession of seven seismic facies units (sfu) comprising seismic-stratigraphical Unit B1 (Maldegem Formation), correlates with grain-size and/or lithological changes in the VR1 cored borehole. However, comparable seismic facies do not always correlate with comparable lithological facies.

(4) convex mounds of medium-amplitude, prograding, hummocky reflectors, (5) a reflection-free interval, (6) medium-amplitude, draping reflectors, and (7) a reflection-free interval with a low-amplitude, discontinuous, draping reflector (see also De Batist & Henriet 1995, Fig. 11). Quite surprisingly and contrary to what has been suggested previously by De Batist & Henriet (1995), this succession of seismic facies does not simply mimic the succession of lithofacies that typifies these deposits. Figure 15 illustrates that seismic facies boundaries sometimes, although not always, coincide with the major lithofacies or grain-size boundaries, and that comparable seismic facies do not always correlate with comparable lithofacies.

In the VR1 well, the Wemmel sands Member at the base of the Maldegem Formation consists of grey glauconitic, slightly clayey fine sands, slightly fining upwards, with two discontinuous thin calcarenite horizons, weathered shell fragments and a sharp but irregular top. These seem to be slowly deposited in a central (to ?inner) shelf environment with an oxygenated bottom under 30 to 20 m palaeowater depth. They are characterized by palynomorph Pco (and ?reworked Pla) associations.

In the VR1 well (Fig. 15), the remainder of the Maldegem Formation can be described as composed of three stacked fining upward cycles of decametric thickness departing from erosion surfaces with sand-filled burrows. The lowermost cycle starts with highly glauconitic clayey sands with glauconite concentrations in burrows ('bande noire' of authors). They are covered by an alternation of bioturbated (mostly *Chondrites*) clayey sands and sandy clays of the Asse Member. They have been deposited in an outer shelf environment under 30–40 m palaeowater depth, with palynomorph Pco associations indicating the presence of a condensed section. These sandy clays grade into a blue-grey bioturbated massive clay – the Ursel clay Member – with pyrite concretions of prodeltaic origin (central to ?outer shelf with 40–50 m palaeowater depth; palynomorph Aar zonation). The second cycle also departs from a burrowed lower surface and consists of moderately clayey sands – the Onderdale sands Member – sometimes showing coarse interlayered bedding and flaser lamination typical of tidal flat sedimentation. It fines upward into a strongly bioturbated blue-green clay – the Zomergem clay Member – of prodeltaic origin, deposited in outer shelf (to ?slope) conditions (more than 50 m palaeowater depth). It belongs to the palynomorph Gin zone. The third cycle is similar in nature and is characterized in its lower part by slow sedimentation in a central (to ?outer) shelf environment, under 40 m palaeowater depth. It shows towards its top a shallowing tendency towards an inner shelf environment of about 15–20 m depth, while further onshore even thin detrital peat layers have been recorded in its burrowed and truncated top (Gulinck 1969*a*; Jacobs 1975).

On the basis of the aforementioned characteristics and of well-log signatures, the Maldegem Formation can be subdivided into three depositional sequences, each composed of transgressive and highstand deposits.

Priabonian

In the VR1 well, the Maldegem Formation is separated from the overlying Zelzate Formation – of Priabonian age – by a burrowed erosional surface. It is equivalent to seismic-stratigraphical Unit P1 of De Batist & Henriet (1995) and thickens considerably in a basinward direction. In the VR1 well, it consists of strongly bioturbated clayey fine sands – the Bassevelde sands Member – with a mottled texture, locally containing carbonised plant remains. A bioturbated (*Chondrites*) intercalated sandy clay layer coarsens upward into glauconitic, bioturbated and mottled, slightly clayey fine sands, containing some shell grit (palynomorph Rpo zone). These units are interpreted as shallow marine (inner shelf environment under 20 m water conditions) and barrier-protected lagoonal, washover and tidal flat deposits (Jacobs & Sevens 1993*a*; Jacobs 1995*b*). Unit P1's seismic facies of (1) continuous, parallel, draping reflectors of varying amplitude, and (2) homogeneous pattern of continuous, parallel, low- to medium-amplitude reflectors, is consistent with this type of environment. Seismic-stratigraphically, the base of Unit P1 stands out as a low-angle coastal onlap surface and can therefore be characterized as a sequence boundary.

In the NE prolongation of the North Hinder Deformation Zone (De Batist 1989), the upper boundary of Unit P1 shows erosional truncation. Here, structurally undisturbed Oligocene strata rest unconformably on folded strata of Unit P1, of Eocene age. The boundary between Units P1 and R1 therefore represents a tectonically enhanced unconformity.

Oligocene

Rupelian

Seismic-stratigraphical Unit R1 of De Batist & Henriet (1995) probably correlates with most of

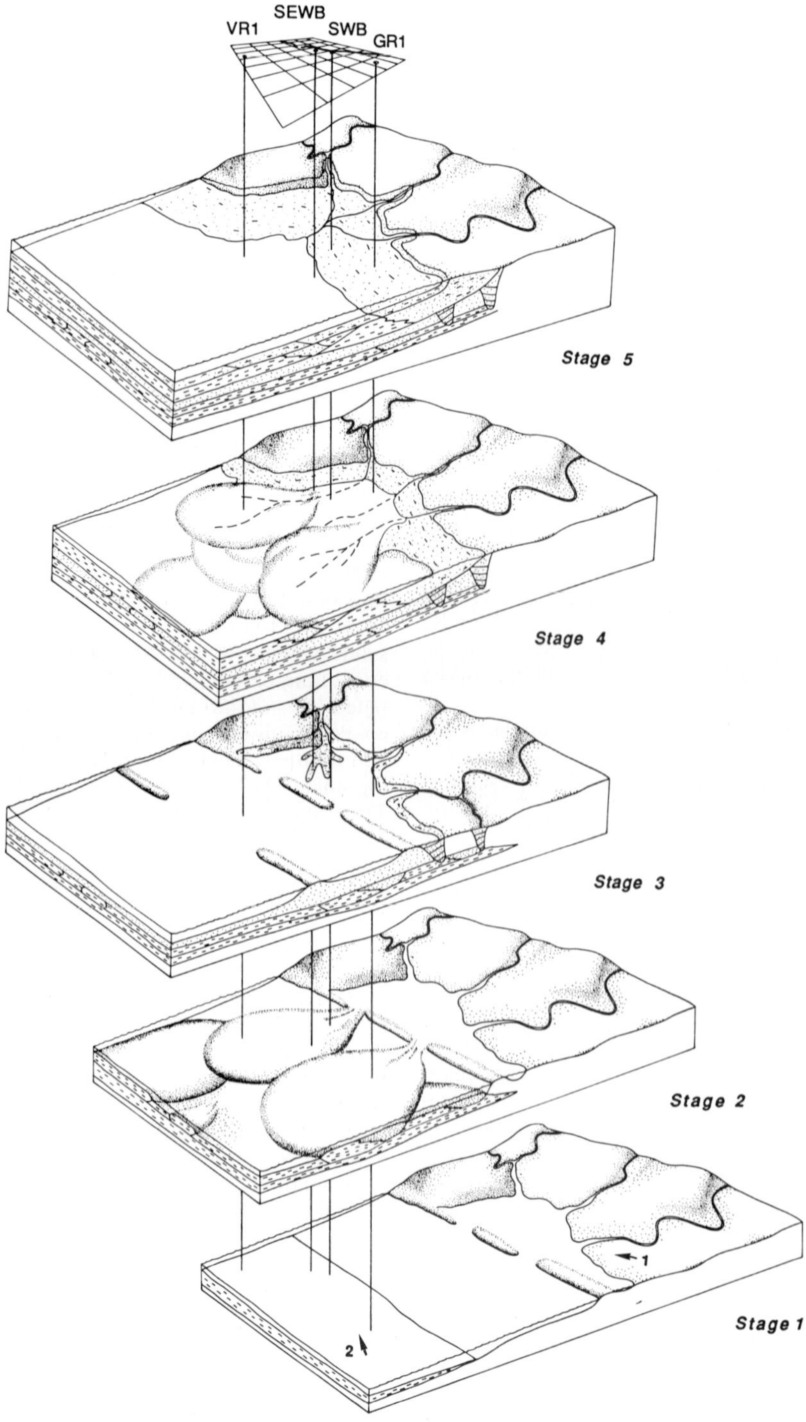

Fig. 16. Evolution of the siliciclastic sedimentary system in the southern North Sea Basin during the Eocene in time and space. Stage 1: early Early Eocene; Stage 2: middle to late Early Eocene; Stage 3: Middle Eocene; Stage 4: Late Eocene; Stage 5: late Late Eocene. Arrows indicate (1) north and (2) main longshore current direction.

the Niel Formation of northwestern Belgium (Steurbaut 1992), consisting of shallow-marine to deltaic/estuarine sands and clays (Jacobs & De Coninck 1977; Steurbaut 1986). The irregular and laterally discontinuous seismic facies of Unit R1, composed of subparallel to wavy reflectors with variable amplitude, is in agreement with these types of deposits. Unfortunately, this unit is not encountered in the VR1 well. Onshore, large 'Big Mac' concretions with a diameter of more than 1 m and a hamburger-like shape characterize the base of the Niel Formation. Macro- and submicroscopical growth patterns supported by geochemical and petrographical evidence suggest that they were formed during relative sea-level highstand when meteoric water was pumped through the sediment aquifers active in these shallow marine deposits (Olivier, pers. comm.).

Offshore, it is overlain by Unit R2, the youngest seismic-stratigraphical unit in the area that was assigned a Palaeogene age by De Batist & Henriet (1995). It has a maximum thickness of about 60 m. It forms the offshore equivalent of the Boom clay Formation of northern Belgium, but possibly also includes the uppermost part of the underlying Niel Formation. The seismic facies of the Boom clay is very characteristic and consists of a very regularly banded pattern of continuous, parallel, high- to medium-amplitude reflectors (see De Batist & Henriet 1995, Fig. 13). On very high-resolution seismic data, these parallel reflections consist of alignments of diffraction hyperbolae, generated by the well-known bands of Boom clay concretions or 'septaria'.

Sedimentation model

On basis of the sedimentary facies, architecture and sequence-stratigraphical characteristics of the sediment series observed in the offshore seismic and borehole data sets, the depositional palaeoenvironments in the Southern Bight of the North Sea Basin during the Eocene (Palaeocene and Oligocene strata were not present in the offshore boreholes) could be reconstructed (Fig. 16).

Stage 1: early Early Eocene (Ypresian)

During early Early Eocene times (Ypresian), distal Kortrijk Formation sediments were deposited on an open offshore mud shelf during a period when the uplift of the Weald–Artois High was initiated causing the separation of the Southern Bight of the North Sea Basin from the opening North Atlantic Ocean. The Kortrijk Formation litho-units form stacked parasequences indicating a constantly rising relative sea level (transgressive systems tract, TST).

Stage 2: middle to late Early Eocene (Ypresian)

In middle to late Early Eocene times, relative sea level fell and a delta complex started to prograde onto the shelf. Heavy minerals indicate that this delta complex was fed by a fluvial drainage system with a southern sediment supply (Jacobs 1995*a*). The Kortemark Member of the Tielt Formation probably forms a complete lowstand systems tract (LST)–TST–HST cycle. The ensuing sea-level fall must have been considerable as indicated by the 'blocky' geophysical log response of the sharp-based LST of the Egem sands Member (Tielt Formation) in the following cycle. After a temporal reinstallation of an offshore mud shelf during a limited period of locally greater water depths, delta sedimentation with lagoonal mud flats and sand shoals is restored in the top portion of the Merelbeke and Pittem clay Members (Gent Formation), with deposition of regressive facies of HST as relative sea level starts to fall again. At the end of Early Eocene times (Vlierzele sands Member of the Gent Formation) tidal influence became prominent with deposition of wave-influenced subtidal sand shoals.

Stage 3: Middle Eocene (Lutetian)

Because of the constant lowering of relative sea level during Middle Eocene times, the sedimentation system shifted landwards towards its most proximal position. High wave energy, longshore currents and large supply of coarse sediment force the delta to retreat. Wave-influenced subtidal and intertidal environments with subtidal gullies and mixed tidal flats (Beernem Member of the Aalter Formation) and submarine coastal barriers protecting a lagoon open to the sea (with storm deposits of the Oedelem Member) give rise to stacked HST parasequences, indicating constant shoreline regression and shallowing of the basin.

Stage 4: Late Eocene (Lutetian–Bartonian)

In Late Eocene times, after a major hiatus, prominent but stepwise overall relative sea-level rise induces cyclic sedimentation from a tidal sand flat environment (sandy deposits of the Maldegem Formation) towards a prodelta environment (predominantly clayey Maldegem Formation deposits). Distal muddy delta fans alternate with proximal sandy sediments, displaying the progradational pulses of the delta lobes. Aggradation produces (at least three) stacked plurimetric sedimentary (para)sequences of TST–HST nature. They display a very simple, parallel layered (planar) geometry throughout the basin, indicating the almost total absence of intrabasinal relief. Each sequence appears to depart from a basal erosion surface, which indicates a repetition of transgressive pulses.

Stage 5: late Late Eocene (Priabonian)

After a new but minor relative sea-level drop, responsible for the development of thin detrital

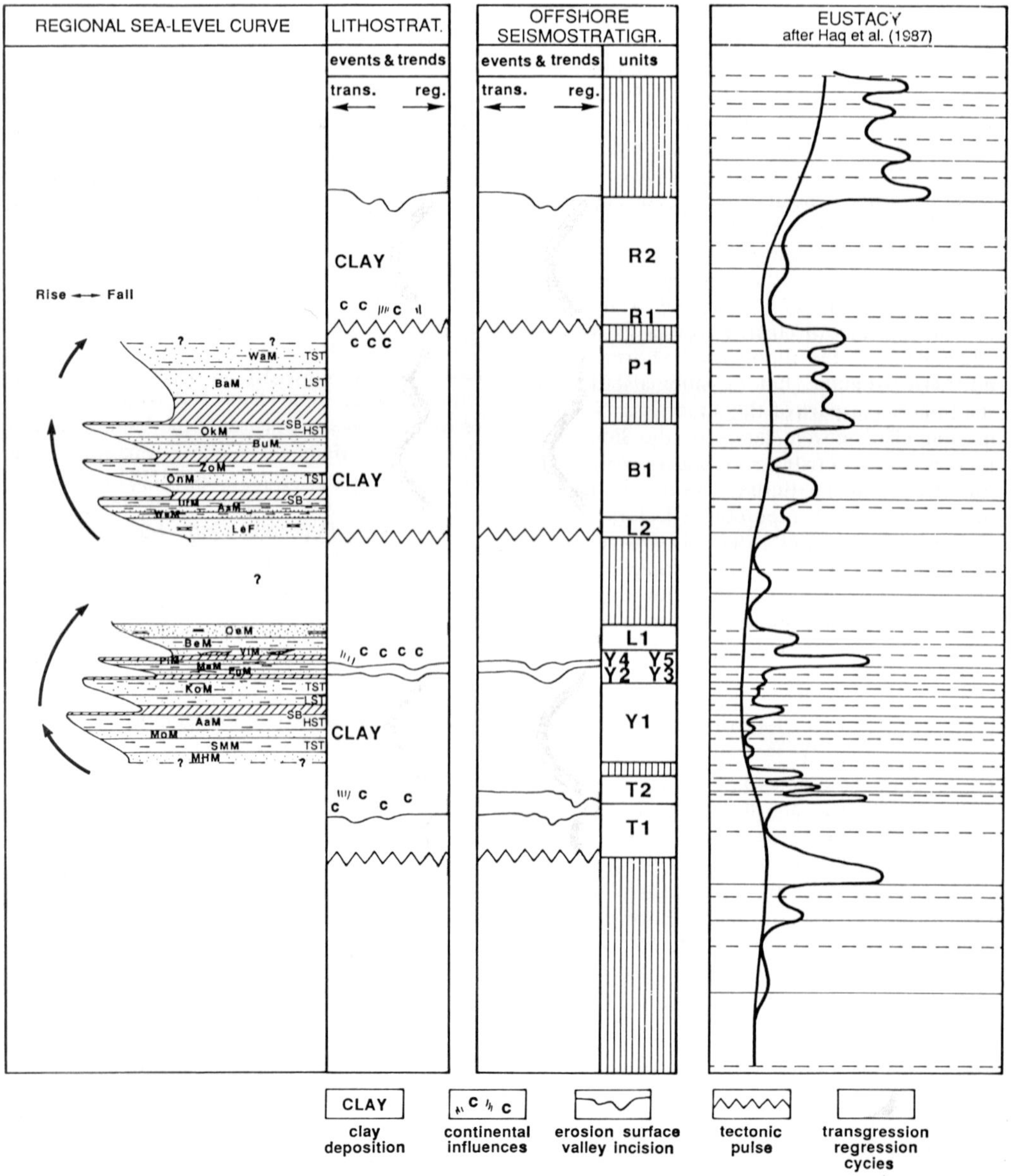

Fig. 17. Synoptic correlation diagram for the Belgian Basin. Seismic stratigraphy illustrates the major 'events and trends'. Eocene schematical regional relative sea-level curve displays two asymmetric second-order transgressive–regressive cycles with nine 'lithostratigraphical' third-order sequences.

peat layers and burrows backfilled with peaty sand in the top of the Maldegem Formation, stacked LST–TST (para)sequences develop that constitute the Uppermost Eocene Zelzate Formation. The sandy tidal flat and lagoonal sedimentation is reinstalled in slightly shallower-water conditions compared to Stage 4.

Regional relative sea-level curve

A schematic relative sea-level curve has been constructed for the Eocene (Fig. 17) on the basis of the detailed lithofacies analysis of the offshore wells and of the geometrical data derived from the offshore seismic profiles, and using additional onshore outcrop data. It should be kept in mind that chronostratigraphical calibration should be made with caution as biostratigraphical control is still in progress and lacks precision.

In the Eocene, two asymmetric second-order transgressive–regressive cycles can be recognized, comprising nine 'lithostratigraphical' sequences of third-order. The lowest sequence of the lower cycle is characterized by a widespread relative sea-level rise (stacked TST) but followed by an overall fall: sediment texture in the four remaining lithosequences is gradually coarsening upward from massive, fine clays towards medium sands. Loss of accommodation space is shown by considerable shallowing of the basin and progradation, because of high supply of coarse sediment.

A major hiatus is interpreted as an important sea-level fall. It means that the ensuing pulsating relative sea-level rise must have been of comparable height (or even greater). This rise is responsible for a deepening of the basin, depositing three stacked fining upward sequences of overall transgressive nature and great lateral continuity. A similar sedimentary pattern, however, might be caused by pulsating variations in sediment flux and gradual rise, but this is highly unlikely because it would suppose the sediment flux variations to have the same amplitude over time. The last sequence of this upper cycle gradually coarsens and shallows upward due to relative sea-level lowering.

Conclusions

Interpretation of an extensive high-resolution seismic grid has allowed 13 seismic-stratigraphical units to be identified within the Palaeogene section in the offshore Belgian Basin, their geometries to be defined on a regional scale and a well-documented seismic-stratigraphical model to be developed. Detailed analysis of four offshore cores has allowed these seismic-stratigraphical units to be correlated with the classical Palaeogene lithostratigraphy of northwestern Belgium.

When attempting to interpret the observed seismic-stratigraphical and lithostratigraphical units in a sequence-stratigraphical way, it should be kept in mind that the basic concepts of sequence stratigraphy have been developed for typical shelf-slope–basin sections along an Atlantic-type continental margin. Their application to margins in a 'ramp-type' setting, such as the Palaeogene southern North Sea shelf, is therefore not always straightforward. In this study, some important differences from the general sequence-stratigraphy model for passive margins could be observed, which impeded sequence-stratigraphical interpretation:

(i) geometries are not very obvious due to the low gradient and the distance from the shelf break; consequently, sequences, systems tracts and parasequences are stacked in a quasi-parallel and conformable way, separated by low-angle unconformities;
(ii) due to the low gradient, ravinement during transgression (even on a parasequence scale) can be quite severe and can completely erode underlying lowstand deposits and even large parts of previously deposited sequences, thus 'cannibalising' sequence boundaries;
(iii) sequences tend to be mainly composed of transgressive and highstand systems tracts, while lowstand deposits are mostly only preserved as incised channel fills;
(iv) because of the low gradient, discrete changes in sea level, sediment input or subsidence can create strong lithofacies shifts – this is particularly true for the entire Late Ypresian and Lutetian section in the Belgian Basin.

M.D.B. is Senior Research Assistant of the Belgian National Fund for Scientific Research. R. Birchall (University College of North Wales), and I. Olivier and B. Van Bavinchove (both University of Gent) graduated on event stratigraphy projects in the Belgian Tertiary. Their MSc theses with P.J. provided vital information on crucial horizons. The authors gratefully acknowledge the support of the Belgian Science Policy Office and the Ministry of Education of the Flemish Community, which enabled most of the data to be acquired. Financial support was also granted through projects by the Federal and Regional Ministries of Economic Affairs. Logistic support was offered by the Belgian Management Unit of the Mathematical Model of the North Sea and the Schelde Estuary. We would also like to thank the captains and the crews of the various research vessels from which the data were acquired. The Belgian Geological Survey kindly provided the cores of the wells, and the National Fund for Scientific Research the necessary laboratory analysis equipment. The authors greatly appreciate the many suggestions from various colleagues, especially J. M. Vilain, D. Michoux and B. Lambert of Total-France for their valuable micropalaeontological contributions, and G. Postma and N. Vandenberghe for a critical review of a first version of this paper.

References

BALSON, P. S. 1989. Neogene deposits of the UK sector of the southern North Sea (51°–53° N). *In*: HENRIET, J. P. & DE MOOR, G. (eds) *The Quaternary and Tertiary Geology of the Southern Bight, North Sea*. Belgian Geological Survey, Brussels, 89–95.

CAMERON, T. D. J., CROSBY, A., BALSON, P. S., JEFFERY, D. H., LOTT, G. K., BULAT, J. & HARRISON, D. J. 1992. *United Kingdom Offshore Regional Report: the Geology of the Southern North Sea*. HMSO for the British Geological Survey, London.

——, LABAN, C. & SCHÜTTENHELM, R. T. E. 1989. Upper Pliocene and Lower Pleistocene stratigraphy in the Southern Bight of the North Sea. *In*: HENRIET, J. P. & DE MOOR, G. (eds) *The Quaternary and Tertiary Geology of the Southern Bight, North Sea*. Belgian Geological Survey, Brussels, 97–110.

COSTA, L. I. & MANUM, S. B. 1988. The description of the interregional zonation of the Paleogene (D1–D15) and the Miocene (D16–D20). *In*: VINKEN, R. (ed.) *The Northwest European Tertiary Basin, Results of the International Geological Correlation Programme Project No. 124*. Geologisches Jahrbuch A, **100**, 321–330.

DE BATIST, M. 1989. *Seismostratigrafie en Struktuur van het Paleogeen in de Zuidelijke Noordzee*. PhD Thesis, Rijksuniversiteit Gent.

—— & HENRIET, J. P. 1995. Seismic sequence stratigraphy of the Palaeogene offshore of Belgium, southern North Sea. *Journal of the Geological Society of London*, **152**, 27–40.

——, DE BRUYNE, H., HENRIET, J. P. & MOSTAERT, F. 1989. Stratigraphic analysis of the Ypresian off the Belgian coast. *In*: HENRIET, J. P. & DE MOOR, G. (eds) *The Quaternary and Tertiary Geology of the Southern Bight, North Sea*. Belgian Geological Survey, Brussels, 75–88.

DE GEYTER, G. 1981. Contribution to the lithostratigraphy and the sedimentary petrology of the Landen Formation in Belgium. *Mededelingen van de Koninklijke Academie voor Wetenschappen, Letteren en Schone Kunsten van België, Klasse Wetenschappen*, **43**, 111–153.

DEPRET, M. 1983. *Studie van de Lithostratigrafie van het Kwartair en van het Tertiaire Substraat te Zeebrugge onder meer met Diepsonderingen*. Belgian Geological Survey, Professional Paper, **201**.

DUPUIS, C., DE CONINCK, J. & ROCHE, E. 1984. Remise en cause du rôle paléogéographique du horst de l'Artois à l'Yprésien inférieur. Mise en évidence de l'intervention du Môle Bray-Artois. *Comptes Rendus de l'Académie des Sciences, Paris*, **298 II**(2), 53–56.

GULINCK, M. 1965. Le passage du Bartonien au Rupélien dans la région Boom-Malines. *Bulletin de la Société Belge de Géologie*, **LXXIV**, 115–120.

——1969*a*. Coupe résumée des terrains traversés au sondage de Kallo et profil géologique NS passant par Woensdrecht-Kallo-Halle. *Mémoires Explicatives des Cartes Géologiques et Minières de la Belgique*, **11**, 3–7.

——1969*b*. Le passage Oligocène–Eocène dans le sondage de Kallo et le Nord de la Belgique. *Mémoires du Bureau de Recherches Géologiques et Minières*, **69**, 193–195.

HENRIET, J. P., DE BATIST, M., Van VAERENBERGH, W. & VERSCHUREN, M. 1988. Seismic facies and clay tectonic features of the Ypresian clay in the southern North Sea. *Bulletin van de Belgische Vereniging voor Geologie*, **97**, 457–472.

——, —— & VERSCHUREN, M. 1991. Early fracturing of Palaeogene clays, southernmost North Sea: relevance to mechanisms of primary hydrocarbon migration. *In*: SPENCER, A. M. (ed.) *Generation, Accumulation and Production of Europe's Hydrocarbons*. European Association of Petroleum Geologists, Special Publication, **1**, 217–227.

HOUTHUYS, R. 1990. *Vergelijkende Studie van de Afzettingsstruktuur van Getijdenzanden uit het Eoceen en van de Huidige Vlaamse Banken*. Aardkundige Mededelingen, Leuven University Press, **5**.

JACOBS, P. 1975. *Bijdrage tot de Litostratigrafie van het Boven-Eoceen en het Onder-Oligoceen in Noordwest België*. PhD Thesis, Rijksuniversiteit Gent.

——1978. *Litostratigrafie van het Boven-Eoceen en van het Onder-Oligoceen in Noordwest België*. Belgische Geologische Dienst, Professional Paper, **151**.

——1995*a*. Eocene sediment supply in Western Belgium as determined through heavy mineral distribution. *Contributions to Tertiary and Quaternary Geology*, **32**, 35–52.

——1995*b*. Eocene to early Oligocene deltas in the Southern North Sea Bight, Belgium. *In*: OTI, M. & POSTMA, G. (eds) *Geology of Deltas*. Balkema, Rotterdam. 139–152.

—— & DE CONINCK, J. 1977. Sedimentologische en paleontologische kenmerken van het Eo-Oligoceen te Waasmunster. *Natuurwetenschappelijk Tijdschrift*, **59**, 157–183.

—— & SEVENS, E. 1988. *Sedimentation around the Eo-Oligocene boundary in the Belgian Basin*. International Association of Sedimentologists 9th European Regional Meeting, Excursion Guidebook, Belgian Geological Survey, 48–50.

—— & —— 1993*a*. Eocene siliciclastic continental shelf sedimentation in the Southern Bight North Sea, Belgium. *In*: *Progress in Belgian Oceanographic Research*. Royal Academy of Belgium, Brussels, 95–118.

—— & —— 1993*b*. Middle Eocene sequence stratigraphy in the Balegem Quarry (Western Belgium, Southern Bight North Sea). *Bulletin van de Belgische Vereniging voor Geologie*, **102**, 203–213.

——, ——, DE BATIST, M. & HENRIET, J. P. 1990. Grain size-, facies and sequence analysis of West Belgian Eocene continental shelf deposits. *Zentralblatt für Geologie und Paläontologie, Teil I*, **8**, 931–955.

KING, C. 1990. Eocene stratigraphy of the Knokke borehole (Belgium). *In*: *The Knokke Boring (11E/138)*. Mémoires Explicatives des Cartes Géologiques et Minières de la Belgique, **29**, 67–102.

LAGA, P. & VANDENBERGHE, N. (èds) 1990. The Knokke well (11E/138) with a description of the Den Haan (22W/276) en Oostduinkerke (35E/142) wells. *Mémoires Explicatives des Cartes Géologiques et Minières de la Belgique*, **29**.

LERICHE, M. 1912. L'Eocène des bassins parisien et belge (Livret-guide de la réunion extraordinaire de la Société géologique de France). *Bulletin de la Société Géologique de France*, **XII** (4), 692–724.

——1922. *Les Terrains Tertiaires de la Belgique*. Congrès Géologique International, Livret-guide pour la XIIIe Session, Excursion A4, Bruxelles.

——1928. Sur la répartition des facies lagunaires et fluviatile du Landénien, dans les Bassins belge et parisien. *Bulletin de la Société Belge de Géologie*, **38**, 69–91.

MARÉCHAL, R. & LAGA, P. 1988. *Voorstel Lithostratigrafische Indeling van het Paleogeen*. Belgian Geological Survey, Brussel.

MITCHUM, R. M. JR, VAIL, P. R. & SANGREE, J. B. 1977. Stratigraphic interpretations of seismic reflection patterns in depositional sequences. *In*: PAYTON, C. E. (ed.) *Seismic Stratigraphy – Application to Hydrocarbon Exploration*. American Association of Petroleum Geologists, Memoir, **26**, 117–134.

MOURLON, M. 1888. Sur l'existence d'un nouvel étage de l'Eocène moyen dans le bassin franco-belge. *Bulletin de l'Académie Royale de Belgique*, **XVI**(3), 252–276.

O'BRIEN, N. R. & SLATT, R. M. 1990. *Argillaceous Rock Atlas*. Springer, New York.

POWELL, A. J. 1988. A modified dinoflagellate cyst biozonation for latest Palaeocene and earliest Eocene sediments from the central North Sea. *Review of Palaeobotany and Palynology*, **56**, 327–344.

—— & VAIL, P. R. 1988. Eustatic control on clastic deposition. II: sequence and system tract models. *In*: WILGUS, C. K., HASTINGS, B. S., KENDALL, C. G. ST. C., POSAMENTIER, H. W., ROSS, C. A. & VAN WAGONER, J. C. (eds) *Sea-level Change – An Integrated Approach*. Society of Economic Paleontologists and Mineralogists, Special Publication, **42**, 125–154.

POSAMENTIER, H. W., JERVEY, M. T. & VAIL, P. R. 1988. Eustatic controls on clastic deposition I. Conceptual framework. *In*: WILGUS, C. K., HASTINGS, B. S., KENDALL, C. G. ST. C., POSAMENTIER, H. W., ROSS, C. A. & VAN WAGONER, J. C. (eds) *Sea-level Change – an Integrated Approach*. Society of Economic Paleontologists and Mineralogists, Special Publication, **42**, 109–124.

RUTOT, A. 1882. Résultats de nouvelles recherches dans l'Eocène supèrieur de la Belgique. IV Résolution de la question du Tongrien et du Wemmelien. Création du système Asschien. *Annales de la Société Royale Malacologique de Belgique*, **XVII**, Bulletin des sèances, clxxxi–clxxxv.

——1883. Résultats de nouvelles recherches dans l'Eocène supèrieur de la Belgique. II. Constitution géologique des collines tertiaires comprises entre Bruges et Eecloo. *Annales de la Société Royale Malacologique de Belgique*, **XVII**, Bulletin des sèances, clxxvii–clxxix.

STEURBAUT, E. 1986. Late Middle Eocene to Middle Oligocene calcareous nannoplankton from the Kallo well, some boreholes and exposures in Belgium and a description of the Ruisbroek Sand Member. *Mededelingen van de Werkgroep Tertiaire en Kwartaire Geologie*, **23**, 49–83.

——1990. Calcareous nannoplankton assemblages from the Tertiary in the Knokke borehole. *Mémoires Explicatives des Cartes Géologiques et Minières de la Belgique*, **29**, 47–62.

——1992. Integrated stratigraphic analysis of Lower Rupelian deposits (Oligocene) in the Belgian Basin. *Annales de la Société Géologique de Belgique*, **115**, 287–306.

—— & NOLF, D. 1986. Revision of Ypresian stratigraphy of Belgium and northwestern France. *Mededelingen van de Werkgroep Tertiaire en Kwartaire Geologie*, **23**, 115–172.

TUCKER, M. 1993. Carbonate diagenesis and sequence stratigraphy. *Sedimentology Review*, **1**, 51–72.

VAIL, P. R., MITCHUM, R. M. JR. & THOMPSON, M. S. 1977. Seismic stratigraphy and global changes of sea level, part 4: global cycles of relative changes of sea level. *In*: PAYTON, C. E. (ed.) *Seismic Stratigraphy – Application to Hydrocarbon Exploration*. American Association of Petroleum Geologists, Memoir, **26**, 63–81.

VANDENBERGHE, N. 1978. *Sedimentology of the Boom Clay (Rupelian) in Belgium*. Verhandelingen van de Koninklijke Academie voor Wetenschappen, Letteren en Schone Kunsten van België, Klasse Wetenschappen, **147**.

—— & VAN ECHELPOEL, E. 1987. Field guide to the Rupelian stratotype. *Bulletin van de Belgische Vereniging voor Geologie*, **96**, 325–337.

——, LAGA, P., STEURBAUT, E., HARDENBOL, P. & VAIL, P. 1996. Sequence stratigraphy of the Tertiary at the southern border of the North Sea Basin in Belgium. *In*: DE GRACIANSKY, P. CH., HARDENBOL, J., JACQUIN, T., VAIL, P. R. & FARLEY, M. B. (eds.) *Mesozoic–Cenozoic Sequence Stratigraphy of Western European Basins, 2*. Society of Economic Paleontologists and Mineralogists, Special Publication.

VANDENBROECK, E. 1893. Coup d'oeil synthétique sur l'Oligocéne belge et observations sur le Tongrien supérieur du Brabant. *Bulletin de la Société Belge de Géologie*, **7**, 208–303.

VAN WAGONER, J. C., MITCHUM, R. M. JR, POSAMENTIER, H. W. & VAIL, P. R. 1987. Seismic stratigraphy interpretation using sequence stratigraphy. Part II: Key definitions of sequence stratigraphy. *In*: BALLY, A. W. (ed.) *Atlas of Seismic Stratigraphy. Vol. 1*. American Association of Petroleum Geologists, Studies in Geology, **27**, 11–14.

——, POSAMENTIER, H. W., MITCHUM, R. M., VAIL, P. R., SARG, J. F., LOUTIT, T. S. & HARDENBOL, J. 1988. An overview of the fundamentals of sequence stratigraphy and key definitions. *In*: WILGUS, C. K., HASTINGS, B .S., KENDALL, C. G. ST. C., POSAMENTIER, H. W., ROSS, C. A. & VAN WAGONER, J. C. (eds) *Sea-level Change – an Integrated Approach*. Society of Economic Paleontologists and Mineralogists, Special Publication, **42**, 39–45.

VERSCHUREN, M. 1992. *An Integrated 3D Approach to Clay Tectonic Deformation*. PhD Thesis, University of Gent.

ZIEGLER, P. A. 1982. *Geological Atlas of Western and Central Europe*. Shell Internationale Petroleum Maatschappij, Elsevier, Amsterdam.

Tidal sedimentation in Inner Hebrides half grabens, Scotland: the Mid-Jurassic Bearreraig Sandstone Formation

D. MELLERE[1] & R. J. STEEL[2]

[1] *Statoil Research Centre, Arkitekt Ebbels veg 10, Rotvoll, N-7005 Trondheim, Norway*
[2] *Department of Geology and Geophysics, University of Wyoming, Laramie, USA*

Abstract: The Bearreraig Sandstone Formation (Late Toarcian–Bajocian), on the islands of Skye and Raasay, in NW Scotland, forms part of the marine infill of one of the Hebridean rift basins that developed during the early stages of the opening of the North Atlantic. The Formation, up to 250 m thick in south Skye, and from 150 m to 180 m thick in north Skye and Raasay, is a spectacular succession of medium to coarse-grained, tidally generated, cross-bedded sandstones, and subordinate shales. The sandstones sharply overlie the Dun Caan Shales and Raasay Ironstone and are abruptly blanketed by the Garantiana Clay Member.

The succession was deposited under the influence of a series of significant sea-level fluctuations during a 12 Ma interval. During regressive phases of the basin coastline, sedimentation occurred largely at the mouth of a tidal-dominated delta which evolved into a macrotidal estuary during each transgressive phase. Five major facies associations have been recognized: (1) shale/siltstone-dominated successions containing sharply based, strongly bioturbated sandstone beds interpreted as prodelta deposits; (2) medium to coarse sandy successions of thickening-upward, small to very large-scale planar cross-stratified sets, representing dune fields in the tidally dominated delta-front environment; (3) delta-plain, very fine sandstones with roots and fluvial channels; (4) estuarine, tidal channel-fill deposits which overlie erosion surfaces up to 10 m deep and consist of both bioturbated and non-bioturbated, large-scale, tabular or trough cross-bedded medium/coarse sandstones (sets 2–10 m thick) often deformed by water-escape structures and slumping; (5) bioturbated, thinning-upward tabular sandstone beds, interpreted as transgressive shelf deposits.

Palaeocurrent data indicate that during the early stage of deposition of the Bearreraig Formation, tidal-dominated sedimentation occurred separately in the north Skye/Raasay and in the south Skye sub-basins. During a late stage of deposition, the sub-basins merged, and sedimentation occurred uniformly throughout the region. This change in palaeogeographical configuration is associated with a regionally extensive unconformity. Tectonic activity, particularly at an early stage in the form of tilting and block rotation, is believed to have enhanced tidal circulation within the fault-constrained sub-basins.

The Mesozoic basins of the Hebrides form a series of linked extensional, westerly tilted half grabens (Steel 1971, 1977; Brewer & Smithe 1984; Stein 1988; Earle *et al.* 1989; Stein & Blundell 1990; Morton 1992*a*) (Fig. 1), controlled in their position and extent by the extensional reactivation of Caledonian thrusts. The Aalenian–Bajocian Bearreraig Sandstone Formation was deposited across depositional 'blocks' which were tilting slightly to the west-northwest during sedimentation. The northern block was bounded to the east by the North Raasay Fault, whereas three blocks further to the southeast were bounded successively by the Beinn na Leac and South Scalpay/Camasunary fault system (Fig. 1). Three of the tilted blocks probably had exposed basement 'highs' supporting local drainage, whereas the sub-basin bounded by the North Raasay and Beinn na Leac faults probably had only subdued uptilting of Triassic/Jurassic strata along its crest.

The striking differences in thickness, in facies and in sequential organization of the Bearreraig Sandstone Formation within and between the four sub-basins is probably attributable to the differential subsidence history of these linked extensional basins. The purpose of this paper is to record and emphasize the differing styles and evolution of the alternating deltaic and estuarine sedimentation during high-frequency regressions and transgressions across the province. We have constructed a sedimentological and stratigraphical model for the Bearreraig Sandstone Formation. Details of the tectonic effects on the succession within and between the tilted blocks will be presented elsewhere.

Stratigraphy

The Bearreraig Sandstone Formation, a prominent part of the Jurassic succession in the

From De Batist, M. & Jacobs, P. (eds), 1996, *Geology of Siliciclastic Shelf Seas*,
Geological Society Special Publication No. 117, pp. 49–79.

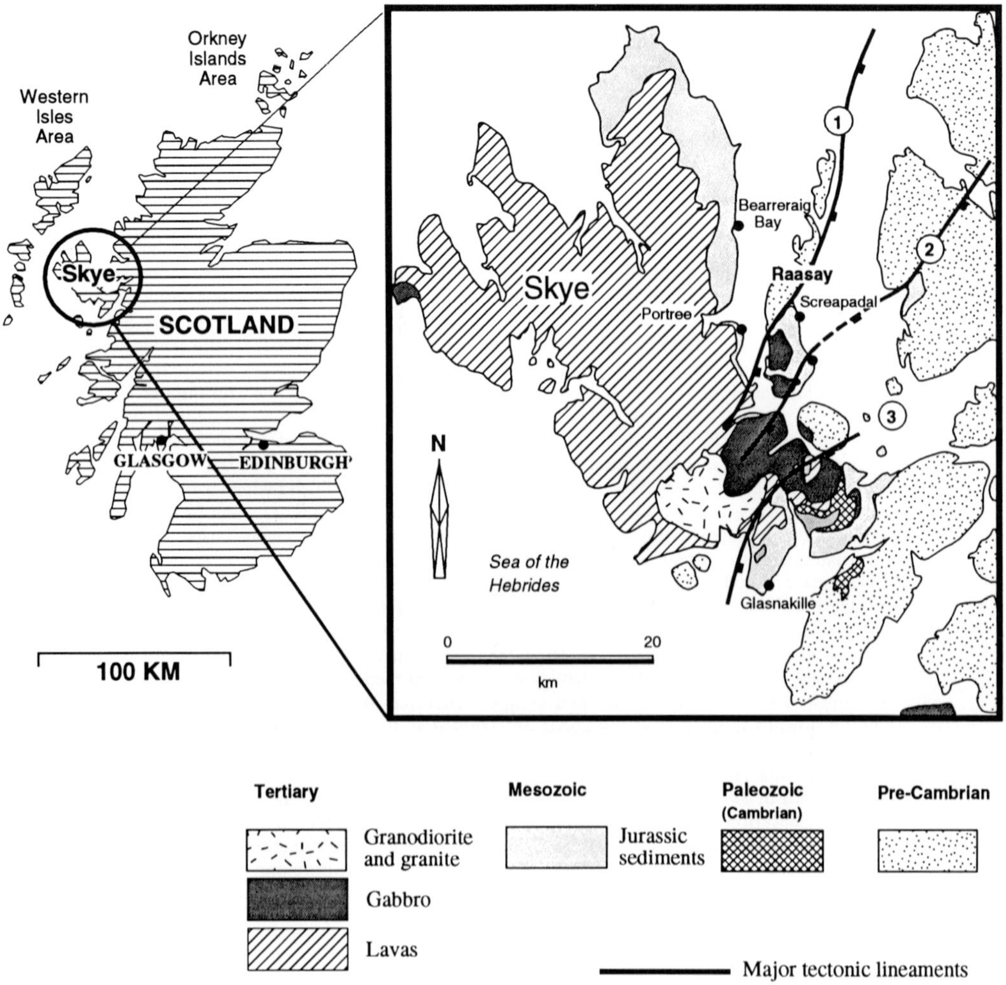

Fig. 1. Location of the study area and schematic geological map of Skye and Raasay showing also the distribution of the Jurassic deposits across the depositional blocks bounded by: (1) the North Raasay Fault; (2) Beinn na Leac Fault and (3) South Scalpay/Camasunary Fault.

Hebridean basins, signalled a major change in sedimentation style compared to earlier Jurassic deposition. The Formation is variable both in thickness, ranging from 38 m in Mull (Morton 1965) to up to 250 m in south Skye, and in facies. On Raasay and in south Skye the Formation consists mainly of coarse-grained cross-bedded sandstones whereas in north Skye, cross-stratified to bioturbated sandstone and shale dominate. The interval has been interpreted in terms of tectonic rejuvenation and renewed lithospherical extension (Morton 1987, 1989). The Formation sharply overlies the Raasay Ironstone Formation (Fig. 2), a widespread unit of shales and intercalated chamositic oolite sandstone beds. The contact is a stratigraphical hiatus, locally evidenced by a minor angular unconformity (Morton 1987, 1989). The upper bounding surface of the Bearreraig Sandstone Formation is defined by the contact with the dark, fine-grained shales of the Garantiana Clay Member of the Bearreraig Sandstone Formation (Morton 1965) which grade upwards into the Cullaidh Shale Formation (Harris & Hudson 1980; the basal 'Oil Shale' of the previous nomenclature) of the Great Estuarine Group (Fig. 2).

The Bearreraig Sandstone Formation was extensively examined by Morton (1965, 1976, 1983) and Morton & Dietl (1989) who have provided an excellent biostratigraphical framework. Morton (1983) also made a first study on the facies and palaeoenvironment of the

	Stages	Ammonite Zones	Ammonite Subzones	GREAT ESTUARINE GROUP
JURASSIC				Oil Shale
	BAJOCIAN	*Garantiana*		Garantiana Clay
		Subfurcatum		
		Humphresianum		
		Sowerbyi	*Sauzei*	**BEARRERAIG SANDSTONE FORMATION**
			Laeviuscula	
			Trigonalis	
			Discites	
	AALENIAN	*Concavum*		
		Murchisonae		
		Scissum		
		Opalinum		
	TOARCIAN	*Levesquei*	*Aalensis*	Dun Caan Shale
				RAASAY IRONSTONE

Fig. 2. General stratigraphy of the Bearreraig Sandstone Formation and correlation with the Jurassic ammonite zones (biozonation from Morton 1965).

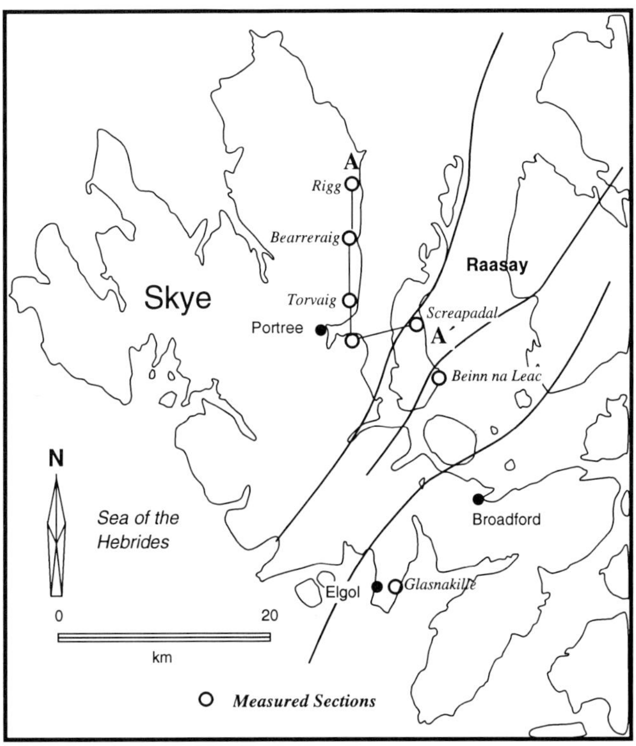

Fig. 3. Schematic structural map of Skye and Raasay with localities of measured sections and trace of the cross-section in Fig. 4.

cross-bedded sandstones recognizing three main facies associations: (1) cross-bedded sandstones deposited as tidal sand waves; (2) massive sandstones to inner shelf shales, interpreted as offshore sand bars; and (3) offshore shales and sandy limestones.

We attempt here to build upon this work, showing more of the details of the tidally generated system, based on sections from Raasay, northeast Skye and Glasnakille (southeast Skye) respectively (Fig. 3).

The facies of the Bearreraig Sandstone Formation

The sedimentary facies recognized within the Bearreraig Sandstone Formation are grouped into five main associations: (1) coarsening-upward, mudstone-siltstone successions (prodelta deposits); (2) small to large-scale, coarsening-upward cross-bedded sandstone (tidally dominated delta-front deposits); (3) very fine sandstones with roots and fluvial channels (delta-plain deposits); (4) cross-stratified sandstones infilling surface of marked incision (estuarine channel-fill deposits); (5) bioturbated sandstones in upward-thinning units (transgressive shelf deposits). Associations 1–3 relate to regressive phases of tide-dominated delta construction, whereas associations 4 and 5 accumulated during transgressive phases and estuary development.

A cross-sectional panel illustrating the lateral and vertical facies relationships through the studied localities on Raasay and north Skye is shown in Fig. 4.

Facies Association 1: coarsening-upward shaley successions (prodelta to lower delta-front deposits)

This relatively fine-grained association becomes more prominent from the southeastern to the northwestern area, where it reaches its maximum thickness in the Bearreraig Bay. Complete successions are seldom preserved, owing to the scouring and truncating effect of overlying estuarine channels. The succession has two major facies components with a gradational boundary between them: strongly bioturbated shales and siltstones of prodelta origin and distal mouth bar siltstones and sandstones of lower delta front origin. In Raasay the coarsening-upward prodelta deposits are directly overlain by the small-scale dunes of Facies Association 2.

Strongly bioturbated interbedded siltstones and sandstones: prodelta to shelf deposits. This facies characterizes the middle and/or basal parts of the succession in Bearreraig Bay and on Raasay. It consists of strongly bioturbated siltstones with interbeds of strongly bioturbated very fine sandstone (Fig. 5). The latter can be organized into 5–15 m thick, coarsening- and thickening-upward units. Individual sandstone beds are 5–120 cm thick, and display a tabular to broadly lenticular geometry. Bed bases are generally very sharp with occasional scours. Bed tops are normally diffuse due to intense and pervasive bioturbation and an upward gradation into siltstone. Bioturbation is so intense with large forms of *Thalassinoides* and *Planolites* traces, that original internal sedimentary structures are mostly obliterated. Only in a few places were normal grading, climbing-ripple lamination and parallel lamination recognized. In some beds on southeast Raasay, parallel lamination is overlain by current-ripple lamination. Generally there is no evidence of wave reworking. The interbedded siltstones and shales are also extensively bioturbated, and they occur between the sandstone beds in 0.2–3 m thick intervals or as up to 15 m thick units below sandier intervals.

In the Bearreraig Bay, at some 100 m from the base and below the channellized tidally dominated deposits of Facies Association 4 (Fig. 6), the uppermost part of the facies interval is typically represented by very fine-grained sandstones with a high-density but low-diversity ichnofossil assemblage dominated by small forms of *Planolites* burrows with characteristic mud walls (Fig. 6B).

Interpretation. The sharp-based sandstones indicate discrete emplacement of sand, either suddenly or after an episode of sea-floor scouring. The parallel lamination indicates deposition from rapid flows in the upper flat-bed field of the upper flow regime. These flows could be either turbidity currents or powerful flows moving on the bottom under the influence of storm-generated pressure gradients. The lack of any storm evidence such as hummocky or swaley cross-stratification and wave ripple lamination suggests that the deposits were emplaced as sediment gravity flows. This facies also resembles the lower part of the Chungo Member in the Wapiabi Formation in Canada (Rosenthal & Walker 1987) interpreted as turbidity deposits. The presence of a low-diversity, locally high abundance of *Thalassinoides* and *Planolites* and the record of few brackish-water palynomorphs (Riding *et al.* 1991), indicate a depositional environment close to a delta, probably on the slopes of prograding mouth bars. A more proximal depositional setting (possibly lower delta front) is inferred for the strongly

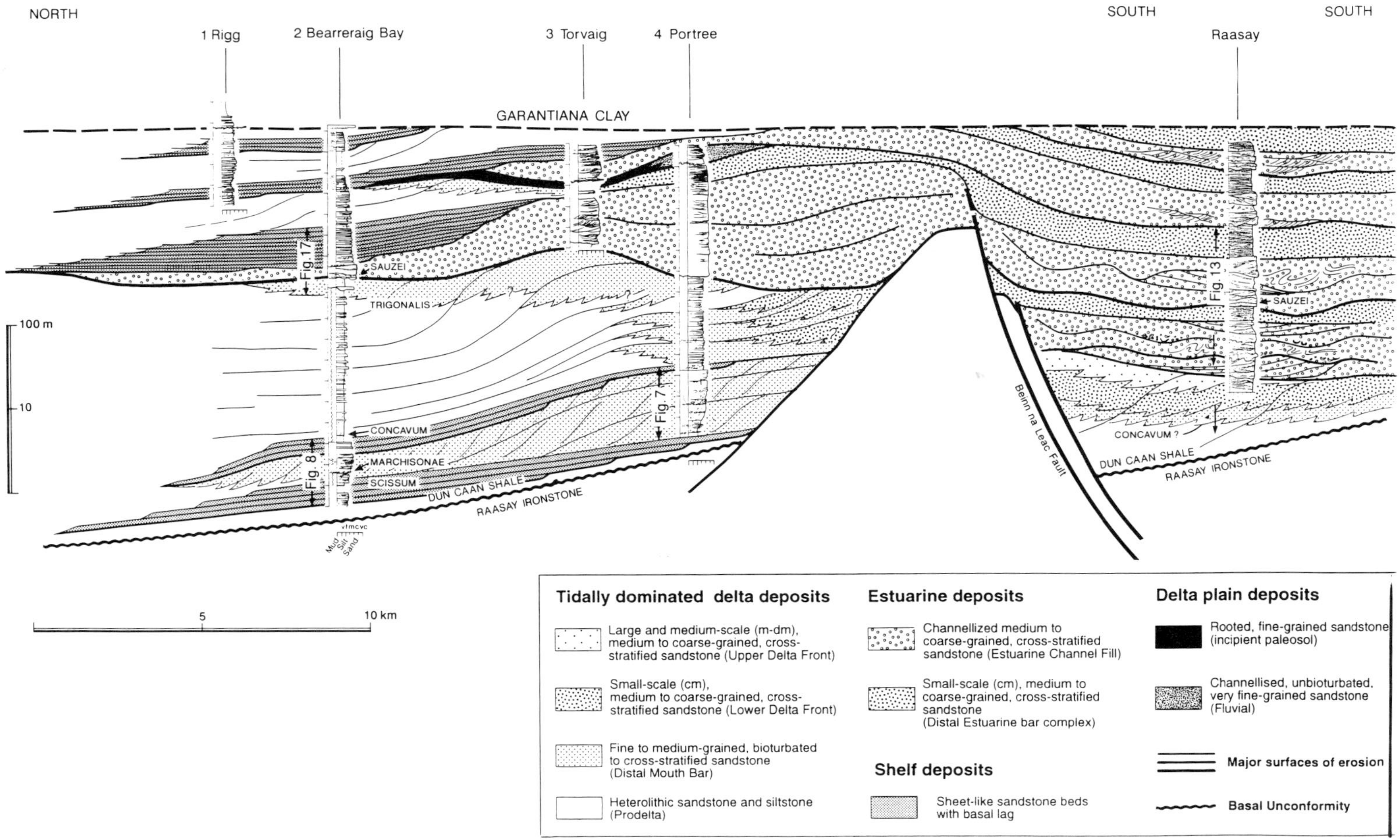

Fig. 4. Correlation panel through the sections (location in Fig. 3) measured in north Skye and in Raasay (Screapadal). The sections lie within two different structural blocks. Within each block the sedimentary units were traced physically laterally. Correlation between the blocks was possible using the biozones defined by Morton (1965). In the legend, we used the same symbol for the small-scale dunes of the lower delta front and of the distal estuarine bar complexes. Lithologically the small-scale dunes are identical. Their attribution to deltaic or estuarine setting is based on their vertical thickening (deltaic) or thinning-upward (estuarine) trends.

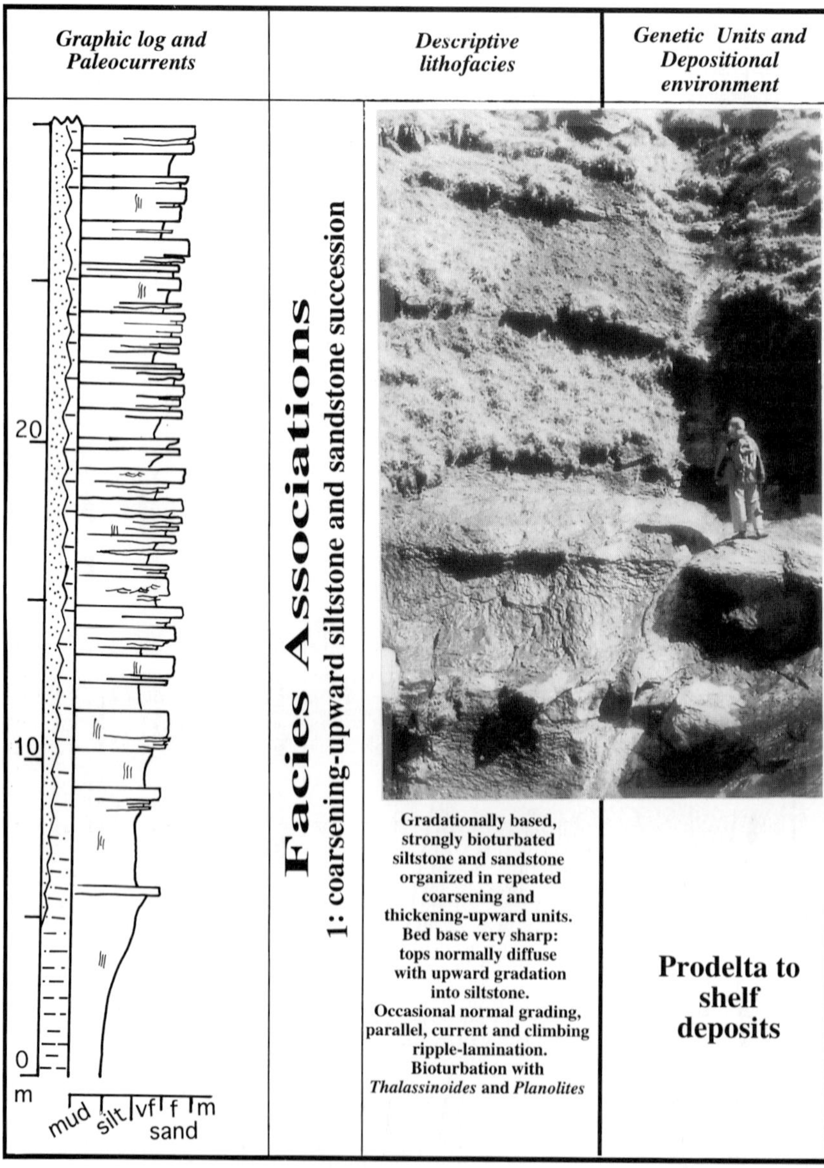

Fig. 5. Measured section at Beinn na Leac (Raasay) and photograph of sharp-based sandstone beds with siltstone-mudstone background (prodelta–shelf deposits).

bioturbated sandstones that underlie the tidal channel in the Bearreraig Bay (Fig. 6). The low diversity and diminutive size of the ichnofossils strongly suggest stressed conditions, probably induced by salinity, oxygenation and nutrient availability. A similar ichnofacies assemblage was described by Beynon & Pemberton (1992) in the Cretaceous Grand Rapids Formation of Alberta and was interpreted as being associated with a brackish-water environment. Stable oxygen isotope studies on concretions within the same sandy bioturbated level in the Bearreraig Bay, indicating a late phase of brackish-water influence (Wilkinson 1991), also support the present interpretation.

Bioturbated, rippled-laminated and cross-stratified sandstones: distal mouth-bar deposits. This facies association is characterized by a thickening- and coarsening-upward profile. It is present

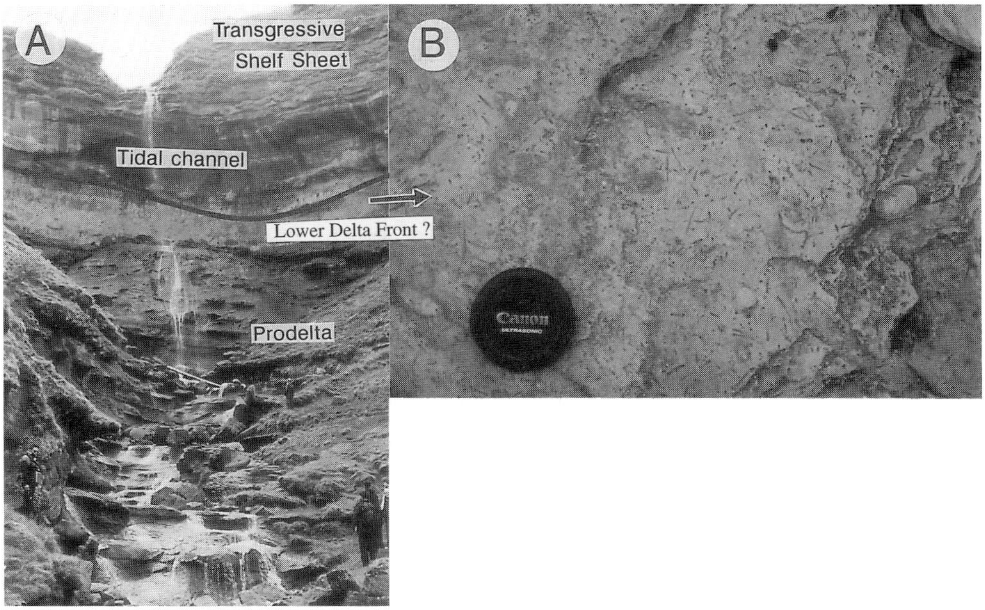

Fig. 6. (**A**) View of the succession at the Bearreraig bay. The prodelta deposits of heterolithic sandstones and siltstones (Facies Association 1) grade upwards into bioturbated sandstones of possibly lower delta-front environment. The deposits are sharply eroded by a tidal channel complex (Facies Association 4). The uppermost part of the section shows a fining-upward succession of shelf deposits (Facies Association 5). (**B**) Details of the abundant but low-diversity ichnofacies association of the bioturbated interval with *Planolites* traces.

in the basal part of the succession at Portree, in the Bearreraig Bay, at Glasnakille and on Raasay. The facies association grades from silty, fine-grained bioturbated to rippled-laminated and tabular cross-stratified sandstones, organized into well developed coarsening- and thickening-upward units, up to 15 m thick. At Portree the association occurs at the base of the formation and consists almost exclusively of bioturbated sandstone and siltstones (Fig. 7). Bioturbation is pervasive with *Anchonichnus*, *Thalassinoides* and *Chondrites* traces. Belemnites, *Pecten*, oyster, *Cardium* and other molluscs are widespread throughout the interval or concentrated within strongly cemented nodular beds, with cemented nodules up to 50 cm in diameter. At 14 m from the base oysters and belemnites are concentrated in a 5 cm thick horizon (Fig. 7).

In the section at Bearreraig Bay the facies occurs some 15 m from the base (Fig. 8) and overlies a sandstone layer characterized by a marked concentration of belemnites (Fig. 8C). The facies consists of bioturbated sandstone and siltstones at the base, which grade upwards into planar tabular and wedge cross-strata. Bioturbation is pervasive in the basal interval with *Anchonichnus* and *Chondrites* traces. Climbing ripple lamination is the only sedimentary structure detectable. The cross-bedded sandstones show well developed, bundled foresets and finer-grained double drapes along the set and foreset boundaries. There is a cyclical lateral change in thickness of the bundled foresets, from 15–20 to 50 cm.

Interpretation. This facies is interpreted to have been deposited on a shelf to subtidal depositional environment as suggested by the sedimentary structures recognized within the uppermost level of the coarsening-upward interval at Portree and within the cross-stratified sandstone in the Bearreraig Bay. The fine-grained double drapes along the cross-bedded sets are distinctive features of tidal deposits (de Raaf & Boersma 1971; Reineck & Singh 1973; Visser 1980; Clifton 1983), reflecting suspension deposition during slack-water periods of the tidal cycle. Also, tidal bundling is common, representing the repeated and cyclic occurrence of material deposited on the slipface of a bedform during the dominant and the subordinate tidal current stages (Visser 1980). Repetitive change in thickness (laterally) of the bundles within the Bearreraig cross-bedded deposits may be attributed to ebb-flood and neap-spring tidal cycles (see also Boersma 1969; Visser 1980;

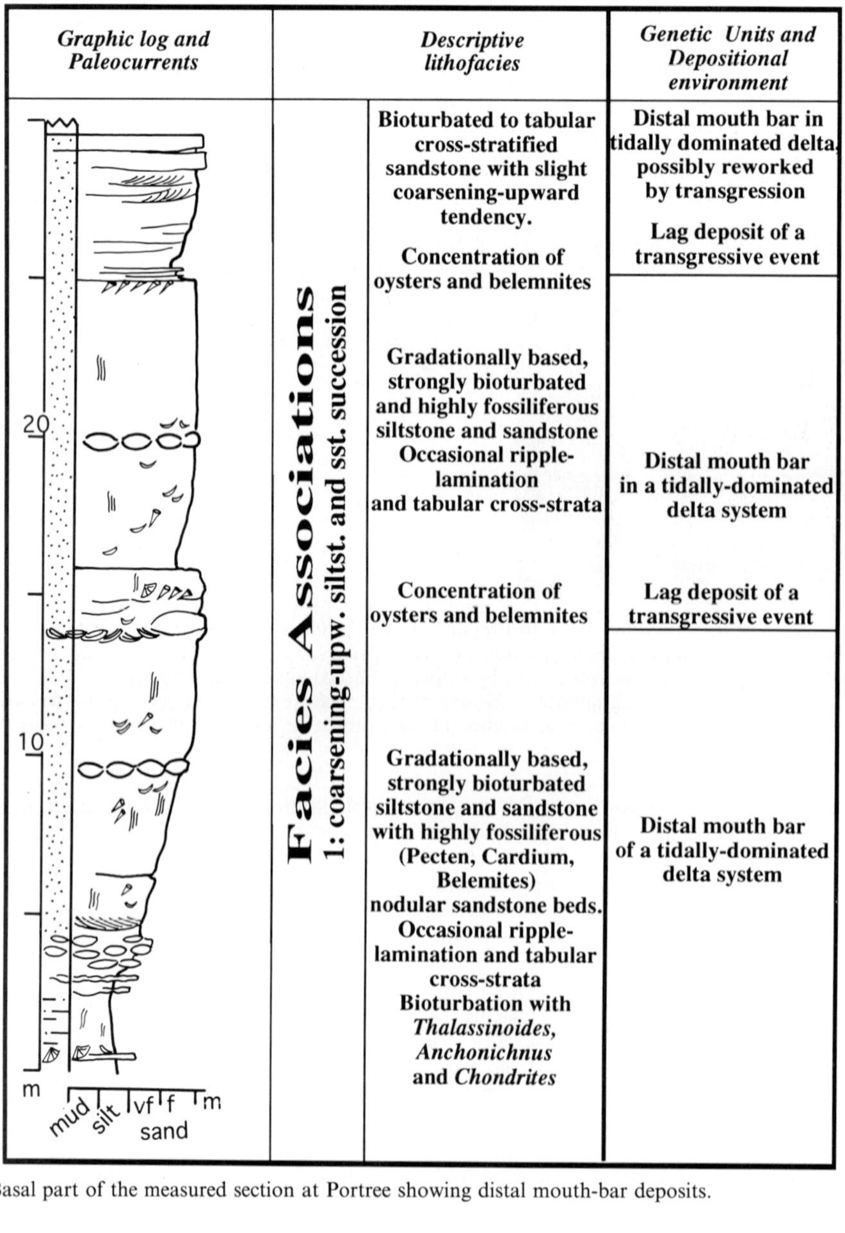

Fig. 7. Basal part of the measured section at Portree showing distal mouth-bar deposits.

Boersma & Terwindt 1981; Kreisa & Moiola 1986).

The vertical passage from bioturbated and more heterolithic, tabular cross-bedded sandstones to cleaner cross-strata reflects a decrease in water depth with a consequent change in hydrodynamic regime from low-energy to moderate- and high-energy conditions. The coarsening-upward sandbodies are interpreted as prograding distal mouth bars of a tidally dominated delta. Similar facies have been described by Maguregui & Tyler (1991) in western Venezuela. The mouth bar in the Bearreraig Bay is interpreted to have been deposited in a more offshore position with respect to that at Portree, as indicated by the strong degree of bioturbation and higher shale content. The depositional area was only mildly influenced by tidal currents and waves, although moments of higher-energy regime, (associated

Fig. 8. General views (**A**, **B**) of the basal part of the succession in Bearreraig Bay. The strata here are interpreted as part of a tidally dominated mouth-bar complex which downlaps tabular to lenticular, bioturbated sandstones interpreted as transgressive shelf deposits (Facies Association 5). One of the transgressive beds (arrowed) has a marked basal lag of belemnites, overlying a ravinement surface (**C**).

possibly with transgressive phases) may have been responsible for accumulation of shells within the tabular, more cemented beds.

Facies Association 2: thickening- and coarsening-upward planar cross-stratified sandstones (tidal-dominated delta-front deposits)

The uppermost part of the coarsening-upward deltaic facies association culminates with thick intervals of planar tabular cross-stratified sandstones. This association dominates the basal part of the Bearreraig Sandstone Formation on Raasay and on south Skye, where it can reach up to 80 m in thickness. The association is composed of dunes (tabular to wedge, planar cross-bedded sandstones), further classified on the basis of their thickness. The classification adopted here follows Ashley *et al.* (1990), where 'dune' is used to refer collectively to both megaripples and sandwaves. Within the facies association, three classes of bedforms have been distinguished: small (5–40 cm thick), medium to large (0.4–3 m thick) and very large (3–10 m thick). These bedforms are themselves organized into coarsening- and thickening-upward units.

Small-scale dunes. The small-scale dunes consist of 5–40 cm thick (average 15 cm) tabular to broadly undulating cross-stratified sets of coarse to very coarse sandstones and occur at the base of the delta-front facies association. They are organized commonly into aggradational and slightly progradational intervals up to 20 m thick (Fig. 9). On Raasay, the small-scale dunes overlie, with a gradual but very rapid transition, the coarsening-upward packages of prodelta to shelf deposits described above, and pass upwards into medium-scale cross-stratified sandstone. Generally, the passage between the

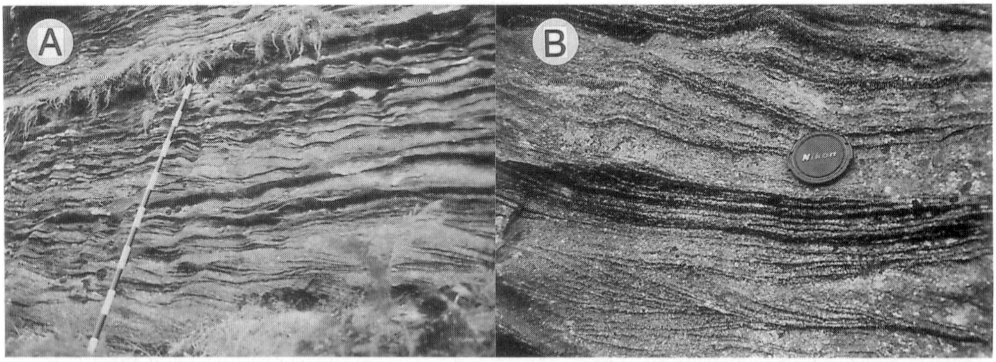

Fig. 9. General view (**A**) and details (**B**) of the small-scale dunes in Raasay. The basal surfaces of the cross-bedded sets are broadly lenticular, in places erosive. The cross-bedded foresets are typically marked by double sandstone laminae.

prodelta fine-grained sediments and the small-scale dunes is not exposed.

Grain-size increases rapidly upwards, where the sediments become among the coarsest in the Bearreraig Sandstone Formation, with quartz granule concentrations along foresets and bottomsets. Foresets, typically bundled, contains double drapes (Fig. 9B) consisting of coarse- and fine-grained sandstone. Topsets are usually truncated. Bottomsets are tangential to the basal surface of the sets. The system is completely shale-free. The basal bounding surface of the cross-bedded sets is broadly undulating and lenticular, in places clearly erosive, scouring into the underlying beds with scours up to 70 cm deep. Bioturbation is generally absent, but it can be very intense in the lower part of the coarsening-upward interval, by the transition with the underlying prodelta deposits.

Palaeocurrents are mostly unidirectional (towards the north) and are consistent with the direction of the overlying medium and larger-scale dunes.

Interpretation. The tabular sets of cross-strata and the double laminae along sets and foresets indicate migrating two-dimensional dunes that fluctuated in migration speed and asymmetry (see also Allen 1980; Visser 1980; Boersma & Terwindt 1981; Rubin 1987). A subtidal depositional environment is suggested: more precisely, it is believed that the small-scale dunes were deposited in widespread lobate dune fields, influenced primarily by tidal currents in the area of transition between the delta-front and the prodelta regions. This depositional interpretation is consistent also with the presence of bioturbation in the lowermost level of the cross-stratified interval: the increased tidal-current strength upwards would have inhibited both the biological activity and the deposition of mud drapes in the sand deposits. The low-angle broad scours, which interrupted the migration of the dunes, are interpreted as having been produced by shallow channelling, probably formed by tidal action.

Large- to medium-scale, tabular to wedge-shaped planar cross-strata. The tabular and wedge-shaped planar cross-stratified sandstone facies (Fig. 10) occurs at the top of the coarsening- and thickening-upward profile of the cross-bedded facies association. Intervals of tabular planar cross-strata are more common than wedge-shaped sets, although the latter dominate locally (Fig. 10B). Cross-bedded sets are 40 to 300 cm thick and are developed in fine to coarse-grained sandstone. Sets are composed of centimetre-thick, normally graded, planar foresets. Reactivation surfaces and pause planes are particularly common in the thickest cross-beds. Topset laminae are usually truncated sharply. In places, however, sigmoidal cross-bedding (Fig. 10A) with preserved topsets grades downcurrent into steeper tabular planar foresets. The dip of the foresets can exceed the angle of repose and over-steepened and overturned cross-bedding is quite common. Bottomset laminae typically exhibit tangential (sigmoidal) contacts with the lower bounding surface of the set. Except at a few localities (notably on the Elgol road, in south Skye where shale constitutes 10–15% of the facies, see Fig. 10C), the cross-bedded sandstones of the Bearreraig Sandstone Formation are usually clay-free. Although shale interbeds are very rare, thin double drapes (usually silty very fine sandstone) can still be detected separating sets and foresets. Foresets are also typically bundled with cyclical repetition of

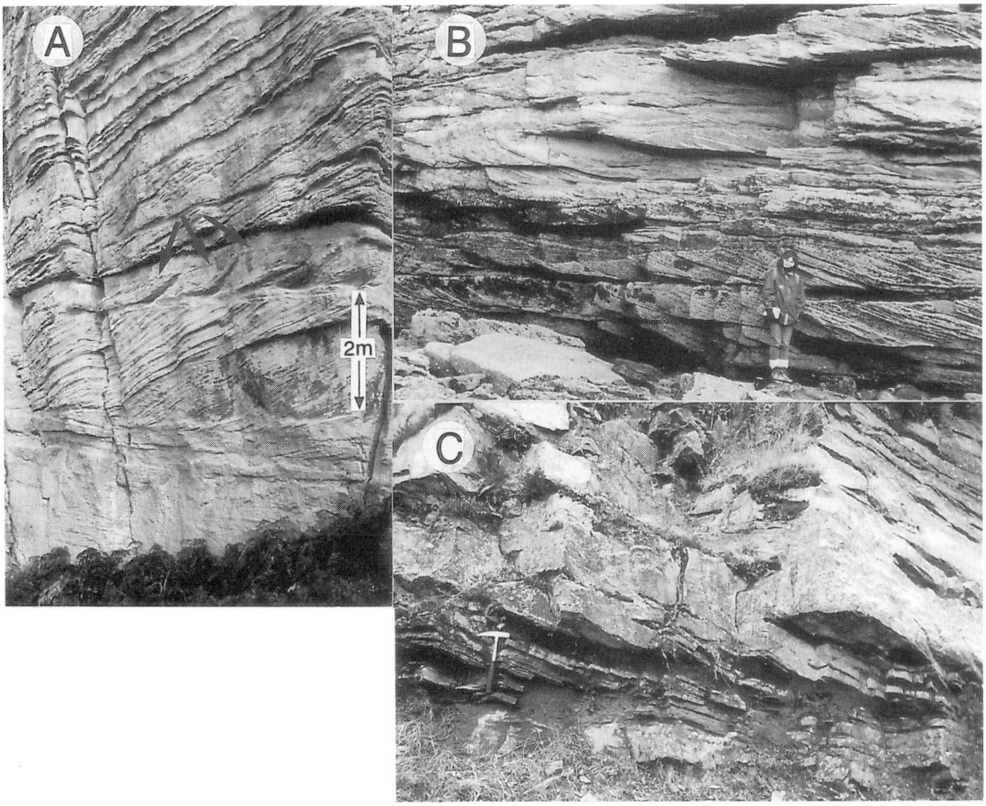

Fig. 10. Details of the large and medium-scale dunes in tidally dominated upper delta front. (**A**) Details from an outcrop at Screapadal, in Raasay. The large-scale dune in the lower part of the photograph is here infilling a broad channel. It is overlain by a medium-scale dune displaying sigmoidal tidal bundles (indicated by arrows). (**B**) Medium-scale dunes from Glasnakille. The cross-bedded sandstones are dominated by tabular planar sets, though intervals of wedge-planar sets can also be present. (**C**) Mudstone lenses and shales in the bottomsets of large bedforms along the Elgol road, in south Skye.

thicker and thinner sandstone units, the latter characteristically double draped. On Raasay, in the lower part of the succession, sets of quartzitic and hybrid arenites (i.e. a mixture of intrabasin bioclasts and terrigenous sandstone) alternate cyclically with sets of more cemented biocalcarenites. The planar tabular cross-beds are generally separated by 3–10 cm thick intervals of undulating, more cemented and rippled-laminated, 2–5 cm thick sandstone beds, rhythmically alternating with thin siltstone and shale layers.

Bioturbation is very slight or absent. Occasional *Thalassinoides* are observed. Although palaeocurrent directions may differ (Morton 1983) between the different basins, they are usually strictly unimodal: towards the north in north Skye, Raasay and in the middle–upper part of the succession at Glasnakille (a subordinate mode towards the south can also be present); towards the south in the lower part of the succession at Glasnakille. In south Skye, herring-bone cross-stratification has been documented.

Interpretation. The planar cross-bedded sandstone facies is interpreted in terms of the migration of two-dimensional dunes (Ashley *et al.* 1990) under the influence of tidal currents. The large scale of the sets, double drapes, reactivation surfaces and herring-bone cross-stratification are all characteristic of tidally influenced deposits (Visser 1980; Allen 1980; Clifton 1983; Allen & Homewood 1984; Smith 1988; Dalrymple *et al.* 1992). Although mudstone is very rare, the double fine laminae along the cross-bed foresets are believed to be related to bidirectional flows in a subtidal setting (see also de Raaf & Boersma 1971; Reineck & Singh 1973; Visser 1980; Boersma & Terwindt 1981; Smith 1988). Horizontal rippled-laminated sandstone and fine-grained sandstone and siltstone (or thin drapes of shale) may represent a similar style of deposition but on a flat surface, commonly the surface over which the dunes were migrating. Reactivation surfaces are interpreted to have been formed by the migration of large bedforms under asymmetrically reversing currents (Boersma 1969).

Very large-scale cross-bedded sandstones. This facies is present in the lowermost part of the section in Glasnakille and on south Raasay. It consists of coarse-grained, planar tangential cross-bedded sandstones in very large-scale sets, extending tens of metres basinwards with very gentle dip angles (10–15°). At Glasnakille, sets are up to 8 m thick (Fig. 11A) and are composed

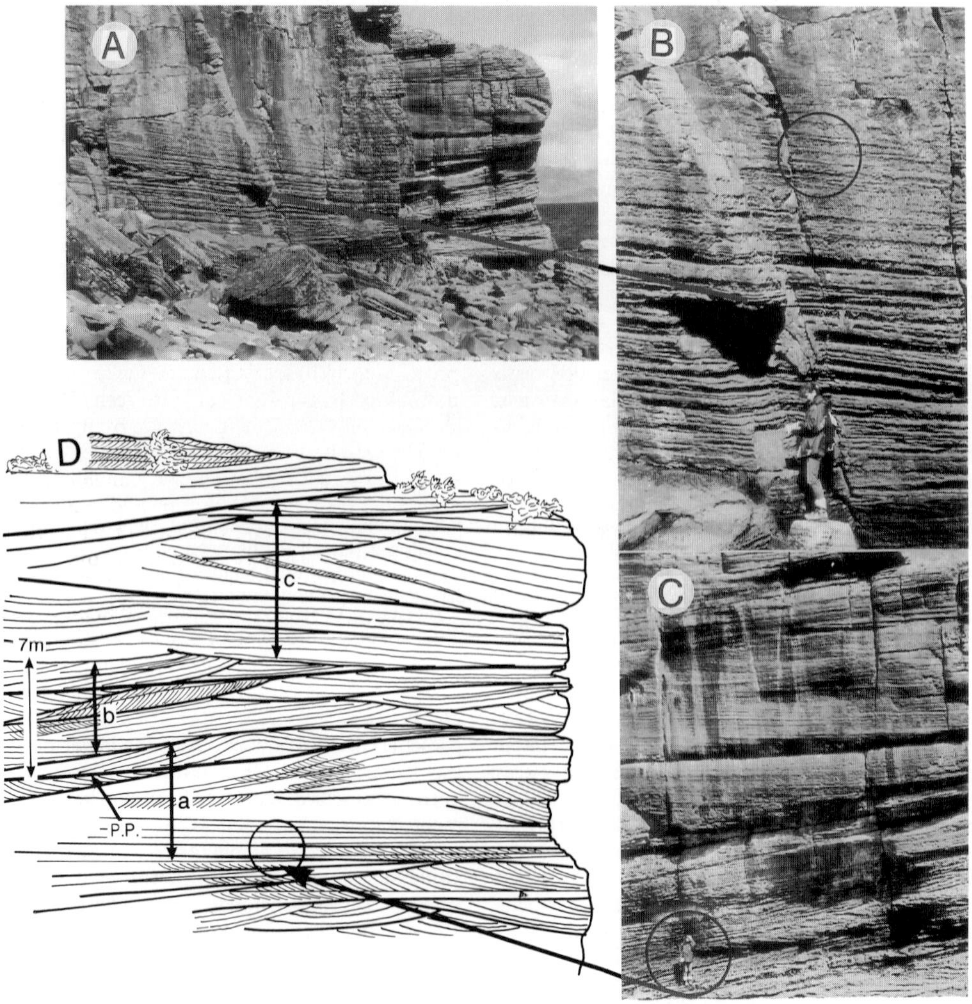

Fig. 11. Views and field sketch of one of the very large-scale dunes at Glasnakille, south Skye. (**A**) The basal cross-bedded sandstone set is some 8 m thick. It is composed of gently (up to 10°) southwards dipping, ebb-oriented foresets. (**B**) Detail of the foresets. Through most of the thickness of the very large-scale sets, foresets are composed of homogeneous, thinly planar laminated sets up to 20 cm thick, separated by double siltstone drapes. In the uppermost part, small flood-oriented dunes are superimposed on the large foresets (circle). (**C**) The large dune (**A**) migrated over trough cross-beds (person in the circle as scale). (**D**) Migration of the large dune was not continuous but punctuated by pause planes (pp) which became a stable platform for dunes oriented 45–90° (southwest-west) with respect to the large set. In this outcrop the large-scale dunes seem to change their polarity upwards: the ebb-oriented large-scale dune (a) is overlain by alternated ebb and flood-oriented cat-back dunes (b) and again by larger ebb-oriented dunes (c).

internally of homogeneous, thinly laminated sets up to 20 cm thick, separated by double siltstone/very fine sandstone drapes (Fig. 11B), usually darker, finer-grained and more cemented than the main sandstones of the sets. In a few places sets are separated by true double mud drapes up to 2 cm thick. Foreset laminae (1–5 cm thick) are homogeneous or normally graded.

The very large dunes are composite: in their lower levels and along the bottomsets the large foresets contain ripples (climbing ripples along the bottomsets) migrating in a direction opposite to the direction of migration of the large bedform. The superimposed bedforms become larger towards the crest, where the large foresets contain reverse, 20–40 cm thick tabular cross-beds (for example at Glasnakille, Fig. 11B). The stoss side of the very large bedform is more complex, with smaller bedforms (8–100 cm thick) commonly superimposed, some of them migrating in a reverse direction. Growth of the large bedform is punctuated by pause and reactivation surfaces (Figs 11C and 11D). Pause planes appear to have generated stable platforms for the migration of smaller bedforms oriented 45–90° with respect to the largest ones. The large bedforms migrated over small to medium-scale, planar and trough cross-bedded sets, 10–40 cm thick. The uppermost levels of the very large-scale set are made of medium-scale trough cross-beds with a general thinning-upward tendency of the sets.

In the outcrops of Glasnakille (Fig. 11) the large-scale dunes change their polarity upwards in the section. A southward-oriented dune, up to 6 m thick, is overlain by a 7 m interval of alternating southward and northward-oriented large dunes, 1–3 m thick. The southward-oriented foresets are scoured by a unit with basal bounding surfaces characteristically dipping northwards. The upper part of the outcrop again records southward dominance of very large bedforms. The smaller ubiquitous superimposed forms were more dynamic and show a bigger dispersal palaeocurrent pattern.

At Screapadal, and in other localities in Raasay, very large bedforms of this style, with foresets up to 10 m high and dipping northwards, infill longitudinal channels. The very large dunes of Glasnakille appear to be unconfined.

Shell fragments occur throughout the association. Bioturbation is very slight to absent and tends to be concentrated along foreset boundaries.

Interpretation. The large-scale dunes are strongly asymmetric, a feature which can provide a simple qualitative indicator of the direction of the local net bed-load transport (Johnson *et al.* 1982): the lee face orientation is consistent with the orientation of the medium and small-scale dunes. Tidal influence in these deposits is demonstrated by abundant high-angle cross-strata, superimposition of small-scale bedforms oriented in a direction opposite to the large-scale foresets, indicating reverse flows, and double fine-grained laminae separating the sets.

Although regional considerations suggest hinterlands to the south and open basinal conditions to the north (Steel 1977; Morton 1992*a*, *b*; Harris 1992), indicating north as the dominant ebb-current direction, the southward main palaeocurrent vector at Glasnakille is also believed to have been generated by ebb-tidal currents.

The large-scale cross-bedded sandstones of this facies resemble the class IV sand waves of Allen (1980), which have a stacked ebb and flood cross-stratification produced by superimposed small dunes with reversed polarity during each semi-diurnal cycle. The large dunes described here are remarkably similar in internal structure to the progressive ebb-dominated and the cat-back ebb-dominated larger dunes of Van Veen (1935) and Berné *et al.* (1993). As described by Berné *et al.* (1993), the ebb-oriented dunes in the entrance of the Gironde estuary have ebb-dominated foresets, but their surface is covered with flood-oriented superimposed small dunes. This latter is analogous to what is described above for the lowermost part of the outcrop at Glasnakille (Fig. 11). As in the Gironde estuary the main internal structure of the large dunes seems to be related to long-term evolution of net bed-load transport direction rather than to a semi-diurnal reversal, and the stratification related to the small dunes is most likely to be oblique or perpendicular to the axis of the tidal ellipse rather than parallel to it, as in the Allen model. The polarity of the small superimposed dunes indicates an oblique orientation relative to the asymmetrical large dunes and suggests that the net sand transport is oblique or parallel to, rather than perpendicular to, the crest of the large bedforms (see also Bokuniewicz *et al.* 1977; Berné *et al.* 1993). It is also possible that these 45–90° oriented superimposed dunes are infilling swatchway channels (Robinson 1960). The cat-back dunes are dominated by ebb-oriented foresets but they are covered by a cap of flood-oriented bedforms, as was observed also in the analogous middle part of the outcrop at Glasnakille (Fig. 11). The observed bounding surfaces, scouring the ebb-oriented foresets and dipping along a flood orientation, are likely to be related to a change in overall hydrodynamic conditions, rather than to local processes like movement of superimposed bedforms. They seem to record a change in asymmetry of the large bedforms due to the progressive development of flood-oriented flows and, as a consequence, the passage from ebb-dominated progressive dunes to cat-back dunes.

The presence of rippled bottomsets indicates backflow ripples produced in front of the larger current bedform. Since no wave action has been recognized within this facies or the adjacent ones, the most likely process for the bedform inversion may have been the fortnightly variation of tidal amplitude and/or seasonal change in river discharge (see also Berné *et al.* 1993).

Vertical arrangement and depositional setting of the coarsening- and thickening-upward cross-bedded facies association

The cross-bedded facies association is vertically organized into 15–30 m thick, coarsening- and thickening-upward intervals, and these are followed by a slight fining-upward tendency. A profile with several of these coarsening-upward units can be observed in the lowermost part of the succession along the shore at Glasnakille (Fig. 12). Here successive thickening-upward units are capped by a thinning-upward interval (generally up to 2 m thick) of laterally persistent, partially cemented, massive to tabular cross-stratified, broadly lenticular sandstone beds (Fig. 12). The massive character is probably due to a high carbonate content. The coarsening-upward intervals may also be locally eroded at their top by a channel.

The general coarsening- and thickening-upward trend of the cross-bedded facies association, its vertical relationship with underlying prodelta deposits and with overlying channels suggest that the association represents the deposition of the upper reaches of a tidally dominated delta front. Similar facies, developed within embayments and tidally dominated deltas, have been described by Maguregui & Tyler (1991) in western Venezuela. Build-up and seaward migration of the Bearreraig tidally dominated mouth bars was not continuous, but was punctuated by stages of lower sediment supply and weaker tidal regime, as demonstrated by the small-scale fining- and thinning-upward motifs within the larger thickening-upward trend. The laterally extensive, sheet-like, partially cemented beds (sometimes with overlying mud drapes) which occur at the top of the coarsening-upward units, are interpreted as representing longer periods of low clastic input and high marine productivity. They probably originated from the winnowing action of storm-enhanced, sheet-like flows, followed by considerable delay prior to renewed sand transport, as indicated by the overlying mud drapes. They can represent hard grounds, similar to the layers capping the ebb-tidal delta deposits in Roda (Yang & Nio 1989) or more generally condensed sections and flooding surfaces at the top of a prograding unit. The repetition of such late-stage flooding events, together with the thinning-upward tendency of the beds, indicates that despite the general regressive dominance of the Beareraig delta front, the mouth-bar systems were periodically reached by transgressive phases.

Facies Association 3: delta-plain deposits

Facies Association 3 lies above the coarsening-upward, tidally dominated deltaic deposits of Facies Association 2 and below the thinning-upward channel-fill succession of Facies Association 4. Upper and lower bounding contacts are sharp, often marked by calcite- or iron-cemented horizons. The association consists of two facies: fine-grained sandstones with root traces, and channellized, unbioturbated very fine-grained trough cross-stratified sandstones.

Fine-grained sandstone with root traces. This facies was recognized at Glasnakille and Torvaig. At Glasnakille a 50 cm thick horizon with roots was found above a succession of tidally dominated delta-front cross-stratified sandstones and immediately below fining-upward channellized deposits. At Torvaig (Fig. 4), some 30 m from the base, a 30–50 cm thick rooted sandstone horizon immediately underlies the sharp base of a bioclastic thinning-upward unit of cross-stratified medium-grained sandstones of Facies Association 4.

Channellized, unbioturbated, very fine-grained trough cross-stratified sandstones. This facies was recognized only in the uppermost part of the section at Portree (Fig. 4), where it scours into small-scale, cross-stratified sandstones of possible tidally dominated lower delta-front origin. Load casts up to 70 cm deep and iron cementation characterize the lower bounding surface. The facies, up to 5 m thick, consists of very fine sandstones with trough cross-strata infilling a broad channel. There is a sharp contrast in grain-size and colour between the white, calcite-cemented, medium-grained, small-scale

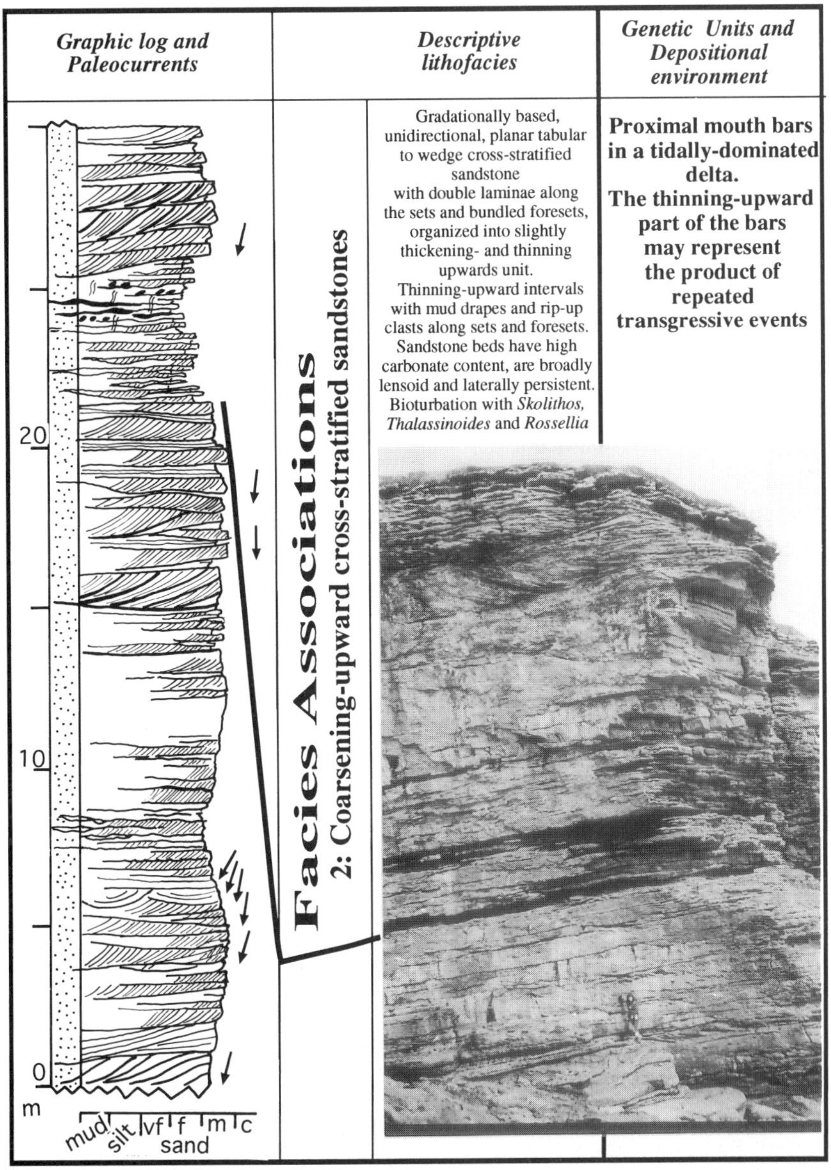

Fig. 12. Representative measured section at Glasnakille and photograph of the outcrops interpreted in terms of tidally dominated delta front. Note the thickening (due to deltaic progradation) and thinning-upward (possibly related to flooding or transgressive events) tendency of beds.

cross-stratified sandstones of the underlying deposits and the very fine-grained buff sandstones of the channel fill. No sign of bioturbation was found in the deposits. The upper bounding surface and the overlying sediments are not exposed.

Interpretation. Interpretation of this association is based on the stratigraphic position of the deposits (generally immediately underlying the tidally dominated channel fills of Facies Association 4), the presence of root traces, and the character of the cross-strata which is absolutely different from the underlying delta-front and overlying estuarine cross-stratified associations.

The root horizons reflect harsh physical and chemical conditions associated with possible brackish-water environment and subaerial exposure. The sharp boundary both with the underlying delta-front deposits and with the overlying tidal cross-bedded sandstones suggests that the sediments were deposited immediately prior to or during a relative fall of sea level. The latter caused plant growth and incipient palaeosols and a subsequent subaerial incision. The very fine-grained unbioturbated sandstones at Portree can be interpreted as small crevasse channels of fluvial origin, deposited at a time when other localities recorded subaerial exposure. Most of the deposits related to this stage of sea-level fall were subsequently eroded and overlain by transgressive tidal-channel deposits.

Facies Association 4: Erosively based, massive to cross-stratified sandstones (the channel-fill deposits)

Channel deposits form the thickest units within the succession and dominate entirely the sections of eastern Raasay, although some units with strong erosional bases have been seen also at Glasnakille and in the Bearreraig Bay. Channel-fill units have a lower contact which is erosional and abrupt, scouring deeply into underlying subtidal, small-scale dunes and/or coarsening-upward packages (Figs 13 and 14). At Screapadal (Fig. 13B), the basal erosional relief of some of the lowermost channels reaches up to 15 m. Here the channels are multi-storied and laterally juxtaposed, typically forming belts extending for several kilometres along the axis of the basin and more or less parallel to the main palaeoflow vector. The lithology of the infill of the erosional relief is a basal coarse-grained lag followed by fine to coarse-grained, trough to tabular cross-bedded sandstone with minor shales in the form of mud drapes and rare horizons of rip-up clasts. The large- and medium-scale trough and tabular cross-beds of the channel fill are, in places, intensively and spectacularly deformed by

Fig. 13. (**A**) One of the tidal channels on Raasay (base marked by continuous line), scouring deeply into underlying medium-scale dunes. Most of the channel fill is here represented by convolute bedded sandstone (indicated by arrow). (**B**) View of the outcrops of Screapadal (sheep in the circle as scale). The bases of the major channels are indicated by arrows. The cliff is up to 150 m thick.

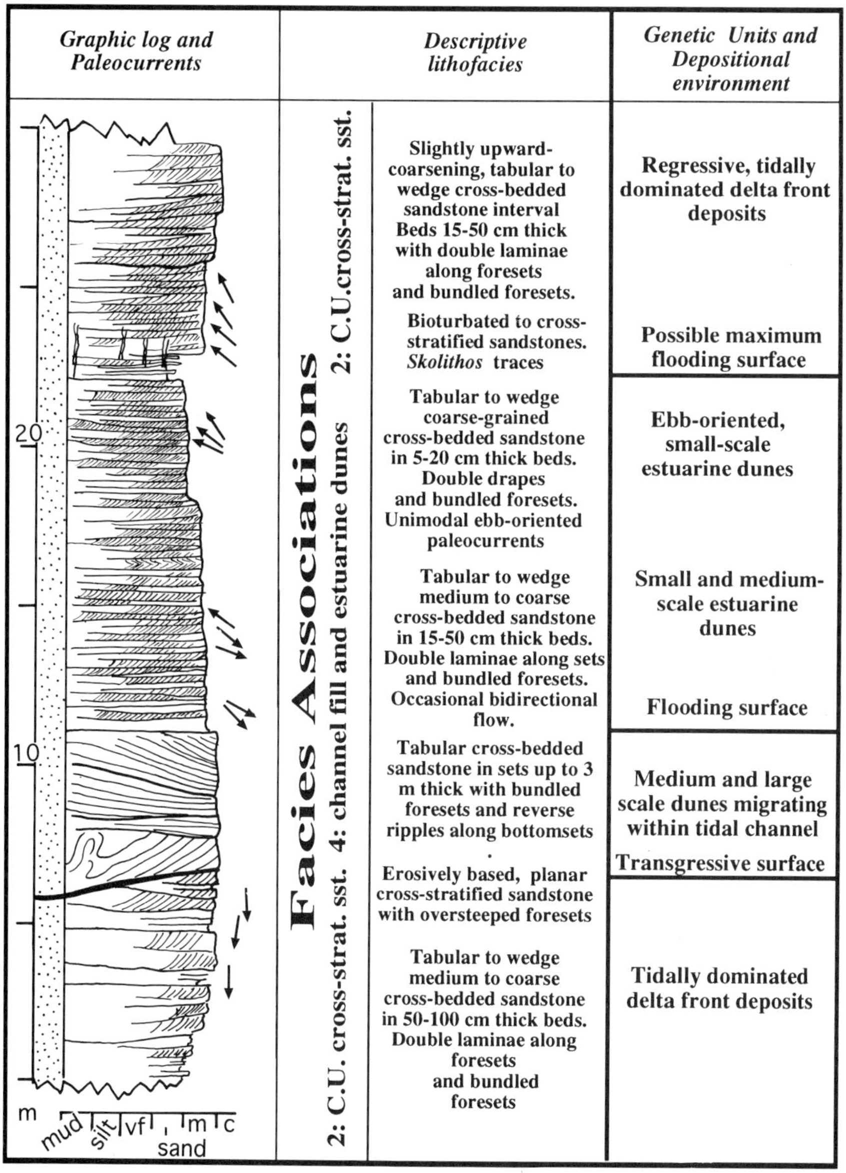

Fig. 14. Sketch of a measured section at Glasnakille illustrating the facies and the vertical relationships between channellized deposits, thinning-upward small-scale dunes and thickening-upward tidally dominated delta-front deposits.

water-escape structures (Fig. 13A). Convolute bedding can occur throughout the association.

Channel morphology and internal organization change through the succession, both on east Raasay and at Glasnakille. Whereas the lowermost channels are usually deeply incised and characterized by large- to medium-scale tabular cross-beds and frequent deformation structures, the channels in the upper part of the succession are relatively shallow, 70–200 cm deep, and occur as multi-lateral and multi-story lenticular bodies, some metres to a few tens of metres wide. Bedforms are systematically oriented northwards on Raasay and southwards on southern Skye, with more palaeocurrent variability in the Bearreraig Bay.

In all the channel-fill units observed there is no evidence of subaerial exposure.

The basal lag. The basal erosion surface has significant relief and is, in most places, overlain by medium to coarse-grained sandstone with a concentration of bivalves, belemnites and small pebbles (up to 2 cm in diameter). Shell debris is often concentrated in gutter casts with up to 50 cm of relief cut into the underlying cross-bedded dune field. The basal deposits (lower 50–100 cm) can be intensively bioturbated with shell fragments either dispersed throughout or concentrated at the top of the basal 10 cm. Where this lag is not present, the erosional surface is commonly overlain directly by trough and planar cross-beds (Fig. 14).

Convolute beds. One of the spectacular features of the channel infills is the intense degree of water-escape and other deformation structures, often associated with massive sandstone intervals. At Screapadal, structureless intervals with only minor trough cross-bedding and deformed beds form units up to 5 m thick (Fig. 13A). At some levels, cross-laminae initially show irregular small-scale (a few centimetres) contortions. The scale of convolutions and degree of deformation increase rapidly upwards, from centimetre-scale up to 1 m. In some cases, the original medium and large-scale cross-stratification (in sets up to 3 m thick) is almost completely obliterated. The traces of folds are evidenced by brown, bioclastic, more cemented beds. Superimposed flood-current dunes, which sank into the underlying sediment, locally produce sand-in-sand load balls.

The trough and planar cross-stratified sandstones within the channels. Trough cross-bedded sandstones are common, especially high in the succession. They consist of medium to coarse-grained trough cross-beds, 0.5–3 m thick, typically multilateral and multi-storey and with a thinning-upward tendency of the sets. Thin finer-grained, double drapes occur along bundled foresets. In the lowermost channels of the succession, the trough cross-bedded sandstones are not present; here the infill consists almost entirely of planar to wedge-shaped tabular cross-beds and convolute beds. The planar cross-stratified sandstones are similar, in form and size, to the medium and large-scale cross-stratified sandstones previously described within the tidal delta-front facies association and interpreted as fields of two-dimensional dunes. The dunes here, however, migrated within major channels (Fig. 14), and tend to display a general thinning-upward set tendency. As mentioned above, a very large dune, up to 10 m thick, fills one of the channels at Screapadal.

Interpretation. The basal deposits are interpreted as channel lags. The gutter casts at the channel bases record vortex and scouring of sinuous dunes during the phase of channel cutting, when large amounts of sand were transported through the system. It is believed that frequent convolution and water-escape structures were formed during deposition, since the axial planes of the folds have a preferential direction of inclination, similar to the foresets of the tabular cross-beds. Deformation can be initiated by creep of sediment induced by channel migration and undercutting (Ovenshine *et al.* 1976), wave-induced liquefaction (Dalrymple 1979), rise and fall of the water table through the sediment, open-grain packing during deposition by avalanching on any dune lee face, or by tectonic events (see Bartsch-Winkler & Ovenshine 1984). Rapid changes in water level, together with excavation of bluffs by channel migration are believed to have been the major local causes which enhanced liquefaction within the channel-fill deposits. However, considering the scale of deformation, and the active tectonic region in which deposition occurred (Morton 1989), it is believed that earthquakes may have had an important role in triggering the soft-sediment deformation.

The trough cross-beds, characterizing the channel fill in the upper part of the succession, were formed by the migration of three-dimensional dunes. Double drapes along the sets and bundled foresets suggest a subtidal depositional environment. The largest dunes in the channel at Screapadal resemble the largest dunes recorded near the mouth of Bahia Blanca estuary (Aliotta & Perillo 1987) in the central and southern margin of the main channel.

The cross-stratified sandstone between the channels. The major channels of the channel belt are separated or underlain by up to 20 m thick semi-continuous aggrading or slightly thinning-upward intervals of generally medium-scale dunes (Fig. 15A); this continuity is broken by numerous small channels, 50–100 cm thick, up to 10 m wide, usually infilled by massive sandstones or by small-scale dunes (Fig. 15B).

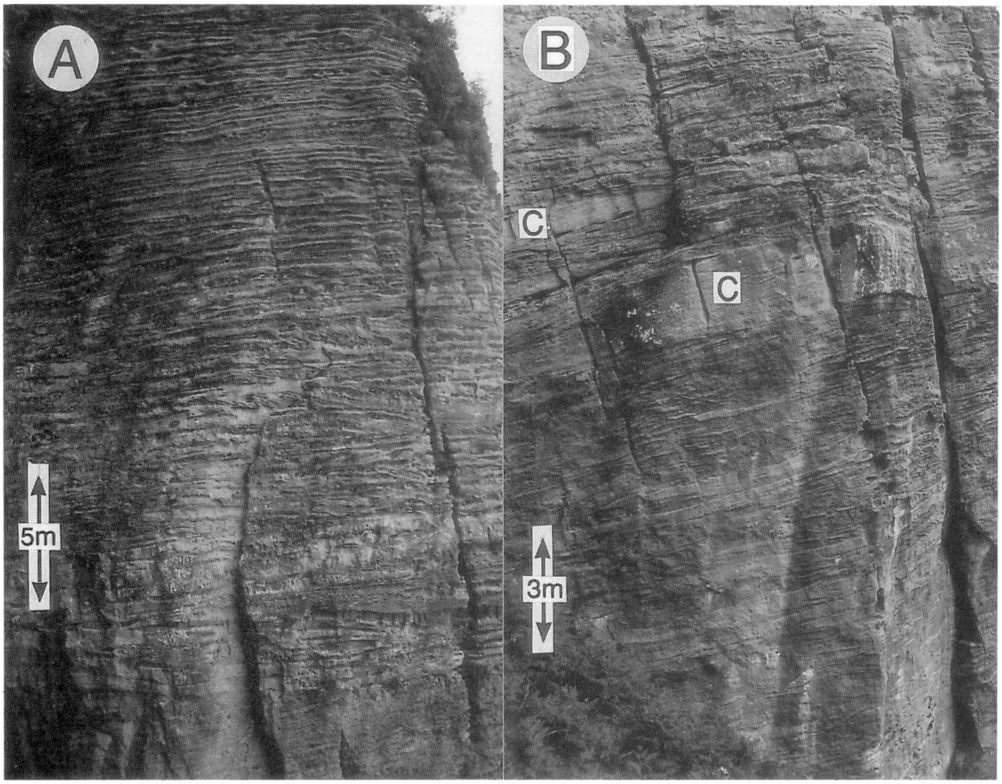

Fig. 15. Outcrops south of Screapadal on Raasay. (**A**) Continuous bar chains in units up to 30 m thick. The medium-scale dunes are overlain by smaller dunes. Although the palaeocurrent vector is generally oriented northwards (ebb orientation), a few opposite sets can be seen (arrow). (**B**) Small channels, C, often interrupt the continuity of the dunes.

The small-scale cross-stratified sandstone at the top of the channel-fill deposits. Small-scale, coarse-grained dunes, analogous to those described in the cross-bedded facies association, may also occur at the top of the channel-fill deposits, both on Raasay and at Glasnakille (Fig. 14 and uppermost part of Fig. 15A) where they usually form up to 30 m thick, aggradational to fining and thinning-upward intervals. Their basal bounding surface is usually sharp, marked by a clear change in hydrodynamic conditions with superimposition of small-scale bedforms over large-scale and deformed cross-beds of the channel fill. The facies can be overlain by the basal part of the small-scale dunes of Facies Association 2 (Fig. 14) or alternatively, may be eroded by a subsequent channelling episode (Fig. 16). Palaeocurrents show a wider dispersal as compared to the channel-fill deposits, although they are consistent with the main northward palaeocurrent vector.

Depositional setting of the channel fills and associated cross-bedded sandstones

Ebb tide-dominated estuary

The deep channels and the associated cross-stratified sandstones of the Bearreraig Sandstone Formation are believed to have been deposited within an estuary. This hypothesis is based on the scale of incision at the base of channels, on the infill architecture (discussed further below) and on the presence of transgressive lags, in places up to 1 m thick, at the base of the channels. Both channels and associated dunes are similar to the channel network and dune fields described by Dalrymple *et al.* (1978, 1990) from the macrotidal Bay of Fundy, and to the large dunes at the mouth of the Gironde estuary (Berné *et al.* 1993). For the large composite bedforms migrating within the channels in the Bearreraig Sandstone Formation, the presence of double laminae along foreset beds,

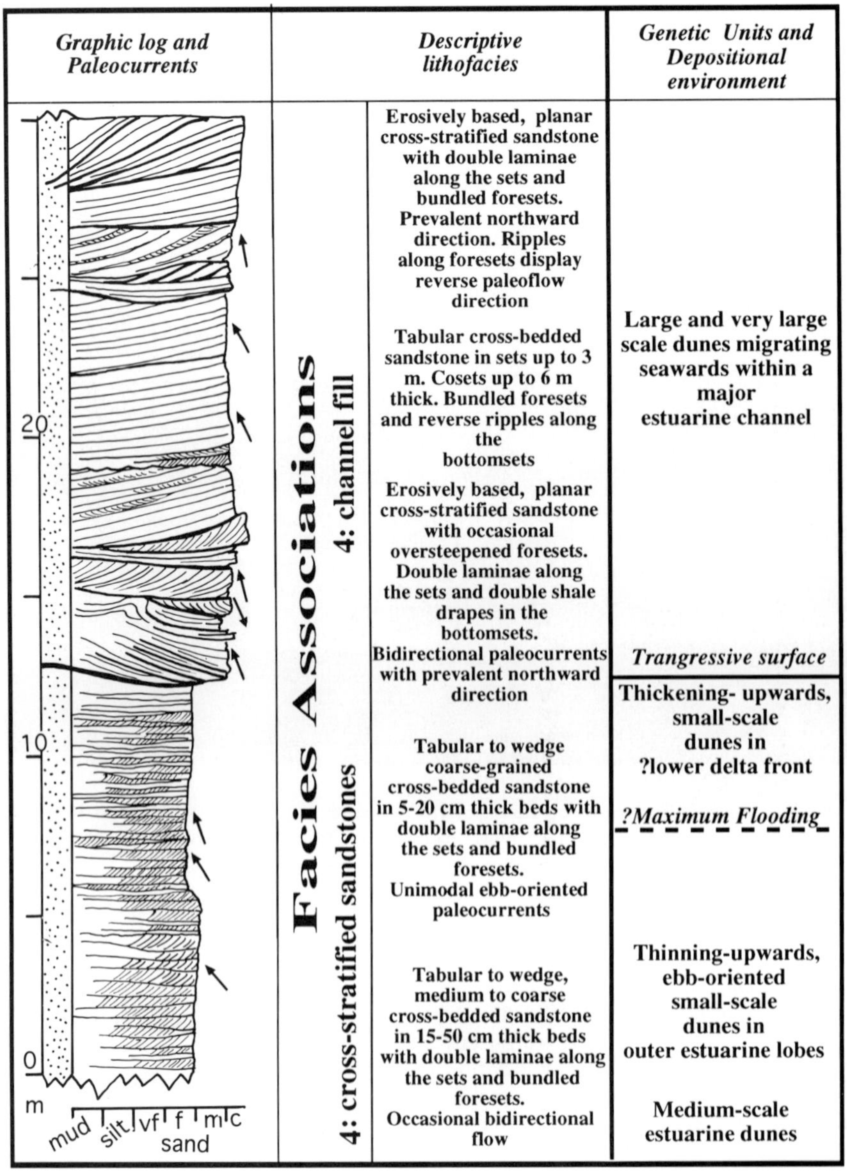

Fig. 16. Sketch of a measured section on Raasay (Beinn na Leac) illustrating the facies and the vertical relationships between thinning-upward small-scale dunes and overlying channellized deposits.

bundled foresets together with the lack of shale interlayers, indicate a largely subtidal depositional setting (Klein 1970; Visser 1980; de Raaf & Boersma 1971; Clifton 1983; Dalrymple *et al.* 1990). The scour depth of the channels confirms this.

The lack of barrier island and lagoonal sediments suggests deposition in a wide-mouthed, tidal-dominated estuarine setting (definition from Boyd *et al.* 1992 and Dalrymple *et al.* 1992). The absolute abundance of cross-bedded deposits and the very large size of individual bedforms, strongly suggests, moreover, a macrotidal regime.

The palaeocurrent vector within the estuarine channel-fill deposits is oriented almost constantly northwards (regional offshore position) and indicates a dominance of ebb currents at the estuary mouth. This ebb dominance at estuary mouths is fairly unusual (Dalrymple *et al.* 1992), but not unknown (Allen 1991; Allen & Posamentier 1993).

Estuarine mouth deposits

The very large dunes which migrated within the channels are believed to represent deposition at the estuary mouth, which is dominated by a system of channels and large and medium-scale

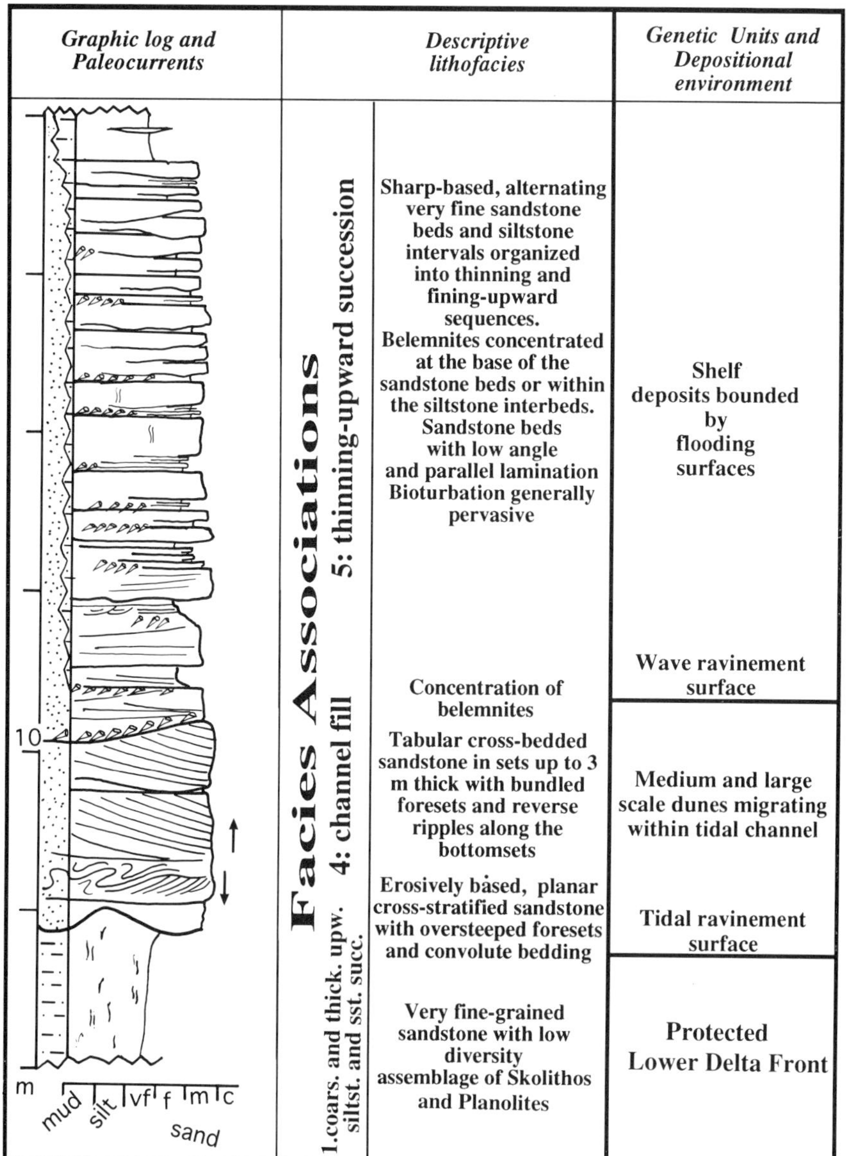

Fig. 17. Measured section from the middle part of the Bearreraig Bay succession. Lower delta front sandstones with stressed ichnofacies are overlain by channellized estuarine cross-stratified sandstones. The bounding surface is interpreted to represent a tidal ravinement surface. The upper bounding surface of the cross-bedded sandstones may be interpreted as a wave ravinement surface.

dunes. The largest dunes at Screapadal have a preserved relief of 6–8 m. Considering that the height/depth ratio in estuarine settings varies between 1/6 and 1/10, with an average of 1/7 (Yalin 1964; Bokuniewicz *et al.* 1977; Rubin & McCulloch 1980), the water depth at which such a bedform was deposited is likely to have been 30–50 m. Modern dunes, with amplitudes up to 10 m and kilometres in wavelength are known from tidally dominated environments (Berné *et al.* 1992, 1993), as well as in the fossil record (Allen & Homewood 1984; Mutti *et al.* 1985; Smith 1988). The stacking of sandwave cosets and the presence of low-angle master-bedding planes, may indicate that the very large dunes represent portions of linear sand banks (Houbolt 1968; Harris 1988).

Between the channels, the system of large and medium-scale dunes probably represents a tidal sand-bar complex which occupied the seaward reaches of the zone of tidal energy maximum (see also Dalrymple & Zaitlin 1989; Dalrymple *et al.* 1990).

The small-scale dunes are suggested to have been deposited in an unconfined setting, probably outside the estuary mouth, as indicated by the wider range of dip directions recorded within these deposits compared with those found in the more inshore tidal channels. We are not suggesting that all the small-scale dunes were located seaward of the sand-bar complex, but particularly those closely associated with underlying channels or those that fine upwards. It is likely that small dunes also occurred headward of the estuary mouth, in the central estuarine zone approaching the zone of tidal energy maximum (Dalrymple *et al.* 1990, 1992), eventually associated with upper-flow regime sand flats.

Facies Association 5: bioturbated sandstones in upward-thinning units (shelf deposits)

This facies association is restricted to northwest Skye (Bearreraig Bay), where it occurs immediately overlying channel-fill deposits (cf. Figs 6A and 17), and at the base of the succession underlying the coarsening-upward succession interpreted as tidally dominated distal mouth bars (Fig. 18). It consists of sharp-based, very fine to fine-grained bioturbated sandstones 5–110 cm thick, interbedded with 3–40 cm thick shales or bioturbated sandy mudstones, and occurs as intervals 4–8 m thick displaying an overall thinning and fining-upward tendency of the sandstone beds.

Sandstone beds form laterally continuous sheets and discontinuous lenticular bodies, mimicking hummocky cross-stratification (Fig. 18B). Bed bases are sharp, locally with small tool marks. Belemnites are often concentrated at the base of the beds (see also Figs 7A and 7C). Internal sedimentary structures are generally completely obliterated by intense bioturbation. Occasional low-angle stratification can be detected. Bed tops often display straight-crested symmetrical ripples, though usually they are very diffuse, due to intense bioturbation and gradation into siltstones. The lack of lithological contrast makes identification of specific traces difficult, although a few *Thalassinoides* were recognized. The lenticular beds split laterally into thinner heterolithic units of sandstone and shale. Boundaries are always accentuated by shale layers. The interbedded mudstones and sandy shale are extensively bioturbated.

Interpretation. The depositional environment for this association appears to have been a

Fig. 18. General view (**A**) and details (**B**) of the transgressive shelf deposits containing sharp-based bioturbated sandstones in upward-thinning units (Facies Association 5). The sandstones have shale interbeds. In places the beds are lenticular, mimicking hummocky morphology, and have wave ripple-laminated tops.

flat-lying area of relatively quiet sand and shale deposition, prone to bioturbation. At times, sand was rapidly emplaced and later completely or almost totally reworked by bioturbation. This suggests depths above storm wave-base, but greater than fair-weather wave-base. The scoured and sharp-based soles of the beds result from initiation of high-velocity, competent flows (Hunter & Clifton 1982), and the gutter casts, commonly associated with the base of the beds, are believed to record periods of localized erosion in a generally cohesive muddy substratum, probably by oscillatory wave scour (Duke 1985; Plint & Norris 1991). The environment was presumably a prodelta-shelf area.

The belemnites concentrated at the base of the beds are interpreted as storm lags. The sharp character of some of the basal bounding surfaces and the presence of basal lags suggest that these surfaces have been generated by wave ravinement processes (Swift 1968; Nummedal & Swift 1987). The overall upward thinning of the facies of the succession and the stratigraphic context as argued below, suggest a transgressive setting for Facies Association 5.

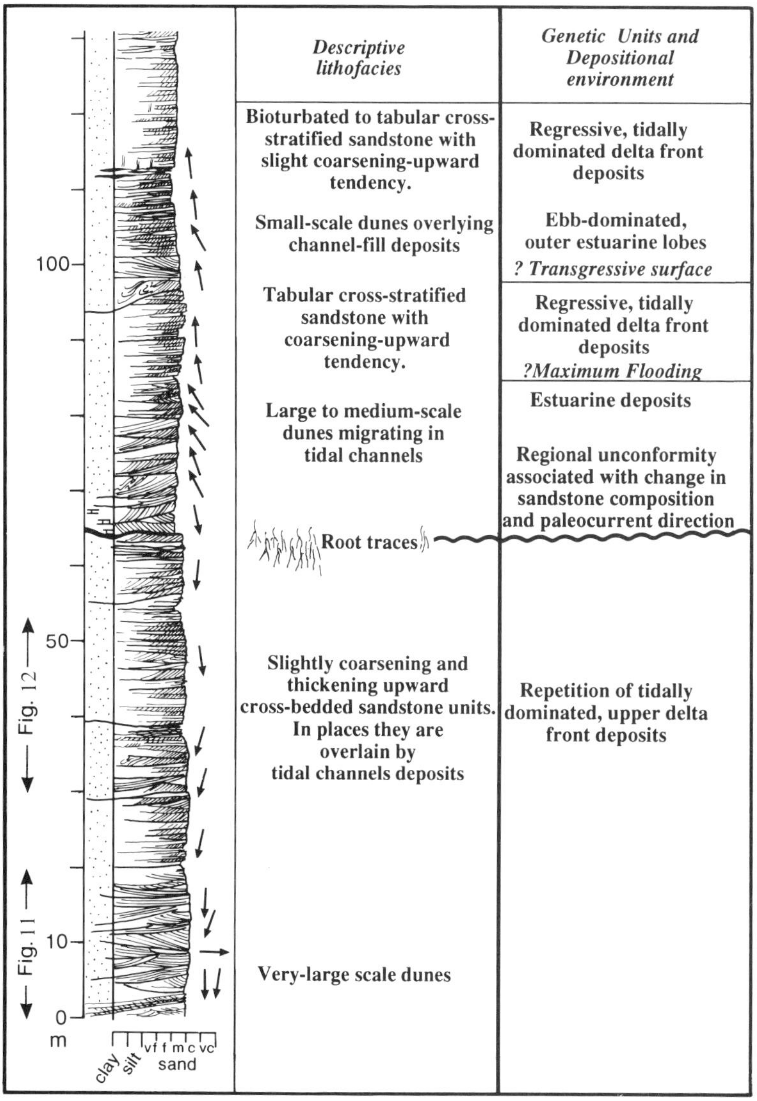

Fig. 19. Measured section at Glasnakille.

Time trend of sedimentation at the main localities

The succession has been particularly studied at three main localities: north Skye (the representative section in Bearreraig Bay), east Raasay (Screapadal) and south Skye (Glasnakille) (see Fig. 3 for location). The section of Glasnakille is shown in Fig. 19. The sections of Bearreraig Bay and Screapadal are shown in the correlation panels of Figs 4 and 20.

North Skye: Bearreraig Bay

The lower part of the Bearreraig Sandstone Formation consists of shales, siltstones and shelf sandstone sheets (Facies Association 5) deposited in a transgressive setting. The overlying bioturbated to cross-bedded tidally influenced mouth-bar deposits record a regressive phase. The interval then shows progressive deepening and flooding of the system by offshore shales.

The middle part of the succession (40 to 100 m in Figs 4 and 20) represents a major change in depositional style, a period characterised by distal deltaic sediments. A series of major thickening-upward sandstone units within the generally shaley package, suggests that there were at least three progradation and retreat episodes of the delta system during deposition.

The upper part of the succession is marked by a pronounced, erosive surface (some 100 m from the base) cut into prodelta/lower delta-front deposits. This major erosion surface denotes a change in depositional environment and sedimentary regime from deltaic to estuarine sedimentation. The erosion surface is a tidal ravinement surface onto which there developed a bed-load-dominated, estuarine-channel system broadly oriented to the northeast. As in the lowermost part of the succession, the estuarine deposits are abruptly capped by a thinning-upward package of bioturbated to ripple-laminated sandstones (Facies Association 5) recording a progressive transgression of the estuarine system. The sandstone beds of Facies Association 5 can therefore be interpreted as transgressive sheets, eroded products of the underlying deposits as the estuarine mouth shifted landward in response to rising sea level (e.g. Swift *et al.* 1991). The uppermost levels of the Bearreraig Sandstone in Bearreraig Bay are characterized by two more progradational episodes of delta-front deposits. Tidal activity appears to have decreased here as cross-bedded sandstones are no longer present. The contact with the overlying Garantiana Clay Member is sharply marked by a fossiliferous granule to small-pebble conglomerate.

Raasay

The Bearreraig Sandstone Formation on Raasay is up to 150 m thick (Fig. 13B) and is dominated, except for the lowermost 30–40 m by estuarine channel-fill sandstones separated by intervals of aggrading to slightly prograding small-scale dunes regarded as the initial progradation of proximal mouth-bar deposits (Fig. 4).

The upper half of the succession, corresponding to the Sauzei ammonite zone (Morton 1965), is characterized by a marked change in the composition of the cross-bedded sandstones which pass from brown to white hybrid-arenites to markedly white quartz-arenites. This change corresponds to a renewed episode of channel-belt incision and large-scale dune migration, and is correlative with the main unconformity recognized in the Bearreraig Bay, some 100 m from the base.

The contact with the overlying Garantiana Shale is sharp and marked below a very coarse-grained sandstone layer.

South Skye: Glasnakille

The measured succession in Glasnakille is up to 250 m thick (Fig. 19) and represents the thickest section of the Bearreraig Sandstone Formation in the study region (see also Morton 1965, 1983). The base of the succession is sharp, as can be seen along the Elgol road near Kilmarie some kilometres northward of the measured section of Fig. 19. As on Raasay, tide-dominated delta facies and estuarine cross-bedded sandstone dominate the succession.

The basal part of the succession consists of a monotonous repetition of small, medium and very large-scale dunes (tidal sand-bar complexes) organized into coarsening-upward units (tidally dominated proximal delta mouth bars). The great thickness of the sandstones and the lack of interbedded shales indicate constant and vigorous sand input, punctuated by minor flooding events at the top of each coarsening-upward cycle. As on Raasay, the succession records a change in composition from hybrid-arenites to quartz-arenites. This change occurs at about 80 m above the base, and is associated with a change in colour of the deposits, palaeocurrent direction, sedimentary patterns (from tide-dominated delta facies association to estuarine channel), a rooted horizon and a

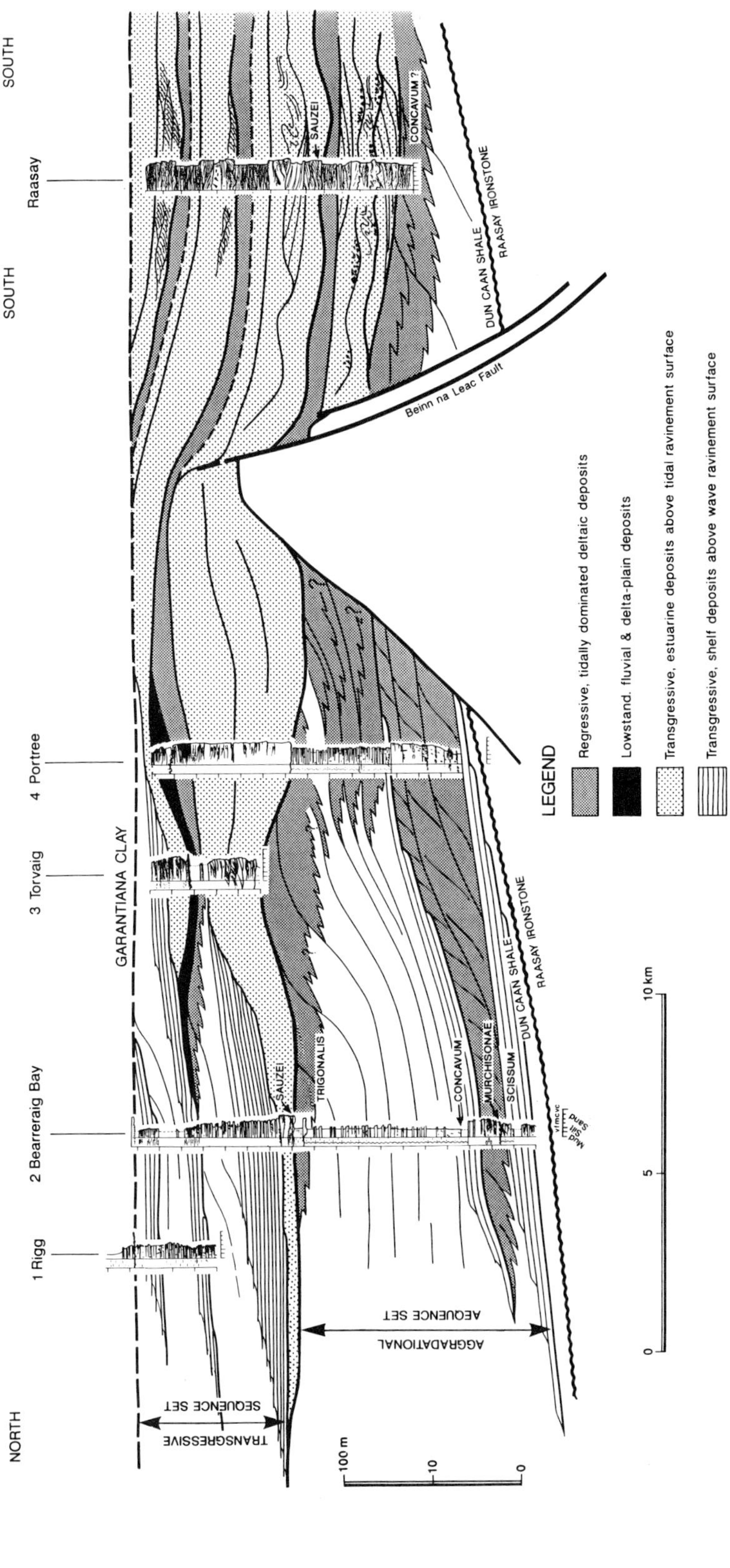

Fig. 20. Sequence stratigraphical framework for the Bearreraig Sandstone Formation across Raasay and north Skye.

marked mappable unconformity (Fig. 19). Channelled estuarine tabular and trough cross-bedded sandstones and small-scale dunes oriented northwards replace tide-dominated, delta-front sandstones showing southward oriented currents.

Vertical and lateral facies relationships

Sequence stratigraphical framework

The facies associations described above are interpreted as the proximal–distal–lateral components in a macrotidal estuary/tidally dominated delta system. We use the term 'estuary' in the sense of Dalrymple *et al.* (1992) only to describe transgressive settings. Where tidally generated facies show clear progradational trends, we refer to this as having originated from a tide-dominated delta.

The Bearreraig system evolved through an early (Aalenian–early Bajocian) stage which saw the development and vertical aggradation of two relatively muddy, tide-dominated deltaic sequences. This was followed by a later (Bajocian) phase of overall transgression with the development of more extensive estuarine deposits. On Raasay only the lowest strata record the muddy deltaic sequences, the succession otherwise being dominated by estuarine deposits. The overall low-order aggradational–transgressive cycle is punctuated by higher-order depositional sequences, each defined by a basal bounding surface (unconformity in the most proximal areas, paraconformity in the distal reaches), and composed of a transgressive systems tract (with estuarine and shelf deposits) and a highstand systems tract (progradational deltaic deposits). Although the formation was deposited across three different structural blocks, a landward–basinward facies relationship is believed to have been developed within each of them. The cross-section of Fig. 20 shows a sequence stratigraphical framework for the Bearreraig Sandstone succession, somewhat more complex than the single, low-order tectonically related genetic sequence described by Morton (1989).

Sequence sets

The Bearreraig Sandstone Formation consists of lower and upper parts, each distinguished on the basis of the dominant facies association, internal stacking architecture and lithological composition. The two parts, separated by a major erosion surface, can be recognized in each sub-basin.

The upper part consists of three transgressive–regressive sequences arranged in a landward-stepping architecture; it is thus referred to as a transgressive sequence set (see also Van Wagoner *et al.* 1990) and is volumetrically dominated by transgressive deposits (Fig. 20). The lower part has two main transgressive–regressive sequences, but here arranged in an aggradational or seaward-stepping architecture; it is thus referred to as a regressive sequence set (Fig. 20). Each of the sequences within the sequence sets is some 20–40 m thick, has a markedly incised lower boundary (at least in the landward reaches), and has a transgressive to regressive thickness ratio which increased landwards.

Although there is good evidence of only one of these sequence boundaries being subaerially exposed, as demonstrated by the rootlet horizons at Glasnakille and Torvaig, and by the fluvial deposits in the uppermost part of the section at Portree (Figs 4 & 20), the near-symmetry of many of the sequences and the incised lower boundaries make it inappropriate to refer to them as parasequences. Simple or high-frequency sequences is a more correct description (see also Cant 1991).

The lower sequence set is bounded by a regional unconformity (Morton 1987, 1989) associated with the Raasay Ironstone Formation. Morton (1987) noted the importance of this boundary and made it a sequence boundary in his stratigraphic scheme. Above the unconformity there are three progradational tide-dominated delta units, separated by transgressive shelf deposits. The uppermost deltaic unit is truncated by the regional unconformity and has little preservation of its uppermost sands. In spite of the lack of preservation of the uppermost sediments, the lower part of the formation can be seen to have an overall aggradational to progradational stacking pattern. Apart from the stacking pattern and the muddy deltaic depositional environments, the lower sequence set also has a characteristic brown-coloured, carbonate-rich sandstone composition, in contrast to the quartz-arenite white sandstones in the upper part of the Formation.

The upper part of the Formation is a transgressive sequence set and overlies the marked erosional surface referred to above. The latter is associated with the maximum northward progradation of the Bearreraig deltaic system (Fig. 20) and is believed to represent a time (Sauzei ammonite subzone) of major erosion and valley incision in the upstream part of the system. This erosional surface is overlain first by rootlet sandstones and then by repeated sandy estuarine units. The base of the channellized estuarine deposits is interpreted as a tidal ravinement surface (sensu Allen & Posamentier 1993) on which there are very large-scale estuarine dunes.

The sequence set displays an overall retrogradational stacking pattern and is abruptly overlain by the Garantiana Clay Member.

Facies partitioning

Figure 20 also illustrates the differing partitioning of estuarine and deltaic deposits during the transgressive–regressive phases of the Bearreraig system. Transgressive estuarine deposits are progressively thicker landwards within the transgressive systems tracts in the north Skye block, and are particularly thickly stacked on Raasay. The volume partitioning of the tidally influenced deltas shows an opposite pattern. Deltaic lobes and prodelta deposits progressively increase in thickness offshore and down-dip on individual blocks. On Raasay the regressive deposits, here represented by slightly thickening-upward, small-scale dunes, immediately overlie thinning-upward small-scale dune successions, considered to have been deposited during transgressive stages. Most of the regressive deltaic deposits appear to have by-passed the Raasay sub-basin (at least in the presently exposed areas) to be transported into a widely developed deltaic-slope system in the wider north Skye sub-basin.

Another interesting aspect of the succession is the volume of 'shelf' sands lying above the lowermost (or master) wave ravinement surface, and associated with a landward-onlapping complex of ravinement surfaces. The shelf deposits form narrow wedges, extending up to 10 km down-dip and up to 25 m in thickness, and characteristically located near the faulted basin margin of the north Skye sub-basin. The thickness of the transgressive deposits is thought to have been enhanced by a high gradient of ravinement. Despite the 'high' gradient setting, the rapid flattening of the shelf (and of the basinward continuity of the ravinement surface) has allowed sand to accumulate. The transgressive sand is thick simply because the landward translation of the ravinement was slow. In practice, there are always clear signs of erosional surfaces and thin lags within the transgressive shelf tabular beds, suggesting that changes in the rate of rise of sea level have periodically generated basinward-extending erosion surfaces. Repeated upward-fining motifs (tens of centimetres to a few metres thick) characterize these lithosomes, and classic parasequences are absent.

Sediment dispersal and palaeogeographic reconstruction: tectonic control on sedimentation

Outcrop palaeocurrent data have been combined with regional mapping to reconstruct the palaeogeography of the Bearreraig Sandstone Formation. Palaeocurrents were generally constrained by the axis of the trough cross-stratification and by the dip of the tabular cross-beds. In most of the measured sections on Raasay and north Skye, palaeocurrent indicators show little variability, with the main vector oriented northwards. In south Skye (Glasnakille, Fig. 19) the basal part of the succession shows palaeocurrents oriented southwards, whereas the upper part shows a palaeocurrent vector to the north. This change occurs above the major unconformity recognized in the Glasnakille section at 80 m from the base. These marked differences in palaeocurrent directions and composition are believed to be caused by a fault-related tilting of the basin axis associated with uplift and erosion of marginal areas which become new source areas for deltaic and estuarine deposits.

An attempt at palaeogeographical reconstruction during the earliest stage of sedimentation of the Bearreraig Sandstone Formation is shown in Fig. 21A. The structural framework is based on the maps of Steel (1977) for the Triassic succession and of Harris (1992) for the Great Estuarine Group. Palaeohighs separate the north Skye and Raasay sub-basins and the Raasay and Glasnakille sub-basins. At a later stage (Fig. 21B), during the time of deposition of the main estuarine channellized deposits in the Bearreraig Bay, estuarine sedimentation was widespread throughout the study region. Palaeocurrent vectors are conformably oriented northwards, although minor southeastward flood directions were recorded in the Bearreraig Bay and at Torvaig. The palaeogeography shows a basin open to the north, and a drowning of the previous palaeohighs.

The strong tide-dominated character of the Bearreraig Sandstone Formation marks out these deposits from others of the same age in the northern North Sea, which are mainly dominated by waves and only subordinately by tidal currents (Brent Group; see Graue *et al.* 1987; Fælt & Steel 1990).

The presence of strong currents able to create and move bedforms up to 10 m high is believed to be a direct consequence of the particular tectonic setting of the Middle Jurassic Hebridean sub-basins, with a series of relatively narrow, slightly tilting blocks where confinement enhanced tidal currents. The details of the

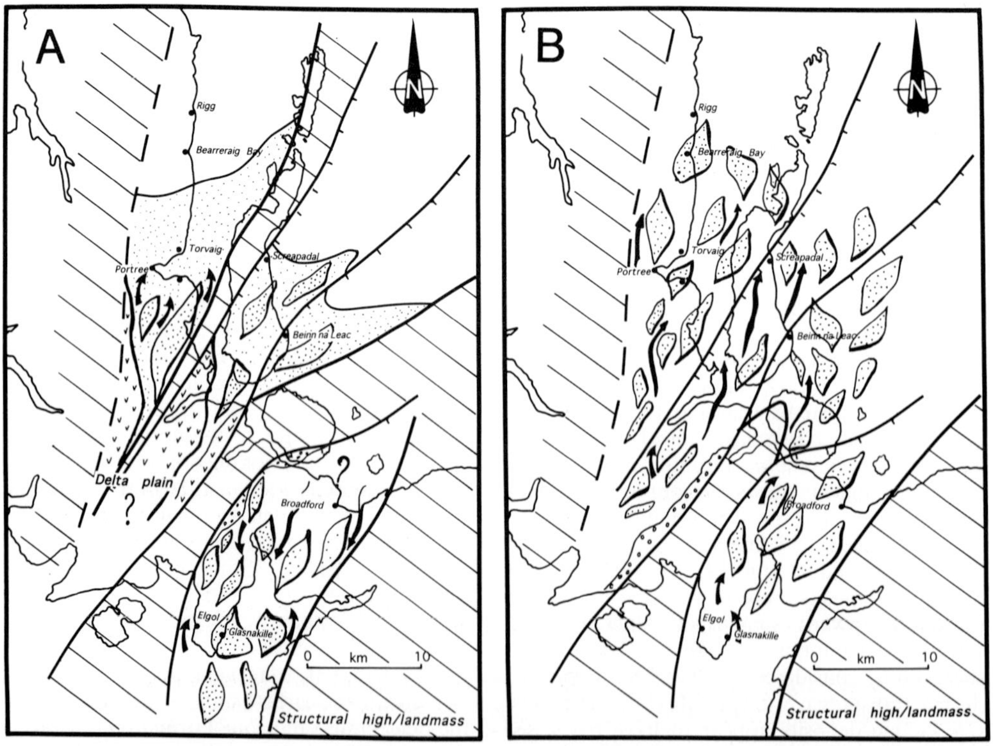

Fig. 21. Palaeogeographical reconstructions for the initial stage ((**A**) *Scissum-Trigonalis* ammonite zones) and the late stage ((**B**) *Sauzei-Subfurcatum* zones) of deposition of the Bearreraig Sandstone Formation.

relationship between sedimentation patterns, tilt episodes and sequence stratigraphy is beyond the scope of this work. Narrow confined basins enhancing tidal currents are not exclusive to extensional tectonic settings. In the Baronia deposits of the Eocene south Pyrenean foreland basin, the emergence of a blind thrust produced a constriction in the basin which acted to amplify tidal activity (Mutti *et al.* 1985).

Conclusions

The facies analysis, stratigraphic correlation and palaeogeographic interpretation outlined here lead to the following conclusions.

(1) Facies analyses demonstrate that the Bearreraig Sandstone Formation on Raasay and at Glasnakille consists primarily of coarsening- and thickening-upward tidally dominated deltaic and thinning-upward estuarine cross-bedded deposits.

(2) The tidally dominated deltaic deposits show two distinct facies and grain-size distributions: the prodelta deposits and the distal mouth bars consist of siltstone and very fine-grained, bioturbated sandstones; the proximal mouth bars consist entirely of medium-grained cross-stratified sandstone.

(3) The estuarine deposits are represented by deep and wide channel belts with associated bar chains. They record deposition within an ebb-dominated macrotidal setting.

(4) The correlation panel presented here shows that the Bearreraig Sandstone Formation represents a large-scale regressive to transgressive development. It can be further subdivided into eight transgressive–regressive cycles. The regressive phases recorded repeated progradation of a tide-dominated delta system. During the transgressive phases the delta was transformed into a macrotidal estuary. The latest stage of transgression produced a thick succession of transgressive shelf sandstones.

(5) Previous work, palaeocurrent analysis and palaeogeographical reconstructions indicate that the Formation was deposited on at least three distinct fault blocks, subject to syn-depositional tilting and separated by

structural highs. During the early stage of deposition a deltaic system, open to the south, was established in south Skye, while in Raasay and in north Skye a northward tidal delta system was active. During a later stage, palaeogeography was more uniform: the previous tectonically separated provinces of Glasnakille, north Skye and Raasay were probably unified into a very large tidally dominated system open to the north.

(6) This change in palaeogeographical conditions seems to have been associated with a regional unconformity and a change in palaeocurrent directions. Block rotation and tilting is likely to have been responsible for the southward closure of the basin in south Skye, its opening to the north, and the formation of the regional unconformity

(7) The active tilt-block setting of the narrow, Mid-Jurassic Hebridean basins is believed to have enhanced tidal current circulation.

The authors wish to thank Enterprise Oil for initiating funding and following up this study, and particularly Mike Whyatt for his enthusiastic support. The paper has benefited from the comments of the reviewers G. Postma and R. Dalrymple. The penetrating observations of Bob Dalrymple were much appreciated and resulted in significant changes to the original manuscript. Responsibility for facts and interpretations rests, nevertheless, with the authors. Financial support by Statoil during the last stage of the preparation is gratefully acknowledged.

References

ALIOTTA, S. & PERILLO, G. M. E. 1987. A sand wave field in the entrance to Bahia Blanca estuary, Argentina. *Marine Geology*, **76**, 1–14.

ALLEN, G. P. 1991. Sedimentary processes and facies in the Gironde estuary: a recent model for macrotidal estuarine systems. *In*: SMITH, D. G., REINSON, G. E., ZAITLIN, B. A. & RAHMANI, R. A. (eds) *Clastic Tidal Sedimentology*. Canadian Society of Petroleum Geologists, Memoir, **16**, 29–40.

—— & POSAMENTIER, H. W. 1993. Sequence stratigraphy and facies model of an incised valley fill: the Gironde estuary, France. *Journal of Sedimentary Petrology*, **63**, 378–391.

ALLEN, J. R. L. 1980. Sand waves: a model of origin and internal structure. *Sedimentary Geology*, **26**, 281–328.

—— & HOMEWOOD, P. 1984. Evolution and mechanics of a Miocene tidal sand wave. *Sedimentology*, **31**, 63–81.

ASHLEY, G. M. *et al.* 1990. Classification of large-scale subaqueous bedforms: a new look at an old problem. *Journal of Sedimentary Petrology*, **60**, 160–172.

BARTSCH-WINKLER, S. & OVENSHINE, A. T. 1984. Macrotidal subartic environment of Turnagain and Knik Arms, Upper Cook Inlet, Alaska: sedimentology of the intertidal zone. *Journal of Sedimentary Petrology*, **54**, 1221–1238

——, CASTAING, P., LE DREZEN, E. & LERICOLAIS, G. 1993. Morphology, internal structure, and reversal of asymmetry of large subtidal dunes in the entrance to Gironde estuary (France). *Journal of Sedimentary Petrology*, **63**, 780–793.

BERNÉ, S., DURAND, J. WEBER, O. 1992. Architecture of moder tidal dunes (sand waves), Bay of Bourgneuf, France. *In*: MIALL, A. D. TYLER, N. (eds) *The Three-dimensional Facies Architecture of Terrigenous Clastic Sediments and its Implications for Hydrocarbon Discovery and Recovery*. Society of Economic Paleontologists and Mineralogists, Concepts in Sedimentology and Paleontology, **3**, 245–260.

BEYNON, B. M. & PEMBERTON, S. G. 1992. Ichnological signature of a brackish water deposit: an example from the lower Cretaceous Grand Rapids Formation, cold Lake Oil Sands area, Alberta. *In*: PEMBERTON, S. G. (ed.) *Application of Ichnology to Petroleum Exploration – A Core Workshop*. Society of Economic Paleontologists and Mineralogists, Core Workshop, **17**, 199–221.

BOERSMA, J. R. 1969. Internal structure of some tidal megaripples on a shoal in the Westerschelde estuary, The Netherlands. Report of a preliminary investigation. *Geologie en Mijnbouw*, **48**, 409–414.

—— & TERWINDT, J. H. J. 1981. Neap-spring tide sequences of intertidal shoal deposits in a mesotidal estuary. *Sedimentology*, **28**, 151–170.

BOKUNIEWICZ, H. J., GORDON, R. B. & KASTENS, K. A. 1977. Form and migration of sand waves in a large estuary, Long Island. *Marine Geology*, **24**, 185–199.

BOYD, R., DALRYMPLE, R. & ZAITLIN, B. A. 1992. Classification of clastic coastal depositional environments. *Sedimentary Geology*, **80**, 139–150.

BREWER, M. D. & SMITHE, D. K. 1984. MOIST and the continuity of crustal reflector geometry along the Caledonian–Appalachian orogen. *Journal of the Geological Society of London*, **141**, 105–120.

CANT, D. J. 1991. Geometric modelling of facies migration: theoretical development of facies successions and local unconformities. *Basin Research*, **3**, 51–62.

CLIFTON, H. E. 1983. Discrimination between subtidal and intertidal facies in Pleistocene deposits, Willapa Bay, Washington. *Journal of Sedimentary Petrology*, **53**, 353–369.

DALRYMPLE, R. W. 1979. Wave-induced liquefaction: a modern example from the Bay of Fundy. *Sedimentology*, **26**, 835–844.

—— & ZAITLIN, B. A. 1989. *Tidal sedimentation in the macrotidal, Cobequid Bay–Salmon River estuary, Bay of Fundy*. 2nd International Research Symposium on 'Clastic Tidal Deposits', Field Guide. Canadian Society of Petroleum Geologists.

——, KNIGHT, R. J. & LAMBIASE, J. J. 1978. Bedforms and their hydraulic stability relationships in a tidal environment, Bay of Fundy, Canada. *Nature*, **275**, 100–104.

——, ——, ZAITLIN, B. A. & MIDDLETON, G. V. 1990. Dynamics and facies model of a macrotidal sand-bar complex, Cobequid Bay–Salmon river estuary (Bay of Fundy). *Sedimentology*, **37**, 577–612.

——, ZAITLIN, B. A. & BOYD, R. 1992. Estuarine facies models: conceptual basis and stratigraphic implications. *Journal of Sedimentary Petrology*, **62**, 1130–1146.

DE RAAF, J. F. M. & BOERSMA, J. R. 1971. Tidal deposits and their sedimentary structures. *Geologie en Mijnbouw*, **50**, 479–503.

DUKE, W. 1985. Hummocky cross stratification, tropical hurricanes and intense winter storms. *Sedimentology*, **32**, 167–194.

EARLE, M. M., JANKOWSKY, E. J. & VANN, I. R. 1989. Structural and stratigraphic evolution of the Faeroe–Shetland Channel and Northern Rockall Trough. *In*: TANKARD, A. J. & BALKWILL, H. R. (eds) *Extensional Tectonics and Stratigraphy of North Atlantic Margins*. American Association of Petroleum Geologists, Memoir, **46**, 461–469.

FÆLT, L. M. & STEEL, R. J. 1990. A new paleogeographic model for the Brent delta system. Discussion. *Journal of the Geological Society of London*, **147**, 1085–1090.

GRAUE, E., HELLAND-HANSEN, W., JOHNSEN, J. *et al.* 1987. Advance and retreat of Brent Delta System, Norwegian North Sea. *In*: BROOKS, J. & GLENNIE, K. (eds) *Petroleum Geology of North West Europe*. Graham & Trotman, London, 915–937.

HARRIS, J. P. 1992. Mid-Jurassic lagoonal delta systems in the Hebridean basins: thickness and facies distribution patterns of potential reservoir sandbodies. *In*: PARNELL, J. (ed.) *Basins of the Atlantic Seaboard: Petroleum Geology, Sedimentology and Basin Evolution*. Geological Society, London, Special Publication, **62**, 111–144.

—— & HUDSON, J. D. 1980. Lithostratigraphy of the Great Estuarine Group (Middle Jurassic). Inner Hebrides. *Scottish Journal of Geology*, **16**, 231–250.

HARRIS, P. T. 1988. Large-scale bedforms as indicators of mutually evasive sand transport and the sequential infilling of wide-mouthed estuaries. *Sedimentary Geology*, **57**, 273–298.

HOUBOLT, J. J. H. C. 1968. Recent sediments in the southern Bight of the North Sea. *Geologie en Mijnbouw*, **47**, 245–273.

HUNTER, R. E. & CLIFTON, H. E 1982. Cyclic deposits and hummocky cross-stratification of probable storm origin in Upper Cretaceous rocks of Cape Sebastian area, southwestern Oregon. *Journal of Sedimentary Petrology*, **52**, 127–144.

JOHNSON, M. A., KENYON, N. H. & BELDERSON, R. H. 1982. Sand transport. *In*: STRIDE, N. H. (ed.) *Offshore Tidal Sands. Processes and Deposits*. Chapman and Hall, London, 58–94.

KLEIN, G. 1970. Depositional and dispersal dynamics of intertidal sand bars. *Journal of Sedimentary Petrology*, **40**, 1095–1127.

KREISA, R. D. & MOIOLA, R. J. 1986. Sigmoidal tidal bundles and other tide-generated sedimentary structures of the Curtis Formation, Utah. *Geological Society of America Bulletin*, **97**, 381–387.

MAGUREGUI, J. TYLER, N. 1991. Evolution of middle Eocene tide-dominated deltaic sandstone, Lagunillas Field, Maracaibo Basin, western Venezuela. *In*: MIALL, A. D. TYLER, N. (eds) *The Three-dimensional Facies Architecture of Terrigenous Clastic Sediments and its Implications for Hydrocarbon Discovery and Recovery*. Society of Economic Paleontologists and Mineralogists, Concepts in Sedimentology and Paleontology, **3**, 232–244.

MORTON, N. 1965. The Bearreraig Sandstone Series (Middle Jurassic) of Skye and Raasay. *Scottish Journal of Geology*, **1**, 189–216.

——1976. Bajocian (Jurassic) stratigraphy in Skye, western Scotland. *Scottish Journal of Geology*, **12**, 23–33.

——1983. Paleocurrents and palaeo-environment of part of the Bearreraig Sandstone (Middle Jurassic) of Skye and Raasay, Inner Hebrides. *Scottish Journal of Geology*, **19**, 87–95.

——1987. Jurassic subsidence history in the Hebrides, NW Scotland. *Marine and Petroleum Geology*, **4**, 226–242.

——1989. Jurassic sequence stratigraphy in the Hebrides, NW Scotland. *Marine and Petroleum Geology*, **6**, 243–260.

——1992*a*. Late Triassic to Middle Jurassic stratigraphy, palaeogeography and tectonics west of the British Isles. *In*: PARNELL, J. (ed.) *Basins of the Atlantic Seaboard: Petroleum Geology, Sedimentology and Basin Evolution*. Geological Society, London, Special Publication, **62**, 53–68.

——1992*b*. Dynamic stratigraphy of the Triassic and Jurassic of the Hebrides Basin, NW Scotland. *In*: PARNELL, J. (ed.) *Basins of the Atlantic Seaboard: Petroleum Geology, Sedimentology and Basin Evolution*. Geological Society, London, Special Publication, **62**, 97–109.

—— & DIETL, G. 1989. Age of the Garantiana Clay (Middle Jurassic) in the Hebrides Basin. *Scottish Journal of Geology*, **25**, 153–159.

MUTTI, E., ROSELL, J., ALLEN, G. P., FONNESU, F. & SGAVETTI, M. 1985. The Eocene Baronia tide dominated delta-shelf system in the Ager Basin. *In*: MIALL, M. D. & ROSELL, J. (eds) *6th European Regional Meeting, International Association of Sedimentologists, Excursion Guidebook*, 579–600.

NUMMEDAL, D. & SWIFT, D. J. P. 1987. Transgressive stratigraphy at sequence-bounding unconformities: some principles derived from Holocene and Cretaceous example. *In*: NUMMEDAL, D., PILKEY, O. H. & HOWARD, J. D. (eds) *Sea Level Fluctuation and Coastal Evolution*. Society of Economic Paleontologists and Mineralogists, Special Publication, **41**, 241–260.

OVENSHINE, A. T, LAWSON, D. E. & BARTSCH-WINKLER, S. R. 1976. The Placer River Formation: intertidal sedimentation caused by the Alaska earthquake of March 27 1964. *Journal of Research, United States Geological Survey*, **4**, 151–162.

PLINT, A. G. & NORRIS, B. 1991. Anatomy of a ramp-margin sequence: facies, palaeogeography and sediment dispersal patterns in the Muskiki and Marshybank formations, Alberta Foreland Basin. *Bulletin of Canadian Petroleum Geology*, **39**, 18–42.

REINECK, H. E. & SINGH, I. B. 1973. *Depositional Sedimentary Environments – With Reference to Terrigenous Clastics*. Springer, Berlin.

RIDING, J. B., WALTON, W. SHAW, D. 1991. Toarcian to Bathonian (Jurassic) palinology of the Inner Hebrides, northwest Scotland. *Palynology*, **15**, 115–179.

ROBINSON, A. H. W. 1960. Ebb-flood channel systems in sandy bays and estuaries. *Geography*, **45**, 183–199.

ROSENTHAL, R. L. P. & WALKER, R. G. 1987. Lateral and vertical facies sequences in the Upper Cretaceous Chungo Member, Wapiabi Formation, southern Alberta. *Canadian Journal of Earth Science*, **24**, 771–783.

RUBIN, D. M. 1987. *Cross-bedding, Bedforms, and Palaeocurrents*. Society of Economic Paleontologists and Mineralogists, Concepts in Sedimentology and Paleontology, **1**.

—— & MCCULLOCH, D. S. 1980. Single and superimposed bed forms: a synthesis of San Francisco bay and flume observations. *Sedimentary Geology*, **26**, 207–231.

SMITH, D. G. 1988. Tidal bundles and mud couplets in the McMurray Formation, northeastern Alberta, Canada. *Bulletin of Canadian Petroleum Geology*, **36**, 216–219.

STEEL, R. J. 1971. New Red Sandstone movement on the Minch Fault. *Nature*, **234/51**, 158–159.

——1977. Triassic rift basins of northwest Scotland – their configuration, infilling and development. *In*: FINSTAD, F. & SELLEY, R. C. (eds) *Proceedings Mesozoic Northern North Sea Conference 7*. Norwegian Petroleum Society. 1–18.

STEIN, A. 1988. Basement controls upon Hebridean basin development, NW Scotland. *Basin Research*, **1**, 107–119.

—— & BLUNDELL, D. J. 1990. Geological inheritance and crustal dynamics of the northwest Scottish continental shelf. *Tectonophysics*, **173**, 455–467.

SWIFT, D. J. P. 1968. Coastal erosion and transgressive stratigraphy. *Journal of Geology*, **76**, 444–456.

——, PHILLIPS, S. & THORNE, J. A. 1991. Sedimentation on continental margins, IV: lithofacies and depositional systems. *In*: SWIFT, D. J. P., OERTEL, G. F., TILLMAN, R. W. & THORNE, J. A. (eds) *Shelf Sand and Sandstone Bodies – Geometry, Facies and Sequence Stratigraphy*. International Association of Sedimentologists, Special Publication, **14**, 89–152.

VAN VEEN, J. 1935. Sandwaves in the North Sea. *International Hydrography Review*, **12**, 21–29.

VAN WAGONER, J. C., MITCHUM, R. M., CAMPION, K. M. & RAHMANIAN, V. D. 1990. *Siliciclastic Sequence Stratigraphy in Well Logs, Cores and Outcrops: Concepts for High-resolution Correlation of Time and Facies*. American Association of Petroleum Geologists, Methods in Exploration, **7**.

VISSER, M. J. 1980. Neap-spring cycles reflected in Holocene subtidal large-scale bedform deposits: a preliminary note. *Geology*, **8**, 543–546.

WILKINSON, M. 1991. The concretions of the Bearreraig Sandstone Formation: geometry and geochemistry. *Sedimentology*, **38**, 899–912.

YALIN, M. S. 1964. Geometrical properties of sand waves. *American Society of Civil Engineers*, Proceedings, Hydraulics Division Journal, **90**, 105–119.

YANG, C. S. & NIO, S. D. 1989. An ebb-tide delta depositional model – a comparison between the modern Eastern Scheldt tidal basin (southwest Netherlands) and the Lower Eocene Roda Sandstone in the southern Pyrenees (Spain). *Sedimentary Geology*, **64**, 175–196.

Estuarine and shallow-marine sedimentation in the Upper Cretaceous–Lower Tertiary west-central Patagonian Basin (Argentina)

L. A. SPALLETTI

Centro de Investigaciones Geológicas, calle 1-644, La Plata, 1900, Argentina

Abstract: The Upper Cretaceous–Palaeocene Paso del Sapo–Lefipán Basin of west-central Patagonia (southern Argentina) is an intracratonic depocentre located to the east of the Andean magmatic arc. The depression is related to uplifted basement blocks and was generated by a complex system of strike-slip faults.

The sedimentary infill comprises two stratigraphical units: the Paso del Sapo and the Lefipán Formations. The Campanian–Maastrichtian Paso del Sapo Formation (145 m thick) is mainly composed of quartz-rich sandstones and conglomerates associated with minor heterolithic intervals, mudstones and coal beds. The Maastrichtian–Palaeocene Lefipán Formation (200 m thick) is composed in its lower part of gypsiferous mudstones and shales with isolated cross-bedded and plane-bedded sandstone bars. The middle and upper sections of this unit are distinguished by cross-bedded sandstone multi-storeys associated with several coquina beds.

Sediments from Paso del Sapo Formation show features of fluvial and tidally influenced systems. Sandstones and conglomerates were deposited as subtidal and intertidal estuarine bars. Heterolithic sections, mudstones and coal beds represent the more restricted inter- to supratidal marginal estuarine deposition.

Most of the Lefipán Formation was formed under open marine conditions. Offshore fine-grained transgressive deposits accumulated during basin starvation. Later on, sandstone multi-storeys and coquinas were deposited in a wave and tidally influenced lower to upper shoreface environment.

The sedimentary record of the K-T basin is in the range of a second-order eustatically controlled cycle. Based on facies arrangement and stratal geometries, the estuarine deposits of the Paso del Sapo Formation and the basal section of the Lefipán Formation are interpreted as a retrogradational systems tract. Later on, during a highstand period, the middle and upper Lefipán Formation was deposited as a progradational systems tract.

During the late Cretaceous, the south of South America was subjected to relative calm tectonic conditions and to a generalized transgressive process (Uliana & Biddle 1988). Thus, in the west of the Argentine Patagonia there was a remarkable embayment (Figs 1a and 1b) in which the accumulation of dominantly siliciclastic deposits corresponding to the Paso del Sapo Formation (PSF) and Lefipán Formation (LF) occurred (Fig. 1c).

The Cretaceous–Tertiary basin developed on a continental crustal substrate, to the east of the magmatic arc located along the Pacific margin of South America. However, in the sedimentary materials no supplies are known from the magmatic arc, but clastic contributions occur from highlands adjacent to the depocentres, constituted by Palaeozoic and Jurassic plutonic and volcanic rocks (Spalletti *et al.* 1989). In the backarc region, a complex system of strike-slip faults was responsible for the conformation and regional development of the depressions filled with Cretaceous–Tertiary sediments, as well as for the positive relief generation, with basement blocks, which surrounded such basins.

The PSF and the LF are exposed in extensive cliffs along the Chubut River Valley (Fig. 1c) where almost complete studies of siliciclastic lithofacies, facies associations and sediment-body geometry are possible. The aim of this paper is to present the sedimentological analysis of these units in order to determine the processes responsible for the depositional features and to depict a dynamic conceptual model, stressing the interaction between continental and marine environments.

General features of the Cretaceous–Tertiary sedimentary rocks

The Paso del Sapo Formation (PSF) is as thick as 145 m and is composed of quartz arenites and granule conglomerates, in which cross-bedding and plane-bedding are very frequent and constitute thick and amalgamated lithosomes. Siltstones, carbonaceous mudstones, coals and heterolithic intervals are subordinately found. Debris of carbonized leaves and petrified trunks are commonly found.

From De Batist, M. & Jacobs, P. (eds), 1996, *Geology of Siliciclastic Shelf Seas*, Geological Society Special Publication No. 117, pp. 81–93.

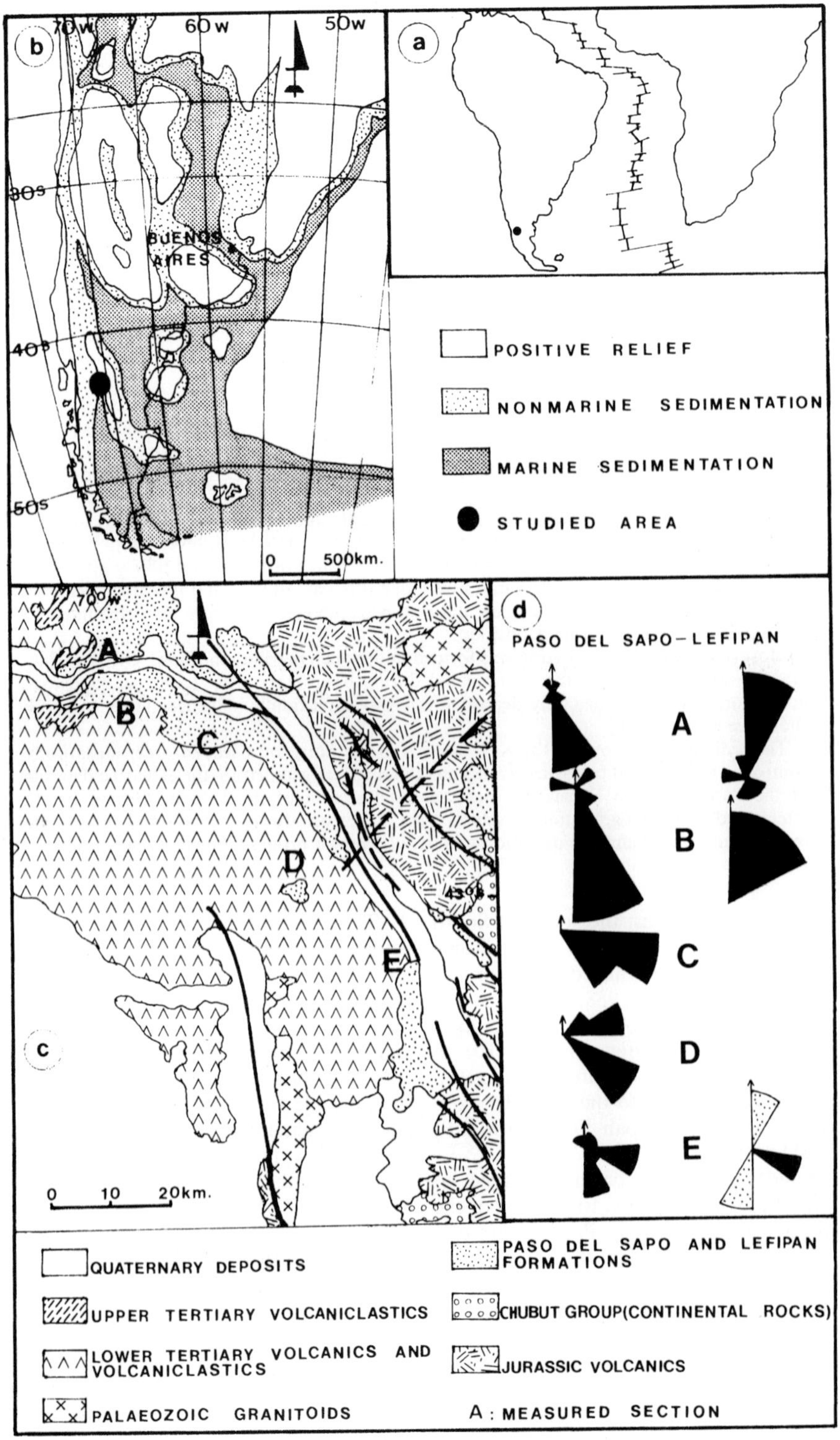

Fig. 1. (**a**) Location of the studied area in the plate tectonic reconstruction for the Late Cretaceous (from Scotese 1991). (**b**) Late Cretaceous palaeogeographical reconstruction (70 Ma) of southern South America (modified from Uliana & Biddle 1988). (**c**) Geologic map and general stratigraphy of the studied area, showing the outcrop belt of the Paso del Sapo and Lefipán Formations. Capital letters indicate the location of the measured sections.
(**d**) Palaeocurrent trends of the Paso del Sapo and Lefipán Formations. Dark pattern represents the orientation of current-induced cross-bedding; the stippled pattern represents the orientation of wave-formed ripple marks.

The Lefipán Formation (LF) is approximately 200 m thick. Its lower section is characterized by the prevalence of carbonaceous and gypsiferous siltstones and mudstones, as well as heterolithic (wavy-bedded) sections. In an evident upward coarsening arrangement, as medium and upper levels are reached, multi-episodic bodies of quartz-rich arenites with intercalations of biorudites and grainstones (coquina beds) become progressively more common. The LF is rich in marine invertebrate fossils and in sandstone levels with bioturbation and mottled structures.

The average detrital mode of the Cretaceous–Tertiary sandstones is quartz (72%), feldspar (15%), rock fragments (13%). That of mudstones is quartz (45–65%), kaolinite (10–40%) and illite (5–15%). The PSF mudstones are rich in kaolinite, whereas those of the LF also have authigenic glauconite and illite. The compositional information suggests that the detrital components come from plutonic and local volcanic terranes (continental block provenance) subjected to strong weathering processes (Iñiguez *et al.* 1988).

On the basis of palaeopalynological information, Papú *et al.* (1988) have demonstrated that the PSF was accumulated during the Campanian–Maastrichtian in brackish environments and tropical to subtropical conditions. As regards the LF, the marine invertebrate fauna shows that its deposition began in the Maastrichtian (with doubts about the Campanian) and finished in the Palaeocene (Medina *et al.* 1990). On the other hand, Baldoni (1992) points out that the palynoflora of the lower part of the LF corresponds to the Maastrichtian and suggests humid palaeoclimatic conditions.

Facies and architectural analysis

The studies of the PSF and LF were based on the survey of five principal sections (Fig. 1c) to 1:100 scale). In each locality lithosomes (or architectural elements) were also characterized with a two-dimensional-type analysis (Spalletti 1994; Miall 1994). Likewise, complementary studies on lithosome geometry and lithology were made on isolated exposures of both stratigraphical units.

In each of the sections, observational lithofacies were defined from the texture, composition and primary sedimentary structures (Fig. 2). As can be seen in Table 1 Miall's code (1978) was followed for lithofacies designation. This code was adapted to the lithologies and structures which were present in the measured sections (Figs 3, 4 and 5). Lithofacies were grouped in facies associations and in sequences (fining, coarsening, thinning and thickening upwards) of different vertical scale.

For the architectural studies, the lateral and vertical scale of lithosomes, their external geometry and the nature of bounding surfaces were taken into account (Muñoz *et al.* 1992; García Gil 1993). Each lithosome was also characterized by the lithofacies associations,

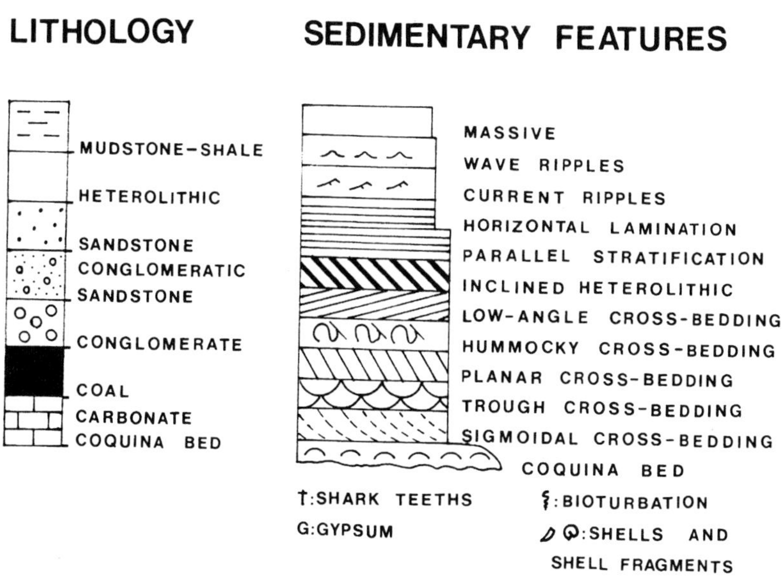

Fig. 2. Legend for lithological columns of measured sections (Figs 3,4 and 5).

Table 1. *Lithofacies codes, based on the combination of lithology (texture–composition) and sedimentary structures*

	Gravel	Gravel-sand	Sand	Heterolithic sand-mud	Mud	Carbonate	Coal	Interpretation
Trough cross-bedding	Gt	SGt	St	(Drapes)	(Drapes)	–	–	Coarse and medium-grained transverse (3D) megaripples
Planar cross-bedding	Gp	SGp	Sp	(Drapes)	(Drapes)	–	–	Coarse and medium-grained transverse (2D) megaripples
Low-angle cross-bedding	Gl	SGl	Sl	–	–	–	–	Scour fills, antidunes
Hummocky cross-bedding	–	–	Shcs	–	–	–	–	Proximal tempestites. Storm-oscillatory flows and unidirectional flows and unidirectional 'steady' currents
Epsilon cross-bedding	–	SGϵ	–	–	–	–	–	Lateral accretion, point bar deposits
Inclined heterolithic cross-bedding	–	–	–	Hihs	–	–	–	Tidally influenced high-sinuosity channels. Lateral accretion mixed-load point bar deposits
Parallel stratification	Gh	SGh	Sh	–	–	–	–	Planar bed flow (upper flow regime)
Horizontal lamination	–	–	–	–	Fh	–	Ch	Laminar flow (lower flow regime)
Imbrication	Gi	–	–	–	–	–	–	Upper flow regime. Longitudinal and lateral bars
Wave ripples	–	–	Swr	Hwr	–	–	–	Wave ripples (oscillatory flow)
Current ripples	–	–	Scr	Hcr	–	–	–	Current ripples (lower flow regime)
Mass	Gm	–		–		CQm		Concentrated fast currents (upper flow regime).
(Bioturbated?)			Sm		Fm		Cm	Unknown flow regime, bioturbation

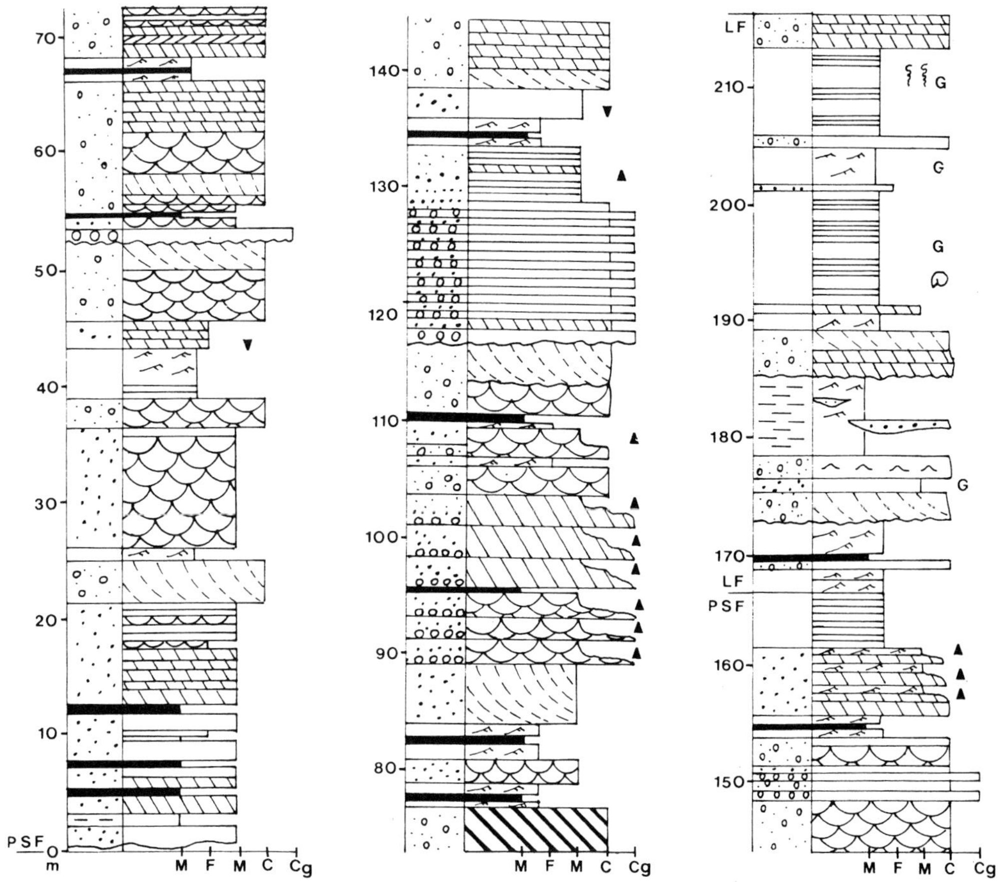

Fig. 3. Schematic representation of the Paso del Sapo (PSF) and Lefipán (LF) Formations at the Pirola Section (location B in Fig. 1c).

the internal sequences, the lower-hierarchy bounding surfaces and the orientation of the directional structures (Miall 1988; Miall & Tyler 1991). In Table 2 a summary of the main recognized architectural elements is given. These elements were objectively defined and were designated with a number and a letter: the numbers indicate the external geometry of each lithosome and its dominant lithology, and the letters give details about the scale, position in the sequence and internal features (especially sedimentary structures).

Some lithosomes are present both in the PSF and the LF. This is the case of those which are interpreted as channel-lag deposits (1a), channel fills and subtidal bars (2b, 2d) and subtidal sand bars (3a, 3b). Others appear exclusively in the PSF, such as the palaeochannels (2a), large bars with lateral accretion features (2c), the bars composed of intertidal and subtidal heterolithic deposits (3c) and the marginal tidal fine-grained deposits (4a). The most typical bodies of the LF are those attributed to storm events and foreshore environments (1b, 1c, 1d), together with open marine subtidal bars (3d) and the fall-out and combined fall-out and traction fine-grained deposits (4b, 4c).

Palaeocurrent trends

The study of palaeocurrents, based on the cross-bedding orientation, shows for the PSF a very uniform trend of the vectors towards the south-southeast (Fig. 1d). In the LF the orientation is somewhat more variable, since in the north region the foresets are oriented towards the northeast, whereas in the south they point to the east and southeast. In the last region, the general orientation of the wave-ripple crests (NNE–SSW) is transversal to that of the current-induced cross-bedding (Fig. 1d).

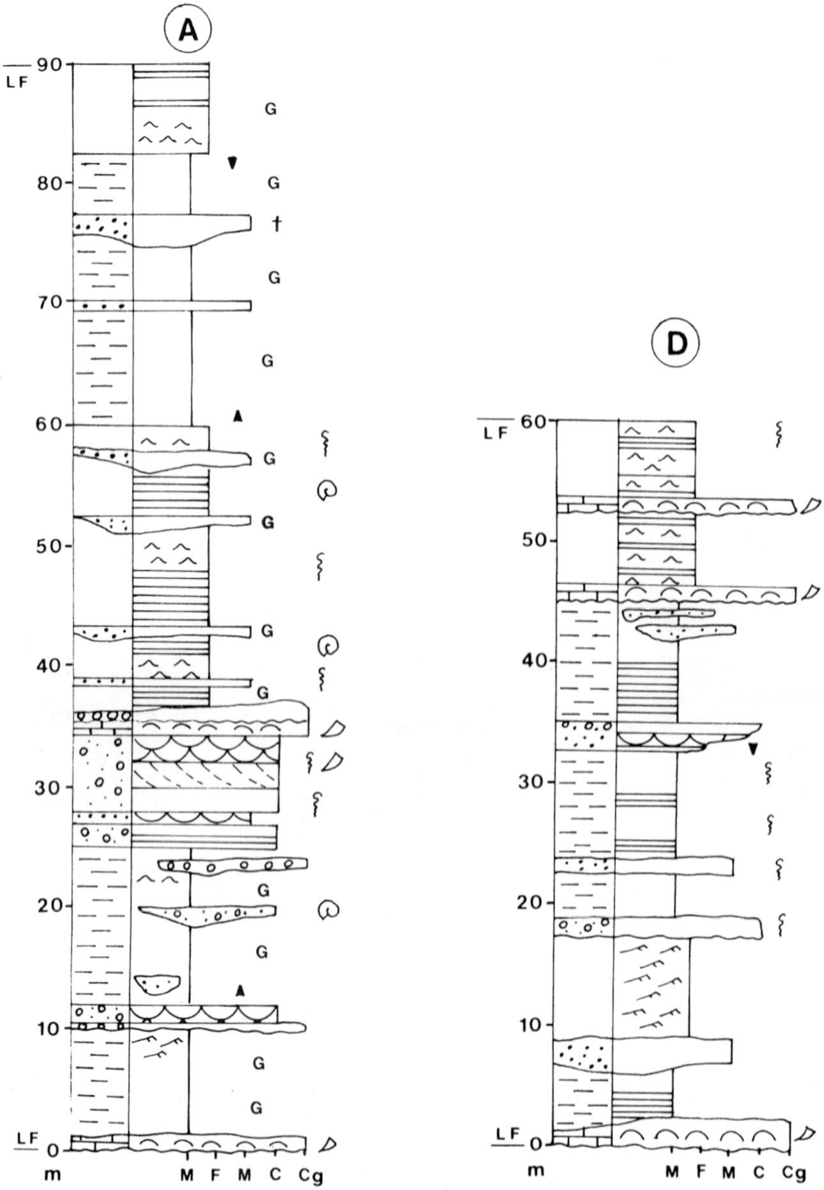

Fig. 4. Schematic representations of the Lefipán Formation (LF) at Los Fortines (A) and section on the road to Colán Conhué (D). Refer to Fig. 1c for location.

Diagnostic features of the Paso del Sapo and Lefipán Formations

As shown in Tables 1 and 2 and in Figs 3 and 5 the Paso del Sapo Formation is characterized by features that are typical of both fluvial and tidal-influenced environments (Smith 1987, 1988; Terwindt 1988; Eberth & Miall 1991; Shanley *et al.* 1992; Lanier *et al.* 1993; Richards 1994). The main features can be listed as follows:

(a) amalgamated cross-stratified sandstone bodies (sand bars);
(b) trough and planar cross-bedded units (3D and 2D dunes);
(c) sigmoidal beds; sand bundles with mud and carbonaceous drapes;

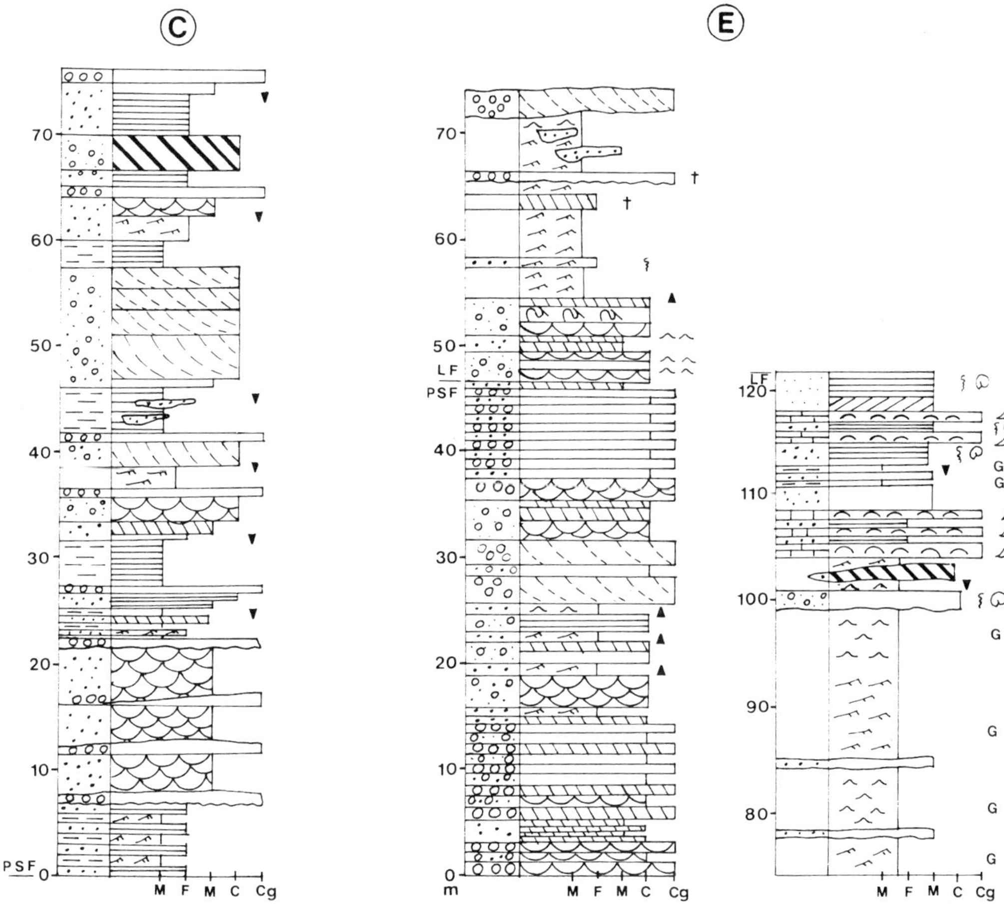

Fig. 5. Schematic representations of the Paso del Sapo (PSF) and Lefipán (LF) Formations at San Martín (C) and Ibarra (E) sections. Refer to Fig. 1c for location.

(d) isolated herringbone cross-stratification;
(e) reactivation surfaces in cross-bedded sets;
(f) frequent lateral accretion sand bodies with cross-bedded subsets;
(g) parallel stratified (upper flow regime) sandstones between large sand bars;
(h) fine to medium current-rippled and wave-rippled sandstones;
(i) decimetre-thick lenticular and tabular beds of conglomerate lags (lithosome type 1a);
(j) subordinated carbonaceous mudstones and coals with abundant comminuted plant fragments;
(k) fine-grained intervals composed of heterolithic (flaser, wavy and lenticular) couplets;
(l) presence of large-scale inclined heterolithic stratification (IHS); mud and carbonaceous debris draping small-scale cross-bedded sandstones;
(m) very common plant debris in drapes and in fine-grained intervals; light bioturbation; reworked trunks in sandstones.

The main features of the Lefipán Formation (Figs 3, 4 and 5; Tables 1 and 2) suggest a transition from tidal flats to open marine settings, influenced by both fair-weather and storm conditions (Terwindt 1988; Deynoux *et al.* 1993; Jennette & Pryor 1993; Brenchley *et al.* 1993; Willis & Moslow 1994; Ricketts 1994). They can be synthesized as follows:

(a) abundant mudstones, shales and heterolithic sections (type 4b and 4c lithosomes);
(b) flaser, lenticular and especially wavy bedding in heterolithic intervals; the traction bed is composed of fine-grained current-rippled sandstone;

Table 2. *Lithosome characterisation*

Type	Geometry	Scale	Facies association	Remarks	Origin	Stratigraphic unit
1a	Tabular	dm	Gm, Gi	Overlain by an erosion surface	Basal channel lag	PSF, LF
1b	Tabular	dm	Gm, Gi	Top of sandstone bed	Storm gravel bed	LF
1c	Tabular	dm-m	Sh, Gh, SGh, Sm	Accretion-bedding. Bioturbation common	Beach, foreshore	LF
1d	Tabular Lensoid	dm	CQm	Uncommon parallel lamination. Bioturbated. Massive	Storm carbonate bed	LF
2a	Lensoid	m	Gh, SGh, Sh, Gl, SGl, Sl	Concave-up basal contact, symmetric infill	Channel	PSF
2b	Lensoid	m	SGt, St	Concave-up basal contact, 3D dunes	Channel + subtidal bars	PSF, LF
2c	Lensoid Tabular	m	SGϵ, Gt, St	Planar to concave-up basal contact	Attached large bars, lateral accretion + small bars	PSF
2d	Tabular Lensoid	m	Hihs, (st, Src)	Planar to concave-up basal contact	Channel, tidal creek and point bar	PSF, LF
3a	Mound	m	St, Sp, Sl, Sh, SGt, SGp, Gt, Gp, Gl, Gh, Swr, Scr	Large scale, compound or single sets. Sigmoids, clay drapes and reactivation surfaces. Thick and normal graded foresets. Planar basal contacts	Large subtidal sand bars. 2D and 3D aggrading megaripples	PSF, LF
3b	Mound Tabular	dm	SGp, Sp, Scr	Small scale sigmoids, tangential base, mud drapes. Planar basal contact	Subtidal sand bars. 2D megaripples, reverse flows (ebb-flood)	PSF, LF
3c	Mound Tabular	dm	St (Ht, Hp)	Sigmoids, heterolithic drapes. Planar basal contact	Subtidal to intertidal sand bars	PSF
3d	Mound Tabular	dm-m	St, Sp, Sm, Shcs	No mud. Planar basal contact. 'Coastal parallel' palaeocurrent trends	Subtidal, open marine nearshore sand bars: shallow storms and longitudinal currents	LF
4a	Tabular	dm-m	Hcr (Hwr), F, C	Traction (ripples and small dunes)-suspension couplets	Inter- to supratidal marginal estuary	PSF
4b	Tabular	dm-m	F	Gypsiferous mudstones and shales. Fall-out deposits	Lagoon (gypsiferous mudstones) and offshore (dark shales)	LF
4c	Tabular	dm-m	Hwr, Swr	Heterolithlic (wavy and lenticular), oscillatory and translational flows	Open marine. Fair-weather and storm waves	FL

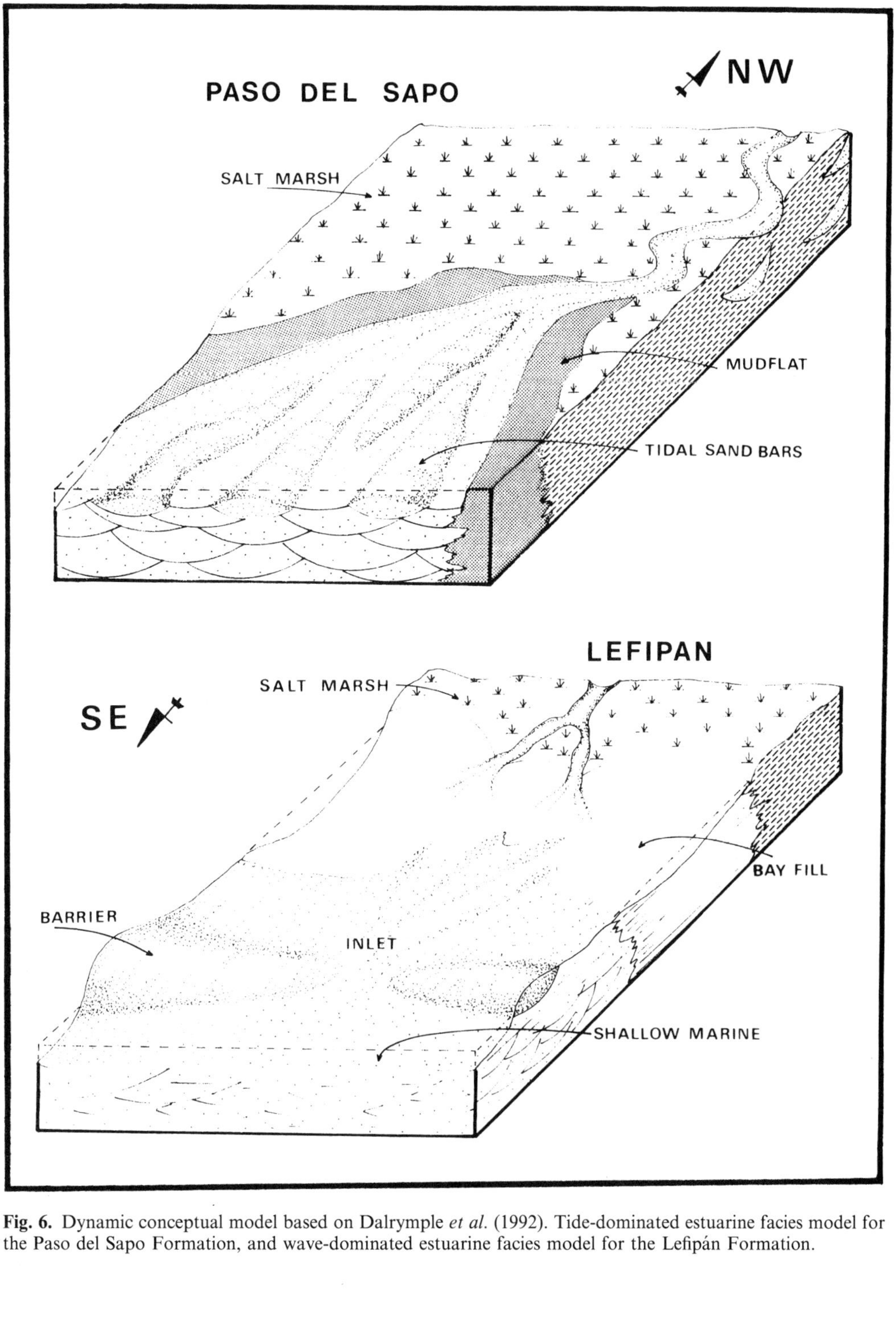

Fig. 6. Dynamic conceptual model based on Dalrymple *et al.* (1992). Tide-dominated estuarine facies model for the Paso del Sapo Formation, and wave-dominated estuarine facies model for the Lefipán Formation.

(c) gypsum veins and carbonized plant debris in fine-grained deposits;
(d) fine-grained current-rippled and wave-rippled sandstones associated with fine-grained intervals;
(e) amalgamated lenticular sandstone bodies (plane-bedded, trough, tabular-planar sets, and massive bioturbated beds) intercalated in fine-grained successions; wave-ripples may be present towards the top of cosets;
(f) light to intense bioturbation; vertical, diagonal, branching and truncated burrows in sandstones;
(g) sections (up to 8 m thick) with upward coarsening arrangements (mudstone grading upward into medium sandstone);
(h) subordinated wave-ripples and HCS in sandstone beds;
(i) frequent coquina beds and shelly sandstones (molluscs dominant);
(j) uncommon thin beds of granule conglomerate (type 1a and 1b lithosomes).

Dynamic conceptual model and controls on the depositional system

The Cretaceous–Tertiary units of the studied region constitute an ephemeral geological system of both discrete thickness and duration. The deposits show features of fluvial and tidal action with subordinated influence of fair-weather and storm waves. The sand accumulations are always present, though they are much more remarkable towards the lower sedimentary succession (PSF). In these accumulations, palaeoflow evidences are predominantly parallel to the elongation of the sedimentary bodies.

The vertical arrangement of the PSF–LF (Figs 3 and 5) is fining upwards. In the lower portion fluvial deposits prevail. These deposits are later replaced by tide-influenced facies, whereas the last stage was characterized by open marine sediments. These features suggest that the succession occupied a drowned valley along a transgressive system with facies translation landward as transgression proceeded. The Cretaceous–Tertiary depocentre can be interpreted as an inlet of the sea reaching into a river system; that is a river mouth which has been flooded by the sea.

According to the ideas proposed by Leckie & Singh (1991), Boyd *et al.* (1992), Dalrymple *et al.* (1992) and Richards (1994), the mentioned characteristics are adequately framed in a conceptual model of an estuarine environment (Fig. 6). In this model the sediment supply was important, although this supply could never compensate the relative rise in sea level.

In the case of the Paso del Sapo Formation, the deposits of its lower part suggest an internal estuarine environment (Nichols & Biggs 1985; Cooper 1993), with predominant fluvial action and with slightly brackish overbank deposits. The rest of the unit has been deposited in a mixed estuarine to marine environment with prevailing tidal action (Boyd *et al.* 1992; Dalrymple *et al.* 1992). Among the tractional deposits, bodies formed in tidal meanders and tidal sand bars are identified. The fine-grained deposits were generated in a marginal mud flat environment under intertidal to supratidal conditions.

The Lefipán Formation has been deposited under a major marine influence. In the lower part, a tide and wave-dominated estuarine environment is inferred (Dalrymple *et al.* 1992; Leithold 1994). In this environment, tidal and bay-fill deposits together with isolated tidal sand bars were accumulated. In the uppermost sections of the LF, shallow marine platform deposits associated with upper shoreface (nearshore) sandstones and bioclastic carbonates are observed.

The estuarine deposits have recently been considered to be essential in sequence stratigraphy interpretations (Leckie & Singh 1991; Shanley *et al.* 1992; Dalrymple *et al.* 1992; Allen & Posamentier 1993). In this regard, by following the nomenclature of Galloway (1989), it is possible to recognize a retrogradational systems tract in the lower half of the Cretaceous–Tertiary sedimentary fill, and a progradational systems tract in its upper half (Fig. 7). This record is in the range of a second-order eustatically controlled cycle. The unconformity of the base of the Paso del Sapo Formation corresponds to the transgressive surface, and the boundary between both tracts is the maximum marine flooding surface which is located near the base of the offshore mudstones which represent the climax of the regional expansion of the Lefipán Formation. As shown in Fig. 7, the retrogradational systems tract comprises both fluvial-dominated internal estuarine facies and tidal-dominated mixed estuarine facies of the Paso del Sapo Formation, as well as the bay-fill and tidal-flat deposits of the lower sections of the Lefipán Formation. The progradational systems tract involves most of the Lefipán Formation (Fig. 7); it includes in its lower part the offshore deposits and in its upper part those of the upper shoreface environment.

Conclusions

(1) In west-central Patagonia, excellent exposures of the Upper Cretaceous–Palaeocene Paso del Sapo (PSF) and Lefipán (LF)

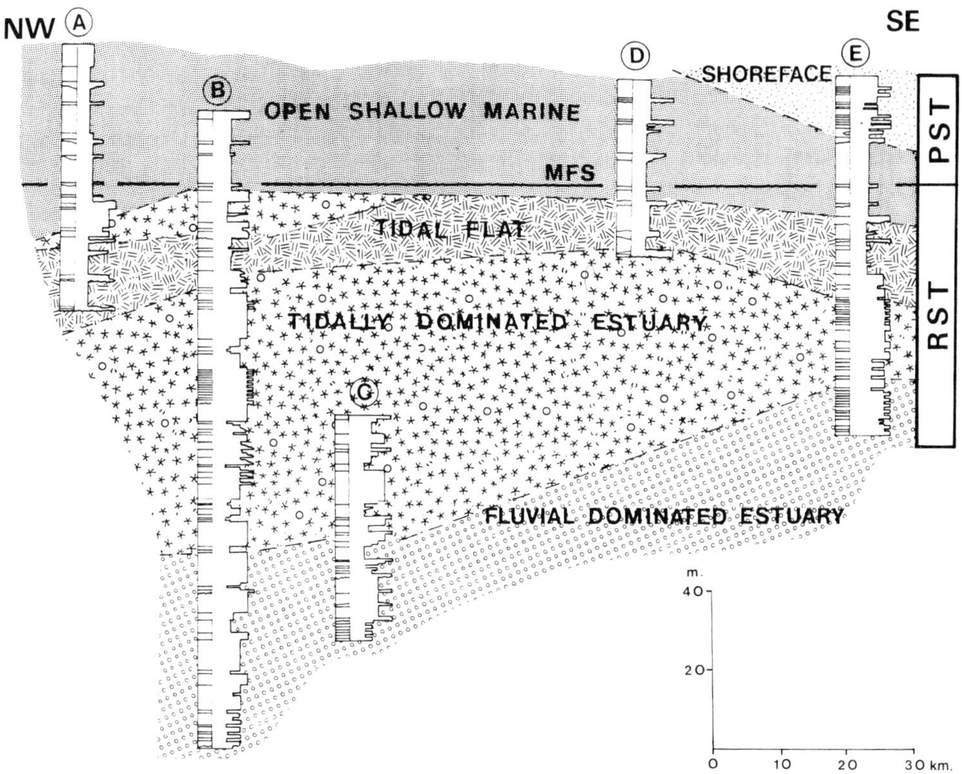

Fig. 7. Northwest–southeast schematic cross-section of the Paso del Sapo and Lefipán Formations showing main depositional environments and systems tracts. Locations are shown in Fig. 1c. RTS, retrogradational systems tract; PST, progradational systems tract; MFS, maximum flooding surface.

formations allow detailed analysis of facies and two-dimensional lithosome types.

(2) The PSF starts with fluvial deposits associated with slightly brackish overbank deposits, which are later replaced by tide-influenced sand bars and marginal inter- and supratidal mud-flat deposits. Fine-grained tidal and bay-fill sediments together with isolated sand bars are typical of the lower LF. Owing to maximum marine flooding, offshore mudstones and shales developed in the middle and upper sections of the LF. A shallowing-up trend is seen towards the top of this unit, where nearshore sandstones and bioclastic carbonates prevail.

(3) Results of the present study indicate that features of the PSF and LF are adequately framed in a model of estuarine sedimentation. The PSF evolved from a fluvial-dominated (internal) to a tidal-dominated (mixed) estuary. However, a tide and wave-dominated estuarine environment is proposed for its lower part; most of the LF was formed under offshore and nearshore open marine conditions.

(4) From the sequence stratigraphical perspective, this study allowed the recognition of a lithogenetic unit that follows a long-term eustatic fluctuation. Two successive (retrogradational and progradational) systems tracts, separated by a marine flooding surface of regional extent, are recognized. The retrogradational tract comprises the fluvial-dominated and tidal-dominated estuarine facies of the PSF and the lower LF. The progradational tract involves the upward shallowing cycle composed of both the offshore and nearshore deposits of the middle and upper LF.

References

Allen, G. P. & Posamentier, H. W. 1993. Sequence stratigraphy and facies model of an incised valley fill: the Gironde estuary, France. *Journal of Sedimentary Petrology*, **63**, 378–391.

BALDONI, A. 1992. Palynology of the lower Lefipán Formation (Upper Cretaceous) of Barranca de los Perros, Chubut Province, Argentina. Part I. Cryptogam spores and gymnosperm pollen. *Palynology*, **16**, 117–136.

BOYD, R., DALRYMPLE, R. W. & ZAITLIN, B. A. 1992. Classification of clastic coastal depositional environments. *Sedimentary Geology*, **80**, 139–150.

BRENCHLEY, P. J., PICKERILL, R. K. & STROMBERG, S. G. 1993. The role of wave reworking on the architecture of storm sandstone facies, Bell Island Group (Lower Ordovician), eastern Newfoundland. *Sedimentology*, **40**, 359–382.

COOPER, J. A. G. 1993. Sedimentation in a river dominated estuary. *Sedimentology*, **40**, 979–1017.

DALRYMPLE, R. W., ZAITLIN, B. A. BOYD, R. 1992. Estuarine facies models: conceptual basis and stratigraphic implications. *Journal of Sedimentary Petrology*, **62**, 1130–1146.

DEYNOUX, M., DURINGER, P., KHATIB, R. & VILLENEUVE, M. 1993. Laterally and vertically accreted tidal deposits in the Upper Proterozoic Madina-Kouta Basin, southeastern Senegal, West Africa. *Sedimentary Geology*, **84**, 179–188.

EBERTH, D. A. & MIALL, A. D. 1991. Stratigraphy, sedimentology and evolution of a vertebrate-bearing, braided to anastomosed fluvial system, Cutler Formation (Permian-Pennsylvanian), north-central New Mexico. *Sedimentary Geology*, **72**, 225–252.

GALLOWAY, W. E. 1989. Genetic stratigraphic sequences in basin analysis. I: architecture and genesis of flooding-surface bounded depositional units. *AAPG Bulletin*, **73**, 125–142.

GARCIA GIL, S. 1993. The fluvial architecture of the Upper Buntsandstein in the Iberian Basin, central Spain. *Sedimentology*, **40**, 125–143.

IIGUEZ, A. M., MERODIO, J. C. & SPALLETTI, L. A. 1988. Mineralogía y geoquímica de las Formaciones Paso del Sapo y Lefipán (Cretácico–Terciario), provincia del Chubut. *Asociación Geológica Argentina Revista*, **43**, 13–23.

JENNETTE, D. C. & PRYOR, W. A. 1993. Cyclic alternation of proximal and distal storm facies: Kope and Fairview Formations (Upper Ordovician), Ohio and Kentucky. *Journal of Sedimentary Petrology*, **63**, 183–203.

LANIER, W. P., FELDMAN, H. R. & ARCHER, A. W. 1993. Tidal sedimentation from a fluvial to estuarine transition, Douglas Group, Missourian–Virgilian, Kansas. *Journal of Sedimentary Petrology*, **63**, 860–873.

LECKIE, D. A. & SINGH, C. 1991. Estuarine deposits of the Albian Paddy Member (Peace River Formation) and lowermost Shaftesbury Formation, Alberta, Canada. *Journal of Sedimentary Petrology*, **61**, 825–849.

LEITHOLD, E. L. 1994. Stratigraphical architecture at the muddy margin of the Cretaceous Western Interior Seaway, southern Utah. *Sedimentology*, **41**, 521–542.

MEDINA, F. A., CAMACHO, H. H. & MALAGNINO, E. C. 1990. *Bioestratigrafía del Cretácico superior – Paleoceno Marino de la Formación Lefipán, Barranca de los Perros, Río Chubut, Chubut.* V Congreso Argentino Paleontología Bioestratigrafía, **I**, 137–142.

MIALL, A. D. 1978. Lithofacies types and vertical profile models in braided river deposits: a summary. *In*: MIALL, A. D. (ed.) *Fluvial Sedimentology*. Canadian Society of Petroleum Geologists, Memoir, **5**, 597–604.

——1988. Architectural elements and bounding surfaces in fluvial deposits: anatomy of the Rayenta Formation (Lower Jurassic), southwest Colorado. *Sedimentary Geology*, **55**, 233–262.

——1994. Reconstructing fluvial macroform architecture from two-dimensional outcrops: examples from the Castlegate Sandstone, Book Cliffs, Utah. *Journal of Sedimentary Research*, **B64**, 146–158.

—— & TYLER, N. (eds) 1991. *The Three-dimensional Facies Architecture of Terrigenous Clastic Sediments and its Implications for Hydrocarbon Discovery and Recovery*. Society of Economic Paleontologists and Mineralogists, Concepts in Sedimentology and Paleontology, **3**.

MUÑOZ, A., RAMOS, A., SANCHEZ-MOYA, Y. & SOPEÑA, A. 1992. Evolving fluvial architecture during a marine transgression: Upper Buntsandstein, Triassic, central Spain. *Sedimentary Geology*, **75**, 257–281.

NICHOLS, M. M. & BIGGS, R. B. 1985. Estuaries. *In*: DAVIS, R. A. (ed.) *Coastal Sedimentary Environments*. Springer, New York, 77–186.

PAPU, O. H., VOLKHEIMER, W. & SEPULVEDA, E. G. 1988. *Masulas de Salviniaceae del Cretácico Tardío de Nordpatagonia y sur de Mendoza, Argentina. Su Importancia Bioestratigráfica y Paleoambiental.* V Congreso Geológico Chileno **III**, H67-H81.

RICHARDS, M. T. 1994. Transgression of an estuarine channel and tidal flat complex: the Lower Triassic of Barles, Alpes de Haute Provence, France. *Sedimentology*, **41**, 55–82.

RICKETTS, B. D. 1994. Mud-flat cycles, incised channels, and relative sea-level changes on a Paleocene mud-dominated coast, Ellesmere Island, Artic Canada. *Journal of Sedimentary Research*, **B64**, 211–218.

SCOTESE, C. R. 1991. Jurassic and Cretaceous plate tectonic reconstructions. *Palaeogeography, Palaeoclimatology, Palaeoecology*, **87**, 493–501.

SHANLEY, K. W., MCCABE, P. J. & HETTINGER, R. D. 1992. Tidal influence in Cretaceous fluvial strata from Utah, USA: a key to sequence stratigraphic interpretation. *Sedimentology*, **39**, 905–930.

SMITH, D. G. 1987. Meandering river point bar lithofacies models: modern and ancient examples compared. *In*: ETHRIDGE, F. G., FLORES, R. M. & HARVEY, M. D. (eds) *Recent Developments in Fluvial Sedimentology*. Society of Economic Paleontologists and Mineralogists, Special Publication, **39**, 83–91.

——1988. Modern point bar deposits analogous to the Athabasca Oil Sands, Alberta, Canada. *In*: DE BOER, P. L., VAN GELDER, A. & NIO, S. D. (eds) *Tide-influenced Sedimentary Environments and Facies*. Reidel, Dordrecht, 417–432.

SPALLETTI, L. A. 1994. Evolución de los ambientes fluviales en el Triásico de la Sierra Pintada (Mendoza, Argentina): análisis sobre la influencia de controles intrínsecos y extrínsecos al sistema depositacional. *Asociación Argentina de Sedimentología Revista*, **1**, 125–142.

——, DEL VALLE, A., MANASSERO, M. J. & MATHEOS, S. D. 1989. Procedencia y ambiente tectónico de las areniscas cretácico – terciarias del sector norte de la Patagonia Argentina. *In*: SPALLETTI, L. A. (ed.) *Contribuciones de los Simposios sobre Cretácico de América Latina, A: Eventos y Registro Sedimentario*, 149–163.

TERWINDT, J. H. J. 1988. Palaeo-tidal reconstructions of inshore tidal depositional environments. *In*: DE BOER, P. L., VAN GELDER, A. & NIO, S. D. (eds) *Tide-Influenced Sedimentary Environments and Facies*. Reidel, Dordrecht, 233–263.

ULIANA, M. A. & BIDDLE, K. T. 1988. Mesozoic–Cenozoic paleogeographic and geodynamic evolution of southern South America. *Revista Brasileira de Geociencias*, **18**, 172–190.

WILLIS, A .J. & MOSLOW, T. F. 1994. Stratigraphic setting of transgressive barrier-island reservoirs with an example from the Triassic Halfway Formation, Wembley Field, Alberta, Canada. *AAPG Bulletin*, **78**, 775–791.

Late Quaternary sedimentation and high-resolution sequence stratigraphy of the east Texas shelf

J. B. ANDERSON[1], K. ABDULAH[1], S. SARZALEJO[1], F. SIRINGAN[2] & M. A. THOMAS[3]

[1] *Department of Geology and Geophysics, Rice University, Houston, TX 77251, USA*
[2] *National Institute of Geological Sciences, University of the Philippines, Diliman, Quezon City, The Philippines*
[3] *Geology Research, Shell Development Company, Houston, TX 77001, USA*

Abstract: An ongoing study utilizes outcrop-scale seismic data and lithofacies data from cores and platform borings collected from the east Texas continental shelf to test assumptions and models that relate sedimentary facies patterns and sequence stratigraphy. The current data base consists of nearly 15 000 km of high-resolution seismic data and lithological data from hundreds of sediment cores and platform borings. The Texas shelf is ideally suited for this work because sediment supply during the Pleistocene was high enough to keep pace with the relatively rapid rise and fall in sea level, thus producing sequences that can be imaged on high-resolution seismic records. Furthermore, sediment supply, shelf gradient, and the degree of diapirism and faulting vary across the shelf, so the relative roles of these agents in controlling the overall packaging of facies into systems tracts can be evaluated.

Strong contrasts exist between low-sediment-supply (Trinity/Sabine) and high-sediment-supply (Brazos and Colorado) fluvial systems and their associated systems tracts. The predominant difference is that sediment delivered to the shelf by the Trinity/Sabine fluvial system was, for the most part, deposited within the incised fluvial valleys. Only during lowstands did the Trinity/Sabine system deliver sediment directly to shelf-margin deltas and slope minibasins. Reincision, associated with fifth-order eustatic fluctuations, flushes the valley of sediments deposited during the previous transgression and highstand, thus large quantities of sediment are delivered from this valley system during lowstands. Transgressive shelf sand bodies occur adjacent to the Trinity and Sabine incised valleys and are scattered widely across the shelf. Backstepping parasequences, the product of the episodic nature of glacial eustatic sea-level rise, characterize the incised valley fill. The highstand systems tract is thin to absent in interfluve areas.

The Brazos and Colorado rivers have much larger sediment supplies than the Trinity and Sabine rivers. They have filled their incised valleys with fluvial deposits and abandoned them to occupy several more shallow valleys. The result of this fluvial avulsion has been the sequestering of a significant part of the sediments delivered by these rivers to fluvial valleys on the shelf. Both the Colorado and Brazos rivers constructed large shelf-margin deltas during the lowstand, but these deltas differ in terms of their overall morphology and sediment facies. The ancient Colorado delta is sandy and it directly sourced two slope fans during the maximum lowstand. The Brazos shelf-margin delta consists mainly of mud and there is little evidence of significant bypass during the lowstand. During initial transgression, the Brazos/Colorado shelf-margin deltas backstepped onto the outer shelf. Rapid transgression and associated erosion removed the delta-plain beds. Continued transgression led to decapitation of sandy fluvial and deltaic facies, reworking the sands into widespread shelf sand bodies, and further backstepping of the deltas. During the previous highstand, the Brazos and Colorado rivers constructed fluvial-dominated deltas on the shelf.

The shelf-margin deltas of the area show a complex pattern of progradation and aggradation that varies from one delta to the next. This complexity is due predominantly to the different response of systems with different sediment supplies to fifth-order eustatic fluctuations. Before we understood these autocyclic effects, attempts at relating seismic stratigraphic changes to the oxygen isotope curve, our proxy for sea-level change, were unsuccessful.

Sequence stratigraphy is proving to be an invaluable tool for global stratigraphical correlation (Vail *et al.* 1977), although important questions remain concerning the degree of eustatic control on sequence development, particularly for higher order glacial eustatic cycles. We are now in a phase in which the method is being tested for its value in predicting the distribution of sedimentary facies within a sequence stratigraphic context (e.g. Posamentier & Vail 1988; Sarg 1988; Van Wagoner *et al.* 1990).

Sequence stratigraphy has its roots in the analysis of conventional seismic data, which has

From De Batist, M. & Jacobs, P. (eds), 1996, *Geology of Siliciclastic Shelf Seas*, Geological Society Special Publication No. 117, pp. 95–124.

resolution on the order of several tens of metres (Vail *et al.* 1977). Direct observation of acoustic properties and features that are indicative of lithofacies rarely is possible with these data. Therefore, correlation of conventional seismic records with outcrop and well-log data involves a huge jump in scale.

The long-term objective of our research is to test and refine models that predict lithofacies distribution patterns from seismic stratigraphic relationships. The approach we are using involves the use of high-resolution (outcrop-scale resolution) seismic reflection data and oil company platform borings and cores to map stratigraphic boundaries and to characterize sedimentary facies bounded by these surfaces. Approximately 15 000 km of high-resolution seismic data, several hundred industry platform boring descriptions, and several hundred sediment cores provide the data base for mapping sedimentary facies on the east Texas continental shelf and upper slope. The sedimentary facies targeted were deposited during the last glacial eustatic cycle (Stage 5e to present). These facies include incised fluvial valley fill deposits, fluvial meanderbelts, fluvial- and wave-dominated deltas, tidal inlet/delta systems, coastal lithosomes, shelf sand bodies, shelf-margin deltas, and slope fans and sediment gravity flow aprons. The objective of the study was to see how these depositional systems responded to various forcing agents, particularly changing sea level.

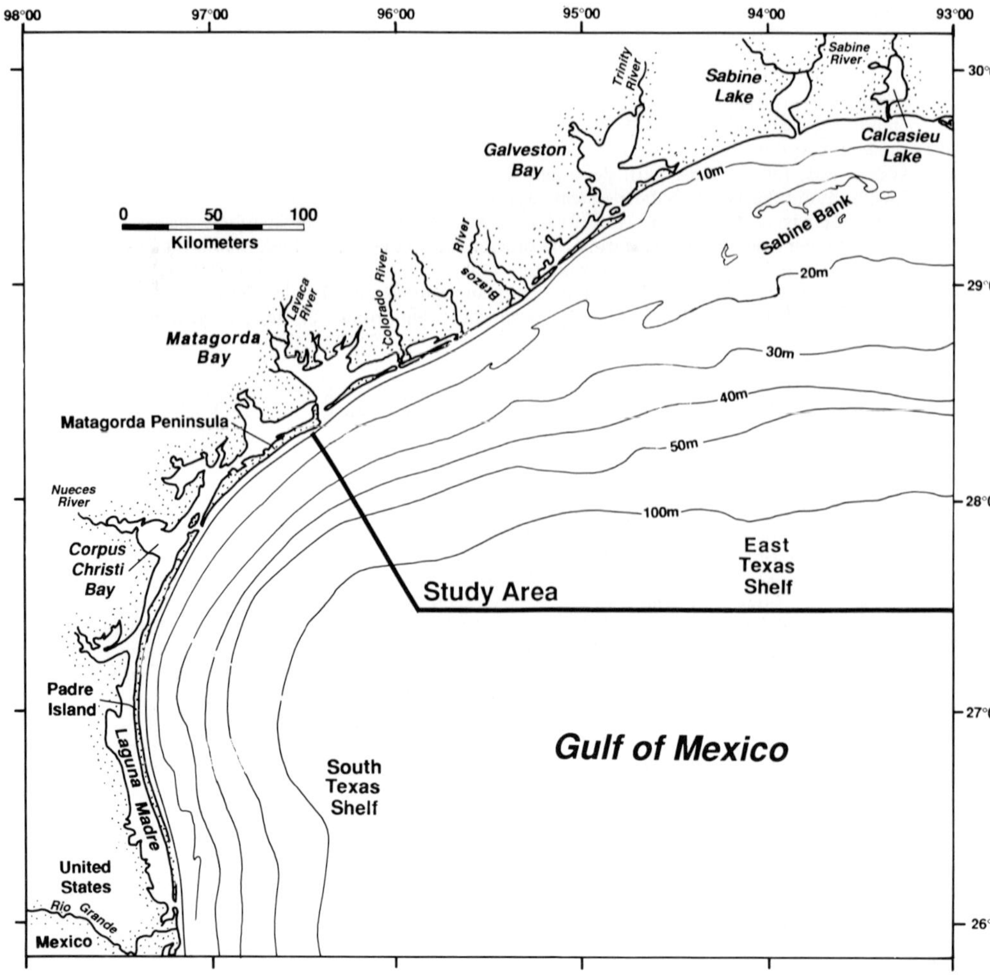

Fig. 1. The Texas continental shelf is characterized by a wide range in shelf gradient, climate, and sediment yields from different rivers. This results in a gradation from more fluvial-dominated depositional systems on the east Texas shelf to more storm-dominated depositional systems on the south Texas shelf.

There are those who argue that the lessons learned from analyses of Pleistocene strata are not applicable to older strata because the frequency of eustatic rise and fall is much higher for the Pleistocene than for the remainder of the Cenozoic and Mesozoic. We view this as an important factor only in regulating the thickness and extent of sequences; the magnitudes of Pleistocene sea-level changes have been great enough to impact the entire continental shelf. Another assumption in studies of this type is that the processes controlling facies architecture (i.e. fluvial input, storm influence, etc.) are at work regardless of the rate of sea-level rise and fall. Indeed, we should consider ourselves fortunate to be able to image the sedimentary record of this very unique geological time period; sea level is rising and falling fast enough to provide sequences that are thick enough to be preserved but thin enough to be imaged using high-resolution seismic techniques. The key is to find the right areas in which to conduct such studies.

The east Texas shelf is located between the strongly fluvial-dominated Mississippi Delta (Berryhill *et al.* 1987; Coleman & Roberts 1988), where delta-lobe shifting dominates sedimentation on the shelf, and the storm-dominated south Texas shelf. The shelf gradient more than doubles (from 0.5 to 2.0 m/km) from north to south along the Texas shelf (Fig. 1). In addition, the east Texas shelf is characterized by a strong climatic gradient, ranging from humid in the east to semi-arid in the southwest portion of the state. This climatic gradient is manifested in the sediment yields of the rivers in the area (Morton & McGowen 1980). Sediment supply to the shelf by these rivers varies widely. There is also a wide range in the bedload versus suspended load of these rivers. Subsidence and sedimentation rates on the Texas shelf have been sufficient for the glacial eustatic rises and falls to produce sufficiently thick packages of strata to be imaged using high-resolution seismic reflection data. Also, tectonic influences on sedimentation (salt tectonics and growth faulting) vary considerably within the study area. Tectonism decreases from east to west across the outer shelf. Finally, there is a wealth of industry

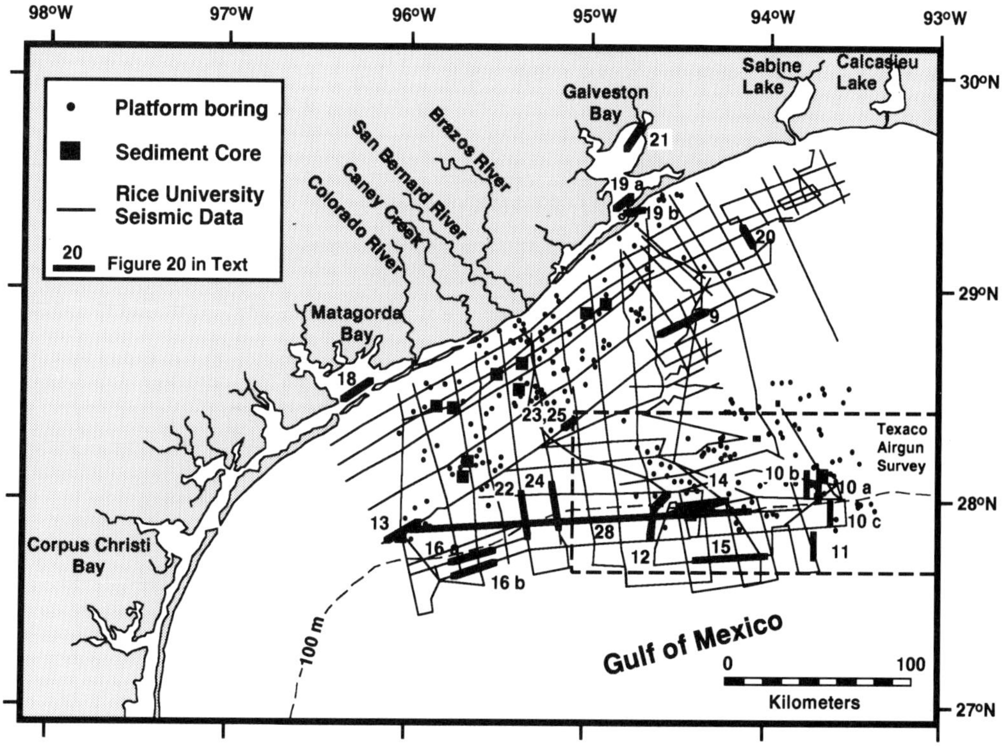

Fig. 2. Locations of seismic track lines and oil company platform borings within the study area. Circles designate sites where only lithological descriptions are available and squares designate sites for which we have core. The boxed area marks a dense grid of small airgun seismic records, acquired by Texaco and donated to Rice University. Heavy lines indicate line segments used in different figures in the text.

platform boring descriptions and sediment cores (literally hundreds) available that provide ground truth for seismic facies interpretations and provide material for chronostratigraphical analysis. Most of these drill sites penetrated enough (average 100 m) to record the stratigraphical interval being examined. All of these factors combined make the east Texas shelf a natural laboratory for investigating how sedimentation on the shelf responds to various forcing agents.

Methods

Three seismic sources were employed in the survey: a Uniboom (300–2400 Hz), which provides sub-bottom penetration to depths of approximately 200 m; a 15 in^3 watergun (40–2000 Hz), which provides data to depths of approximately 700 m and a 50 in^3 bubble-free airgun (20–500 Hz), which provides data to depths of approximately 1.5 s. In addition, Texaco contributed a data set of 40 in^3 and 10 in^3 airgun data, and 3.5 kHz data (Fig. 2). The majority of our seismic data were acquired digitally using the Elics Delph1 and Delph2 acquisition systems. Only minimal processing was undertaken, including band-pass filtering and vertical stacking.

Approximately 300 oil company platform borings and over 200 sediment cores provide lithological ground truth for seismic facies analyses (Fig. 2). Descriptions of platform borings include general lithology, stiffness of units, and presence or absence of fossils. Fugro-McClelland engineers provided ten cores from the continental shelf. These cores contained carbonate and carbonaceous material for radiocarbon dating and benthic and planktonic foraminifera for palaeobathymetric work. Benthic foraminifera from one of these cores were used to generate an oxygen isotope record for the study. In addition, some descriptions include well-log data. On the upper slope, the seismic grid covers the core sites of Shell Production Company's 1965–67 Eureka Project, which was based on the study of several 330 m cores. Shell has kindly provided detailed biostratigraphical and lithological data from these cores.

When dealing with high-resolution (high-frequency) data, seismic reflection character (and therefore seismic facies) is dominantly controlled by the scale of sedimentary structures within the deposit, the nature of the bounding surfaces, and the lateral and stratigraphic variability that creates impedance contrasts (Anderson *et al.* 1994). In this study, seismic facies were defined using parameters such as reflection configuration, external form, and facies association.

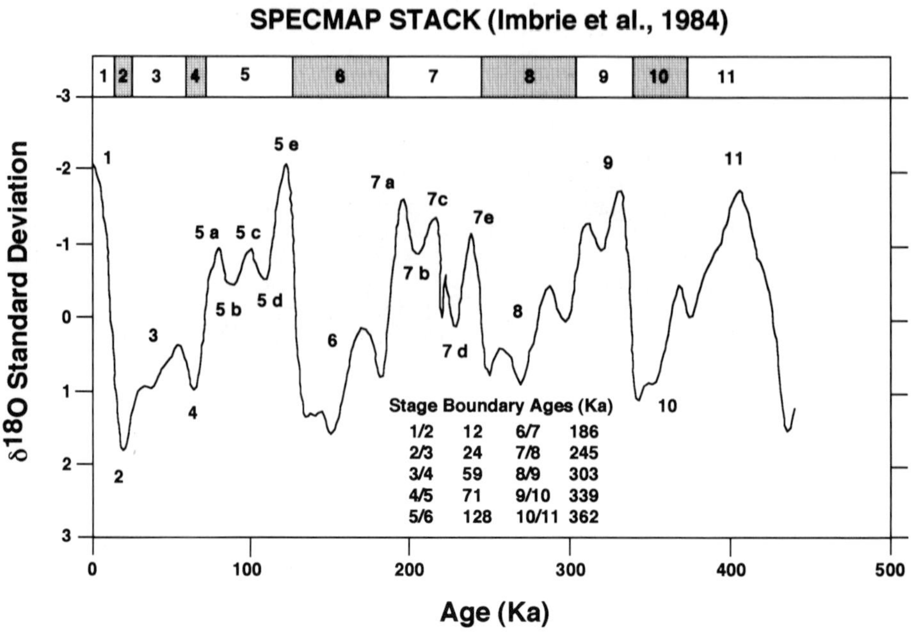

Fig. 3. The oxygen isotope curve (SPECMAP curve of Imbrie *et al.* 1984) is used as a proxy for sea level.

This methodology is based, in large part, on earlier studies by Mitchum *et al.* (1977) and Sangree *et al.* (1978).

The oxygen isotope curve (SPECMAP STACK of Imbrie *et al.* 1984) is used as a proxy to sea-level change during the Pleistocene (Fig. 3). We recognize the limitations of this curve (ice volume versus temperature effects), but research to date has shown a good correlation between the oxygen isotope curve and the sea-level curve for the past 20 000 years (Fairbanks 1989).

Distribution of sedimentary facies on the shelf

Seismic data, platform boring descriptions, and sediment cores were used to examine the distribution of major depositional systems on the continental shelf and upper slope. The results are presented in a series of five maps (Figs 4 to 8) depicting the distribution of depositional systems during the last highstand (130 000 to 80 000 years BP), early lowstand (~80 000 to 22 000 years BP), maximum lowstand (~22 000 to 15 000 years BP), transgression (~15 000 to 4000 years BP), and the present highstand setting (~4000 years BP to present). The following discussion describes the factors influencing sedimentation and lithofacies distribution during each of these time intervals.

Previous highstand (~130 000 to 80 000 years BP)

During the previous highstand, sea level was approximately 5 m above present and the shoreline was located approximately 20 km north of the present shoreline; these former shoreline deposits are known as the Ingleside Barrier. The Trinity, Brazos, and Colorado rivers constructed highstand deltas on the continental shelf (Fig. 4). The Colorado River formed a sandy, lobate delta. In contrast, the Brazos River deposited a more elongate, muddy delta. Three phases of progradation of the Brazos delta are recognized.

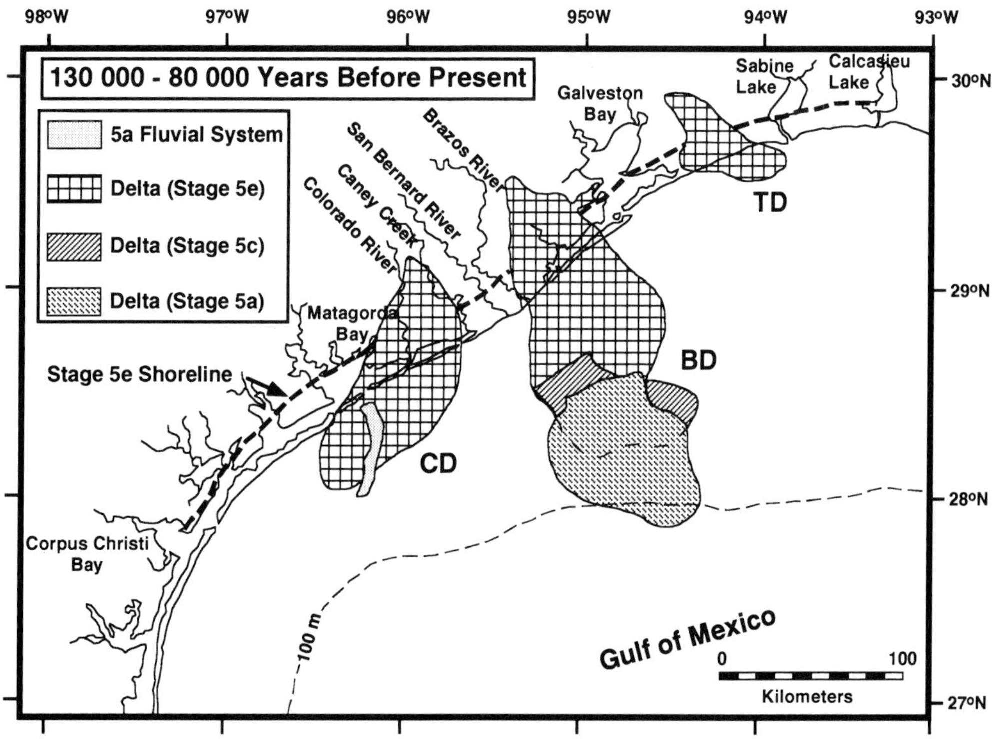

Fig. 4. Map showing the distribution of depositional systems associated with the previous highstand (~130 000 to 80 000 years BP). BD = Brazos delta, CD = Colorado delta, and TD = Trinity delta. Location of Trinity delta from Graf (1966).

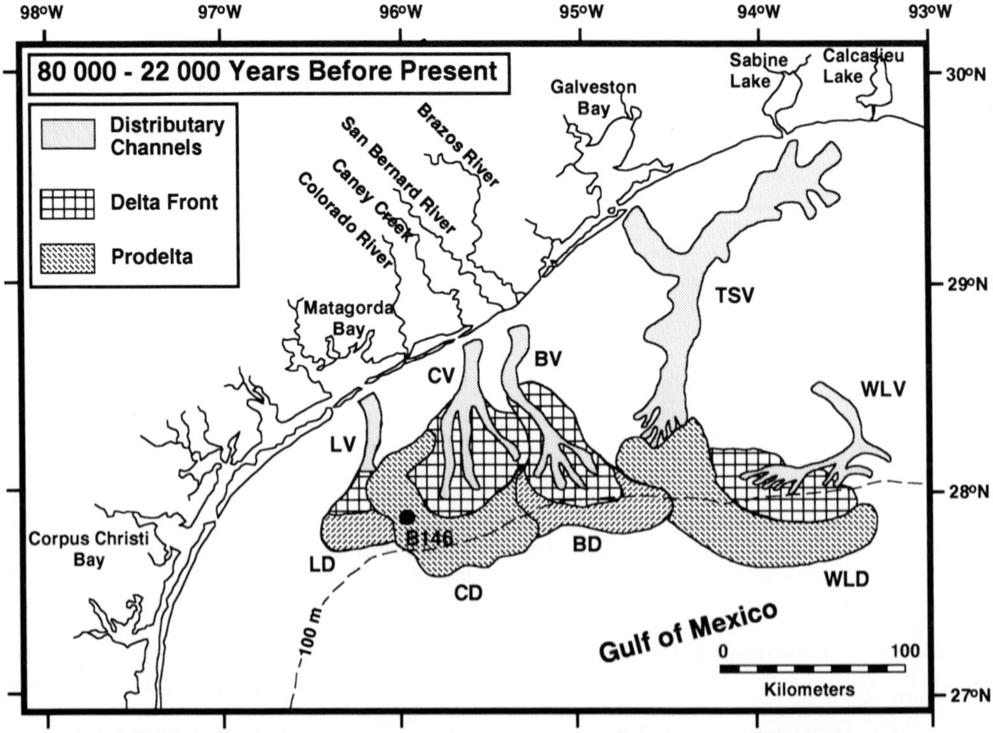

Fig. 5. Map showing the distribution of early lowstand (~80 000 to ~22 000 years BP) depositional systems on the east Texas shelf. LV = Lavaca valley, CV = Colorado valley, BV = Brazos valley, TSV = Trinity/Sabine valley, WLV = western Louisiana valley, LD = Lavaca delta, CD = Colorado delta, BD = Brazos delta, and WLD = western Louisiana delta. Also shown is the location of core B146.

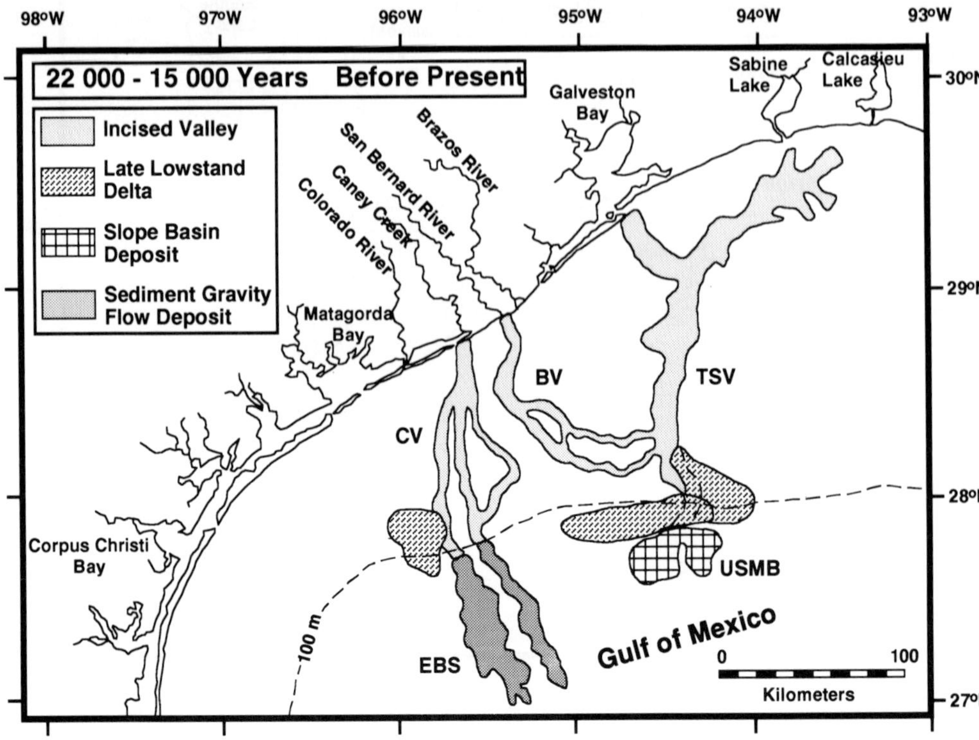

Fig. 6. Map showing the distribution of depositional systems of the maximum lowstand–early transgression (~22 000 to ~15 000 years BP). CV = Colorado valley, BV = Brazos valley, TSV = Trinity/Sabine valley, EBS = East Breaks Slide, and USMB = upper slope minibasin.

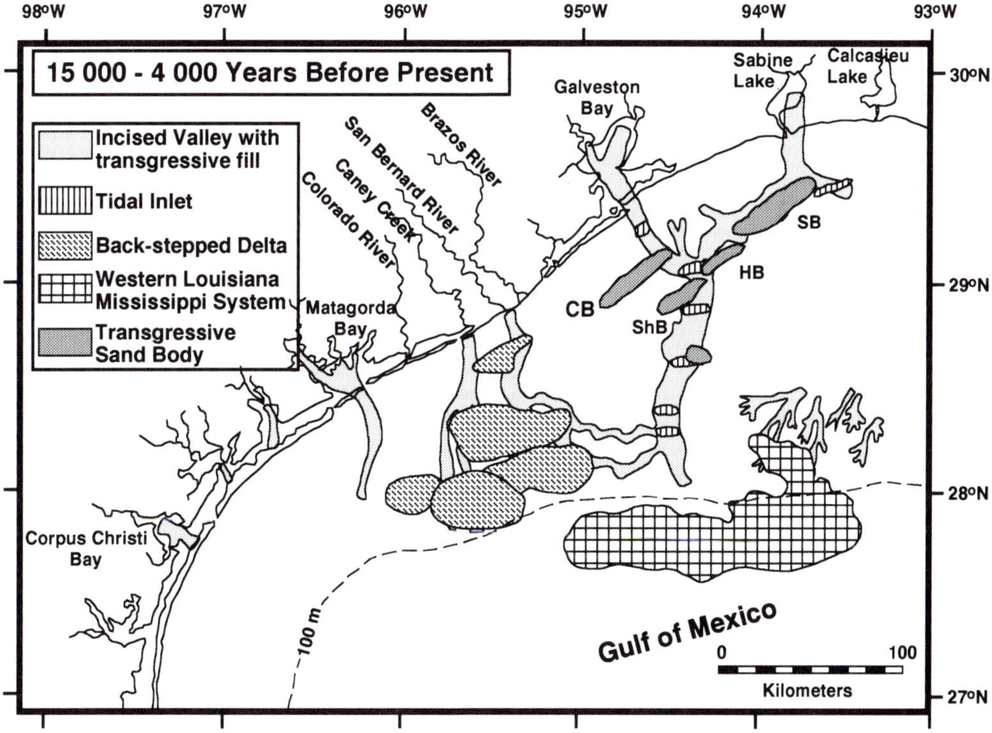

Fig. 7. Map showing the distribution of transgressive (~15 000 to 4000 years BP) depositional systems of the east Texas continental shelf. HB = Heald bank, SB = Sabine bank, ShB = Shepard bank, and CB = Curtis bank.

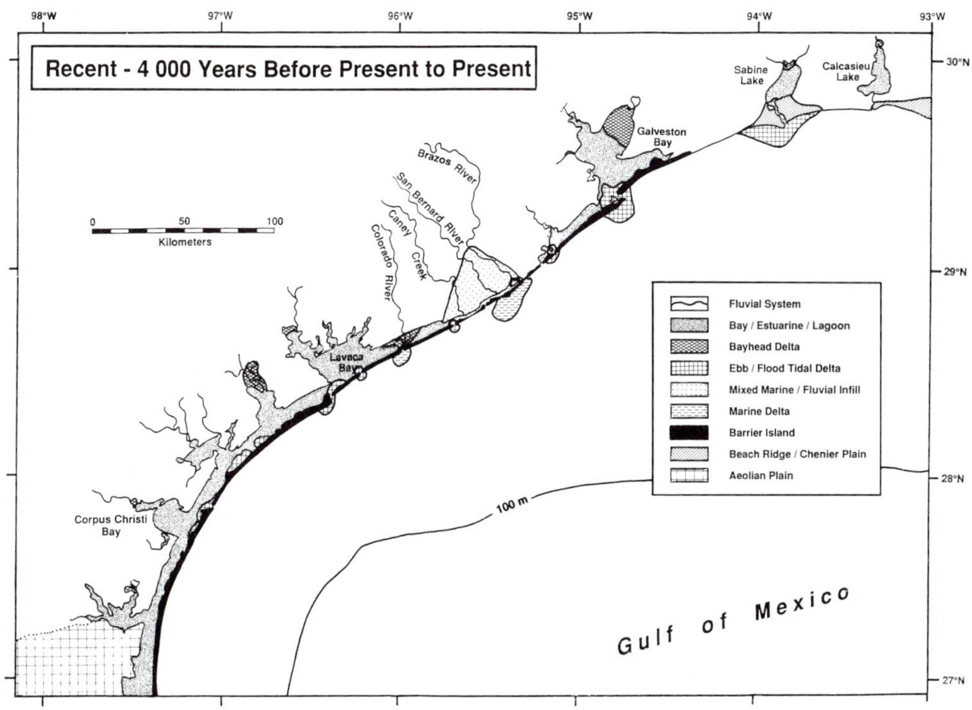

Fig. 8. Map showing the depositional systems of the present highstand (~4000 years BP to present).

Tentatively these are assigned to Stages 5e, 5c, and 5a, respectively. Minor transgressions (5b and 5d) caused backstepping of the deltas. The size and shape of the offshore Trinity delta is inferred from onshore distributaries, thus the Trinity offshore delta is not as well mapped as the Brazos and Colorado deltas.

Early lowstand systems tract (~ 80 000 to 22 000 years BP)

Prominent features of the previous lowstand include incised fluvial valleys, shelf-margin deltas, lowstand slope fans, and other sediment gravity flow deposits. Figure 5 shows the incised valleys and lowstand deltas formed during the early lowstand. Several rivers, including the Sabine, Trinity, Brazos, Colorado, and Lavaca rivers, incised the shelf at this time. One other fluvial incision, originating in western Louisiana but with unknown origin, is also shown. This incision is referred to as the western Louisiana fluvial valley (WLV) and its delta is the western Louisiana delta (WLD).

The Sabine and Trinity incised valleys merge approximately 35 km offshore and extend southward to the shelf break. The rivers remained confined to a single valley on the outer shelf. The Lavaca River incised valley was mapped only on the inner shelf, but shows a relatively straight course. These rivers have relatively small sediment yields compared to the other rivers of the region. The smaller yields are attributed to their smaller drainage basins, relative to the Brazos and Colorado rivers. Unlike the Trinity/Sabine and Lavaca rivers, the western Louisiana, Brazos, and Colorado rivers formed extensive distributary valleys on the mid- and outer shelf (Fig. 5). Initial incision of some of these valleys is believed to have occurred during oxygen isotope Stage 5d (~110 000 years BP), based on seismic stratigraphical analysis (Thomas & Anderson 1994).

Fluvial terraces along the flanks of the Trinity/Sabine valley are interpreted as Stage 5c and 5a deposits that are cut by Stage 5b and Stage 2 incisions (Thomas & Anderson 1994) (Fig. 9). Thomas & Anderson (1991) correlated Stage 5c fluvial deposits with onshore fluvial terraces (Deweyville Terraces) that were mapped by Bernard (1950) and Aten (1983).

The western Louisiana fluvial system constructed a sizeable shelf-margin delta, originally mapped by Suter & Berryhill (1985) (Fig. 5). Sarzalejo *et al.*'s (1993) detailed work on this delta showed that the western Louisiana fluvial delta prograded across the shelf shortly after the Stage 5e highstand, and initially reached the shelf edge approximately 80 000 years BP. Four different delta lobes were active during the early lowstand.

Figure 10 shows portions of seismic profiles illustrating the seismic facies associated with this prograding, fluvial-dominated delta. The main feeder channel was crossed by several seismic profiles. It is characterized by a chaotic-complex fill seismic facies with laterally migrating reflectors (Line 17, Fig. 10b). A platform boring through this portion of the valley penetrated a coarsening upward sequence ranging from sand to silty sand and then sandy silt. This channel sequence rests sharply on clay and is capped by fossiliferous silty clay. A ravinement surface separates the channel from the upper fossiliferous unit. Seismic Line 18 (Fig. 10c) illustrates the chaotic fill seismic facies of the distributary channels. A platform boring through one of these channels penetrated sands at the level of

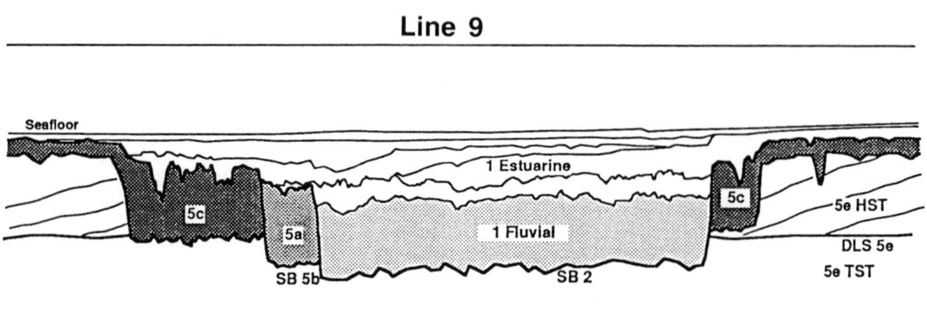

Fig. 9. Interpreted seismic line 9 crosses the Trinity River incised valley on the middle shelf (Fig. 2) and illustrates multiple stages of valley incision. Note that the older Stage 5c and 5a fluvial terrace deposits virtually filled the valley whereas the Stage 1 fluvial deposits are confined to the lower third of the valley. Flood tidal facies within the estuarine section prograde landward, oblique to the valley axis (northwest) (from Thomas & Anderson 1994). SB = sequence boundary, DLS = downlap surface, TST = transgressive system tract, HST = highstand systems tract.

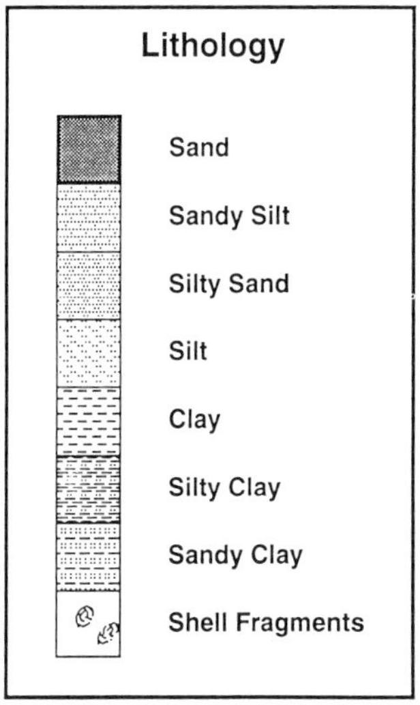

Fig. 10. (**a**) Lithological units that will be used in cores and borings on subsequent seismic sections. (**b**) Seismic line 17 crosses the main feeder channel of the western Louisiana shelf margin delta. Note the chaotic-complex fill and laterally migrating reflectors. A platform boring at approximately this location penetrated a coarsening upward sandy fill resting sharply on clay and capped by fossiliferous silty clay. RV = ravinement surface. (**c**) Seismic line 18 (N) crosses the distributary channel complex of the delta and is used to illustrate this seismic facies, which consists of cross-cutting channels with chaotic fill. A platform boring through one of these channels penetrated sand. (**d**) Seismic line 16 shows amalgamated channels of the distributary mouth bar (DMB) cutting into clinoforms of the prograding prodelta (PD). Platform borings into the distributary mouth bar, such as the one shown, penetrated mainly sands. Borings into the clinoforms penetrated silts and clays. Numerous growth faults occur in this area. Some affect these deltaic facies and others predate these younger deltas. See Fig. 2 for profile locations.

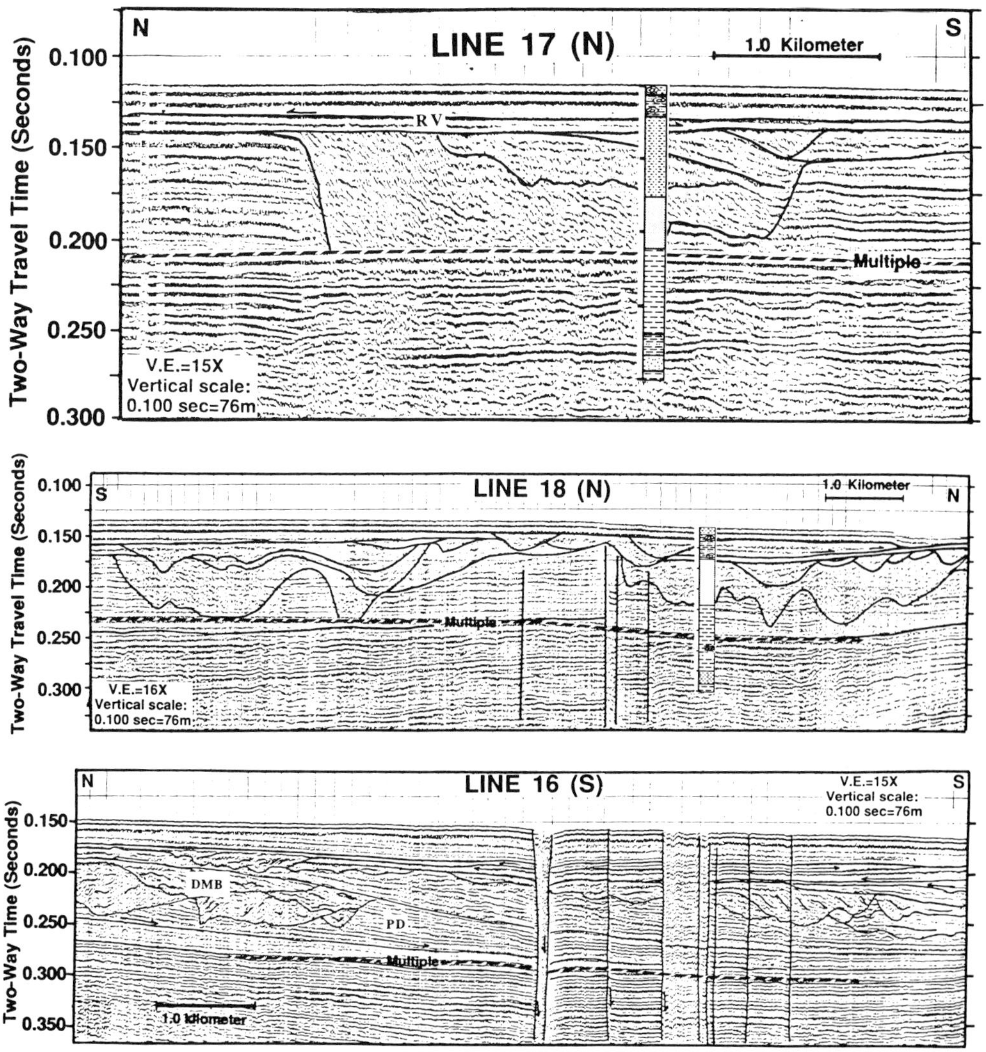

the chaotic seismic facies. The sandy mouth bar complex formed behind a linear trend of salt diapirs. Seismic Line 16 (Fig. 10d) shows the amalgamated channels of the distributary mouth bar cutting into prograding prodelta deposits, which are characterized by continuous reflections. Platform borings into the latter seismic facies penetrated almost exclusively mud with occasional thin sand lenses. The delta downlaps onto a parallel-layered seismic facies; the downlap surface is a regionally defined condensed section. These seismic facies are virtually identical to the ancestral Lagniappe delta on the Alabama–Mississippi continental shelf (Sydow & Roberts 1994). The seismic records from the eastern portion of the study area reveal that a dynamic feedback mechanism exists between salt tectonics, growth faulting, and sedimentation on the outer shelf (Fig. 11).

During the early lowstand, the Brazos and Colorado rivers constructed fluvial deltas roughly equal in size to the western Louisiana delta (Fig. 5). Age constraints for the development of the Brazos shelf-margin delta were obtained using an oxygen isotope record constructed from Site B146 and biostratigraphic data from Site B146 and the Eureka cores. These data indicate that the delta was deposited during Stage 3 (Abdulah & Anderson 1991, 1994).

An oblique progradational reflection pattern characterizes the Brazos shelf-margin delta; virtually no chaotic seismic facies occur within or up-dip of the clinoforms (Fig. 12). This implies that the shelf-margin delta is composed predominantly of fine-grained sediments. Platform borings through the delta penetrated predominantly mud. Seismic Line R92–24 also illustrates the seismic stratigraphy of the delta. Shelf-margin progradation can be seen above the lowest sequence boundary. Rapid transgression, resulting in the erosional ravinement surface, followed the development of this lowstand shelf-margin delta. The maximum-flooding surface above the ravinement surface marks the end of the transgression, and forms the downlap surface for the overlying progradational unit.

Seismic profiles across the Colorado shelf-margin delta record predominantly chaotic seismic facies (Fig. 13). The presence of such facies indicates that the delta has a sandy distributary system similar to that of the western Louisiana delta. Two phases of progradation are indicated by the seismic records (Fig. 13). A platform boring from a channel close to the one imaged in seismic line R90–9 confirms that the chaotic seismic facies consists of sand and gravel, which is also indicated by the log response at this site (Fig. 13).

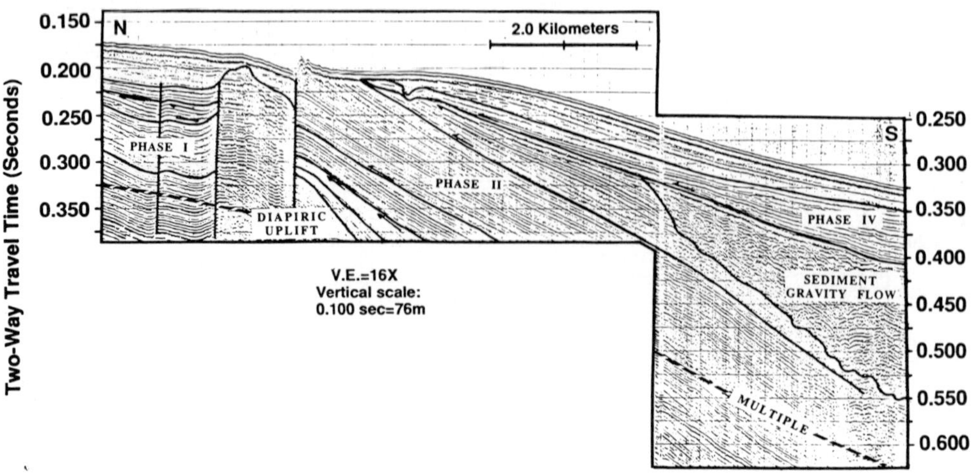

Fig. 11. Growth faults and salt-related structures significantly influenced the dispersal of sediments on the outer shelf and upper slope. Interpreted seismic line 18 provides a good example of these effects. This line also illustrates the different phases of development of the western Louisiana shelf-margin delta. The first phase (Phase I) of early lowstand delta progradation culminated with an episode of diapir and growth fault activity that resulted in uplift and erosion of Phase I deposits. During Phase II sediment bypass occurred on the outer shelf and there was rapid sediment accumulation on the upper slope. Phase III was a period of tectonic uplift and erosion on the shelf, and sediment mass movement on the upper slope, resulting in the 150 ms thick sediment gravity flow deposit on the upper slope. Rapid accumulation and oversteepening during Phase II, and tectonic uplift during Phase III, initiated the mass movement.

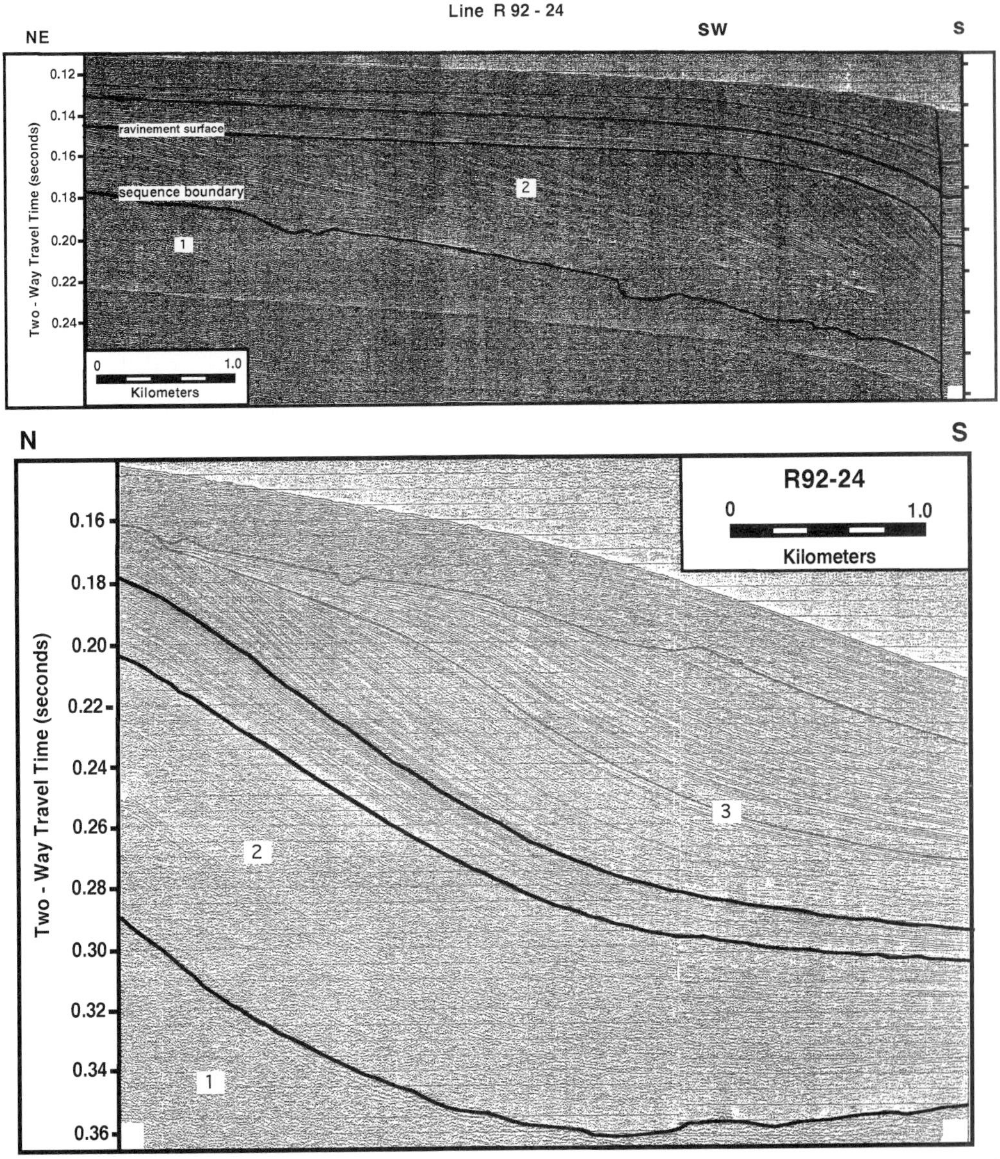

Fig. 12. Seismic line R92-24 is a dip line that crosses the Brazos shelf-margin delta (Fig. 2). The predominance of acoustically laminated seismic facies and platform borings indicate that the delta is composed predominantly of silts and clays; a sandy distributary system does not exist. Sands and gravels are confined to the main feeder channel. This line also illustrates the stratal patterns associated with shelf-edge deposition during the last glacial cycle. Unit 1 is a Stage 5a deposit, Unit 2 is a Stage 3 deposit and Unit 3 is a Stage 2 deposit. Hence shelf margin delta development spans approximately 60 000 years.

Maximum lowstand–early transgression (22 000 to 15 000 years BP)

Figure 6 shows the distribution of major depositional systems on the east Texas shelf and upper slope during the maximum lowstand (22 000 to 15 000 years BP). During the Stage 2 lowstand, the Trinity/Sabine valley was reincised; this was the deepest incision (ranging from −35 to −40 m offshore). This later phase of incision left remnants of older (Stage 5) fluvial terraces along the flanks of the Trinity/Sabine valley at a much higher level than the modern fluvial deposits; the valley virtually was filled with

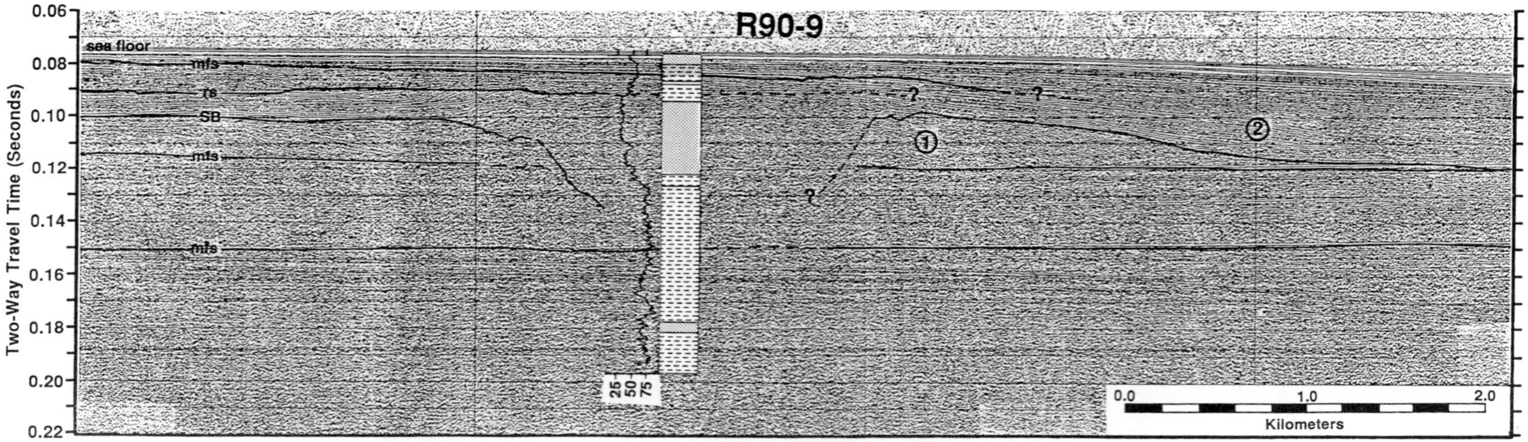

Fig. 13. Seismic line R90-9 illustrates the sandy nature of the Colorado shelf-margin delta (Fig. 2). The chaotic reflection pattern of the valley fill is confirmed by platform borings. Two phases of progradation, labelled 1 and 2, can be identified. Also shown is the lithological description and gamma log for a platform boring acquired near this seismic line.

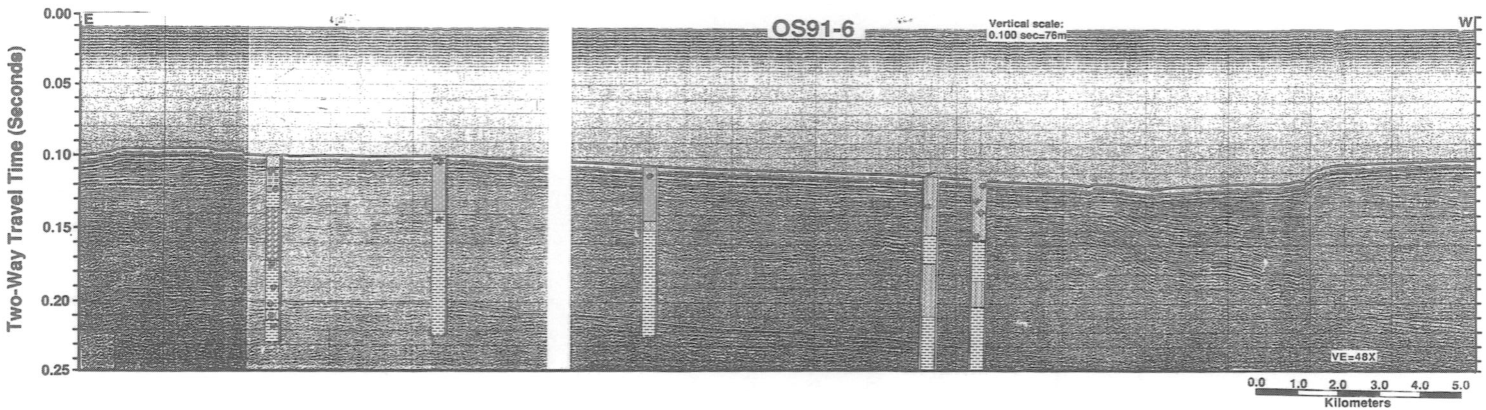

Fig. 14. Seismic line OS91-6 was collected along a transect of platform borings that cross the mouth of the ancestral Trinity/Sabine/Brazos fluvial valley (Fig. 2) . The seismic data provide little evidence of a sandy shelf-margin delta deposit at this location. Platform borings, however, indicate that a sand body averaging 30 m in thickness occurs here. The seismic facies associated with this sand deposit is characterized by strong, discontinuous reflectors and implies broad-scale, planar cross-stratification.

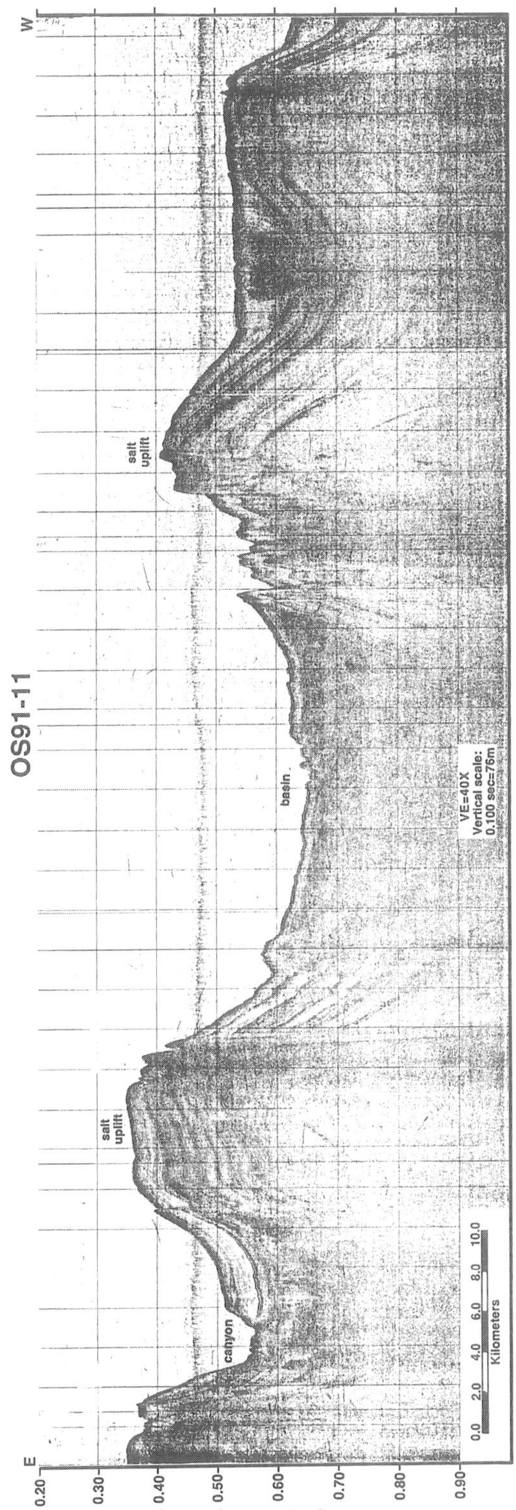

Fig. 15. Seismic line OS91-11 is a strike line from the upper slope (Fig. 2). Three different minibasins, bound by salt diapirs, were crossed in this segment of the line. The minibasins show different amounts of sediment fill and different seismic facies. The central basin occurs down-dip of the mouth of the incised Trinity/Sabine/Brazos valley and is characterized by a mounded chaotic seismic facies interpreted as sediment gravity flow deposits.

fluvial deposits during the previous transgression (Fig. 9). During the Stage 2 incision, these older fluvial deposits were eroded from the valley and transported down-slope. Estimates from the seismic and platform borings indicate that as much as 120 km^3 of sediment was eroded from the valley.

During the Stage 2 lowstand, the ancestral Brazos River cut a new valley around the eastern half of the muddy delta lobe shown in Fig. 5. The incision cut a low area that occurs along the flanks of several diapirs, and merged with the Trinity/Sabine valley on the outer shelf (Fig. 6). The outer portion of the Brazos valley contains a muddy fill, perhaps indicating that there was insufficient time for sandy bedload to be transported to the outer shelf in this suspended-load-dominated river. A relatively small shelf-margin delta formed at the mouth of the Trinity–Sabine–Brazos valley complex, based on the occurrence of an extensive sand body that averages 30 m in thickness (Fig. 14). Seismic profiles across this delta do not show the same chaotic-complex reflector pattern that characterizes the western Louisiana mouth bar deposits (Fig. 10c). Instead, strong, discontinuous reflectors indicate more planar cross-stratification (Fig. 14). The bulk of the sand in this delta probably was derived from erosion of older (Stage 5) fluvial deposits within the Trinity/Sabine incised valley. The sandy mouth bar complex formed behind a linear trend of salt diapirs. This sand body accounts for only a portion of the total sand that was eroded from the Trinity/Sabine valley; the remainder must have bypassed the shelf. Sediment bypass occurred through canyons between diapirs. Seismic profiles across minibasins on the outer shelf and upper slope, and downdip of the Trinity delta, show a mounded chaotic seismic facies interpreted to be sediment gravity flow deposits (Fig. 15). Cores from two of the upper slope minibasins penetrated turbidite sands (Satterfield & Behrens 1990).

The ancestral Colorado River occupied two incised valleys within the western part of the study area that extend to the shelf break. At the shelf margin, these valleys connect with two submarine canyons and elongate slope fans (Fig. 6). Lehner (1969), and later Rothwell *et al.* (1991), first mapped these canyons and fans. Seismic profiles across the canyons record erosional relief in the canyon floors, indicating recent sediment gravity flow processes (Figs 16a and 16b). The seismic records also show that a chaotic reflection pattern, indicative of sandy deposits, characterizes the canyon fill (Fig. 16b). A long core from this region, described by Woodbury *et al.* (1978), penetrated 60 m of contorted sand and clay units.

Transgressive systems tracts (15 000 to 4000 years BP)

The key to understanding the facies relationships of the transgressive systems tract is understanding the nature of sea-level rise. Rates of sea-level rise during the Stage 2 to 1 transgression have ranged from 0.7 cm/year during the first 15 000 years of the transgression to 0.3 cm/year during the last 3500 years (Fairbanks 1989). Anderson & Thomas (1991) argued that a pattern of episodic sea-level rise for the Stage 2 transgression can be recognized globally and was caused by the mass wasting of marine ice sheets. In general, periods of rapid rise in sea level are followed by slow rise and stillstand. While these rapid rises are small in magnitude (probably less than 3 m within a few centuries), they manifest themselves as dramatic landward shifts in facies.

Subsidence within the Trinity incised valley has been somewhere between 0.5 and 1.0 cm/year, based on long-term tidal records collected within the valley over the past 90 years (Swanson & Thurlow 1973). Thus, sedimentation rates of 0.8 to 1.5 cm/year were required to keep pace with the relative rise in sea level. In general, the Trinity, Sabine, and Lavaca rivers were not able to keep pace with the rise in sea level. Their valleys are characterized by continuous flooding and landward shifts in valley-fill facies (Thomas & Anderson 1989, 1994). In contrast, the Brazos and Colorado rivers filled their valleys and shifted to occupy new valleys. For this reason, the Trinity, Sabine and Lavaca rivers are considered relatively low-sediment-supply fluvial systems and the Brazos and Colorado rivers are considered relatively high-sediment-supply systems. The difference between these fluvial systems is evident even today. The Trinity, Sabine and Lavaca river valleys are occupied by estuaries, whereas the Brazos and Colorado rivers extend to the coast and are constructing wave-dominated deltas. The following section underscores the dramatic differences that exist between the relatively low-sediment-supply Trinity/Sabine fluvial system and the relatively high-sediment-supply Brazos and Colorado fluvial systems. Because these systems are so different, the discussion will focus on each separately. Differences in sediment supply of the different rivers and the overall episodic nature of sea-level rise have influenced strongly sedimentation in the interfluve areas as well as the incised valleys.

Figure 7 shows the prominent elements of the transgressive systems tract. These include incised valley-fill deposits, shelf sand bodies, and backstepped deltas.

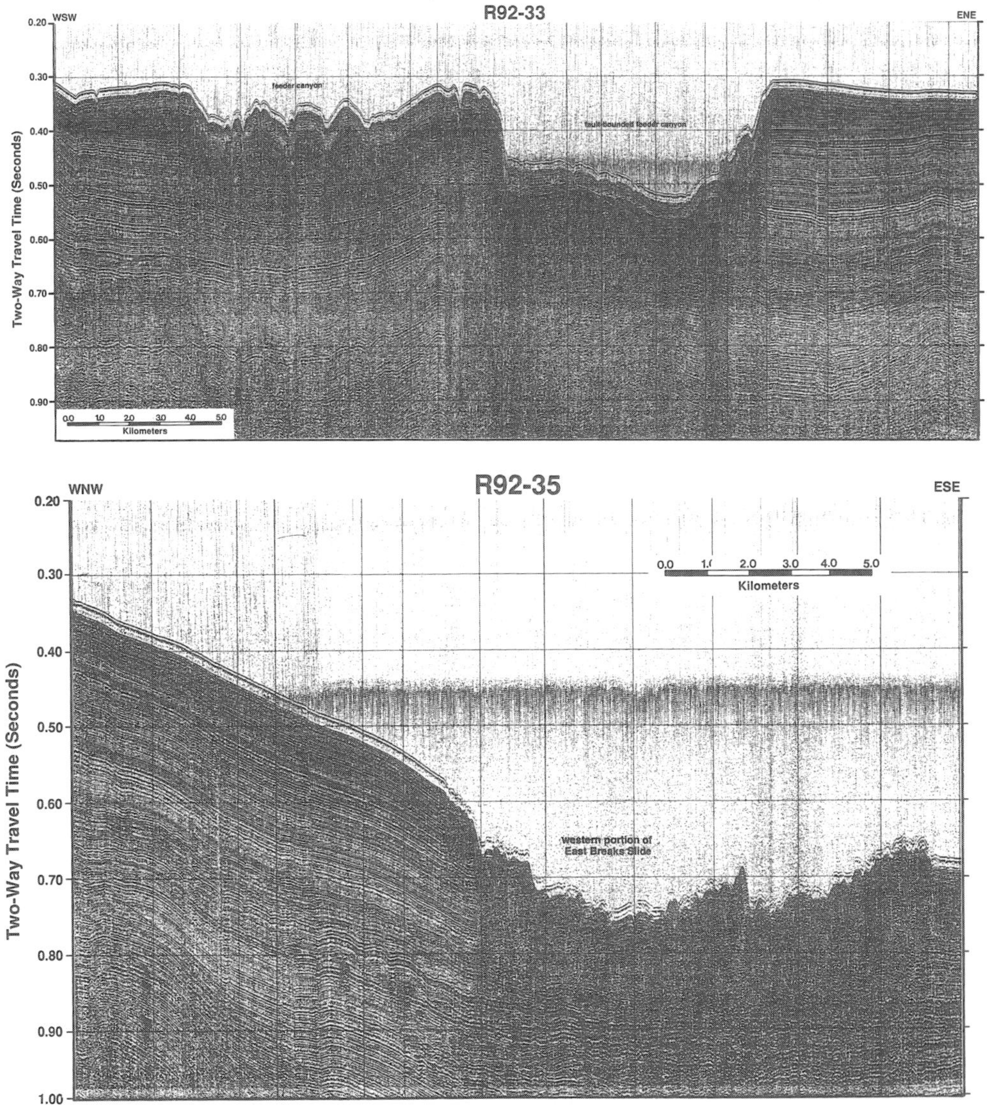

Fig. 16. Two seismic lines collected parallel to the shelf break (Fig. 2) illustrate seismic facies associated with the incised valley–canyon system of the ancestral Colorado River and East Breaks Slide. (**a**) Profile R92-33 crosses the upper fault-bounded feeder canyon. (**b**) Profile R92-35 was collected downslope of line R92–33. It illustrates the chaotic seismic facies within the canyon.

Relatively low-sediment-supply systems

During the construction of the intercoastal barge canal, which extends along the entire Texas coast, closely spaced engineering borings were acquired. A virtually continuous seismic profile along the canal, along with these borings, provides an excellent data set for examining the seismic facies and lithofacies of fluvial valleys that cross the coastal plain. Figure 17 shows a profile across the Trinity River incised valley and illustrates its valley-fill sequences. The Lavaca incised valley is used to illustrate seismic facies associated with relatively low-sediment-supply incised valleys (Fig. 18). The valley-fill deposits shown in Figs 17 and 18 were deposited during the slow rise in sea level of the past few thousand years. This slow rise allowed deposition of a tidal inlet/delta complex. The tidal facies have a discontinuous distribution in

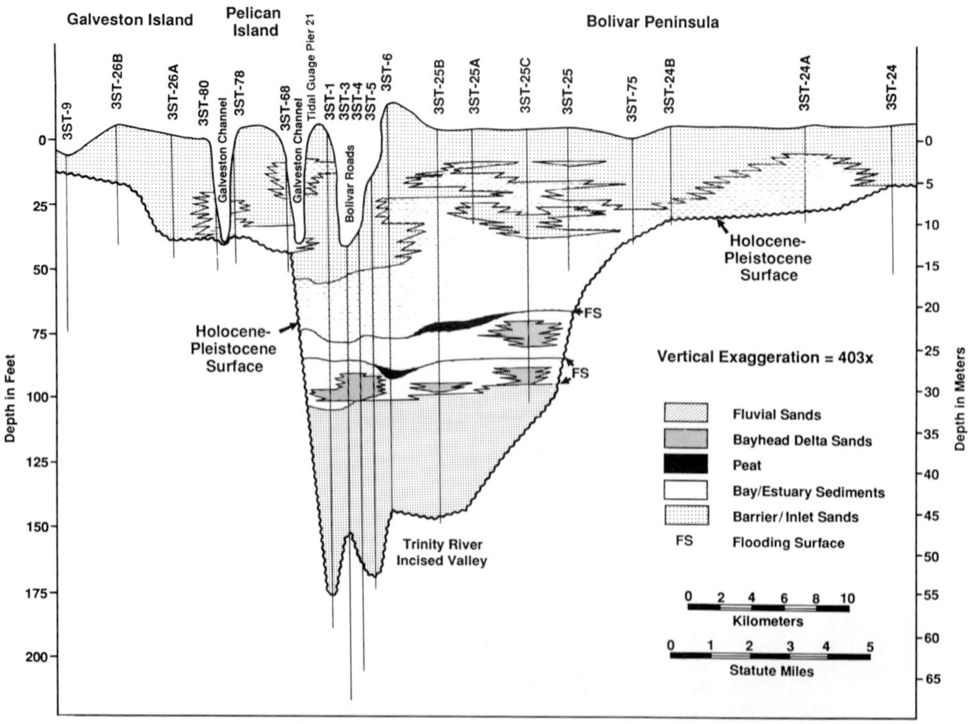

Fig. 17. Example of relatively low-sediment-supply incised valley-fill sequence, the Trinity River incised valley. This profile is based on closely spaced engineering borings collected during construction of the intercoastal barge canal.

offshore portions of these valleys, primarily because more rapid transgression during the interval between 18 000 and approximately 4000 years did not always provide an opportunity for barrier and inlet development.

The sequences shown in Figs 17 and 18 consist of, from bottom to top, fluvial channel, bayhead delta, middle bay, flood tidal delta, tidal inlet, and offshore marine facies. The facies are bound by several types of surfaces, including the lower sequence boundary, the bayline, intermediate flooding surfaces, and ravinement surfaces. Both the seismic facies and bounding surfaces are mappable using good quality, high-resolution seismic data. High-frequency fluctuations in sea level cause repetition of valley-fill facies and bounding surfaces.

The Stage 2 sequence boundary forms the base of the Stage 1 valley fill observed in these examples. This incision averages −35 to −40 m on the inner shelf and decreases with distance from the present shoreline. The depth of incision is fairly constant between different valleys at any given distance from the shelf break. A complex, chaotic reflection pattern characterizes fluvial deposits above the Stage 2 sequence boundary (Fig. 18), and geotechnical borings indicate that the fluvial facies consists of gravels and sands that locally show a fining upward trend (Fig. 17).

The bayline is the fluvial–estuarine contact, which appears as a broad, high-amplitude reflector that shows relief on the order of several metres (Fig. 18). This relief reflects the drowned topography of the former floodplain. The upper bay/bayhead delta facies occurs above the bayline. Horizontal to wavy, subparallel reflectors characterize the bayhead delta facies (Fig. 18). The bayhead delta forms a flat-topped wedge bound at the base by the bayline and at the top by an intermediate flooding surface (Thomas & Anderson 1994). Dip-oriented seismic profiles show very low angle clinoforms that prograde down the valley axis. The clinoforms characterize the delta front. Strike-oriented profiles show individual, lenticular bayhead delta lobes that are stacked in an offsetting pattern due to delta-lobe shifting. McEwen (1969) studied the lithofacies of the modern Trinity bayhead delta. He found the delta to consist of a sandy delta front composed of bioturbated, fine sand. The prodelta consists

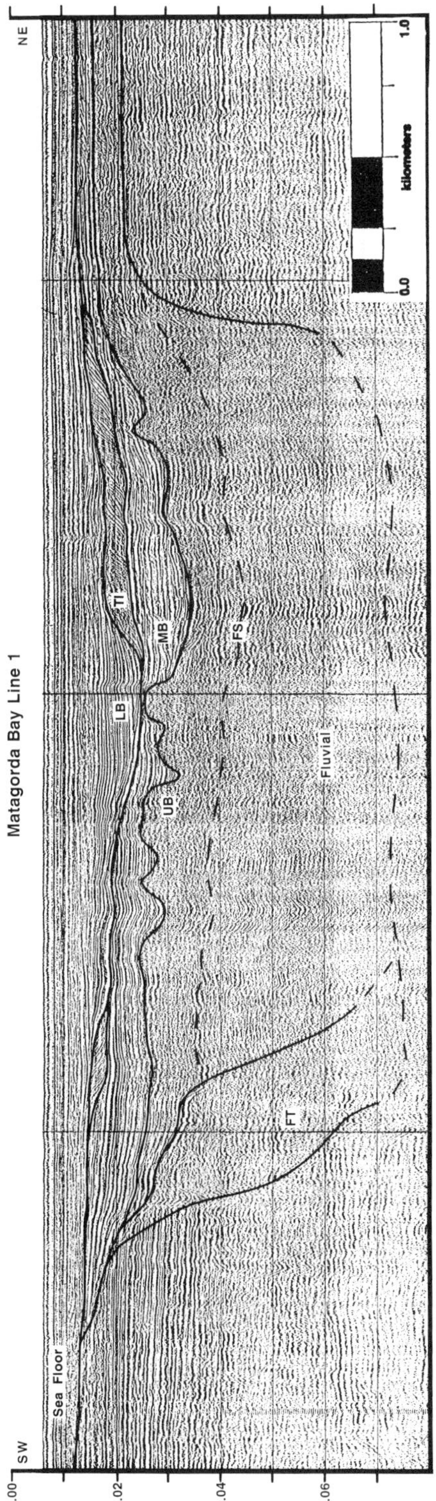

Fig. 18. Uniboom seismic profile across the Lavaca valley showing seismic facies of the complete valley-fill sequence of relatively low-sediment-supply fluvial systems. FT = fluvial terrace, FS = flooding surface (bayline), UB = upper bay seismic facies, MB = mid-bay seismic facies, LB = lower bay seismic facies, and TI = accretionary tidal inlet seismic facies.

(a)

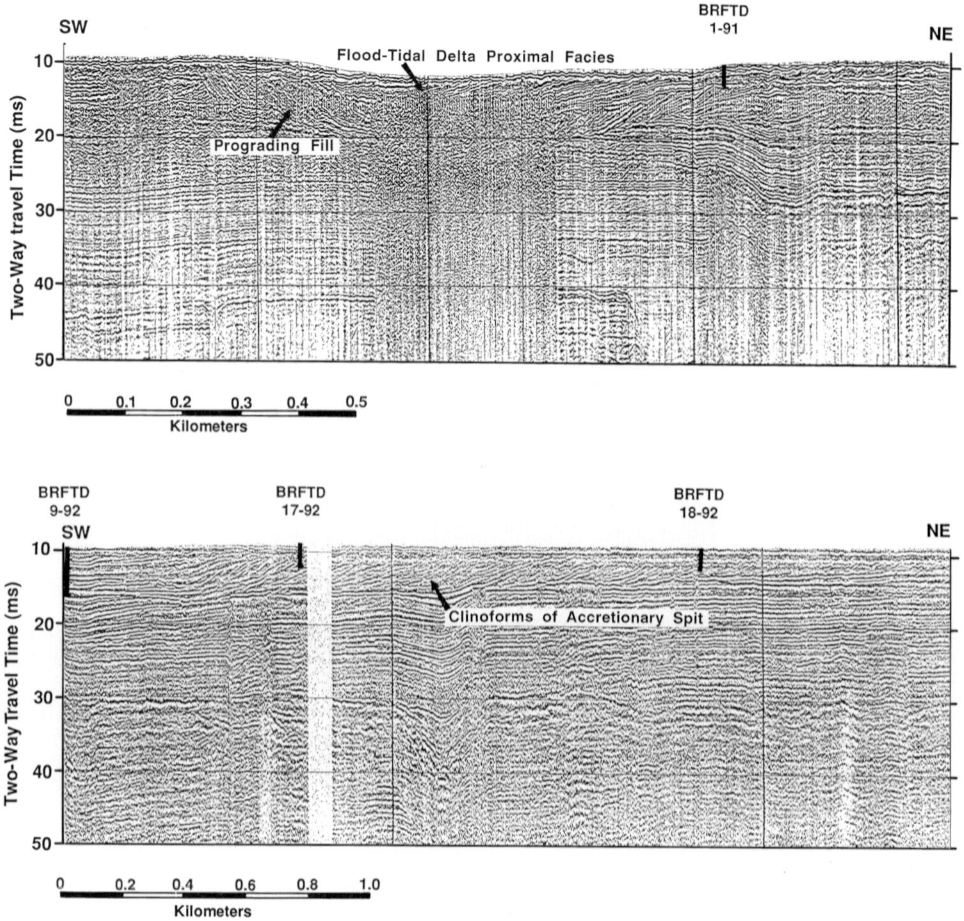

(b)

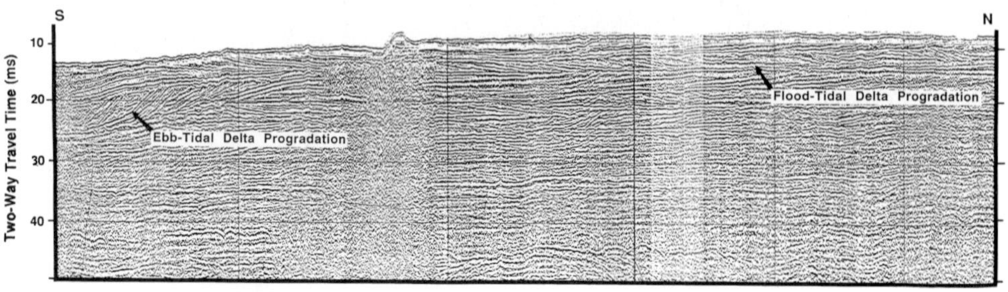

Fig. 19. Uniboom seismic profiles across the Bolivar Roads tidal inlet/flood tidal delta. (**a**) East–west line (in two parts) across the inlet and flood-tidal delta showing accretionary clinoforms. (**b**) North–south line across the ebb-tidal delta and flood-tidal delta showing landward dipping reflectors of the flood-tidal delta and seaward dipping reflectors of the ebb-tidal delta.

predominantly of clay with abundant organic detritus, and Rangia and *Crassostrea* shells. Similar facies occur in the modern Sabine bayhead delta (Anderson *et al.* 1991*a*, *b*).

The middle to lower bay seismic facies is characterized by continuous horizontal reflectors (Fig. 18) and the lithofacies consist of water-saturated, bioturbated muds with a mollusc assemblage dominated by *Ostrea* and *Mulinea*.

The tidal inlet/delta complex can be subdivided into three seismic facies (Siringan & Anderson 1993). The flood tidal delta is characterized by low-angle clinoforms that dip in a landward direction and are oblique to the valley axis (Fig. 19a). Clinoform reflectors generally toplap at the upper surface of the flood tidal delta. Locally, the marine ravinement surface truncates the upper surface of the delta. Landward dipping clinoforms downlap onto a surface that shallows abruptly bayward (Fig. 19b). The tidal inlet is characterized by sigmoidal clinoform reflectors that accrete laterally across the valley as a stacked channel cut-and-fill complex. The basal contact is erosive, with relief on the order of several metres. The depth of the tidal inlet increases toward the modern inlet (Fig. 19a); as spit and barrier development restricted the inlet, it incised to greater depths, cutting into the soft deposits of the Trinity incised valley. Low-angle, seaward dipping reflectors that downlap the marine ravinement surface characterize the ebb tidal delta seismic facies (Fig. 19b).

Lithofacies associated with the modern Bolivar tidal delta complex were studied in detail (Siringan & Anderson 1993) and used to recognize similar facies in the offshore valleys. Interbedded sands and shell hash constitute the tidal inlet. The flood tidal delta consists of interbedded mud and sand, often occurring as thin-bedded tidal couplets. The ebb tidal delta also comprises interbedded sand and mud, but sand is more dominant than in the flood tidal delta and tidal couplets are more rare. Overall, the Bolivar tidal inlet/delta complex is mud-dominated as a result of high influx of fine sediment into the bay (Siringan & Anderson 1993).

In the offshore portions of incised valleys, estuarine facies are truncated by the ravinement surface. The seismic character of the marine facies is marked by horizontal to parallel reflectors. The sea-floor pulse typically obscures the reflectors. Cores from this facies penetrated bioturbated sandy muds. Thickness of the marine facies is greatest (1.5 to 2.5 m) over the incised valley, primarily because of the softer, more easily eroded, valley-fill deposits.

An ongoing investigation of Corpus Christi Bay has revealed a valley-fill facies architecture different from that of the Trinity/Sabine valley. Sediment cores from Corpus Christi Bay sampled more sandy facies, and seismic records show more evidence of tidal influence in the form of large sedimentary structures that typically indicate onshore and westward transport. Corpus Christi Bay, compared to Galveston Bay, has relatively little fluvial input. This difference may cause the contrasts in the facies architecture. Before the Corpus Christi Bay mouth was restricted by the development of north Padre Island, it apparently experienced strong tidal influence.

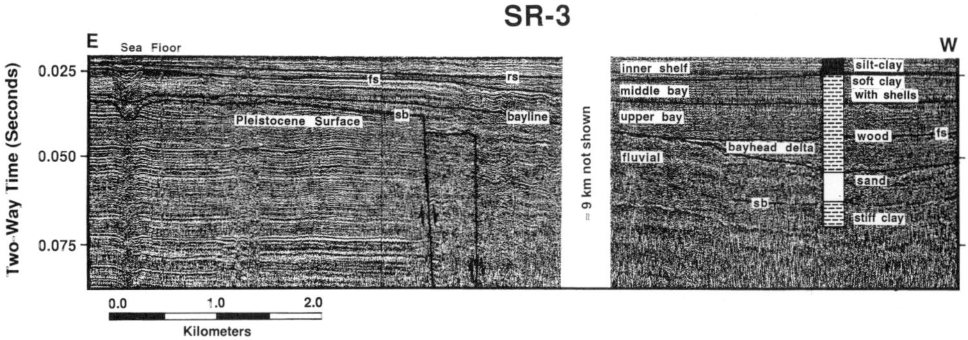

Fig. 20. Seismic line SR-3 crosses the Sabine incised valley and illustrates an incomplete valley-fill sequence. The platform boring provides lithological control for the seismic facies interpretation. The basal fluvial unit is characterized by a chaotic reflector pattern and coincides with fluvial sands in the platform boring. A sharp contact separates the fluvial unit from the overlying bayhead delta unit, which is characterized by wavy reflectors and onlap onto the fluvial surface (the bayline). The middle bay facies is characterized by a strongly laminated reflector pattern. The platform boring penetrated mud with abundant shell material at this level. The inlet/tidal delta facies are missing from this sequence, implying rapid flooding of the bay before a tidal inlet was created by spit and barrier formation. sb, sequence boundary; rs, ravinement surface; fs, flooding surface.

Detailed analyses of the Trinity/Sabine incised valley (Anderson *et al.* 1991*a,b*; Thomas & Anderson 1994) led to the recognition that idealised valley-fill sequences are the exception rather than the rule. Flooding surfaces are often manifested by an absence of one or more of the valley-fill facies. A good example of this is shown in Fig. 20. Fluvial sands are overlain by bayhead delta deposits and middle-bay muds, but the tidally influenced lower-bay facies are missing. Instead, a flooding surface (in this case the ravinement surface) separates middle-bay muds from offshore marine sediments. When this flooding surface is traced up-dip, it exists as a landward shift in the fluvial and bayhead delta facies (Thomas & Anderson 1994). Intermediate flooding surfaces are recognized in cores by the presence of peats that directly underlie bay muds. During periods of slow rise, the bayhead delta progrades, resulting in aggradation of fluvial deposits (Fig. 21). Individual flooding events result in up-dip facies shifts of many tens of kilometres (Thomas & Anderson 1994; Nichol *et al.* 1994) (Fig. 21). Likewise, tidal inlet/delta deposits show a discontinuous distribution within the incised valley (Fig. 7). The tidal inlet/delta facies mark palaeoshorelines where sea level rose slowly, allowing barrier-spit development and, thus, tidal inlet/delta formation. Thomas & Anderson (1994) demonstrated that the flooding surfaces that offset these tidal inlet/delta facies can be traced up-dip and tied to surfaces associated with backstepping fluvial/bayhead delta facies. This stratigraphic relationship indicates that these intermediate flooding surfaces are indeed the product of episodic sea-level rise and not changes in sediment supply. Blum (1990, 1994) has argued that changes in sediment supply, caused by climatic changes, are the principal factor controlling incised valley-fill facies architecture in the up-dip portions of the Colorado River valley.

On the inner shelf, the thickness of fluvial sands within the Trinity/Sabine valley averages 8 m and the thickness of the overlying estuarine and marine deposits averages 25 m (Thomas & Anderson 1991). The estuarine and marine deposits are predominantly fine-grained; the only exceptions are tidal inlet and tidal delta deposits and the distributary mouth bars of bayhead deltas. Thus, as the shoreface advanced across the Trinity/Sabine incised valley, muddy valley-fill facies were exposed to erosion. Fluvial sands will remain in the valley until the next glacial eustatic lowstand when renewed incision flushes these sands down-dip. This mechanism of repeated incision and filling of the valley results in a net bypass of fluvial sands from the shelf; most of this sand reaches the shelf break as a point source.

The Holocene ravinement surface, mapped by Siringan & Anderson (1993, 1994), eroded to depths between −3 to −10 m. The ravinement surface is characterized by erosional steps, similar to those observed on the Adriatic shelf (Trincardi *et al.* 1994). On the interfluve portions of the shelf, a thin (less than one metre thick) unit of marine mud rests directly on the ravinement surface. Even storm beds are noticeably absent (Siringan & Anderson 1994). Thus, coastal lithosomes outside the valley did not escape shoreface erosion. The only sandy deposits on the interfluve portions of the shelf are sand banks that occur along the flanks of the Trinity/Sabine valley (Fig. 7).

A detailed seismic and coring survey of the sand banks that occur along the Trinity/Sabine valley (Sabine, Heald, Shepard, and Curtis banks) (Fig. 7) showed that these banks contain similar stratigraphic and facies relationships. They all consist of a coarsening upward sequence of mud to sand that rests on the ravinement surface.

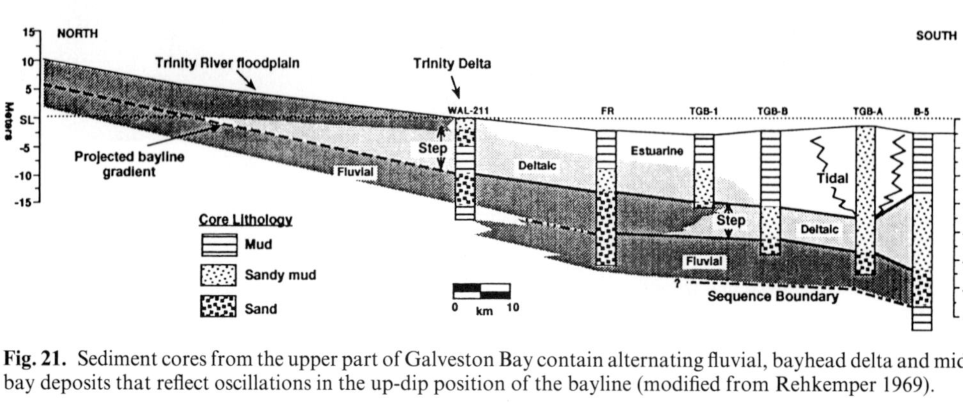

Fig. 21. Sediment cores from the upper part of Galveston Bay contain alternating fluvial, bayhead delta and middle bay deposits that reflect oscillations in the up-dip position of the bayline (modified from Rehkemper 1969).

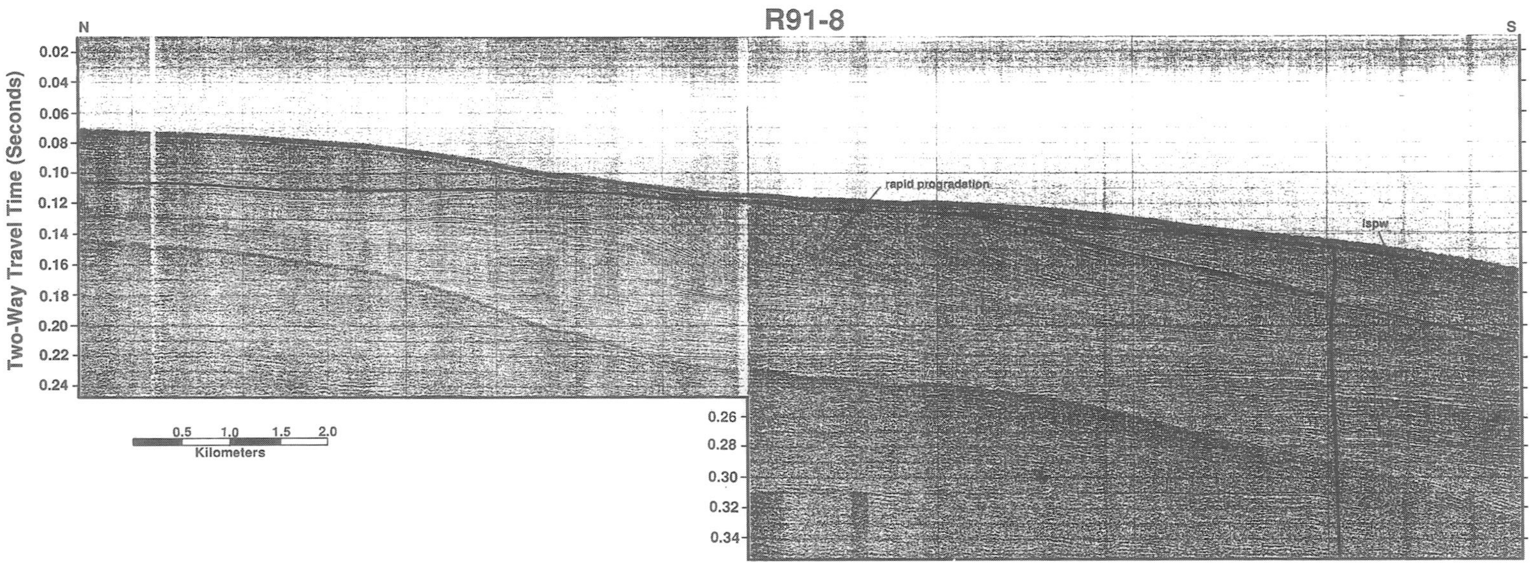

Fig. 22. Seismic line R91-8 crosses the distributary mouth bar and prodelta facies of the ancestral Brazos delta. Transgression resulted in a series of backstepping deltas and erosion of the sandy distributary facies.

The upper, sandy facies of the banks contains offshore marine molluscs that yield relatively young radiocarbon dates, which indicates that these sand bodies were formed in an inner shelf setting. The most likely source of the sand that forms the upper parts of these banks is tidal inlet and tidal delta deposits that were reworked within and along the flanks of the valley.

Relatively high-sediment-supply systems

During transgression, the upper portions of the Brazos and Colorado shelf-margin deltas were eroded and reworked landward. The rise in sea level also forced sedimentation landward and created backstepping deltas (Fig. 22).

The backstepping deltas of the ancestral Brazos and Colorado rivers are associated with fluvial valleys that are incised at progressively shallower depths in an up-dip direction; younger valleys eroded to progressively shallower depths as sea level rose (Abdulah & Anderson 1991). Depths of incision on the inner shelf range from −40 m (Stage 2 valleys) to approximately −12 m (modern valleys). Each valley was back-filled with sediment and abandoned as sea level rose. Seismic facies and platform borings indicate that the valleys are filled primarily with fluvial sands and gravels. Figure 23 shows a seismic profile across one of the Brazos valleys. This profile shows large-scale cross-stratification, which indicates a sand and gravel fill. Note that the valley was eventually abandoned to create the acoustically laminated channel-fill facies. Platform borings from the Brazos and Colorado valleys penetrated mostly sands and gravels.

Seismic profiles show that chaotic seismic facies within fluvial channels and distributary mouth bars are typically truncated by the ravinement surface (Fig. 24). Thus, shoreface erosion removed considerable sand from these valleys and deltas during transgression. Seismic records also provide evidence that the sand eroded from fluvial channels was transported up-dip and to the west (prevailing longshore transport direction) and deposited as shelf sand bodies (Fig. 25). This mechanism of shelf sand body formation is different from the mechanism that produced the sand banks along the Trinity/Sabine incised valley. The size, shape, and facies character of these features is the subject of an ongoing investigation. The existing data suggest that the sand bodies are more extensive, thicker, sandier, and have more rounded shapes than the sand banks that are associated with the Trinity/Sabine valley.

Present highstand (4000 years BP to present)

The depositional systems associated with the present highstand, which began approximately 4000 years BP, are shown in Fig. 8. This period is characterized by a slow rise in sea level that permitted the development of an extensive chain of barrier islands, peninsulas and chenier plains. The higher sediment supply of the Brazos and Colorado rivers relative to the Sabine, Trinity, and Lavaca rivers continues to exert a strong influence on the style of sedimentation in the fluvial valleys. Bays occupy the upper reaches of the Trinity and Sabine valleys (Galveston Bay and Sabine Lake respectively), but the Brazos and Colorado rivers, with their larger sediment supplies, have begun to construct prominent deltaic headlands into the Gulf of Mexico. Perhaps these rivers have reached the point where they will again begin to construct fluvial deltas across the continental shelf.

Sequence stratigraphy

In this paper we examine sedimentation on the continental shelf during one cycle of glacial eustatic sea level, Stage 5e to present. The overall fall in sea level between Stage 5e and Stage 2 occurred over a 100 000 year time span and was punctuated by several fifth-order rises (Fig. 3). These fifth-order eustatic events profoundly influenced shelf sedimentation in the region. Fluvial systems with varying sediment supplies were affected in different ways.

The maximum flooding surface associated with the Stage 5e highstand is a prominent downlap surface that has been mapped throughout the study area. Stage 5 highstand deposits are thin in the eastern portion of the study area. In the Trinity/Sabine area, the Stage 5b and 5d surfaces amalgamate with the Stage 2 sequence boundary on the interfluve areas of the shelf; it is unknown whether or not significant sedimentation occurred on the shelf during these brief highstands. Fifth-order sea-level fluctuations did result in repeated fluvial deposition and incision. Fluvial terraces within the Trinity/Sabine incised valley represent Stages 5d to 5c and 5b to 5a transgressive deposits (Fig. 9). Each reincision flushed the valley of sediments deposited during the previous transgression. Through this mechanism, the Trinity/Sabine rivers formed a significant point source for sediment delivered to the shelf-break and slope. Figure 26 summarizes the proposed distribution of facies of the Trinity/Sabine fluvial system during the last lowstand and subsequent transgression.

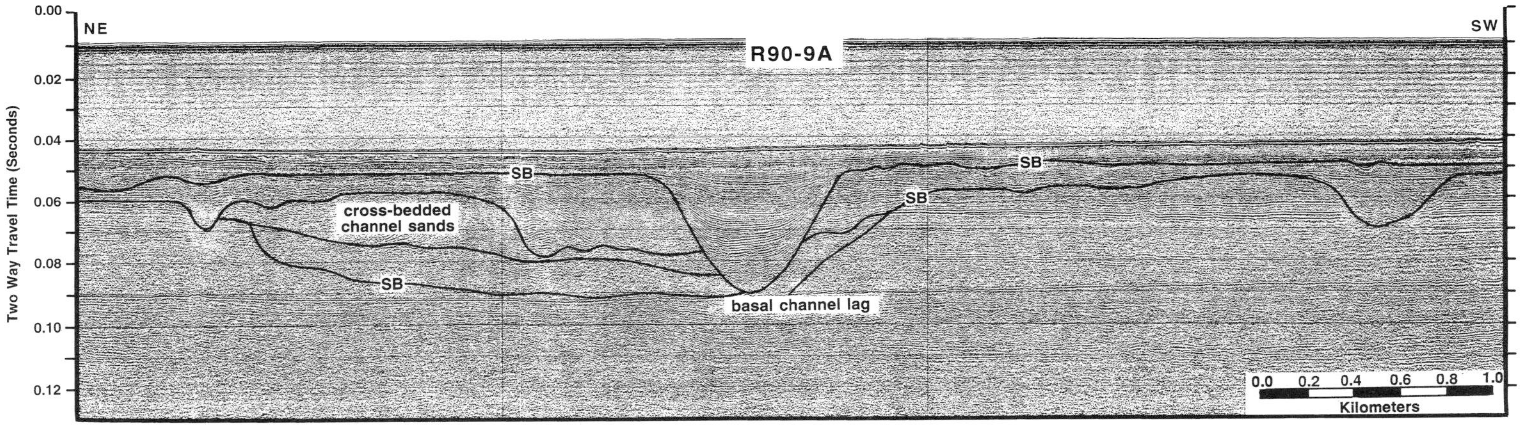

Fig. 23. Seismic line R90-9 crosses the ancestral Brazos incised valley, which is filled almost entirely with fluvial deposits. Crossbeds are seen in portions of this line. The final episode of sedimentation in the valley is marked by acoustically laminated deposits that fill the youngest channel.

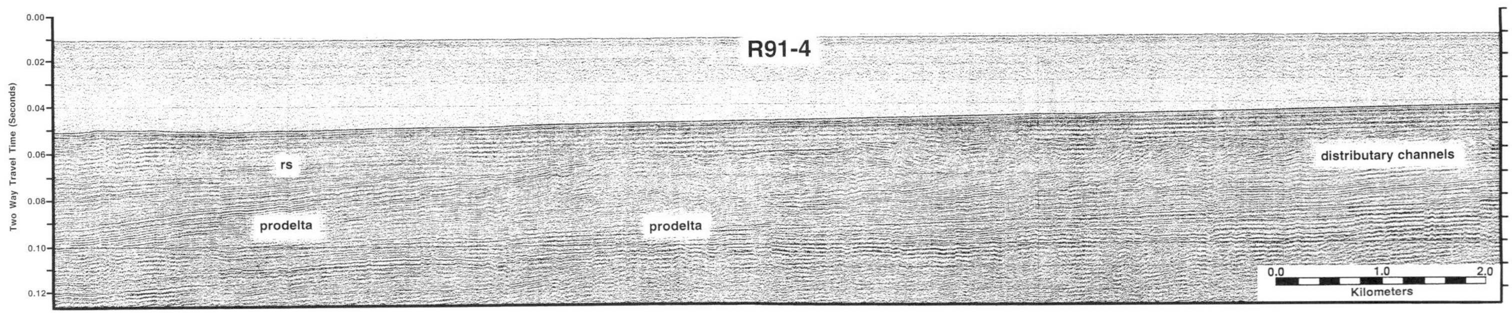

Fig. 24. Seismic line R91-4 shows the complex, chaotic seismic facies interpreted as sandy distributary channels and associated mouth-bar deposits of the Brazos delta. Note that the ravinement surface (rs) truncates these deposits in the up-dip portion of this line.

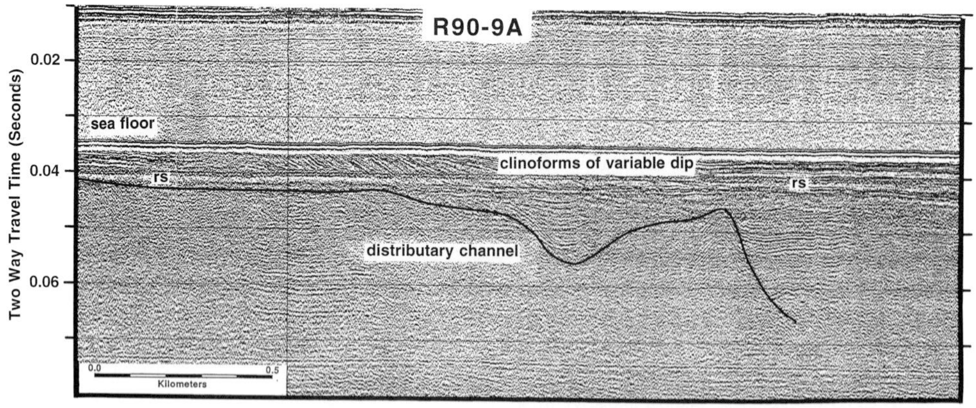

Fig. 25. Seismic line R90-9 illustrates the manner in which sands reworked from fluvial valleys of the ancestral Brazos and Colorado rivers are further reworked during transgression to produce widespread shelf sand bodies. Note that these sands occur above the ravinement surface (rs). The clinoforms indicate progradation to the west.

The thickness of Stage 5 highstand deposits increases dramatically from east to west, a manifestation of the higher sediment supplies of the Brazos and Colorado fluvial systems relative to the Trinity and Sabine rivers. During Stage 5 the Brazos River formed a large delta that advanced across the shelf (Fig. 4); its seaward growth was only slowed during the fifth-order transgressive cycles (Stages 5d to 5c and 5b to 5a). The Stage 5 highstand fluvial deltas of the ancestral Brazos delta reach 60 m in thickness and indicate that this river, which presently forms a wave-dominated delta, is capable of constructing large deltas during prolonged highstands.

During the time interval between 80 000 and 20 000 years BP, sedimentation was focused on the shelf margin (Fig. 5). Different fluvial systems responded differently to the overall fall that occurred during this time. The Trinity/Sabine fluvial system incised to its maximum depth and began to construct a small, sandy delta at the shelf margin. Much of the sediment delivered to the coast by these rivers bypassed the shelf and was transported down-slope through submarine canyons and into slope minibasins (Figs 15 and 26). Meanwhile, the western Louisiana, Brazos and Colorado rivers constructed large shelf-margin deltas. Development of these deltas differs considerably.

The first phase of progradation of the western Louisiana delta is marked by downlap onto the Stage 5e condensed interval. This delta advanced to the shelf margin and later was incised during the Stage 4 lowstand. This event was followed by another phase of progradation (radiocarbon dated as older than 48 000 years), and incision (radiocarbon dated as older than 33 000 years) (Fig. 5). During the maximum lowstand (Stage 2) and early transgression, the western Louisiana delta again prograded to the shelf edge (Fig. 6). This younger shelf-margin delta shows an overall configuration of progradation and aggradation that reflects both delta-lobe shifting and fifth-order eustatic fluctuations; it is impossible to deconvolve the eustatic signal from these deposits.

Figure 27 summarizes the phases of deposition for the high-sediment-supply Brazos/Colorado fluvial/deltaic systems. During highstands, shelf-phase deltas advanced across the shelf (Fig. 27a). During periods of rapid sea-level fall, the main distributary systems cut incised valleys on the inner shelf and carried sediment to the outer shelf to create shelf-margin deltas (Fig. 27b). These deltas aggraded and rapidly prograded in response to higher-order sea-level fluctuations. Such fluctuations probably occurred during $\delta^{18}O$ Stage 3. Another fall in sea level resulted in complete sediment bypassing of the shelf (Fig. 27c), as illustrated by the seismic transect between the ancient Colorado valleys and the East Breaks Slide. The subsequent relative sea-level rise resulted in a second phase of shelf-margin delta development associated with the lowstand prograding wedge (Fig. 27d). Rapid transgression initiated backstepping of shelf-phase deltas and delta-lobe shifting (Fig. 27e).

A regional strike line, collected along the shelf break and crossing the western Louisiana, Brazos and Colorado shelf-margin deltas, illustrates the different lowstand histories of the deltas (Fig. 28). This profile shows that the Colorado and western

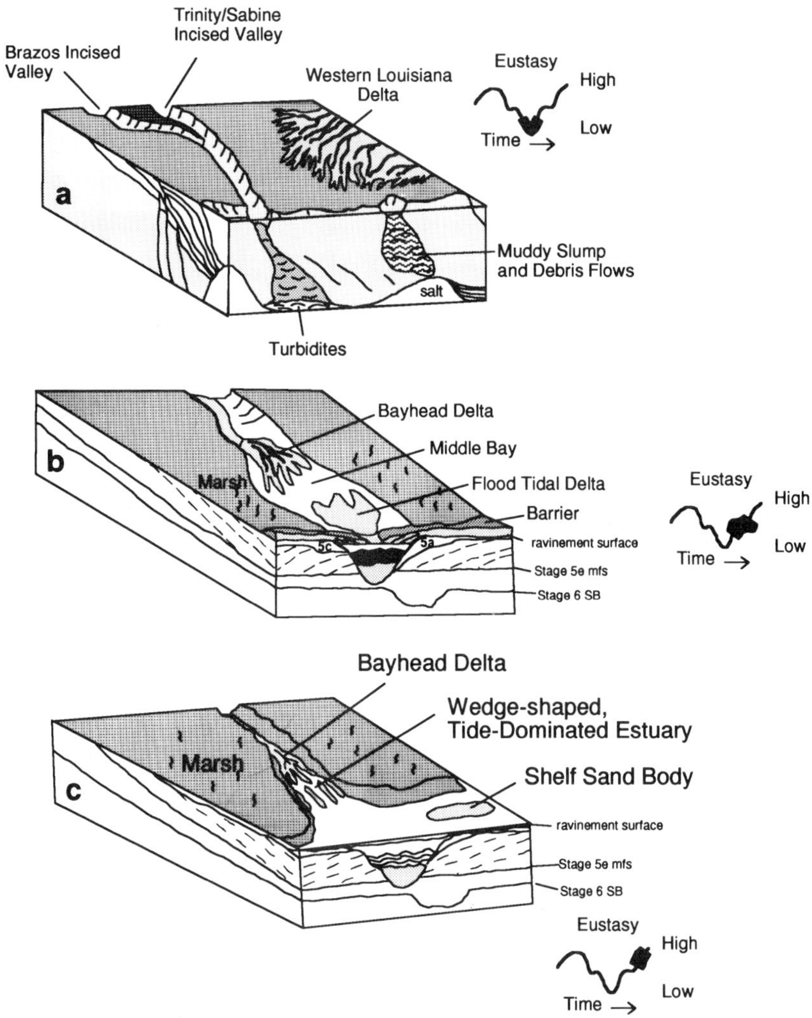

Fig. 26. Block diagrams summarizing facies development for the relatively low sediment supply Trinity/Sabine fluvial-deltaic system during the maximum lowstand and early transgression. See text for discussion.

Louisiana shelf-margin deltas onlap, and therefore are younger than, the Brazos delta. The Brazos delta was the first to prograde across the outer shelf, followed by the western Louisiana delta. The Colorado delta also prograded more slowly across the shelf than the Brazos delta, reaching the shelf break during the latter part of Stage 3. The Colorado incised its sandy delta-front deposits during the last lowstand.

Note that any single dip line across the shelf margin may image the superimposed clinoforms of two different deltas. For example, seismic line R92–24 (Fig. 12) crosses the outer shelf and upper slope at the western end of the western Louisiana delta where it onlaps the older Brazos shelf-margin delta. Two sets of progradational clinoforms can be seen in Line R92–24 (Fig. 12). The lower set, labelled 1 is associated with progradation of the Brazos delta during the eustatic fall from isotope Stage 5 to 4 (Abdulah & Anderson 1994). This lower sequence of clinoforms is bound above by the Stage 4 sequence boundary. The upper set of clinoforms, labelled 2 in Fig. 12, is part of the prograding Stage 3 Brazos delta. The Brazos delta is bound above by a relatively flat erosional surface, which represents the ravinement surface of the Stage 1 transgression. Above this surface,

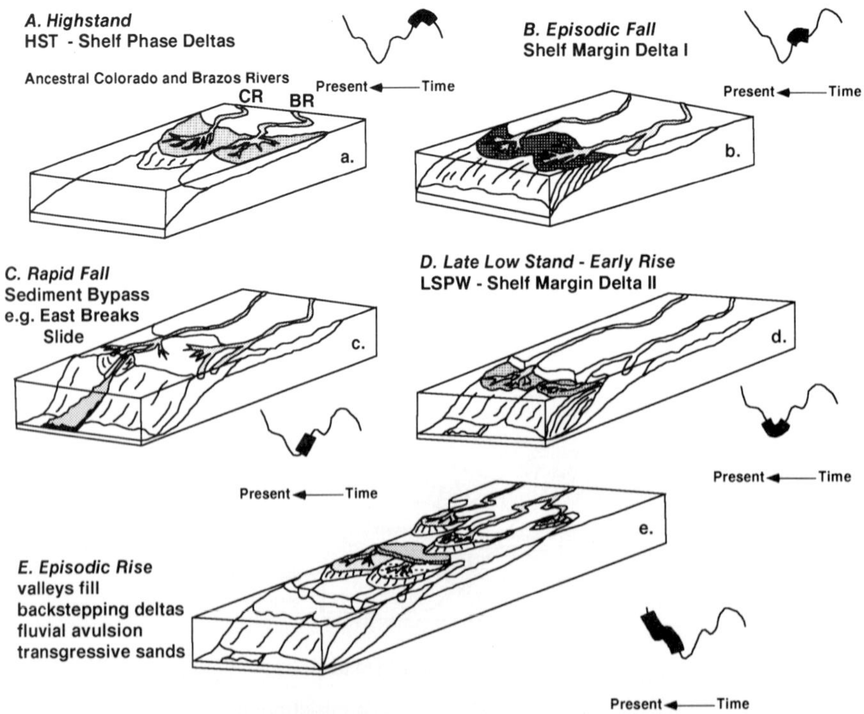

Fig. 27. Block diagrams summarizing the phases of deposition for the Brazos/Colorado fluvial-deltaic systems. See text for discussion.

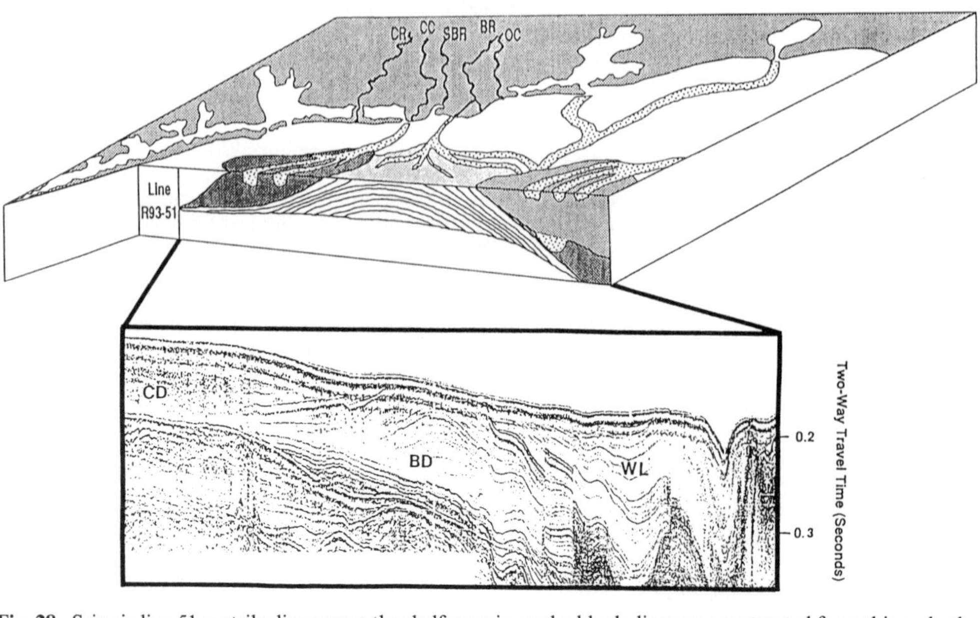

Fig. 28. Seismic line 51, a strike line across the shelf margin, and a block diagram constructed from this and other seismic lines, illustrates the different lowstand histories of the western Louisiana (WLS), Brazos (BD), and Colorado (CD) lowstand deltas.

parallel reflectors represent prodelta deposits from the western Louisiana delta, which eventually prograded into the line of the section and deposited the third and youngest set of prograding clinoforms, labelled 3 in Fig. 12. At the time of this progradation, deposition within the Brazos delta had shifted to the continental shelf, and deposition of the first of a series of backstepping deltas was occurring (Fig. 22). This mixed response of different deltas to sea-level rise is similar to that observed in the Adriatic Basin by Trincardi *et al.* (1994).

The key to understanding the transgressive systems tract of the east Texas continental shelf is recognition of the fact that the overall rise in sea level was episodic in nature (Anderson & Thomas 1991). This episodic nature to the sea-level rise resulted in the discrete nature of preserved coastal lithosomes within incised valleys (Fig. 7), backstepping parasequences within relatively low-sediment-supply fluvial valleys (Fig. 21), and a step-like nature of the Holocene ravinement surface. Had sea level risen in a continuous fashion, more sheet-like facies distribution patterns would be expected on the shelf. The strong influence of the episodic rise in sea level upon stratal patterns on the Texas shelf is similar to that observed on the Spanish continental shelves (Hernandez-Molina *et al.* 1994).

Blum (1990, 1994) has argued that changes in the rate of sediment delivery to the coastal plain are largely controlled by climatic change and not to sea-level change. His conclusion is based on work in the onshore portions of the Colorado fluvial valleys. We do not argue with this point, but our data do show that fifth-order sea-level fluctuations have had a profound influence on sedimentation on the continental shelf. Future studies should focus on trying to better understand the relative roles of eustasy and climate change in controlling sedimentation in the coastal plain to slope setting.

In general, shoreface ravinement has been extremely efficient in removing coastal and estuarine lithosomes from the shelf; these deposits are restricted almost entirely to incised valleys. The ravinement surface is characterized by steps, on the order of two to three metres of relief, that are interpreted to have formed during rapid, episodic rises in sea level (Siringan & Anderson 1991, 1993). The ravinement surface is also deeper within fluvial valleys, as a result of the presence of the more easily eroded material that fills the valleys. Transgressive lag deposits are rarely associated with the ravinement surface. Locally, a double ravinement surface bounds preserved ebb-tidal deltas.

Conclusions

1. Sequence stratigraphic models that predict facies distribution patterns are often oversimplified because glacial eustatic rise and fall is more episodic in nature than these models assume. Subtle changes in sea level, such as the fifth-order fluctuations of the last glacial eustatic cycle, significantly impact facies architecture and distribution patterns on the shelf.
2. Strong contrasts exist between relatively low-sediment-supply (Trinity/Sabine) and high-sediment-supply (Brazos and Colorado) fluvial systems and their associated systems tracts. The predominant difference is that sediment delivered to the shelf by the Trinity/Sabine fluvial system was, for the most part, deposited within the incised fluvial valleys. Reincision of the same valley results in sediment bypass of the shelf. In contrast, the Brazos and Colorado rivers constructed large deltas on the shelf that have shifted in response to the rise and fall in sea level. These rivers have also occupied a number of different fluvial valleys on the shelf. The net effect has been that most of the sediment delivered to the Gulf by the Brazos and Colorado rivers was sequestered on the shelf in fluvial valleys and deltas.
3. The different shelf-margin deltas of the region were active through the late lowstand well into the transgression. These deposits show a complex pattern of progradation and aggradation that varies throughout the area. This complexity is due predominantly to the different responses of systems with different sediment supplies to fifth-order eustatic fluctuations.
4. During lowstands, rivers with high sediment supply and high bedload to suspended load ratios developed extensive, sandy distributary systems that are roughly lobate in shape. Rivers with high sediment supply and high suspended load to bedload ratios developed strike-oriented, elongate shelf-margin deltas composed mainly of mud.
5. Within the eastern part of the study area, salt tectonics and growth faults played a key role in the location of feeder channels and depocentres, as well as increasing the accommodation space available for deposition. A feedback mechanism exists in which sediment loading causes salt migration which, in turn, influences sedimentation. Salt tectonics has had minimal influence on sedimentation in the western part of the study area.

6. A regional strike line across the continental shelf break shows superposition of deltas. Individual dip lines show different progradational histories. This explains why initial attempts to correlate stratal patterns on the shelf to the oxygen isotope curve were unsuccessful. Such analyses are prone to failure unless the locations of major depositional systems on the shelf are first mapped.
7. The transgressive valley fill of the relatively low-sediment-supply Trinity/Sabine incised valley consists of fluvial, bayhead delta, middle bay, tidal inlet/delta, and offshore marine facies. However, the complete succession of these facies rarely occurs. Instead, the valley is characterized by backstepping parasequences bounded by intermediate flooding surfaces. This style of sedimentation is attributed to the episodic nature of sea-level rise during the last transgression. The valley fill of the relatively high-sediment-supply Brazos and Colorado rivers is composed predominantly of fluvial deposits.
8. During transgression, shoreface ravinement of the Trinity/Sabine valleys produced mostly muddy interfluve deposits, with the exception of scattered sand banks that are reworked coastal lithosomes. Avulsion of the Brazos and Colorado rivers during the late Pleistocene–Holocene transgression resulted in the sequestering of a significant volume of sand and gravel on the continental shelf in fluvial channels. During transgression, the fluvial deposits within the upper portions of these channels were reworked into transgressive sand bodies.

We gratefully acknowledge the financial support of Agip, Amoco, Arco, BP, Conoco, Exxon, Marathon, PanCanadian, Shell, Unocal, and Union Pacific. Financial support for this project was also made available by the American Chemical Society – Petroleum Research Fund. We thank Fugro-McClelland and the industry participants who have provided the many platform boring descriptions, gamma logs, and core samples that have greatly enhanced this study. We also wish to thank Stephanie Shipp for her assistance with drafting the figures and editing the text. We also thank Harry Roberts and Richard Davis for their constructive comments on the original version of this paper.

References

ABDULAH, K. C. & ANDERSON, J. B. 1991. Eustatic controls on the evolution of the Pleistocene Brazos–Colorado deltas, Texas. Gulf Coast Section Society of Economic Paleontologists and Mineralogists, Twelfth Annual Research Conference *Coastal Depositional Systems in the Gulf of Mexico*, 1–7.

—— & —— 1994. Contrasting styles of deltaic deposition and the role of eustasy – examples from the Quaternary deltas of the Texas continental shelf. American Association of Petroleum Geologists Hedberg Conference *Applications of Sequence Stratigraphy to Oil Field Development*, Paris, France, 1–11.

ANDERSON, J. B. & THOMAS, M. A. 1991. Marine ice sheet decoupling as a mechanism for rapid, episodic sea-level change: the record of such events and their influence on sedimentation. *Sedimentary Geology*, **70**, 87–104.

——, ABDULAH, K., SIRINGAN, F. & SARZALEJO, S. 1994. Application of high-resolution seismic reflection data in assessing the size, shape, and lithology of near-surface sand bodies. *Proceedings of 26th Offshore Technology Conference*, Houston, Texas, 21–30.

——, SIRINGAN, F. P., SMYTH, W. C. & THOMAS, M. A. 1991*a*. Episodic nature of Holocene sea-level rise and the evolution of Galveston Bay. Gulf Coast Section Society of Economic Paleontologists and Mineralogists, Twelfth Annual Research Conference *Coastal Depositional Systems in the Gulf of Mexico*, 8–14.

——, ——, TAVIANI, M. & LAWRENCE, J. 1991*b*. Origin and evolution of Sabine Lake, Texas-Louisiana. *Gulf Coast Association of Geological Societies, Transactions*, **41**, 12–16.

ATEN, L. E. 1983. *Indians of the Upper Texas Coast*. Academic, New York.

BERNARD, H. A. 1950. *Quaternary Geology of Southeast Texas*. PhD Dissertation, Louisiana State University, Baton Rouge, Louisiana.

BERRYHILL, H. L., SUTER, J. R. & HARDIN, N. S. 1987. *Late Quaternary Facies and Structure, Northern Gulf of Mexico*. American Association of Petroleum Geologists, Studies in Geology, **23**.

BLUM, M. D. 1990. Climatic and eustatic controls on Gulf Coastal Plain fluvial sedimentation: an example from the late Quaternary of the Colorado River, Texas. *In*: ARMENTROUT, J. M. (ed.) *Sequence Stratigraphy as an Exploration Tool: Concepts and Practices in the Gulf Coast*. Gulf Coast Section Society of Economic Paleontologists and Mineralogists, Conference Proceedings, 71–83.

——1994. Genesis and architecture of incised valley fill sequences: a Late Quaternary example from the Colorado River, Gulf Coastal Plain of Texas. *In*: WEIMER, P. & POSAMENTIER, H. W. (eds) *Siliciclastic Sequence Stratigraphy: Recent Developments and Applications*. American Association of Petroleum Geologists, Memoir, **58**, 259–283.

COLEMAN, J. M. & ROBERTS, H. H. 1988. Sedimentary development of the Louisiana continental shelf related to sea-level cycles: Part I – sedimentary sequences. *Geo-Marine Letters*, **8**, 63–108.

FAIRBANKS, R. G. 1989. A 17,000 year glacio-eustatic sea-level record: influence of glacial melting rates on the Younger Dryas event and deep-ocean circulation. *Nature*, **342**, 637–642.

GRAF, C. 1966. *The Late Pleistocene Ingleside Barrier Trend, Texas and Louisiana.* MSc Thesis, Rice University, Houston, Texas.

HERNANDEZ-MOLINA, F. J., SOMOZA, L. & POMAR, R. L. 1994. Late Pleistocene–Holocene sediments on the Spanish continental shelves: Model for very high resolution sequence stratigraphy. *Marine Geology*, **120**, 129–174.

IMBRIE, J., HAYS, J. D., MARTINSON, D. G. *et al.* 1984. The orbital theory of Pleistocene climate: support for a revised chronology of the marine delta O18 record. *In*: BERGER, A. J., IMBRIE, J., HAYS, J., KUKLA, G. & SALTZMAN, B. (eds) *Milankovitch and Climate, Part 1.* Reidel, Dordrecht, 269–305.

LEHNER, P. 1969. Salt tectonics and Pleistocene stratigraphy on continental slope of Northern Gulf of Mexico. *AAPG Bulletin*, **53**, 2431–2479.

MCEWEN, M. C. 1969. Sedimentary facies of the modern Trinity Delta. *In*: LANKFORD, R. R. & ROGERS, J. J. W. (eds) *Holocene Geology of the Galveston Bay Area.* Houston Geological Society, Houston, Texas, 53–77.

MITCHUM, R. M. JR, VAIL, P. R. & SANGREE, J. B. 1977. Stratigraphic interpretations of seismic reflection patterns in depositional sequences. *In*: PAYTON, C. E. (ed.) *Seismic Stratigraphy – Application to Hydrocarbon Exploration.* American Association of Petroleum Geologists, Memoir, **26**, 117–134.

MORTON, R. A. & MCGOWEN, J. H. 1980. *Modern Depositional Environments of the Texas Coast.* Guidebook 20, The University of Texas at Austin, Bureau of Economic Geology.

NICHOL, S. L., BOYD, R. & PENLAND, S. 1994. Stratigraphic response of wave-dominated estuaries to different relative sea-level and sediment-supply histories: Quaternary case studies from Nova Scotia, Louisiana, and eastern Australia. *In*: DALRYMPLE, R., BOYD, R. & ZAITLIN, B. A. (eds) *Incised-Valley Systems: Origin and Sedimentary Sequences.* Society of Economic Paleontologists and Mineralogists, Special Publication, **51**, 265–284.

POSAMENTIER, H. W. & VAIL, P. R. 1988. Eustatic control on clastic deposition II-sequence and system tract models. *In*: WILGUS, C. K., HASTINGS, B. S., KENDALL, C. G. ST. C., POSAMENTIER, H. W., ROSS, C. A. & VAN WAGONER, J. C. (eds) *Sea-level Change – an Integrated Approach.* Society of Economic Paleontologists and Mineralogists, Special Publication, **42**, 125–154.

REHKEMPER, L. J. 1969. Sedimentology of Holocene estuarine deposits, Galveston Bay. *In*: LANKFORD, R. R. & ROGERS J. J. W. (eds) *Holocene Geology of the Galveston Bay Area.* Houston Geological Society, Houston, Texas, 12–52.

ROTHWELL, R. G., KENYON, N. H. & MCGREGOR, B. A. 1991. Sedimentary features of the south Texas continental slope as revealed by side-scan sonar and high-resolution seismic data. *AAPG Bulletin*, **75**, 298–312.

SANGREE, J. B., WAYLETT, D. C., FRAZIER, D. E., AMERY, G. B. & FENNESSY, W. J. 1978. Recognition of continental slope facies, offshore Texas–Louisiana. *In*: BOUMA, A. H., MOORE, G. T. & COLEMAN, J. M. (eds) *Framework, Facies and Oil-trapping Characteristics of the Upper Continental Margin.* American Association of Petroleum Geologists, Studies in Geology, **7**, 87–116.

SARG, J. F. 1988. Carbonate sequence stratigraphy. *In*: WILGUS, C. K., HASTINGS, B. S., KENDALL, C. G. ST. C., POSAMENTIER, H. W., ROSS, C. A. & VAN WAGONER, J. C. (eds) *Sea-level Change – an Integrated Approach.* Society of Economic Paleontologists and Mineralogists, Special Publication, **42**, 155–181.

SARZALEJO, S., ANDERSON, J. B., COTERILL, K. & VAIL, P. R. 1993. *High Frequency Cyclicity in East Texas Outer Shelf and Upper Slope Sequences Revealed by High Resolution Seismic Data and Platform Borings.* American Association of Petroleum Geologists Annual Meeting, New Orleans, Louisiana.

SATTERFIELD, W. M. & BEHRENS, E. W. 1990. A late Quaternary canyon/channel system, Northwest Gulf of Mexico Continental Shelf. *Marine Geology*, **92**, 51–67.

SIRINGAN, F. P. & ANDERSON, J. B. 1991. Facies architecture and evolution of coastal lithosomes on the North Texas Gulf Coast and the occurrence of preserved analogues on the North Texas inner continental shelf. Gulf Coast Section Society of Economic Paleontologists and Mineralogists, Twelfth Annual Research Conference *Coastal Depositional Systems in the Gulf of Mexico*, 247.

—— & —— 1993. Seismic facies architecture and evolution of the Bolivar Roads tidal inlet/delta complex and related deposits, East Texas Gulf Coast. *Journal of Sedimentary Petrology*, **63**, 794–808.

—— & —— 1994. Modern shoreface and inner-shelf storm deposits off the east Texas coast, Gulf of Mexico. *Journal of Sedimentary Research*, **B64**, 99–110.

SUTER, J. R. & BERRYHILL, H. L. JR 1985. Late Quaternary shelf-margin deltas, Northwest Gulf of Mexico. *AAPG Bulletin*, **69**, 77–91.

SWANSON, R. L. & THURLOW, C. L. 1973. Recent subsidence rates along the Texas and Louisiana coasts as determined from tide measurements. *Journal of Geophysical Research*, **78**, 2650–2671.

SYDOW, J. & ROBERTS, H. H. 1994. Stratigraphic framework of a late Pleistocene shelf-edge delta, northeast Gulf of Mexico. *AAPG Bulletin*, **78**, 1276–1312.

THOMAS, M. A. & ANDERSON, J. B. 1989. Glacial eustatic controls on seismic sequences and parasequences of the Trinity/Sabine incised valley, Texas Continental Shelf. *Gulf Coast Association of Geological Societies, Transactions*, **39**, 563–570.

—— & —— 1991. Late Pleistocene sequence stratigraphy of the Texas continental shelf: relationship to $\delta^{18}O$ curves. Gulf Coast Section Society of Economic Paleontologists and Mineralogists, Twelfth Annual Research Conference *Coastal Depositional Systems in the Gulf of Mexico*, 265–270.

—— & —— 1994. Incised valley systems: origin and sedimentary sequences. *In*: DALRYMPLE, R., BOYD, R. & ZAITLIN, B. A. (eds) *Incised-Valley Systems: Origin and Sedimentary Sequences.* Society of Economic Paleontologists and Mineralogists, Special Publication, **51**, 63–82.

TRINCARDI, F., CORREGGIARI, A. & ROVERI, M. 1994. Late Quaternary transgressive erosion and deposition in a modern epicontinental shelf: the Adriatic semi-enclosed basin. *Geo-Marine Letters*, **14**, 41–51.

VAIL, P. R., MITCHUM, R. M., JR, TODD, R. G., *et al.* 1977. Seismic stratigraphy and global changes of sea-level. *In*: PAYTON, C. E. (ed.) *Seismic Stratigraphy – Application to Hydrocarbon Exploration.* American Association of Petroleum Geologists, Memoir, **26**, 49–212.

VAN WAGONER, J. C., MITCHUM, R. M., CAMPION, K. M. & RAHMANIAN, V. D. 1990. *Siliciclastic Sequence Stratigraphy in Well Logs, Cores and Outcrops: Concepts for High-resolution Correlation of Time and Facies.* American Association of Petroleum Geologists, Methods in Exploration, **7**.

WOODBURY, H. O., SPOTTS, J. H. & AKERS, W. H. 1978. Gulf of Mexico continental-slope sediments and sedimentation. *In*: BOUMA, A. H., MOORE, G. T. & COLEMAN, J. M. (eds) *Framework, Facies and Oil-trapping Characteristics of the Upper Continental Margin.* American Association of Petroleum Geologists, Studies in Geology, **7**, 117–137.

Quaternary siliciclastic sequence stratigraphy of western Mediterranean passive and tectonically active margins: the role of global versus local controlling factors

G. ERCILLA & B. ALONSO

CSIC, Instituto Ciencias del Mar, Paseo Juan de Borbón s/n, E-08039 Barcelona, Spain

Abstract: A study of the Quaternary sequence stratigraphy of two siliciclastic Spanish Mediterranean margins with a different geological setting was carried out. This study reveals that the Quaternary stratigraphical architectures of the northern Catalonia passive continental margin and of the northwestern Alboran tectonically active margin are mainly characterized by the stacking of high-frequency lowstand systems tracts. Only the Versilian transgressive and Holocene highstand systems tracts have been recognized. This kind of stratigraphical architecture is the product of the eustatic cyclicity characterized by sea-level changes of high frequency, high amplitude and marked asymmetry, regardless of local geological context.

On the other hand, local factors seem to play an important role in the stratal pattern, its preservation and sedimentary distribution. Thus, tectonics was responsible for strong lateral variations in margin growth pattern, changing from progradational to divergent in the tectonically active margin of the northwestern Alboran. Tectonics also conditioned the preservation of the lowstand systems tracts along and through the margin: the lowstand systems tracts are preserved from shelf in subsiding areas, e.g. northern Catalonia and the central sector of the northwestern Alboran, and from slope in uplifting areas, e.g. eastern and western sectors of the northwestern Alboran. Physiography conditioned the across-margin location of sedimentary depocentres: on the outer shelf and upper slope, when the shelf is relatively wide as occurs in northern Catalonia, and on the base-of-slope when the shelf is relatively narrow as occurs in the northwestern Alboran. Variations in the volume of sediment supply conditioned the along-margin location of sedimentary depocentres, which changed from south to north in northern Catalonia.

The sequence stratigraphy concepts were based on stratal patterns observed in low-resolution seismic records of passive continental margins (Vail *et al.* 1977; Jervey 1988; Posamentier & Vail 1988; Posamentier *et al.* 1988). Attempts are now being made to recognize eustatic cycles in tectonically active areas, and to discern the effects of eustatic sea-level changes from those generated by the tectonism in these active areas (Ramsay 1991; Pirrie *et al.* 1991; Posamentier & Allen 1993). Although tectonism may obliterate the eustatic signal during the most active periods, these attempts show that products of sea-level changes can be identified in many tectonically active basins.

In this paper, we report on an analysis of the role of global versus local controlling factors of two different types of margins in the Spanish Mediterranean (western Mediterranean Sea) (Fig. 1). These margins are the Quaternary passive continental margin in the northern Catalonia Sea (Fig. 1A), and the post-Calabrian tectonically active margin in the northwestern Alboran Sea (Fig. 1B). Although the ages of the strata in the two areas are not exactly coincident, for simplicity we shall refer to them as being of Quaternary age, taking into account that the age of strata on the northwestern Alboran margin dates from the post-Calabrian (0.7 Ma).

The analysis of the two margins is carried out to identify and distinguish between the effects of Quaternary eustatic sea-level cycles and local factors, e.g. tectonism, physiography and sediment supply, in the formation and evolution of the stratigraphical architecture and growth pattern of the two margins. The sequence stratigraphy and growth pattern of these margins were defined by Ercilla *et al.* (1994*a*, *b*, 1995), based on the analysis of discontinuity surfaces, seismic facies, and recent sedimentary facies.

Regional setting

The northern Catalonia and northwestern Alboran margins are siliciclastic margins located in different geological settings of the Spanish Mediterranean Sea (western Mediterranean) (Fig. 1). The two margins, dominated by low-energy waves with weak tidal currents (Flos 1985), have a different tectonic setting and physiographical configuration, but a similar sediment supply from mainly relatively small rivers with drainage basins of several hundreds of square

From De Batist, M. & Jacobs, P. (eds), 1996, *Geology of Siliciclastic Shelf Seas*, Geological Society Special Publication No. 117, pp. 125–137.

kilometres, and a seasonal variable water discharge of only a few cubic metres per second (Institut Cartogràfic de Catalunya 1983).

The northern Catalonia passive continental margin has developed within the southern Mesozoic–Cenozoic Alpine megasuture system in the northeastern Spanish Mediterranean (Duran-Delga & Fontboté 1980). The structural evolution of this continental margin has been mainly controlled by the Pyrenean fault complex

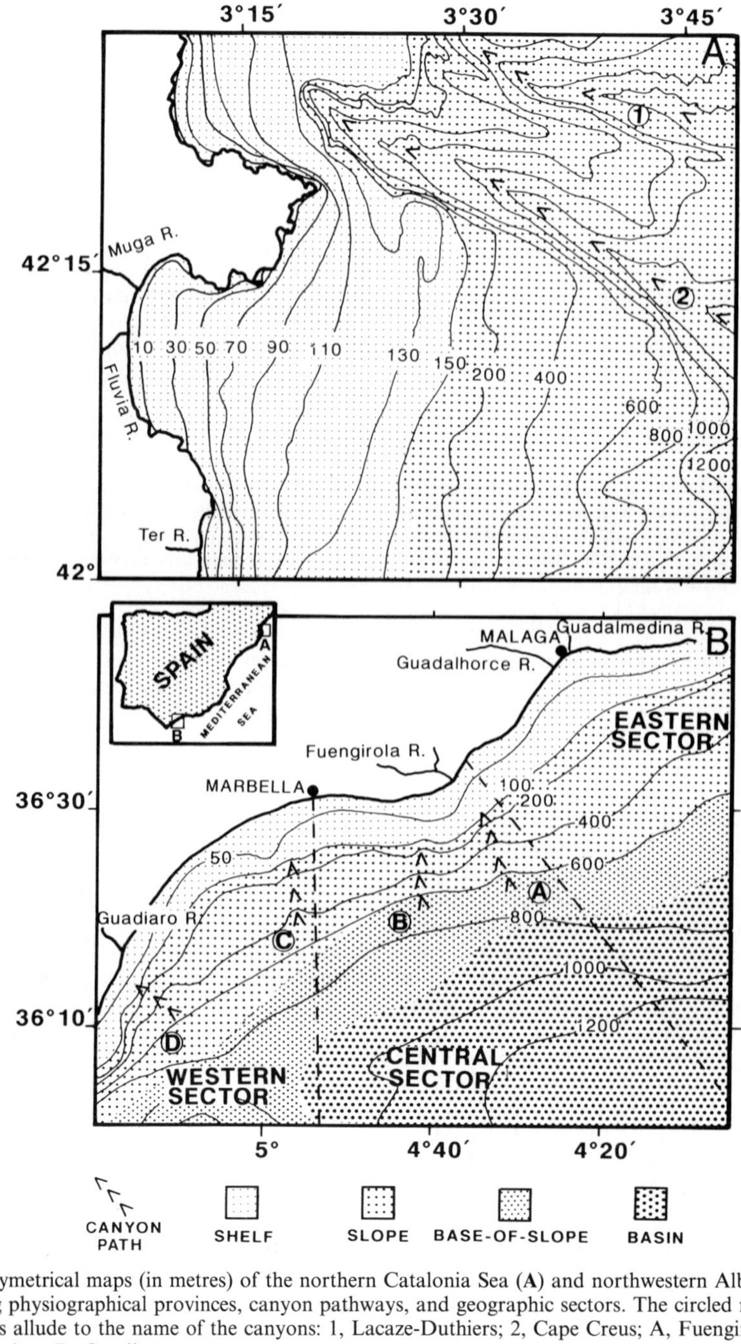

Fig. 1. Bathymetrical maps (in metres) of the northern Catalonia Sea (**A**) and northwestern Alboran Sea (**B**), showing physiographical provinces, canyon pathways, and geographic sectors. The circled numbers and circled letters allude to the name of the canyons: 1, Lacaze-Duthiers; 2, Cape Creus; A, Fuengirola; B, Torre Nueva; C, Baños; D, Guadiaro.

Table 1. *Geometric parameters of the northern Catalonia continental margin*

	Depth (m)	Width (km)	Gradient (°)
Shelf	15–170	6–30	0.10–1.30
Slope	170–1600	23–46	1.30–5.35

from the Oligocene to Quaternary. This complex has caused a system of horsts and grabens that form the present structural configuration of the northern Catalonia continental margin (Got 1973; Stanley *et al.* 1976; Fleta & Escuer 1991; Vázquez & Medialdea, 1992; Medialdea *et al.* 1995). The Quaternary sedimentary evolution has been mainly controlled by sea-level variations, whereas tectonics has been irrelevant (Ercilla *et al.* 1994*b*; Medialdea *et al.* 1995). Table 1 shows the geometric parameters (depth, gradient and width) that define the present physiographical configuration of this continental margin.

The tectonically active margin of the northwestern Alboran Sea is a geologically complex area situated on the inner side of the arc-shaped Betic–Rifian alpine orogenic belt (Banda & Ansorge 1980). The geodynamical evolution of the Alboran Sea is related to the relative motion between Eurasia and Africa (Dewey *et al.* 1989). The Alboran Sea region has been under continuous north–south compression at least since the Tortonian, and probably much earlier (Ott d'Estevou & Montenat 1985; Philip 1987; Sanz de Galdeano 1990; Roest & Srivastava 1991). PostTortonian tectonics modified the architecture of the Miocene basins and margins, formed the structural boundaries of the Alboran Sea, and conditioned its diapiric and volcanic activity (Comas *et al.* 1992). The Quaternary sedimentary evolution has been mainly controlled by the interplay of tectonics and sea-level variations (Alonso & Maldonado 1992; Ercilla *et al.* 1992). To be precise, in the northwestern Alboran margin, the main controlling factors of tectonic origin have been tilting, whose importance has been variable along the margin, and diapirism (Ercilla *et al.* 1992). Table 2 shows the geometric parameters (depth, gradient and width) that define the present physiographical configuration of the Alboran margin. The complexity of the physiographical configuration of this margin leads us to divide this area into three geographic sectors: eastern (Malaga to Fuengirola river mouth), central (Fuengirola river mouth to Marbella) and western (Marbella to Guadiaro river mouth) (Ercilla *et al.* 1992) (Fig. 1B).

Methodology

The analysed data consist of 1400 km of single-channel high-resolution seismic profiles (3.5 kHz, GeoPulse and 15 in^3 sleeve gun) collected on board the 'B/O García del Cid'. Navigation was Maxiram radio-positioning system, satellite (Transit and GPS), and Loran C (Ercilla *et al.* 1994*a*, *b*, 1995). Likewise, a total of 326 cores up to 2.25 m in length were studied. These cores were X-radiographed, and textural and compositional analyses were performed (Ercilla *et al.* 1994*a*, 1995).

Sequence stratigraphy model principles (Posamentier & Vail 1988; Posamentier *et al.* 1988) were used to study the high-resolution stratigraphical architecture of the northern Catalonia and northwestern Alboran margins, and thus to relate the sediments that make up the two margins to the different stages of the Quaternary sea-level cycles. The term 'lowstand systems tract' is here used to define those sediments deposited during the lowering of sea level and lowstand stages, that is to say those sediments deposited after the first drop of sea level due to eustatic changes. The term 'transgressive systems tract' refers to those sediments deposited during the rises of sea level, and the term 'highstand systems tract' to those sediments

Table 2. *Geometric parameters of the northwestern Alboran margin*

	Eastern sector			Central sector			Western sector		
	Depth (m)	Width (km)	Gradient (°)	Depth (m)	Width (km)	Gradient (°)	Depth (m)	Width (km)	Gradient (°)
Shelf	to 105	11	0.07	to 115	11	0.08	to 95	3.5	0.06
Slope	105–600	21	0.55	115–600	10	2.65	95–800	17	0.60
Base-of-slope	600–900	8.5	0.28	600–945	16	0.40	800–945	13	0.31

deposited during the highest position of a sea-level curve. These terms have also been used by other authors in the Mediterranean Sea (e.g. Farrán & Maldonado 1990; Tesson *et al.* 1990; Chiocci 1994; Hernández-Molina *et al.* 1994).

There are not datings, and only the major chronostratigraphical seismic boundaries have been established. In the northern Catalonia continental margin, the lower and upper Quaternary seismic boundaries have been defined by correlation with the boundaries of the Quaternary seismic sequences defined by Got (1973) in the northwestern Mediterranean. In the northwestern Alboran margin, the Calabrian lower boundary has been defined by correlation with the seismic boundary established by Campillo *et al.* (1992) for the Alboran Sea.

Quaternary sequence stratigraphy and growth pattern

The Quaternary sequence stratigraphy of the northern Catalonia continental margin and that of the northwestern Alboran margin are characterized by the stacking of high-frequency depositional sequences formed by incomplete series of systems tracts, except for the most recent depositional sequence, where a complete set of systems tracts has been identified. In the

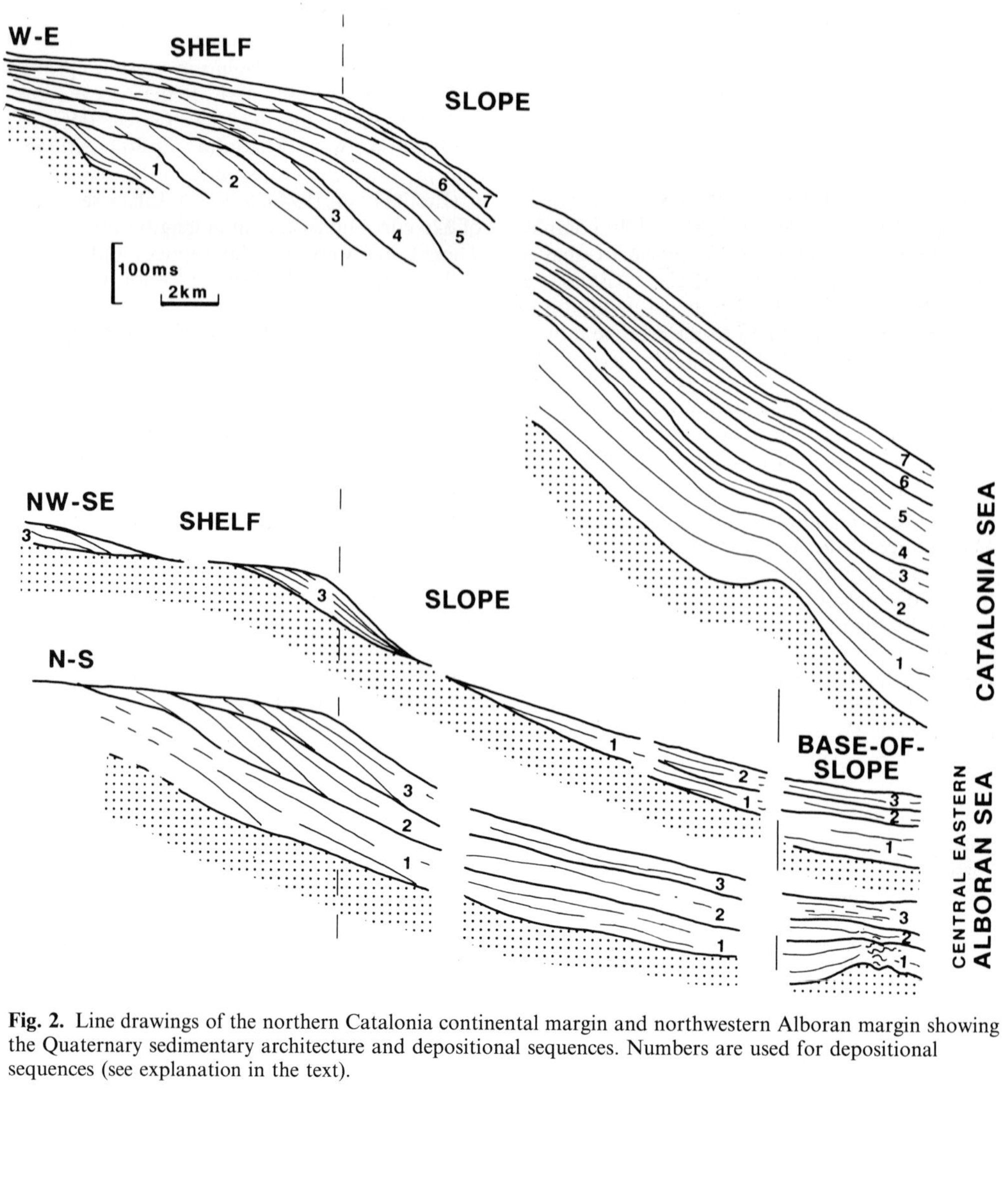

Fig. 2. Line drawings of the northern Catalonia continental margin and northwestern Alboran margin showing the Quaternary sedimentary architecture and depositional sequences. Numbers are used for depositional sequences (see explanation in the text).

following sections we sum up the main features that define the Quaternary sequence stratigraphy and growth patterns of both areas. More details of the two margins can be found in the papers by Ercilla *et al.* (1994*a*, *b*, 1995).

Northern Catalonia continental margin

The Quaternary sequence stratigraphy of the northern Catalonia continental margin is defined by the stacking of seven depositional sequences, called 1 to 7 from older to younger (Ercilla *et al.* 1994*b*) (Fig. 2). By correlation with the boundaries of the Quaternary seismic sequences defined by Got (1973) in the northwestern Mediterranean, depositional sequences 1 and 2 are lower Quaternary in age, and depositional sequences 3 to 7 are upper Quaternary in age. Depositional sequences 1 to 6 are only composed of lowstand systems tracts, whereas the most recent depositional sequence (number 7) includes lowstand, transgressive and highstand systems tracts. The lowstand systems tracts comprise seaward-downlapping deposits (in the sense of Trincardi & Field 1991) on the continental shelf, and parallel deposits on the continental slope (Ercilla *et al.* 1994*b*) (Figs 3A and 4A). The transgressive systems tract has only been sedimentologically recognized for the Versilian transgression. It is represented by relict and palimpsest facies composed of reworked gravels, sands and sandy muds (Fig. 5) deposited with a generalized distribution over the erosional transgressive surface on the shelf (Ercilla *et al.* 1995). The highstand systems tract is characterised by a depositional mound and the Fluviá-Muga and Ter prodeltas, all of them deposited during the Holocene and restricted to the inner-middle continental shelf with a well-defined distribution (Ercilla *et al.* 1995) (Figs 6A and 6B).

This type of stratigraphical architecture suggests that the northern Catalonia continental margin is characterized by the stacking of lowstand systems tracts bounded by erosive surfaces on the continental shelf, and correlative conformity surfaces on the slope (Fig. 2). The erosive surfaces developed due to subaerial erosion during the falls and lowstands, and shoreface erosion during the rises of sea level.

The growth pattern of this continental margin does not show significant sedimentary changes through the Quaternary (Fig. 2). During this time, when depositional sequences 1 to 7 were formed, a vertical stacking of progradational deposits, represented by seaward-downlapping deposits, developed from the middle continental shelf to the upper continental slope. These strata extend seaward into slope-parallel deposits (Fig. 2). This vertical stacking results in the upward and seaward building of the continental margin, with a prevalence of progradational stacking (Figs 2 and 3A). The spatial distribution of the strata for each depositional sequence shows that their sedimentary depocentre shifted in location during the lower and upper Quaternary (Fig. 7A). During the lower Quaternary, the sedimentary depocentres of depositional sequences 1 to 2 developed on the outer continental shelf, as a result of the filling of the La Escala palaeocanyon (Fig. 7A) (Ercilla *et al.* 1994*b*). However, during the upper Quaternary the location of the sedimentary depocentres of depositional sequences 3 to 7 shifted northward to shelf-break and upper continental slope off uplifting coastal areas (Fig. 7B). This location suggests a relatively high volume of sediment supplied by coastal erosion and by ephemeral, seasonal streams in these coastal areas (Medialdea *et al.* 1995).

Northwestern Alboran margin

The Quaternary sequence stratigraphy of the northwestern Alboran margin is defined by the stacking of three depositional sequences, called 1 to 3 from older to younger (Ercilla *et al.* 1994a) (Fig. 2). The two oldest depositional sequences, 1 and 2, are only composed of lowstand systems tracts, whereas depositional sequence 3 also includes transgressive and highstand systems tracts. The lowstand systems tracts comprise shelf-margin deltas on the outer shelf, and parallel deposits, divergent deposits, and the Guadiaro channel-levee complex on the slope and base-of-slope (Figs 3B, 3C, 4B, 4C and 4D). The transgressive systems tract has only been sedimentologically documented for the Versilian transgression. It is composed of reworked gravels and sands spread on the shelf and exposed on the outer shelf as relict facies (Fig. 5B). The highstand systems tract, Holocene in age, is defined by the Guadalmedina–Guadalhorce prodelta confined on the inner-middle shelf, and by hemipelagic deposits on the slope and base-of-slope (Ercilla *et al.* 1992, 1994*a*) (Figs 6C and 6D).

This type of stratigraphical architecture suggests that the northwestern Alboran margin is characterized by the stacking of lowstand systems tracts, bounded by unconformity surfaces on the outer shelf and correlative conformity surfaces on the slope and base-of-slope (Fig. 2). The unconformity surface represents

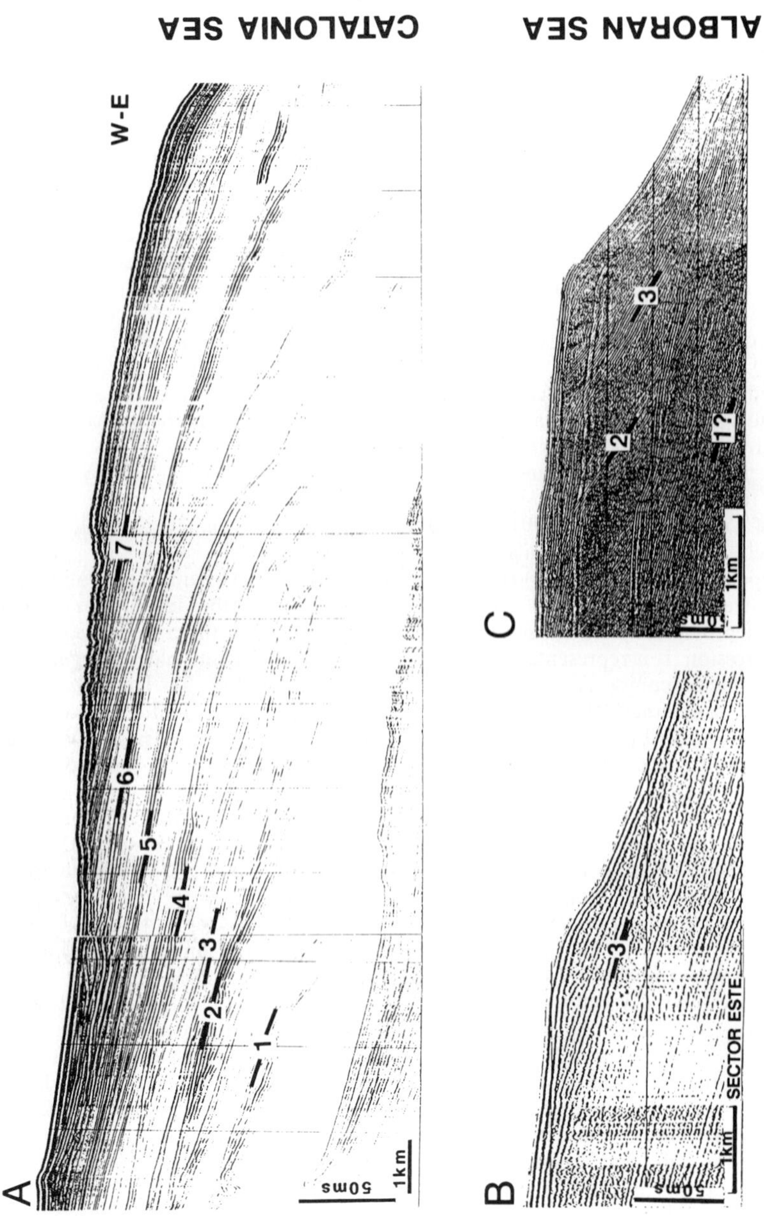

Fig. 3. Lowstand systems tracts: (**A**) seaward-downlapping deposits in the northern Catalonia continental margin; (**B**) shelf-margin deltas in the eastern and western sectors of the northwestern Alboran margin; (**C**) shelf-margin deltas in the central sector of the northwestern Alboran margin. Numbers represent boundaries of depositional sequences with the same number.

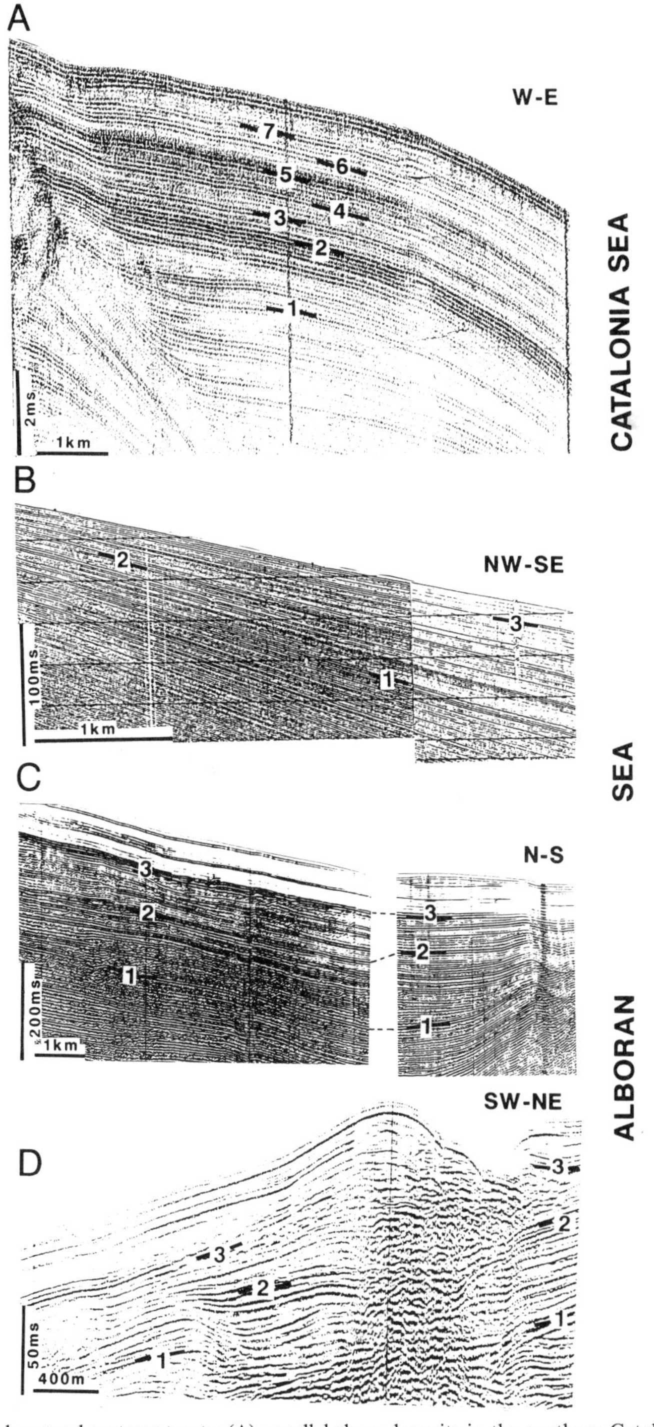

Fig. 4. Quaternary lowstand systems tracts: (**A**) parallel slope deposits in the northern Catalonia continental margin; (**B**) divergent slope deposits in the eastern and western sectors of the northwestern Alboran margin; (**C**) parallel slope deposits in the central sector of the northwestern Alboran margin; (D) the Guadiaro channel–levee complex in the western sector of the northwestern Alboran margin. Numbers represent boundaries of depositional sequences with the same number.

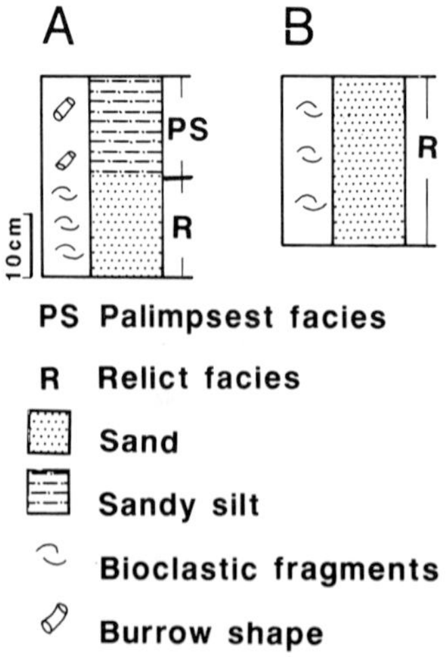

Fig. 5. Versilian transgressive systems tract: (**A**) relict and palimpsest shelf facies in the northern Catalonia continental margin; (**B**) relict shelf facies in the northwestern Alboran margin.

the coincidence between the lowstand erosional surface and the following erosive transgressive surface.

The Quaternary growth pattern displayed by this margin shows noticeable changes from a spatial and temporal perspective. When depositional sequences 1 and 2 developed, progradational shelf-margin delta deposits prograded to the shelf-break and upper slope of the central sector (Figs 2 and 3C). There, progradational strata extend seaward into parallel slope deposits (Figs 2 and 4C). In contrast, depositional sequences 1 and 2 do not occur on the shelf and slope (Figs 2 and 3B), and divergent slope deposits are present in the eastern and western sectors (Figs 2 and 4B).

Major changes in the sedimentary pattern occurred when depositional sequence 3 was formed. These were related to the development of progradational deposits across the entire margin. The progradational deposits are represented by the Guadalmedina–Guadalhorce prodelta and by shelf-margin deltas. The Guadalmedina–Guadalhorce prodelta was constructed on the inner-middle shelf of the eastern sector, and constitutes the only record of Quaternary deposits in this physiographical province (Fig. 6C). The shelf-margin deltas occur on the shelf-break and upper slope of the central sector, and on the shelf-break of the eastern and western sectors (Figs 3B and 3C). In these two geographic sectors, shelf-margin deltas do not have lateral continuity with slope deposits, and represent the only Quaternary record of the shelf-break (Figs 2 and 3B).

The spatial distribution of Quaternary sediments shows a major sedimentary depocentre located at the base-of-slope of the central sector during the development of each depositional sequence (Fig. 7C). The development of this depocentre is related to the presence of morphostructural barriers represented by diapirs (Ercilla *et al.* 1994*a*).

Discussion and conclusions

Comparison of western Mediterranean passive and tectonically active margins: sequence stratigraphy and growth pattern

The Quaternary sequence stratigraphy and growth pattern of the northern Catalonia continental margin and that of the northwestern Alboran margin show several similarities and differences. The major similarity is in the type of systems tract that make up the high-frequency depositional sequences. In both areas, depositional sequences are composed of lowstand systems tracts (Figs 3 and 4), except for the most recent depositional sequence, which also contains transgressive (Fig. 5) and highstand systems tracts (Fig. 6).

The major differences between the two margins refer to the margin stratal patterns and their preservation, and in sedimentary depocentre location. A unique progradational pattern preserved from middle continental shelf is identified in the northern Catalonia continental margin, contrasting with the variable growth pattern and preservation shown by the northwestern Alboran margin (Fig. 2). Here, a progradational pattern conserved from the outer shelf and a divergent pattern preserved from the upper slope are both identified in the northwestern Alboran margin (Fig. 2). The sedimentary depocentres are located on the outer shelf and upper continental slope in the northern Catalonia continental margin, and on the base-of-slope in the northwestern Alboran margin (Fig. 7).

Global controlling factors

Quaternary eustatic sea-level changes have controlled the analogous stratigraphical architecture of both margins, which are mainly composed of stacked lowstand systems tracts, with absence of transgressive and highstand systems tracts between two individual lowstand systems tracts.

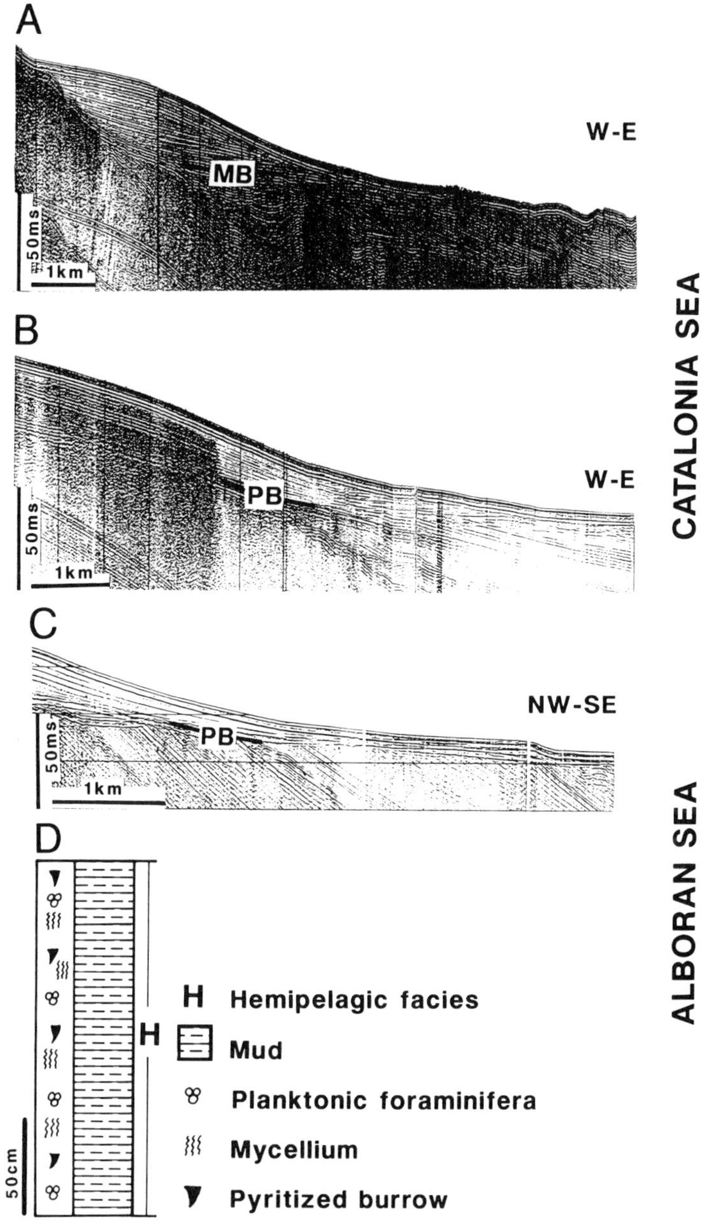

Fig. 6. Holocene highstand systems tract: (**A**) depositional mound in the northern Catalonia continental margin; (**B**) Fluviá-Muga prodelta in the northern Catalonia continental margin; (**C**) Guadalmedina–Guadalhorce prodelta in the northwestern Alboran margin; (**D**) hemipelagic facies in the northwestern Alboran margin. Abbreviations: MB, depositional mound boundary; PB, prodelta boundary.

Quaternary eustasy is characterized by sea-level changes of high frequency (20 to 100 ka), high amplitude (100 m) and marked asymmetry, with sea-level falls and lowstands longer than transgressive and highstands (Chappel & Shackleton, 1986; Williams *et al.* 1988). Eustasy acts as a dominant factor in synchronism with sediment supply and sedimentary processes (Nelson & Kulm 1973; Vail *et al.* 1977; Alonso *et al.* 1989; Ercilla *et al.* 1994*a*). Their combined effects were responsible for the occurrence of periods of active and reduced growth of both margins,

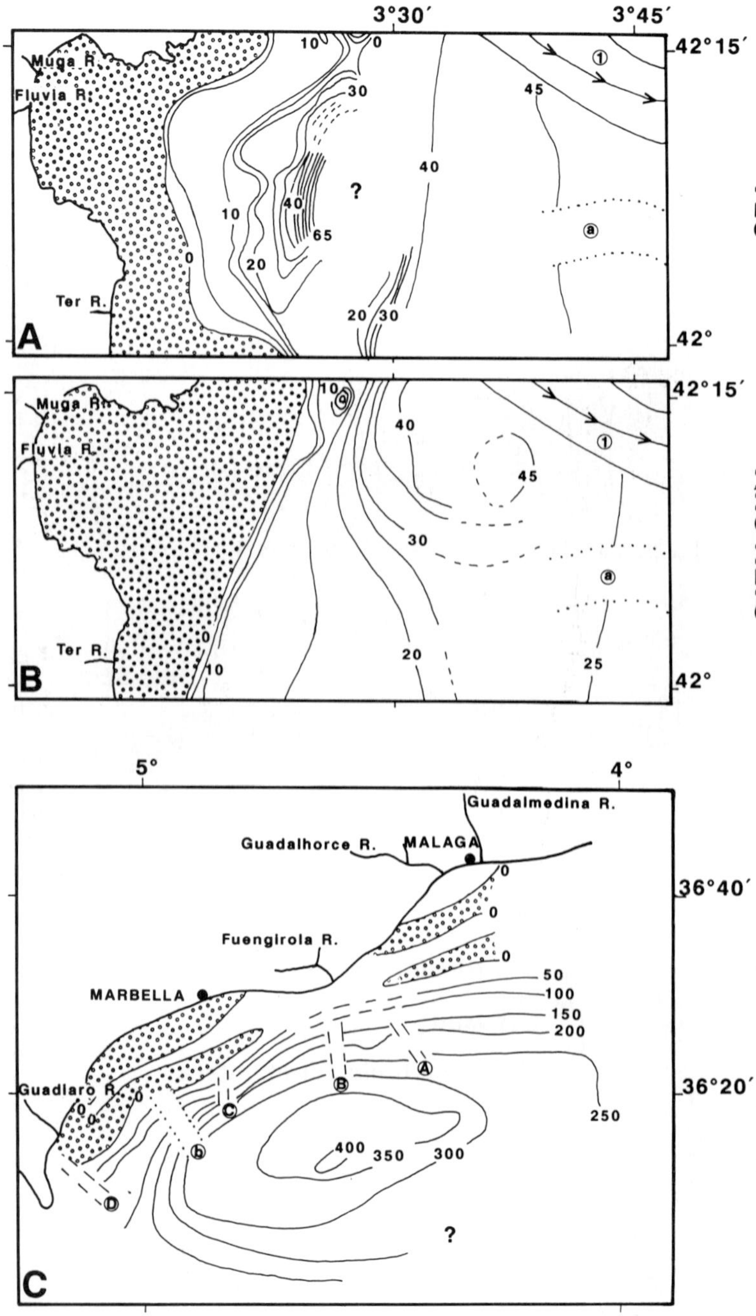

Fig. 7. Representative isopach (milliseconds) maps of: (**A**) depositional sequence 1, lower Quaternary in age, in the northern Catalonia continental margin; (**B**) depositional sequence 6, upper Quaternary in age, in the northern Catalonia continental margin; (**C**) depositional sequences 1 plus 3 in the northwestern Alboran margin. The circled numbers and circled letters allude to the canyons and palaeocanyons: 1, Cape Creus Canyon; a, La Escala palaeocanyon; A, Fuengirola Canyon; B, Torre Nueva Canyon; C, Baños Canyon; b, Estepona palaeocanyon; D, Guadiaro Canyon.

conditioning the poor development of transgressive and highstand systems tracts versus lowstand systems tracts. Therefore, during the falls and lowstand stages, active periods of margin growth occurred due to a combination of the following facts:

(i) these stages were relatively gradual and slower (1.5 mm/year during sea-level fall, according to Williams (1988));
(ii) high volumes of sediment were supplied seaward, because sediment sources moved seaward and the shelf underwent subaerial erosion; and
(iii) triggering of sediment gravity flow processes was favoured by discharge of large quantities of sediment onto the slope.

The combination of these factors favoured sedimentation in all sedimentary environments of the margin, resulting in volumetrically important lowstand systems tracts.

In contrast, during the transgressive and highstand stages the reduced margin development occurred, due to the following factors:

(i) these stages were relatively faster (15 mm/year during sea-level rise, according to Williams (1988));
(ii) a decrease in sediment supply because source areas moved landward, favouring a very low sedimentation on the slope; and
(iii) a predominance of erosive shoreface processes during the rises of sea level, and settling processes from suspension during the highstands.

These combined factors were responsible for the low development of transgressive and highstand systems tracts. Moreover, widening of the shelf due to landward migration of the coastline favoured their development being restricted mainly to shelf, and they were then easily removed during the subsequent falls of sea level. The consequence is that transgressive and highstand systems tracts cannot be recognized in the high-resolution seismic records, because either they are too thin, and below the resolution of the seismic records, or they are absent.

The dominant control exerted by eustatic sea-level changes on the development of the stratigraphical architecture of the northern Catalonia continental margin is in agreement with the Quaternary sedimentary studies of the most passive continental margins in the Mediterranean (Farrán & Maldonado 1990; Tesson *et al.* 1990; Trincardi & Field 1991; Chiocci 1994). Likewise, eustasy, a dominant controlling factor in the sequence stratigraphy growth of the northwestern Alboran margin, has also been observed in other tectonically active Mediterranean margins, such as in the northwestern Aegean margin (Lykousis 1991), where Quaternary sea-level changes developed progradational units reflecting major sea-level lowstands during the glaciation periods.

Local controlling factors

Local tectonics, physiography and sediment supply caused particular modifications to depositional systems on both margins, and were responsible for the differences observed on their sedimentary configuration.

The relative role played by tectonics (tilting) is shown by the development of different growth patterns in the relatively small area of the northwestern Alboran margin. Here, abrupt lateral changes in the stratal patterns are observed: a divergent pattern is identified in the eastern and western sectors where tilting played a prominent role, and a progradational pattern in the central sector, where tilting was less noticeable (Fig. 2). In contrast, and as expected for the passive continental margin of northern Catalonia, the minor role played by tectonics favoured the formation of a uniform progradational pattern along the margin (Fig. 2).

Tectonics, in the sense of subsidence versus uplifting, also influenced the preservation of deposits along and across both margins. Thus, progradational deposits are preserved from shelf, where subsidence is relatively more important than uplifting (Figs 2, 3A and 3C). This occurs in northern Catalonia and in the central sector of northwestern Alboran, where subsidence favoured the creation of accommodation space on the shelf during the successive sea-level changes. In contrast, divergent deposits are preserved from the slope and absent on the shelf, where uplifting is relatively more important than subsidence, as occurs in the eastern and western sectors of the northwestern Alboran margin (Figs 2 and 3B). Here, the tectonic tilting of the margin produced uplifting of the shelf, resulting in a lack of accommodation space and/or erosion of the ancient shelf deposits during the following sea-level variations.

Furthermore, local physiographical configuration and variations in sediment supply resulted in the different locations of the sedimentary depocentres displayed by both areas: on the outer shelf and upper continental slope in the northern Catalonia continental margin (Figs 7A and 7B), and on the base-of-slope in the northwestern Alboran margin (Fig. 7C). The relatively large width of the northern Catalonia continental shelf (up to 30 km) (Fig. 1A), would have caused most of the sediment supplied from onland source areas to pass to the distal margin during falls of sea level, and sediment was then deposited on the shelf-break and upper continental slope. Variations in the volume of sediment supply along-margin conditioned the northward displacement of the sedimentary depocentres during the Quaternary (Figs 7A and 7B). Depocentres with a similar location have been described in the southern Catalonia continental margin, which is characterized by a shelf with a comparably large width (50 km) (Farrán & Maldonado 1990). On the other hand, the relative narrowness (up to 11 km) of the northwestern Alboran shelf (Fig. 1B) contributed to the basinward movement of the sediment supplied by onland sources during sea-level falls. This sediment was mainly funnelled through submarine canyons (Fuengirola, Torre Nueva, Baños, Estepona and Guadiaro), and deposited on the base-of-slope and basin floor (Fig. 7C). The presence of diapiric barriers interrupted further basinward transport of these sediments (Ercilla *et al.* 1994*a*). Depocentres with a similar physiographical location, at the base-of-slope, have also been described in the northeastern Alboran margin (Alonso & Maldonado 1992), where their development during the Quaternary is also related to the narrowness of the shelf (10 km) and to the presence of morphostructural barriers of volcanic origin.

This paper has benefited from reviews by J. B. Anderson. The research was supported by the Spanish Consortium for Ocean Drilling Program (ODP), the Comisión Interministerial de Ciencia y Tecnología (CICYT, Projects GE089-0829, PB91-0080-CO2-02, and AMB95/0196), the Instituto Tecnológico y Geominero de España (ITGE) under the project Mapa geológico de la plataforma continental y zonas adyacentes. Hoja 25, 25E. FIGUERES, and the European Community under the MAST II-MTP project EUROMARGE-NB (MAST2-CT93-0053). G. Ercilla is also especially grateful to Generalitat (Regional Government) of Catalonia for providing two grants. We also thank the officers and crew of the 'B/O Garcia del Cid' for their assistance during cruises, and J. Baraza and J. I. Díaz for their helpful comments.

References

ALONSO, B. & MALDONADO, A. 1992. Plio-Quaternary margin growth patterns in a complex tectonic setting: northeastern Alboran Sea. *Geo-Marine Letters*, **12**(2–3), 137–143.

ALONSO, B., FARRAN, M. & MALDONADO, A. 1989. Estratigrafía sísmica de alta resolución en márgenes continentales pasivos: factores de control durante el Cuaternario. *Revista de la Sociedad Geológica de España*, **2**(3–4), 269–289.

BANDA, E. & ANSORGE, J. 1980. Crustal structure under the central and eastern part of Betic Cordillera. *Geophysical Journal of the Royal Astronomical Society*, **63**, 515–532.

CAMPILLO, A. C., MALDONADO, A. & MAUFFRET, A. 1992. Stratigraphic and tectonic evolution of the western Alboran Sea: Late Miocene to Recent. *Geo-Marine Letters*, **12**(2–3), 165–172.

CHAPPEL, J. & SHACKLETON, N. J. 1986. Oxygen isotopes and sea level. *Nature*, **324**, 137–140.

CHIOCCI, F. L. 1994. Very high-resolution seismics as a tool for sequence stratigraphy applied to outcrop scale – Examples from eastern Tyrrhenian margin Holocene/Pleistocene deposits. *AAPG Bulletin*, **78**, 378–395.

COMAS, M. C., GARCIA-DUEÑAS, V. & JURADO, M. J. 1992. Neogene extensional tectonic evolution of the Alboran Basin from MCS data. *Geo-Marine Letters*, **12**(2–3), 157–164.

DEWEY, J. F., HELMAN, M. L., TURCO, E., HUTTON, D. H. W. & KNOTT, S. D. 1989. *Kinematics of the Western Mediterranean*. Geological Society of London, Special Publication, **45**, 265–283.

DURAN-DELGA, M. & FONTBOTE, J. M. 1980. Le cadre structural de la Méditerranée occidentale. *Mémoires du Bureau de Recherches Géologiques et Minières*, **115**, 67–85.

ERCILLA, G., ALONSO, B. & BARAZA, J. 1992. Sedimentary evolution of the northwestern Alboran Sea during the Quaternary. *Geo-Marine Letters*, **12**(2–3), 144–149.

——, —— & —— 1994*a*. Post-Calabrian sequence stratigraphy of the northwestern Alboran Sea (southwestern Mediterranean). *Marine Geology*, **120**, 249–265.

——, DIAZ, J. I., ALONSO, B. & FARRAN, M. 1995. Late Pleistocene–Holocene sedimentary evolution of the northern Catalonia continental shelf (northwestern Mediterranean Sea). *Continental Shelf Research*, **15**(11–12), 1435–1451.

——, FARRAN, M. ALONSO, B. & DIAZ, J. I. 1994*b*. Pleistocene progradational growth pattern of the northern Catalonia continental shelf (northwestern Mediterranean). *Geo-Marine Letters*, **14**, 41–51.

FARRAN, M. & MALDONADO, A. 1990. The Ebro continental shelf: Quaternary seismic stratigraphy and growth patterns. *Marine Geology*, **95**, 333–352.

FLETA, J. & ESCUER, J. 1991. *Sistemas Sedimentarios de la Cuenca Neogena del Alt Empordà y su Relación con la Tectónica y el Volcanismo*. I Congreso del Grupo Español del Terciario. Vic, Spain. Libro-Guía Excursión, **7**, 128.

FLOS, J. 1985. The driving machine. *In*: MARGALEF, R. (ed.) *Key Environments. Western Mediterranean*. Pergamon, Oxford, 60–69.

GOT, H. 1973. *Etude des Correlations Tectonique-Sédimentation au Cours de l'Histoire Quaternaire du Précontinent Pyrénéo-Catalan*. PhD Thesis, Université de Perpignan.

Hernández-Molina, F. J., Somoza, L., Rey, J. Pomar, L. 1994. Late Pleistocene sediments on the Spanish continental shelves: Model for very high resolution sequence stratigraphy. *Marine Geology*, **120**, 129–174.

Institut Cartografic de Catalunya 1983. *Mapa topogràfic de Catalunya 1 : 250.000*. Institut Cartogràfic de Catalunya, Barcelona.

Jervey, M.T. 1988. Quantitative geological modeling of siliciclastic rock sequences and seismic expression. *In*: Macdonald, D. I. M. (ed.) *Sedimentation, Tectonics and Eustasy. Sea-level Changes at Active Margins*. International Association of Sedimentologists, Special Publication, **12**, 47–69.

Lykousis, V. 1991. Sea-level changes and sedimentary evolution during the Quaternary in the northwest Aegean continental margin, Greece. *In*: Macdonald, D. I. M. (ed.) *Sedimentation, Tectonics and Eustasy. Sea-level Changes at Active Margins*. International Association of Sedimentologists, Special Publication, **12**, 123–131.

Medialdea, J., Maldonado, A., Alonso, B. *et al.* 1995. *Mapa Geológico de la Plataforma Continental Española y Zonas Adyacentes. 1 : 200.000. Memoria y Hojas n° 25 y 25E. Figueres*. Servicio de Publicaciones del Ministerio de Industria y Energía, Madrid.

Nelson, C. H. & Kulm, L. D. 1973. Submarine fans and channels. *In*: Middleton, G. V. & Bouma, A. H. (eds) *Turbidites and Deep Water Sedimentation*. Pacific Section Society of Economic Paleontologists and Mineralogists, Short Course, 29–78.

Ott d'Estevou, P. & Montenat, C. 1985. Evolution structurale de la zone bétique orientale (Espagne) du Tortonien à Holocène. *Comptes Rendus de l'Académie des Sciences*, Paris, **300**, 363–368.

Philip, H. 1987. Plio-Quaternary evolution of the stress field in the Mediterranean zones of subduction and collision. *Annales Geophysicae*, **5**, 301–320.

Pirrie, D., Whitham, A. G. & Ineson, J. T. 1991. The role of tectonics and eustasy in the evolution of a marginal basin: Cretaceous–Tertiary Larsen Basin, Antarctica. *In*: Macdonald, D. I. M. (ed.) *Sedimentation, Tectonics and Eustasy. Sea-level Changes at Active Margins*. International Association of Sedimentologists, Special Publication, **12**, 293–306.

Posamentier, H. W. & Allen, G. P. 1993. Variability of the sequence stratigraphy model: Effects of local basin factors. *Sedimentary Geology*, **86**, 91–109.

—— & Vail, P. R. 1988. Eustatic controls on clastic deposition. II: Sequence and systems tract models. *In*: Wilgus, C. K., Hastings, B. S., Kendall, C. G. St. C., Posamentier, H. W., Ross, C. A. & Van Wagoner, J. C. (eds) *Sea-level Change – an Integrated Approach*. Society of Economic Paleontologists and Mineralogists, Special Publication, **42**, 125–154.

——, Jervey, M. T. & Vail, P. R. 1988. Eustatic controls on clastic deposition I. Conceptual framework. *In*: Wilgus, C. K., Hastings, B. S., Kendall, C. G. St. C., Posamentier, H. W., Ross, C. A. & Van Wagoner, J. C. (eds) *Sea-level Change – an Integrated Approach. Society of Economic Paleontologists and Mineralogists, Special Publication*, **42**, 109–124.

Ramsay, A. T. S. 1991. Sedimentation and tectonics in the Dinantian limestone of South Wales. *In*: Macdonald, D. I. M. (ed.) *Sedimentation, Tectonics and Eustasy. Sea-level Changes at Active Margins*. International Association of Sedimentologists, Special Publication, **12**, 485–512.

Roest, W. R. & Srivastava, S. P. 1991. Kinematics of the plate boundaries between Eurasia, Iberia and Africa in the north Atlantic from the Late Cretaceous to present. *Geology*, **19**, 613–616.

Sanz de Galdeano, C. 1990. Geologic evolution of the Betic Cordilleras in the western Mediterranean, Miocene to the present. *Tectonophysics*, **172**, 107–119.

Stanley, D. J., Got, H., Kenyon, N. H., Monaco, A. & Weilve, Y. 1976. Catalonian, Eastern Betic Balearic Margins: structural types and geologically recent foundering of the Western Mediterranean Basin. *Smithsonian Contributions to the Earth Sciences*, **20**.

Tesson, M., Gensous, B., Allen, G. P. & Ravenne, Ch. 1990. Late Quaternary deltaic lowstand wedges on the Rhône continental shelf, France. *Marine Geology*, **91**, 325–332.

Trincardi, F. & Field, M. E. 1991. Geometry, lateral variability, and preservation of downlapped regressive shelf deposits: eastern Tyrrhenian margin, Italy. *Journal of Sedimentary Petrology*, **49**, 775–790.

Vail, P. M., Mitchum, R. M. Jr & Thompson, S. 1977. Seismic stratigraphy and global changes of sea level, part 4: global cycles of relative changes of sea level. *In*: Payton, C. E. (ed.) *Seismic Stratigraphy – Application to Hydrocarbon Exploration*. American Association of Petroleum Geologists, Memoir, **26**, 63–81.

Vazquez, J.T. & Medialdea, T. 1992. *Evolución Tectónica del Margen Continental entre los Cabos de Bagur y Creus*. III Congreso Geológico de España y VIII Congreso Latinoamericano de Geología. Salamanca, Spain. Actas de las sesiones científicas, **2**, 166–170.

Williams, D. F. 1988. Evidence for and against sea-level changes from the stable isotopic record of the Cenozoic. *In*: Wilgus, C. K., Hastings, B. S., Kendall, C. G. St. C., Posamentier, H. W., Ross, C. A. & Van Wagoner, J. C. (eds) *Sea-level Change – an Integrated Approach*. Society of Economic Paleontologists and Mineralogists, Special Publication, **42**, 31–38.

——, Thunnel, R. C., Tappa, E. & Raffi, I. 1988. Chronology of the Pleistocene oxygen isotope record 0–1.88 m.y. B.P. *Palaeogeography, Palaeoclimatology, Palaeoecology*, **64**, 221–240.

Late Pleistocene-Holocene high-resolution sequence analysis on the Alboran Sea continental shelf

F. J. HERNÁNDEZ-MOLINA[1], L. SOMOZA[2] & J. REY[3]

[1] *Dpto de Estructura y Propiedades de los Materiales, Facultad de Ciencias del Mar, Aptdo 40, Poligono del Rio San Pedro, E-11510 Puerto Real, Cádiz, Spain*
[2] *Marine Geology Division, ITGE Geological Survey of Spain, Rios Rosas 21, E-28003 Madrid, Spain*
[3] *Instituto Español de Oceanografia, IEO, Puerto Pesquero s/n, E-29640 Fuengirola, Málaga, Spain*

Abstract: We present results of a detailed high-resolution seismic stratigraphy analysis of the northwestern margin of the Alboran Sea, between Málaga and the Gibraltar strait, based on a network of high- and very high-resolution seismic profiles. These profiles allow the determination of several scales of depositional sequences that are related to relative sea-level cycles of different order on the continental shelf and slope. The Late Pleistocene–Holocene deposits on these comprise ten seismic units characterizing lowstand, transgressive and highstand deposits. These deposits allow definition of a main Type 1 depositional sequence for the last 20 ka sedimentary record. This sequence is made up of minor depositional sequences produced by high-frequency relative sea-level cycles with a frequency range below the Milankovitch band. In this way, the last eustatic hemicycle was modulated by three types of very high-resolution sea-level cycles: The 'P cycles', being longest, caused the major sedimentary progradations (P_1, P_2, P_3, P_4 and P_5). The smaller eustatic changes of the 'h cycles' created smaller progradations, and the 'c cycles', consisting of short stillstands, resulted in sets of minor progradations within the larger cycles.

The classical sequence-stratigraphical Exxon conceptual model (Vail *et al.* 1977; Van Wagoner *et al.* 1987; Posamentier & Vail 1988; Posamentier *et al.* 1988; Van Wagoner *et al.* 1988; Vail *et al.* 1991) can be applied to very high-resolution seismics of continental shelf Quaternary sequences, although several problems emerge from this application (Suter *et al.* 1987; Boyd *et al.* 1989; Saito 1991; Trincardi & Field 1991; Swift *et al.* 1991; Gensous & Tesson 1992; Gensous *et al.* 1993*a*, *b*; Posamentier & Allen 1993; Tesson *et al.* 1990; 1993; Chiocci 1994; Hernández-Molina *et al.* 1994). The main problem is that the classical conceptual model only partially explains the geometries observed on high-resolution seismic profiles. The classical model considers that the sequence architecture is related to global eustatic cycles that consist of a rapid sea-level fall, a lowstand, a slow sea-level rise, and a highstand (Vail *et al.* 1977). However, some of the aforementioned authors instead recognize asymmetrical cycles punctuated by minor cycles that modulate both regressive and transgressive segment trends. They consider that the sequence architecture is related to relative sea-level cycles which consist of a slow and gradual sea-level fall, a lowstand, a rapid sea-level rise, and a brief highstand (Thomas & Anderson, 1991; Gensous *et al.* 1993*a*, *b*; Tesson *et al.* 1993; Chiocci 1994; Somoza *et al.* 1995; Tesson & Allen 1995) showing major differences from the classical model. Moreover, in most papers dealing with high-resolution stratigraphical analysis, the classical major surfaces such as the sequence boundary (SB), the transgressive surface (TS) and the maximum flooding surface (mfs) are commonly difficult to draw, and sometimes problems emerge with the hierarchy of the surfaces (Posamentier *et al.* 1992; Hunt & Tucker 1992, 1995; Kolla *et al.* 1995). Hernández-Molina *et al.* (1994) proposed that the last lowstand, transgressive and highstand segments are punctuated by several higher-order cycles. The main segments are built up of units which in themselves constitute minor depositional sequences produced by high-frequency relative changes in sea level during the late Pleistocene–Holocene (the last 20 ka). The problem is that, for Quaternary sequences, regressions and transgressions are not rapid simple events, but are modulated by higher-order relative sea-level cycles. The capability for differentiating these higher-order cycles depends on the resolution (in terms of frequency) of the seismic source used (airgun, sparker, boomer, Geopulse, sub-bottom profiles: 3.5 kHz, etc.).

In this paper, we present the results of a detailed high-resolution seismic stratigraphy study

From De Batist, M. & Jacobs, P. (eds), 1996, *Geology of Siliciclastic Shelf Seas*, Geological Society Special Publication No. 117, pp. 139–154.

carried out in the northwestern margin of the Alboran Sea, between Málaga and the Gibraltar strait. This was based on a network of about 5000 km of high- and very high-resolution seismic profiles (sparker 4500 J Geopulse 175 J and 3.5 kHz) (Fig. 1). In the Alboran Basin, previous seismic analyses of the Neogene–Quaternary sequences (Maldonado 1992) have been carried out by Jurado & Comas (1992), who recognized several seismic units using multi-channel seismic records (MCS) and well data, and by Campillo *et al.* (1992), who recognized seismic units on the western Alboran basin based on airgun seismic profiles. The youngest boundary sequence defined by the former authors, based on DSDP site 121, is the Q1 horizon that separates an Upper Pliocene–Lower Pleistocene unit from a Quaternary unit (Sequence 1) which falls within Middle–Late Pleistocene and Holocene age. Although the age of this boundary is not clearly defined at DSDP site 121 (Ryan *et al.* 1973), it has been reported as the top of the Calabrian deposits (Alonso & Maldonado 1992). The post-Calabrian sedimentary record comprises three depositional sequences seen on airgun profiles (Ercilla 1992; Ercilla *et al.* 1994). They defined the two older sequences as lowstand systems tracts and the most recent as also including transgressive deposits. The present paper deals only with this latter sequence using higher-resolution profiles than were gained using the MCS and airgun seismic sources. The high resolution of seismic profiles used, with a resolution of 0.2–0.5 m (pulse length range from 0.2–0.4 ms), allows differentiation of higher-order relative sea-level cycles modulating the lowstand, transgressive and highstand tracts of the Late Pleistocene–Holocene sequence. These could usefully be compared with coeval units in others continental margins.

Seismic stratigraphy analysis

The Late Pleistocene–Holocene sedimentary sequence

The Late Pleistocene–Holocene sedimentary sequence (following Mitchum *et al.* 1977) is separated from underlying Neogene–Quaternary sequences by a strongly reflective, erosional discontinuity which is named the RT unconformity. This basal discontinuity could be correlated with the 3 and 4 boundaries from Ercilla *et al.* (1994). The upper limit of the Late Pleistocene and Holocene sequence corresponds with the present depositional surface of the sea floor and, as a consequence, it cannot be defined because the sequence has not developed yet.

The Late Pleistocene–Holocene depositional sequence (Fig. 2) consists of a sedimentary wedge in infra-littoral and inner shelf areas, and no sedimentation on the outer shelf. Sediment distribution is consistent over all the continental shelf except in those areas where the shelf is very narrow or where submarine canyon heads reach the proximal inner shelf. The isopach map shows that sedimentation over the shelf is directly related to:

(i) the location of main fluvial input in the area (Guadiaro, Guadalmansa, Manilva, Fuengirola, Guadalhorce, etc.); and
(ii) the erosion of littoral sectors located away from fluvial input (Fig. 2).

Structural analysis of the region (Comas *et al.* 1992; Hernández-Molina, 1993; Vazquez *et al.* 1994) shows that the fracture systems do not affect the internal configuration of the Late Pleistocene–Holocene sedimentary deposits; nevertheless they generate important structural highs which control the lateral sedimentary distribution and location of the depocentres. Where subsidence rates show modest magnitudes when considering a very short interval of time, such as the last transgressive segment which spans around 20 000 years in time, the eustatic changes are much more rapid. We assume that the high-order Late Pleistocene–Holocene depositional sequences are not significantly modified by subsidence processes, unlike the lower-order sequences.

Seismic units

The Late Pleistocene–Holocene sequence on the upper slope and continental shelf are built up of ten seismic units (A_1, A_2, B, C, D, E, F_1, F_2, F_3 and F_4) that correspond to sedimentary bodies that are controlled by high-frequency sea-level variations. Each seismic unit is defined by its geometry and seismic facies (Mitchum *et al.* 1977; Sangree & Widmier 1977, 1979):

(i) the type of reflection terminations that are associated with the boundaries of the unit;
(ii) the configuration of the reflection pattern within the unit; and
(iii) the external shape of the unit.

The ten seismic units form the sedimentary record over the continental shelf and have the diagnostic features recorded below.

Upper slope-shelf edge. Unit A is recognized from 120 to 500–600 m depth. Its seismic facies

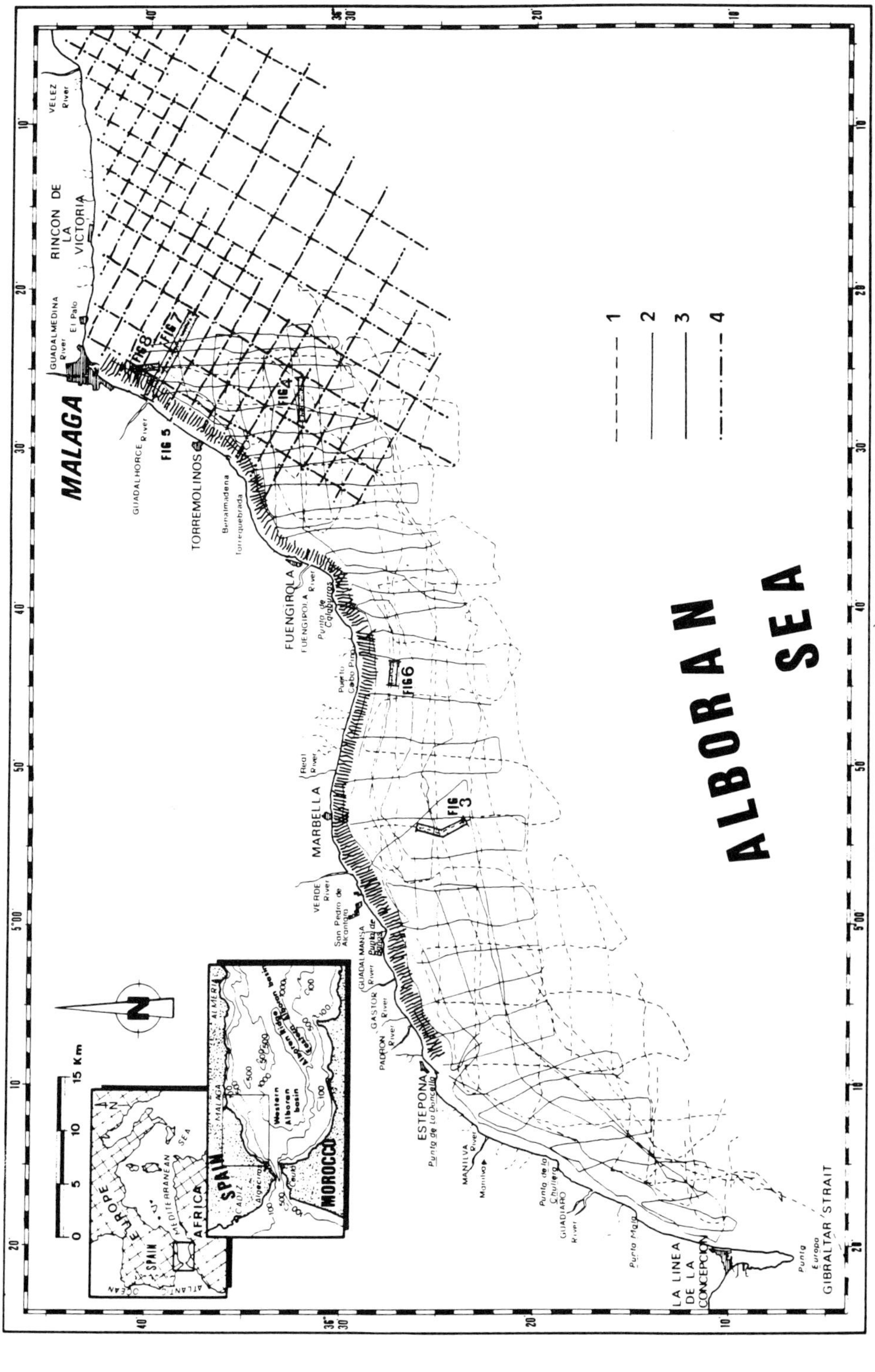

Fig. 1. Location of study area showing the networks of seismic profiles in the northern Alboran Sea continental shelf. Profiles were obtained from four oceanographic cruises: (1) 3.5 kHz and sparker (4500 J), (2) 3.5 kHz, (3) Geopulse, (4) 3.5 kHz and sparker (7000 to 13 000 J).

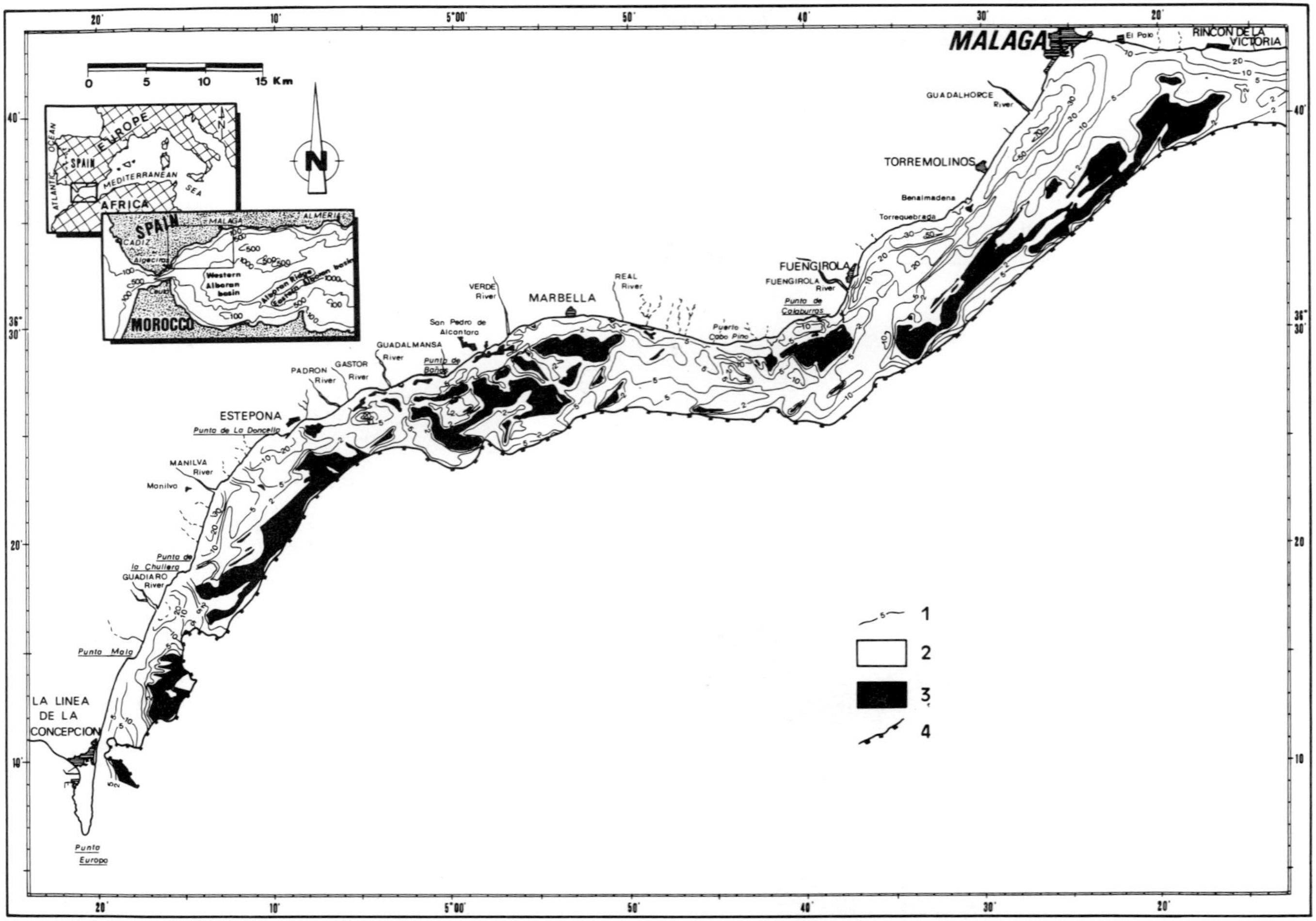

Fig. 2. Late Pleistocene–Holocene depositional sequence distribution over the continental shelf: (1) isopach in metres, (2) Late Pleistocene–Holocene wedge, (3) areas without Late Pleistocene–Holocene sedimentation, and (4) shelf edge.

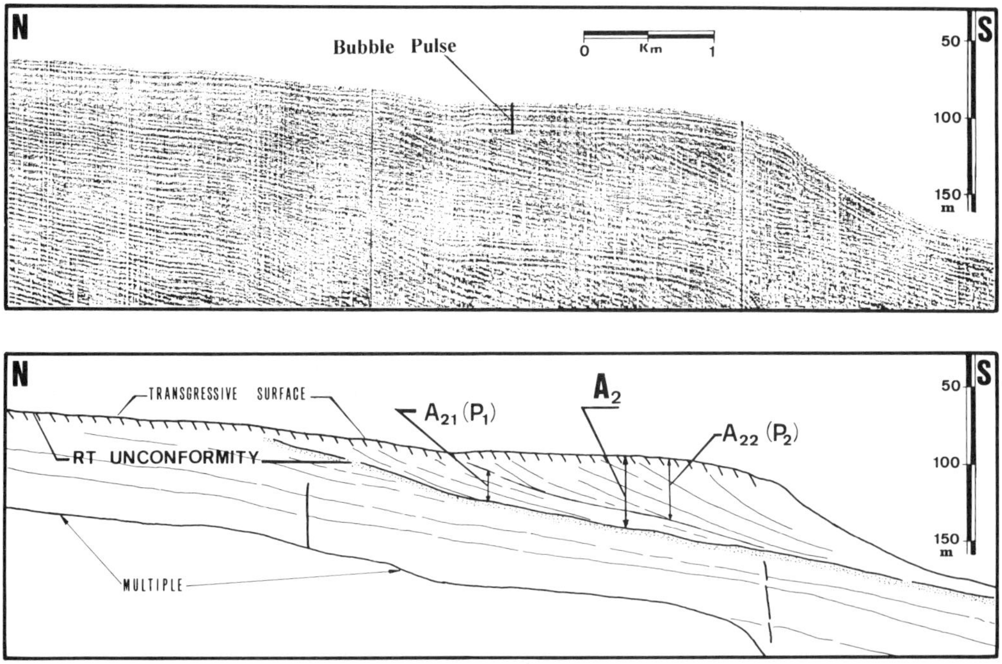

Fig. 3. Sparker (4500 J) profile showing the shelf edge wedge (A_2 seismic unit). This wedge consists of two important progradational bodies: P_1 (sub-unit A_{21}) and P_2 (sub-unit A_{22}). Profile location is in Fig. 1.

is weak, semi-transparent, and is stratified with a progradational and divergent reflection configuration towards the basin. Unit A is composed of two seismic sub-units which grade laterally into each other. Unit A_1 has a transparent seismic facies with progradational-divergent reflection pattern migrating seaward (Figs 3 and 4). It occurs on the middle-upper slope and has a clear progradational configuration. Unit A_2 has an oblique-parallel and progradational

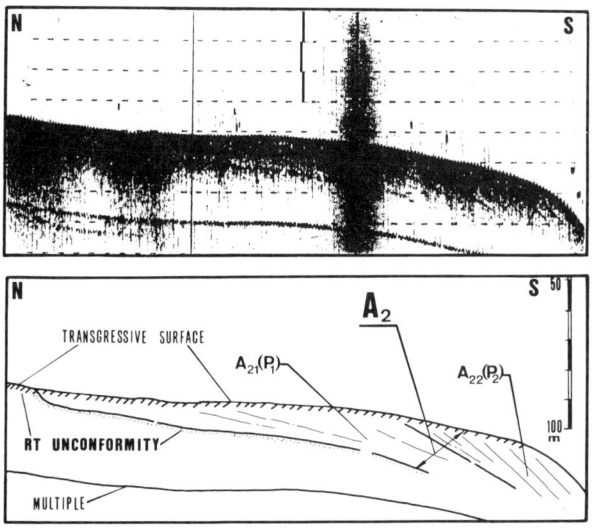

Fig. 4. High-resolution 3.5 kHz seismic profile, across the upper part of the shelf-edge wedge (seismic unit A_2). Two major progradational bodies can be detected: P_1 (sub-unit A_{21}) and P_2 (sub-unit A_{22}). Profile location is in Fig. 1.

reflection configuration that migrates seawards. Within the A_2 Unit it is possible to detect two sub-units (Figs 3 and 4):

(i) A_{21} has an oblique-progradational reflection pattern; and
(ii) A_{22} has an oblique-parallel reflection pattern.

Both sub-units are limited by a downlap surface. The A_2 unit forms the shelf edge (Figs 3 and 4) and both A_{21} and A_{22} sub-units represent two major progradational bodies defined by P_1 (A_{21}) and P_2 (A_{22}) respectively. Also, each of these two major progradational bodies can be divided into minor progradational bodies (h bodies).

Continental shelf. The seismic units covering the continental shelf form a succession of small backstepping and forestepping units. Many such units may be recognized within the upper slope–shelf area, though they have not been labelled separately. Instead each unit has been ascribed the letter B, C or D based on its morphology and internal reflection patterns.

Unit B has weak progradational reflections and lobate morphology. The internal configuration of each unit can be parallel, oblique, or

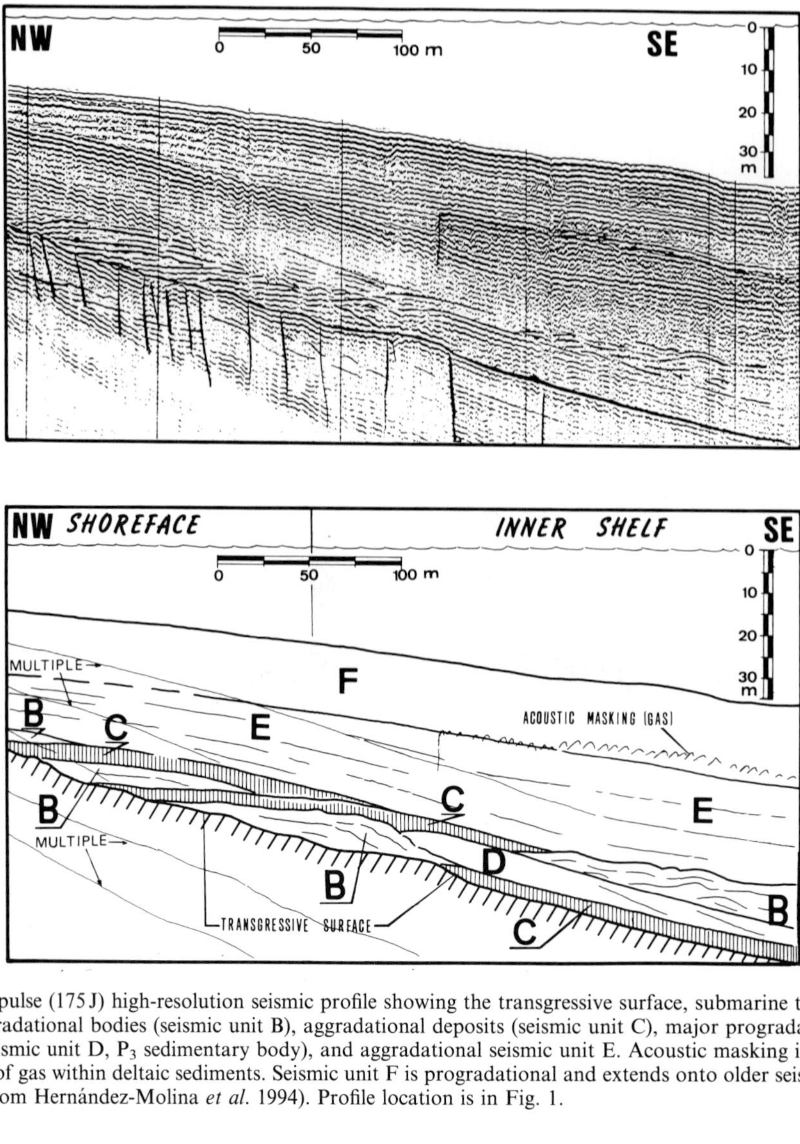

Fig. 5. Geopulse (175 J) high-resolution seismic profile showing the transgressive surface, submarine terraces, minor progradational bodies (seismic unit B), aggradational deposits (seismic unit C), major progradational deposits (seismic unit D, P_3 sedimentary body), and aggradational seismic unit E. Acoustic masking is due to occurrence of gas within deltaic sediments. Seismic unit F is progradational and extends onto older seismic units (modified from Hernández-Molina *et al.* 1994). Profile location is in Fig. 1.

lightly sigmoidal showing downlap terminations seaward and onlap landward. Depending on the type of seismic profile used, unit B shows different seismic facies. In Geopulse high-resolution seismic profiles, unit B is weak, stratified with a progradational-oblique to parallel reflection configuration, or sigmoidal configuration, showing a seaward downlap and a coastal onlap landwards (Fig. 5). In very high-resolution seismic profiles of 3.5 kHz (sub-bottom profiles), unit B is characterized by a weak stratified reflection with a parallel to oblique-progradational reflection configuration (Fig. 6). These seismic units are laterally associated with submarine terraces at 90, 80, 73, 55, 60, 47, 33, and 20 m depth. These sedimentary bodies are interpreted to be beach deposits in the sense of Belknap & Kraft (1985). We define these as h bodies, which show few features of progradation.

Unit C is characterized by weak aggradational reflection patterns with an onlap reflection configuration (Fig. 5). We correlate these units with the aggradational coastal sedimentary h bodies. The geometry, seismic facies, and aggradational configuration of unit C allow us to interpret these units as aggradational beach deposits, perhaps of finer grain-size than the B unit.

Unit D has an oblique-progradational reflection configuration, migrating seaward, and an external wedge morphology. This unit is attributed to a major progradational sedimentary body defined by P3 (Fig. 5).

Inner shelf–littoral wedge. Unit E rests above units B C and D on the inner shelf and under the present littoral wedge; it is especially thick near the major river mouths (Figs 5, 7 and 8). The internal configuration of unit E appears to differ depending on the high-resolution seismic system used: on Geopulse profiles, unit E is weak with aggradational and parallel reflection patterns, whilst in 3.5 kHz profiles, unit E is transparent or shows locally a parallel weak reflection. Unit E is a thick aggradational sedimentary body, and marks the onset of an important change in the dynamics of deposition on the shelf.

Unit F occurs above Unit E on the inner shelf. It appears to have different internal configuration

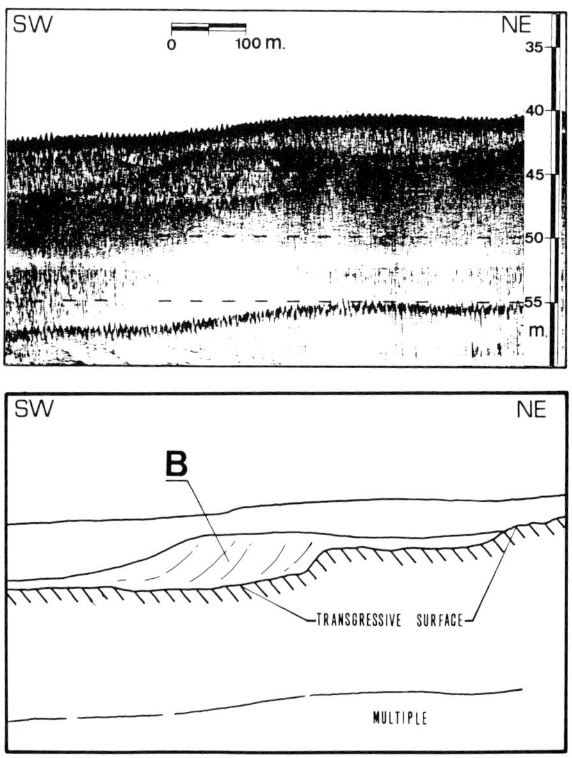

Fig. 6. Progradational unit B on high-resolution 3.5 kHz seismic profile showing oblique or sigmoidal reflection configurations. Profile location is in Fig. 1.

depending on the resolution of the seismic system used. In Geopulse high-resolution seismic profiles (Fig. 5), it has a progradational sigmoidal-oblique reflection configuration and its basal limit is characterized by phase inversion and signal reinforcements. In high-resolution 3.5 kHz seismic profiles, four minor seismic units are identified from bottom to top (Figs 7 and 8).

(i) Unit F1 shows a weak oblique progradational reflection configuration. It forms a downlapping progradational body (P4) with local acoustic masking. It overlies the semi-transparent unit E.
(ii) Unit F2 is semi-transparent with an aggradational internal configuration. We recognize it as an aggradational sedimentary body (defined as one of the h bodies).
(iii) Unit F3 has a weak sigmoidal-oblique reflection configuration. It is an important progradational body (P_5) similar to, but thinner than, P_4 and both overlaps and progrades over older bodies. Unit F_3 erosionally truncates the lower units F_2 and F_1 in the sectors closer to the coastline.
(iv) Unit F_4 is semi-transparent with an aggradational internal configuration, which is similar to F_2, but thinner, and is recognized by us as an aggradational sedimentary h body.

The shelf architecture: high-resolution sequence stratigraphy

Our seismic-sequential stratigraphical study illustrates the late Quaternary deposition on the continental shelf during the last eustatic hemicycle (Fig. 9). The genesis of the seismic units mentioned above can be related to allocyclic processes of base level which have been the main factors controlling accommodation rates. The seismic architecture has been correlated with the main eustatic events (Table 1) and with minor higher-frequency fluctuations, and differentiates between three types of deposits: the lowstand, transgressive and highstand deposits.

Lowstand deposits

The RT discontinuity began to develop as the coastline migrated seawards during the regressive intervals of the last sea-level fall before the Würm glaciation (80 000–20 000 years BP; Duplessy *et al.* 1986).

Unit A on the upper slope (A_1) and on the shelf edge (A_2) comprises lowstand deposits (Fig. 9 and Table 1) that formed during the last glaciation when the sea level was about 120 m below its present position (Fairbridge 1961; Guilcher 1969; Duplessy *et al.* 1981;

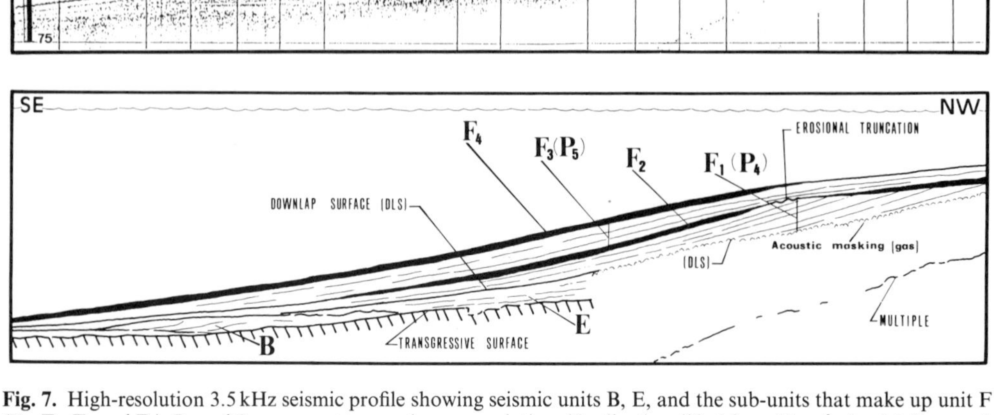

Fig. 7. High-resolution 3.5 kHz seismic profile showing seismic units B, E, and the sub-units that make up unit F (F_1, F_2, F_3 and F_4). P_4 and P_5 represent two major progradational bodies (modified from Hernández-Molina *et al.* 1994). Profile location is in Fig. 1.

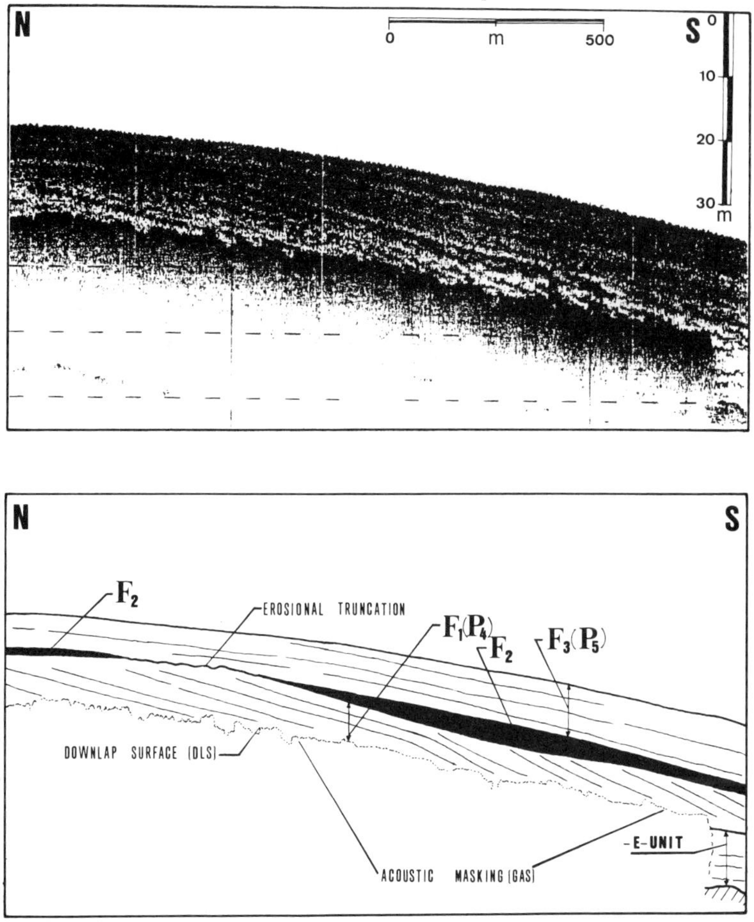

Fig. 8. High-resolution 3.5 kHz seismic profile showing the sub-units that make up seismic unit F. Note the erosional truncation between F_1 and F_3 units. Profile location is in Fig. 1.

Ruddiman & McIntyre 1981; Duplessy *et al.* 1986; Ercilla *et al.* 1992, 1994). Seismic unit A_2 can be separated into two major progradational sedimentary bodies P (P_1 and P_2) and each of these on a minor scale can be separated again into minor sets of smaller progradational sedimentary bodies (h bodies). Of the important major bodies, the lowermost could be correlated with the glacial maximum and eustatic minimum of about 20 000–18 000 years BP. The uppermost major progradational P_2 body may be correlated with a short regressive period of sea-level fall noticed on the Portuguese shelf at 100 to 122 m of water depth, between 14 000 and 15 000 years BP (Dias 1987; Rodrigues *et al.* 1991). During deposition of unit A (20 000–14 000 years BP) turbidite fans developed across the Alboran basin (Bartolini *et al.* 1972) while submarine canyons were active, and fluvial channels undercut the inner shelf.

Transgressive deposits

Seismic units B, C, D and E show a backstepping arrangement that allow them to be recognized as transgressive deposits (Fig. 9 and Table 1). These deposits represent the latest rise of sea level (14 000 to 6500 years BP; Ruddiman & McIntyre 1981; Duplessy *et al.* 1981; Dias 1987; Aloïsi 1986; Perissoratis & Mitropoulos 1989) that produced an accommodation increase and the landward migration of depositional environments (Fig. 9 and Table 1). During transgressive intervals the RT unconformity was reworked, generating a 'ravinement surface'

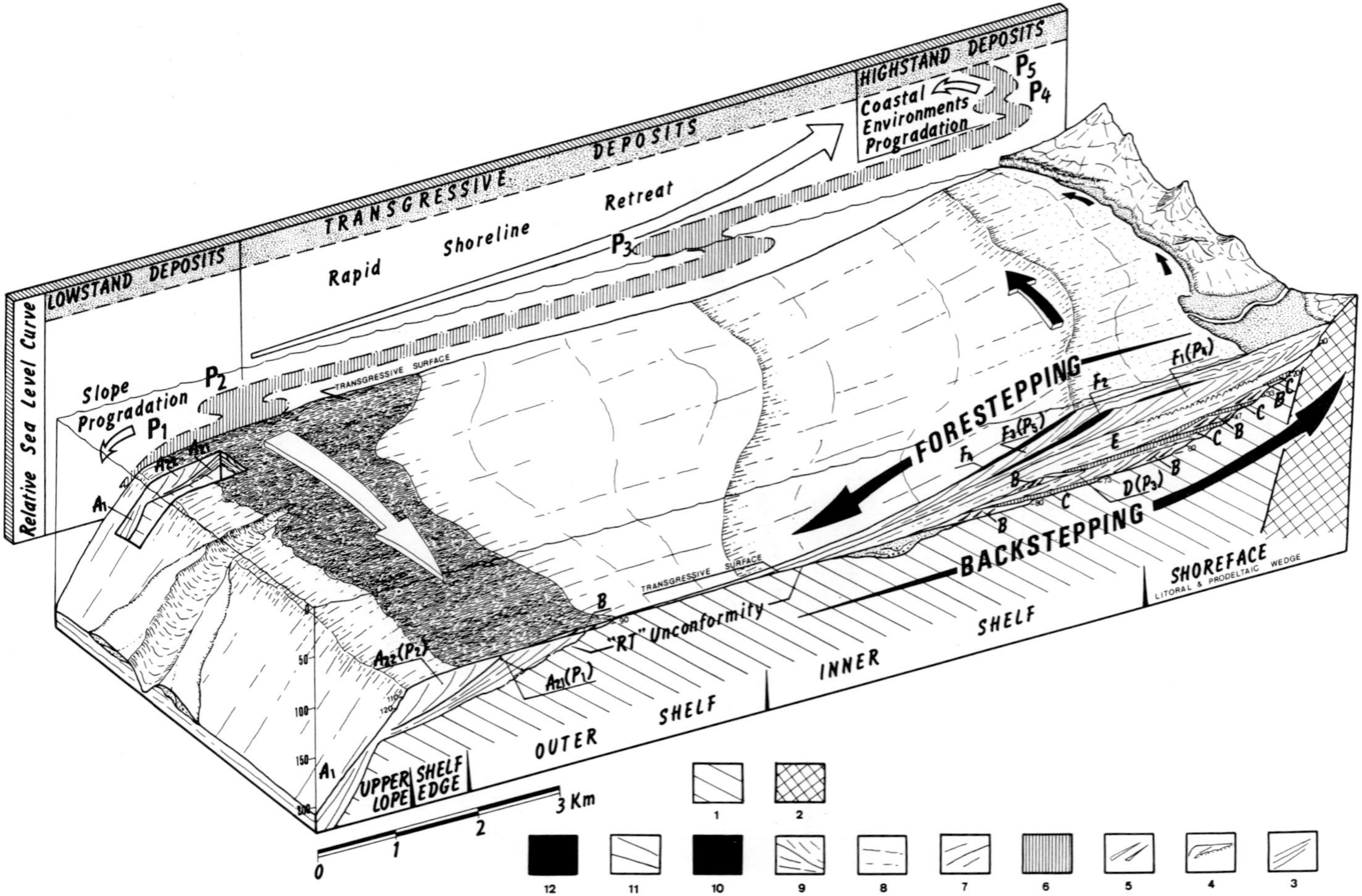

Fig. 9. Sequence stratigraphy model for the northern Alboran Sea continental shelf. (1) Neogene–Quaternary infill, (2) acoustic basement (Internal Subbetic Zone), (3) seismic unit A_1, (4) seismic unit A_2 (A_{21} and A_{22}), seismic unit C (7) seismic unit C (8) seismic unit E (9) seismic unit F_1, (10) seismic unit F_2, (11) seismic unit F_3, and (12) seismic unit F_4. The P major sedimentary bodies are referred to theoretical relative sea-level curve.

Table 1. *Table showing possible correlations between the seismic units, major progradational bodies* (P_1-P_5) *and Late Quaternary relative sea-level curves*

Profile types		Seismic units	Reflection patterns	Bodies	Systems tract
Geopulse profiles	F (3.5 kHz profiles)	IV	Transparent	h	Highstand deposits -HD-
		III	Weak with sigmoidal-oblique reflection configuration	P_5	
		II	Transparent	h	
		I	Weak with oblique-progradational reflection configuration	P_4	
		E	Weak with aggradational and parallel reflection pattern		Transgressive deposits -TD-
		D	Oblique-progradational reflection configuration migrating seaward and wedge morphology	P_3	
		C	Weak with aggradational reflection pattern	h	
		B	Weak with progradational reflection configuration and lobular morphology	h	
3.5 kHz profiles		A_2	Oblique-parallel and progradational reflection configuration migrating seaward	P_{1+2}	Lowstand deposits -LD-
		A_1	Transparent with progradational-divergent reflection pattern		

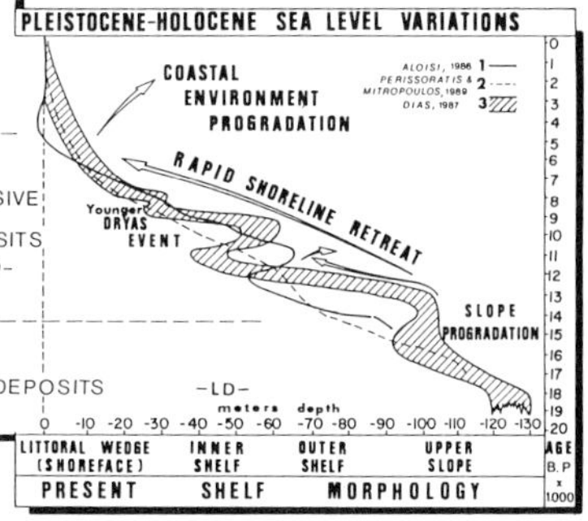

Table modified from Hernández-Molina *et al.* (1994): (1) Aloïsi (1986) for the Western Mediterranean, (2) Perissoratis & Mitropoulos (1989) for the Eastern Mediterranean, and (3) Dias (1987) for the North Atlantic. The seismic units correspond to sedimentary bodies P, h and c. The P bodies are major progradational bodies and are composed of smaller-scale h bodies separated by minor discontinuities. At the smallest scale, the h bodies can be sub-divided into c bodies (not shown in the scheme).

(Stamp 1921; Nummedal & Swift 1987; Suter *et al.* 1987; Thorne & Swift 1991) or 'transgressive surface of erosion' of Vail *et al.* (1991). The incised fluvial channel generated during the former lowstand interval was progressively filled between the end of lowstand and the sea-level rise as observed on other continental margins (Belknap & Kraft 1985).

Seismic units B and C are attributed to progradational and aggradational coastal deposits of ancient crest beach lines, with an onlap reflection configuration onto the RT unconformity. These deposits are correlated with erosive submarine terraces at 90, 80, 73, 55, 60, 47, 33 and 20 m depth (Fig. 9 and Table 1). We suggest that the general rise of the sea level was punctuated by short stillstands that allowed the development of minor terraces and small progradational sedimentary bodies, called h bodies. The retrogradational trend of beach sand deposits is disrupted by seismic unit D a major progradational sedimentary body (P_3) with wedge morphology, at about 60 m depth. It represents the only progradational body between the B and C units with a backstepping configuration. Unit D shows downlap reflection configurations related to a downward shift of the shoreline deposits in relation to a brief eustatic sea-level fall. During the Flandrian transgression, the Younger Dryas climatic event (11 000–10 000 years BP; Bard *et al.* 1987; Kallel *et al.* 1988) represented a phase of polar front advance as a result of global cooling in the northern hemisphere, generating a major regression of sea level. We correlate the development of unit D with the Younger Dryas event (Fig. 9 and Table 1). Minor sets of smaller progradational bodies (h bodies) can be identified within the P_3 main sedimentary body. A minor ravinement surface can be distinguished on the top of the B units (Figs 5 and 6) which can be considered as a higher-order sequence boundary. The ravinement surface of higher-order sequence was reworked, when the sediment supply was low, by a transgressive surface between successive B bodies. Although sea-level rise was generally rapid, forming the aggradational C units, periodic still stands, characterized by high sediment fluxes, created B units which are intercalated between the C units (Fig. 9 and Table 1). This backstepping configuration consisting of aggradational and progradational bodies of higher-order sequences can only be interpreted as the transgressive surface on low-resolution seismics. Furthermore, the backstepping sequence of ravinement surfaces on the tops of the B progradational units can also be confused with the erosional surface of the sequence boundary of a low-order sequence on low-resolution seismics.

It seems there are two orders of stillstand pulses represented by the h bodies and c bodies. We assume that their magnitude depends on the duration and rate of sea-level fall from the inflection point of the stillstand pulse. Major climatic cooling events that gave rise to ice-cap advances in northern latitudes (e.g. Older and

Younger Dryas episodes) are responsible for the major inversion pulses of the Holocene generalized sea-level rise trend (Fig. 9 and Table 1). The short-time periodical pulses which generated the higher-order c bodies are probably related to the rapid decreasing volume of Atlantic water entering the Mediterranean (Zazo *et al.* 1994).

Similar features in the transgressive systems tract have been described in areas of the Mediterranean. In the Adriatic Basin a stack of backstepping marine wedges resting on an erosional transgressive surface marks the first major flooding of the margin following the lowstand (Trincardi *et al.* 1994). Ravinement has also been recognized on top of the backstepping wedges that downlap the maximum flooding surface. In the Tiber Delta a series of retrogradational lagoonal deltas, T1, T2 and T3, within the transgressive systems tract and separated by minor-order flooding surfaces have been reported (Belloti *et al.* 1994). Radiocarbon dating indicates ages between 10 000 and 7750 years BP for the formation of these deltas. In the Tiber Delta, braided fluvial systems became active at 10 000 years BP In the continental shelf of the Tyrrhenian Sea, Chiocci (1994) considers the transgressive systems tract to be formed of two retrogradational parasequences separated by a non-depositional surface, indicative of a flooding surface generated during a rapid sea-level rise between the parasequences. On the continental shelf of the Gulf of Lyon retrogradational bodies within the transgressive systems tract, interpreted as barrier beaches and shoreface facies, indicate radiocarbon ages between 8000 and 10 000 years BP (Tesson *et al.* 1993; Gensous *et al.* 1993*b*). In the Ebro Delta sigmoidal progradational deposits that backstep onto the continental shelf are interpreted as eustatic stillstands (Diaz *et al.* 1990). Radiocarbon dating of these bodies indicates ages of 16 000, 14 300, 12 500, 10 000 and 8000 years BP for them. Several transgressive bodies have been recorded in the above papers, but they do not fall into a hierarchic model.

Seismic unit E, characterized by weak to transparent reflectors (Fig. 9 and Table 1) and capped by the maximum flooding surface (mfs), represents a major change in sedimentation style on the inner shelf. Unit E represents a wedge of marine deposits that several authors (Thorne & Swift 1991; Trincardi *et al.* 1994; Saito 1994) consider to belong to the transgressive deposits and maximum flooding surface. However, we consider that unit E developed at the maximum point of the transgression about 6000–6500 years BP, when sea level reached some 2 to 3 m higher than at present (Fairbridge 1961; Aloïsi 1986; Hoffman 1988; Zazo *et al.* 1994). At this time, the mouths of the rivers of southern Spain were flooded giving rise to estuarine environments that were undersupplied by the limited fluvial input (Hoffman 1988).

Highstand deposits

Seismic unit F represents a thickening upward and outward prograding sigmoidal sediment wedge which correspond to highstand deposits (Fig. 9 and Table 1). These deposits developed from the maximum eustatic event (6.5 ka) to the present (Ruddiman & McIntyre 1981; Aloïsi 1986; Dias 1987; Hoffman 1988; Perissoratis & Mitropoulos 1989; Somoza *et al.* 1991, 1992; Hernández-Molina *et al.* 1994; Zazo *et al.* 1994). Unit F is internally characterized on a minor scale by four seismic units (F_1, F_2, F_3 and F_4) that comprise two major progradational bodies, P_4 (seismic unit F_1) and P_5 (seismic unit F_3), and two minor aggradational bodies, namely h bodies (F_2 and F_4) (Fig. 9). These are related to the filling of previous estuaries and have developed within the present deltaic and littoral prism (Hernández-Molina *et al.* 1994). An acoustic masking above the downlap surface is commonly associated with prodeltaic bodies. Davis (1992) ascribed this phenomenon to the presence of interstitial methane gas within the sediment due to the product of decomposing organic matter. The presence of the erosional truncation boundary between both major progradational bodies P_4 (F_1) and P_5 (F_3), which is well correlated laterally in the study area, obliges us not to interpret these aggradational and progradational bodies as lateral switching of different prodelta lobes. We have interpreted this truncational boundary to have been generated by increasing volume of Atlantic water entering the Alboran Sea, forcing a rapid sea-level rise and causing a drastic increase of the current dynamics on the shelf. A generalized transgressive event from 3000 to 2400 years BP that reached its maximum level during the pre-Roman period, has been reported in the Atlantic and Mediterranean coast of Iberia (Zazo *et al.* 1994). After this transgressive period, major progradation of the coastline took place from 2200 to 1600 years BP.

Discussion and conclusion

The Late Quaternary Alboran continental margin deposits are built up of ten seismic units that are controlled by high-resolution variations of relative sea level. Based on sequence analysis we propose a Type 1 depositional sequence following Vail *et al.* (1984, 1991) that includes lowstand, transgressive and highstand deposits. This depositional sequence originated during part of a general high-frequency sea-level cycle (20 ka) and each of the deposits may be correlated with a particular portion of the last eustatic hemicycle corresponding to part of the last low-frequency (100 ka) sea-level cycle (Mörner 1972; Chappel & Shackleton 1986).

We propose that the main sequence is made up of units that in turn contain minor depositional sequences produced by higher-frequency relative changes in sea level during the Late Pleistocene–Holocene (the last 20 ka). This demonstrates that the Late Pleistocene–Holocene eustatic curve is not a simple one, but lowstand, transgressive and highstand deposits have been modified by three different minor cycles of sea level (P, h and c) with a frequency range below the Milankovitch band. In this way we propose that the last eustatic hemicycle was affected by three types of very high-resolution sea-level cycles (Table 1):

(i) lower-frequency P cycles corresponding to the main periods of stillstands (P_1 to P_5);
(ii) medium-frequency h cycles which affect the previous ones and are correlated with the h bodies; and
(iii) higher-frequency c cycles which are oscillations superimposed on (i) and (ii).

The P cycles, being longest, caused the most significant progradations (P_1, P_2, P_3, P_4 and P_5). The smaller eustatic changes of the h cycles created smaller progradations, and the c cycles, consisting of short stillstands and other changes, resulted in sets of minor progradations within the larger cycles.

Minor depositional sequences developed in relation to the accommodation changes produced by these very high-frequency sea-level cycles have different orders of hierarchy (P, h and c) implying that the different sequence boundary surfaces are also hierarchical. In this way we could recognize in a minor depositional sequence (e.g. a P sequence) a sequence boundary, a transgressive surface, a downlap surface, etc. (Fig. 5).

The major progradational P bodies are themselves composed of smaller-scale h bodies that are bounded by minor discontinuities. Small-scale progradational and aggradational h sedimentary bodies also alternate between the major P sedimentary bodies. On a smaller scale we notice that the h bodies can be sub-divided into very small progradational sets (c bodies).

Our analysis establishes the existence of very high-frequency sea-level cycles during the Late Pleistocene–Holocene. These cycles generated high-frequency depositional sequences which cannot be detected with multi-channel techniques. This leads us to question the exact correlation with the global cycles of a greater period proposed by Haq *et al.* (1987) in the same sense of Boyd *et al.* (1989). The sequence and subsequence time-scales of our Late Pleistocene–Holocene model are related to asymmetrical sea-level cycles comprising a slow sea-level fall, a lowstand, a rapid transgression, and a brief highstand. These cycles are dissimilar to those of the sequence stratigraphy conceptual model.

Thanks for useful comments and constructive criticism made by Dr F. Trincardi (Istituto per la Geologia Marina, CNR, Italy) and Dr John Rees (Coastal Geology Group of British Geological Survey, Nottingham) who reviewed the manuscript.

References

ALOÏSI, J.C. 1986. *Sur un Modèle de Sédimentation Deltaique – Contribution à la Connaissance des Marges Passives.* PhD Thesis, Université de Perpignan.

ALONSO, B. & MALDONADO, A. 1992. Plio-Quaternary margin growth patterns in a complex tectonic setting: Northeastern Alboran Sea. *Geo-Marine Letters*, **12**, 137–143.

BARD, E., ARNOLD, M., MAURICE, P., DUPRAT, J., MOYES, J. & DUPLESSY, J. C. 1987. Retreat velocity of the North Atlantic polar front during the last deglaciation determined by ^{14}C accelerator mass spectrometry. *Nature*, **328**, 791–794.

BARTOLINI, C., GEHIN, C. & STANLEY, D. 1972. Morphology and recent sediments of the western Alborán basin in the Mediterranean Sea. *Marine Geology*, **3**, 159–223.

BELKNAP, D. F. & KRAFT, J. C. 1985. Influence of antecedent geology on stratigraphic preservation potential and evolution of Delawareés barrier systems. *Marine Geology*, **63**, 235–262.

BELLOTI, P., CHIOCCI, F. L., MILLI, S., TORTORA, P. & VALERI, P. 1994. Sequence stratigraphy and depositional setting of the Tiber delta: Integration of high-resolution seismics, well logs, and archeological data. *Journal of Sedimentary Research*, **B64**(3), 416–432.

BOYD, R., SUTER, J. & PENLAND, S. 1989. Relation of sequence stratigraphy to modern sedimentary environments. *Geology*, **17**, 926–929

CAMPILLO, A. C., MALDONADO, A. & MAUFFRET, A. 1992. Stratigraphic and tectonic evolution of the Western Alboran Sea: Late Miocene to Recent. *Geo-Marine Letters*, **12**, 165–172.

CHAPPELL, J. & SHACKLETON, N. J. 1986. Oxigen isotopes and sea level. *Nature*, **324**, 137–140.

CHIOCCI, F. L. 1994. Very high-resolution seismics as a tool for sequence stratigraphy applied to outcrop scale – Examples from eastern Tyrrhenian margin Holocene/Pleistocene deposits. *AAPG Bulletin*, **78**, 378–395.

COMAS, M. C., GARCÍA DUEÑAS, V. & JURADO, M. J. 1992. Neogene tectonic evolution of the Alboran Sea from MCS Data. *Geo-Marine Letters*, **12**, 157–164

DAVIS, M. (ed.) 1992. Methane in marine sediments. *Continental Shelf Research*, **12**(10).

DIAS, J. M. A. 1987. *Dinâmica Sedimentar e Evoluçao Recente da Plataforma Continental Setentrional Portuguesa*. PhD Thesis, Univ. de Lisboa.

DIAZ, J. I., NELSON, C. H., BARBER, J. H. GIRÓ, S. 1990. Late Pleistocene and Holocene sedimentary facies on the Ebro continental shelf. *Marine Geology*, **95**, 333–352

DUPLESSY, J. C., ARNOLD, M., MAURICE, P., BARD, E., DUPRAT, J. MOYES, J. 1986. Direct dating of the oxygen-isotope record of the last deglaciation by ^{14}C accelerator mass spectrometry. *Nature*, **320**, 350–352.

——, DELIBRIAS, G., TURON, J. L., PUJOL, C. & DUPRAT, J. 1981. Deglacial warming of the northeastern Atlantic Ocean: correlation with the paleoclimatic evolution of the European continent. *Palaeogeography, Palaeoclimatology, Palaeoecology*, **35**, 121–144.

ERCILLA, G. 1992. *La Sedimentación Cuaternaria en Márgenes y Cuencas del Maediterráneo Noroccidental y Suroccidental: Modelos de Evolución Sedimentaria y Factores de Control*. PhD Thesis, Universidad Politécnica Cataluña, Universidad de Barcelona and Instituto Ciencias del Mar CSIC.

——, ALONSO, B. & BARAZA, J. 1992. Sedimentary evolution of the northwestern Alboran Sea during the Quaternary. *Geo-Marine Letters*, **12**, 144–149.

——, —— & —— 1994. Post-Calabrian sequence stratigraphy of the northwestern Alboran Sea (southwestern Mediterranean). *Marine Geology*, **120**, 249–265.

FAIRBRIDGE, R. W. 1961. Eustatic changes in sea level. *Physics and Chemistry of the Earth*, **4**, 99–185.

GENSOUS, B. & TESSON, M. 1992. High-resolution sequence stratigraphy of the Late Quaternary Rhone shelf deposits: relations with tectonics and eustacy; reservoirs implications. Symposium on *Sequence Stratigraphy of European Basins*, Dijon, 446–447.

——, TESSON, M. & ARNOLD, M. 1993*a*. Analyse en stratigraphie séquentielle des dépôts pléistocènes de la plate-forme orientale du Rhône (France): confrontation de l'analyse sismique et des données de terrain. *Comptes Rendus de l'Académie des Sciences, Paris*, **317**, 803–810.

——, WILLIAMSON, D. & TESSON, M. 1993*b*. Late Quaternary transgressive and highstand deposits of a deltaic shelf (Rhône Delta, France). *In*: POSAMENTIER, H. W., SUMMERHAYES, R., HAQ, B. U. & ALLEN, J. P. (eds) *Stratigraphy and Facies Associations in a Sequence Stratigraphic Framework*. International Association of Sedimentologists, Special Publication, **18**, 197–211.

GUILCHER, A. 1969. Pleistocene and Holocene sea level changes. *Earth Science Reviews*, **5**, 69–97.

HAQ, B. U., HARDENBOL, J. & VAIL, P. R. 1987. Chronology of fluctuating sea levels since the Triassic. *Science*, **235**, 1156–1167.

HERNÁNDEZ-MOLINA, F. J. 1993. *Dinámica Sedimentaria y Evolución Durante el Pleistoceno Terminal-Holoceno del Margen Noroccidental del Mar de Alborán. Modelo de Estratigrafía Secuencial de Muy Alta Resolución en Plataformas Continentales*. PhD Dissertation, University of Granada.

——, SOMOZA, L., REY, J. POMAR, L. 1994. Late Pleistocene–Holocene sediments on the Spanish continental shelves: Model for very high resolution sequence stratigraphy. *Marine Geology*, **120**, 129–174.

HOFFMAN, G. 1988. *Holozänstratigraphie und Küstenlinieverlagerung an der Andalusischen Mittelmeerküst*. PhD Thesis, Universität Bremen, Berichte aus dem Fachbereich der Geowissenschaften, **2**.

HUNT, D. & TUCKER, M. E. 1992. Stranded parasequences and the forced regressive wedge systems tract: deposition during base-level fall. *Sedimentary Geology*, **81**, 1–9

—— & —— 1995. Stranded parasequences and the forced regressive wedge systems tract: deposition during base-level fall. Reply. *Sedimentary Geology*, **95**, 147–160

JURADO, M. J. & COMAS, M. C. 1992. Well log interpretation and seismic character of Cenozoic sequence in the Northern Alboran Sea. *Geo-Marine Letters*, **12**, 129–136.

KALLEL, N., LABEYRIE, L. D., ARNOLD, M., OKADA, H., DUDLEY, W. C. & DUPLESSY, J. C. 1988. Evidence of cooling during the Younger Dryas in the western North Pacific. *Oceanologica Acta*, **11**(4), 369–375.

KOLLA, V., POSAMENTIER, H. W. & EICHENSEER, H. 1995. Stranded parasequences and the forced regressive wedge systems tract: deposition during base-level fall. Discussion. *Sedimentary Geology*, **95**, 147–160

MALDONADO, A. (ed.) 1992. Alboran Sea. *Geo-Marine Letters*, 12.

MITCHUM, R. M. JR, VAIL, P. R. & THOMPSON, S. III. 1977. Seismic stratigraphy and global changes of sea level, part 2: the depositional sequence as a basic unit for stratigraphic analysis. *In*: PAYTON, C. E. (ed.) *Seismic Stratigraphy – Application to Hydrocarbon Exploration*. American Association of Petroleum Geologists, Memoir, **26**, 53–62.

MÖRNER, N. A. 1972. When will the present interglacial end? *Quaternary Research*, **2**(3), 341–349.

NUMMEDAL, D. & SWIFT, D. J. P. 1987. Transgressive stratigraphy at sequence bounding unconformities – Some principles derived from Holocene and Cretaceous examples. *In*: NUMMEDAL, D., PILKEY, O. H. & HOWARD, J. D. (eds) *Sea-Level Fluctuation and Coastal Evolution*. Society of Economic Paleontologists and Mineralogists, Special Publication, **41**, 241–260.

PERISSORATIS, C. & MITROPOULOS, D. 1989. Late Quaternary evolution of the northern Aegean shelf. *Quaternary Research*, **32**, 36–50.

POSAMENTIER, H. W. & ALLEN, G. P. 1993. Variability of the sequence stratigraphy model: Effects of local basin factors. *Sedimentary Geology*, **86**, 91–109.

—— & VAIL, P. R. 1988. Eustatic control on clastic deposition. II: sequence and system tract models. *In*: WILGUS, C. K., HASTINGS, B. S., KENDALL, C. G. ST. C., POSAMENTIER, H. W., ROSS, C. A. & VAN WAGONER, J. C. (eds) *Sea-level Change – an Integrated Approach*. Society of Economic Paleontologists and Mineralogists, Special Publication, **42**, 125–154.

——, ——, JAMES, D. P. & TESSON, M. 1992. Forced regressions in a sequence stratigraphic framework: concepts, examples, and exploration significance. *AAPG Bulletin*, **76**(11), 1687–1709.

——, JERVEY, M. T. & VAIL, P. R. 1988. Eustatic controls on clastic deposition, I. Conceptual framework. *In*: WILGUS, C. K., HASTINGS, B. S., KENDALL, C. G. ST. C., POSAMENTIER, H. W., ROSS, C. A. & VAN WAGONER, J. C. (eds) *Sea-level Change – an Integrated Approach*. Society of Economic Paleontologists and Mineralogists, Special Publication, **42**, 109–124.

RODRIGUES, A., MAGALHAES, F. & DIAS, J. M. A. 1991. Evolution of the North Portuguese coast in the last 18 000 years. *Quaternary International*, **9**, 145–214.

RUDDIMAN, W. & MCINTYRE, A. 1981. The North Atlantic ocean during the last deglaciation. *Palaeogeography, Palaeoclimatology, Palaeoecology*, **35**, 145–214.

RYAN, W. B. F., HSÜ, K. H. *et al.* 1973. Western Alboran Basin. Site 121. *In*: RYAN, W. B. F., HSÜ, K. J. *et al.* (eds) *Initial Reports of the Deep Sea Drilling Project, Volume 13*. US Government Printing Office, Washington, 43–89

SAITO, Y. 1991. Sequence stratigraphy on the shelf and upper slope in response to the latest Pleistocene–Holocene sealevel changes off Sendai, Northeast Japan. *In*: MACDONALD, D. I. M. (ed.) *Sedimentation, Tectonics and Eustasy. Sea-level Changes at Active Margins*. International Association of Sedimentologists, Special Publication, **12**, 133–150.

——1994. Shelf sequence and characteristic bounding surfaces in a wave-dominated setting: Latest Pleistocene–Holocene examples from the Northeast Japan. *Marine Geology*, **120**, 105–127.

SANGREE, J. B. & WIDMIER, J. M. 1977. Seismic stratigraphy and global changes of sea level, part 9: seismic interpretation of clastic depositional facies. *In*: PAYTON, C. E. (ed.) *Seismic Stratigraphy – Application to Hydrocarbon Exploration*. American Association of Petroleum Geologists, Memoir, **26**, 165–184.

—— & —— 1979. Interpretation of depositional facies from seismic data. *Geophysics*, **44**, 131–160.

SOMOZA, L., ANDRÉS, J. R., REY J. *et al.* 1995. Arquitectura morfo-deposicional del Golfo de Cádiz: Proyecto Golca. *Gaia*, **9**, 79–82

——, GOY, J. L., ZAZO, C., BARDAJI, T., DABRIO, C. J., SILVA, P. & LARIO, J. 1991. Holocene sea-level highstands in the Western Mediterranean (SE Spain). *In*: *A Computer-statistical Analysis of Radiocarbon Dating. Mediterranean and Black Sea Shorelines*. INQUA Commision on Quaternary Shorelines, Subcommission Newsletter, **13**, 61–64.

——, ZAZO, C., GOY, J. L. & DIAZ-DEL-RIO, V. 1992. Sea-level fluctuation cycles in the Mediterranean Spanish coast during the last 7000 years. Applications as forecasting model for future sea-level trend. *In*: SUAREZ, J. L. (ed.) *The Ocean Change: Management Patterns and the Environment*. Universidad de Sevilla 19–26.

STAMP, C. D. 1921. On cycles of sedimentation in the Eocene strata of the Anglo-Franco Belgian basin. *Geological Magazine*, **58**, 108–114/194–200.

SUTER, J. R., BERRYHILL, H. L. & PENLAND, S. 1987. Late Quaternary sea-level fluctuations and depositional sequences, Southwest Lousiana continental shelf. *In*: NUMMEDAL, D., PILKEY, O. H. & HOWARD, J. D. (eds) *Sea-Level Fluctuation and Coastal Evolution*. Society of Economic Paleontologists and Mineralogists, Special Publication, **41**, 199–222.

SWIFT, D. J. P., OERTEL, G. F., TILLMAN, R. W. & THORNE, J. A. (eds) 1991. *Shelf Sand and Sandstone Bodies – Geometry, Facies and Sequence Stratigraphy*. International Association of Sedimentologists, Special Publication, **14**.

TESSON, M. & ALLEN, G. P. 1995. Contrôle tectonique et eustatique haute-fréquence de l'architecture et de la stratigraphie de dépôts de plate-forme péricratonique. Exemple du Golfe du Lion (Méditerranée, France) et des dépôts quaternaires. *Comptes Rendus de l'Académie des Sciences, Paris*, **320**, 39–46.

——, —— & RAVENNE, CH. 1993. Late Pleistocene shelf-perched lowstand wedges on the Rhone continental shelf. *In*: POSAMENTIER, H. W., SUMMERHAYES, R., HAQ, B. U. & ALLEN, J. P. (eds) *Stratigraphy and Facies Associations in a Sequence Stratigraphic Framework*. International Association of Sedimentologists, Special Publication, **18**, 183–196

——, RAVENNE, CH. & ALLEN, G. P. 1990. Application des concepts de l'analyse en stratigraphie séquentielle à un profil sismique haute résolution transverse à la plateforme rhôdanienne. *Comptes Rendus de l'Académie des Sciences, Paris*, **310**, 565–570..

THOMAS, M. A. & ANDERSON, J. B. 1991. Late Pleistocene sequence stratigraphy of the Texas continental shelf: relationship to $\delta^{18}O$ curves. Gulf Coast Section Society of Economic Paleontologists and Mineralogists, Twelfth Annual Research Conference *Coastal Depositional Systems in the Gulf of Mexico*, 265–270.

THORNE, J. A. & SWIFT, D. J. P. 1991. Sedimentation on continental margins, VI: a regime model for depositional sequences, their component systems tracts, and bounding surfaces. *In*: SWIFT, D. J. P., OERTEL, G. F., TILLMAN, R. W. & THORNE, J. A. (eds) *Shelf Sand and Sandstone Bodies – Geometry, Facies and Sequence Stratigraphy*. International Association of Sedimentologists, Special Publication, **14**, 189–255.

TRINCARDI, F. & FIELD, M. E. 1991. Geometry, lateral variability, and preservation of downlapped regressive shelf deposits: eastern Tyrrhenian margin, Italy. *Journal of Sedimentary Petrology*, **49**, 775–790.

——, CORREGGIARI, A. & ROVERI, M. 1994. Late Quaternary transgressive erosion and deposition in a modern epicontinental shelf: the Adriatic semi-enclosed Basin. *Geo-Marine Letters*, **14**, 41–51.

VAIL, P. R., AUDEMARD, F., BOWMAN, S. A., EISNER, P. N. & PEREZ-CRUZ, G. 1991. The stratigraphic signatures of tectonics, eustasy and sedimentology – an overview. *In*: EINSELE, G., RICKEN, W. & SEILACHER, A. (eds) *Cycles and Events in Stratigraphy*. Springer, Berlin, 617–659.

——, HARDENBOL, J. & TODD, R. G. 1984. Jurassic uncorformities, chronostratigraphy and sea-level changes from seismic stratigraphy and biostratigraphy. *In*: SCHLEE, J. S. (ed.) *Interregional Unconformities and Hydrocarbon Accumulation*. American Association of Petroleum Geologists, Memoir, **36**, 129–144.

——, MITCHUM, R. M. JR & THOMPSON, M. S. 1977. Seismic stratigraphy and global changes of sea level, part 4: global cycles of relative changes of sea level. *In*: PAYTON, C. E. (ed.) *Seismic Stratigraphy – Application to Hydrocarbon Exploration*. American Association of Petroleum Geologists, Memoir, **26**, 63–81.

VAN WAGONER, J. C., MITCHUM, R. M. JR, POSAMENTIER, H. W. & VAIL, P. R. 1987. Seismic stratigraphy interpretation using sequence stratigraphy. Part II: Key definitions of sequence stratigraphy. *In*: BALLY, A. W. (ed.) *Atlas of Seismic Stratigraphy. Vol. 1*. American Association of Petroleum Geologists, Studies in Geology, **27**, 11–14.

——, POSAMENTIER, H. W., MITCHUM, R. M., VAIL, P. R., SARG, J. F., LOUTIT, T. S. & HARDENBOL, J. 1988. An overview of the fundamentals of sequence stratigraphy and key definitions. *In*: WILGUS, C. K., HASTINGS, B. S., KENDALL, C. G. ST. C., POSAMENTIER, H. W., ROSS, C. A. & VAN WAGONER, J. C. (eds) *Sea-level Change – an Integrated Approach*. Society of Economic Paleontologists and Mineralogists, Special Publication, **42**, 39–45.

VAZQUEZ, J. T, HERNÁNDEZ-MOLINA, F. J., SOMOZA, L. & REY, J. 1994. Structural setting of the northwestern Alboran Sea continental shelf and its influence in the recent sedimentation. Second International Conference on the *Geology of Siliciclastic Shelf Seas (Southern North Sea and other examples)*, Gent, 73–74.

ZAZO, C., GOY, J. L., SOMOZA, L., *et al.* 1994. Holocene sequence of sea level highstand–lowstand in the Atlantic–Mediterranean linkage coast: forecast for future coastal changes and hazards. *Journal of Quaternary Research*, **10**(3), 1–15.

Late Quaternary transgressive large dunes on the sediment-starved Adriatic shelf

A. CORREGGIARI[1], M. E. FIELD[2] & F. TRINCARDI[1]

[1] *Instituto per la Geologia Marina CNR, via Gobetti 101, I-40129 Bologna, Italy*
[2] *US Geological Survey, 345 Middlefield Road, Menlo Park, CA 94025, USA*

Abstract: The Adriatic epicontinental basin is a low-gradient shelf where the late-Quaternary transgressive systems tract (TST) is composed of thin parasequences of backbarrier, shoreface and offshore deposits. The facies and internal architecture of the late-Quaternary TST in the Adriatic epicontinental basin changed consistently from early transgression to late transgression reflecting: (1) fluctuations in the balance between sediment supply and accommodation increase, and (2) a progressive intensification of the oceanographic regime, driven by the transgressive widening of the basin to as much as seven times its lowstand extent. One of the consequences of this trend is that high-energy marine bedforms such as sand ridges and sand waves characterize only areas that were flooded close to the end of the late-Quaternary sea-level rise, when the wind fetch was maximum and bigger waves and stronger storm currents could form.

We studied the morphology, sediment composition and sequence-stratigraphical setting of a field of asymmetric bedforms (typically 3 m high and 600 m in wavelength) in 20–24 m water depth offshore the Venice Lagoon in the sediment-starved North Adriatic shelf. The sand that forms these large dunes derived from a drowned transgressive coastal deposit reworked by marine processes. Early cementation took place over most of the dune crests limiting their activity and preventing their destruction. Both the formation and deactivation of this field of sand dunes occurred over a short time interval close to the turn-around point that separates the late-Quaternary sea-level rise and the following highstand and reflect rapid changes in the oceanographic regime of the basin.

Modern large-scale bedforms on continental shelves can form in response to oceanic currents, storm flows or tidal currents that impinge on the sea floor (Flemming 1988; Belderson *et al.* 1982; Harris 1988). Examples of active large bedforms also come from estuaries (Berné *et al.* 1993), bays (Berné *et al.* 1991) and semi-enclosed epicontinental seas (Field *et al.* 1981; Kuijpers *et al.* 1993). The recognition and sedimentological interpretation of bedforms that are inactive and drowned on modern continental margins can provide relevant stratigraphical information for reconstructing major changes in the oceanographic regime that took place during distinctive stages of the late-Quaternary sea-level rise; this is particularly important in epicontinental settings where the late-Quaternary relative sea-level rise determined not just the landward translation of the coastline but also major increases in basin extent and coastline length accompanied by possible changes in the oceanographic regime (Trincardi *et al.* 1994).

The North Adriatic epicontinental shelf (Fig. 1 inset) is sediment-starved and subject to a high-energy oceanographic regime (Cavaleri & Stefanon 1980; Mosetti 1985; Malanotte-Rizzoli 1994). This shelf is less than 30 m deep and presents a variety of bedforms that have a patchy distribution (Mosetti 1966; Brambati & Venzo 1967; Colantoni *et al.* 1979; Cavaleri & Stefanon 1980; Newton & Stefanon 1982; Correggiari *et al.* 1992). The bedforms in the North Adriatic are variable in size, are composed of sandy sediment, and rest on older alluvial lowstand or backbarrier transgressive deposits. Although several authors have pointed out the occurrence of bedforms in the Adriatic, no detailed inspection has been carried out to examine the morphology and sediment composition of any of the bedform fields.

This paper focuses on the geometry, sediment character and stratigraphical position of a small field (less than 100 km^2) of large asymmetric bedforms that occur between 20 and 24 m of water, offshore the Venice Lagoon (Figs 1 and 2). This field of large bedforms is one of the best developed of the North Adriatic and is located where the regional contour bends abruptly from ENE–WSW to about N–S (see the 20 m contour on Fig. 1). The bedforms are superimposed on one of the larger-scale reliefs that characterize the North Adriatic (Fig. 3). In this paper we call large dunes the asymmetric bedforms in the study area (see discussion in Berné *et al.* 1993) and shore-parallel mounds the underlying relief. Given the lack of current-meter data on this field of bedforms we prefer not to use a nomenclature (e.g. sand waves) that implies the knowledge of sedimentological process or a

From De Batist, M. & Jacobs, P. (eds), 1996, *Geology of Siliciclastic Shelf Seas*, Geological Society Special Publication No. 117, pp. 155–169.

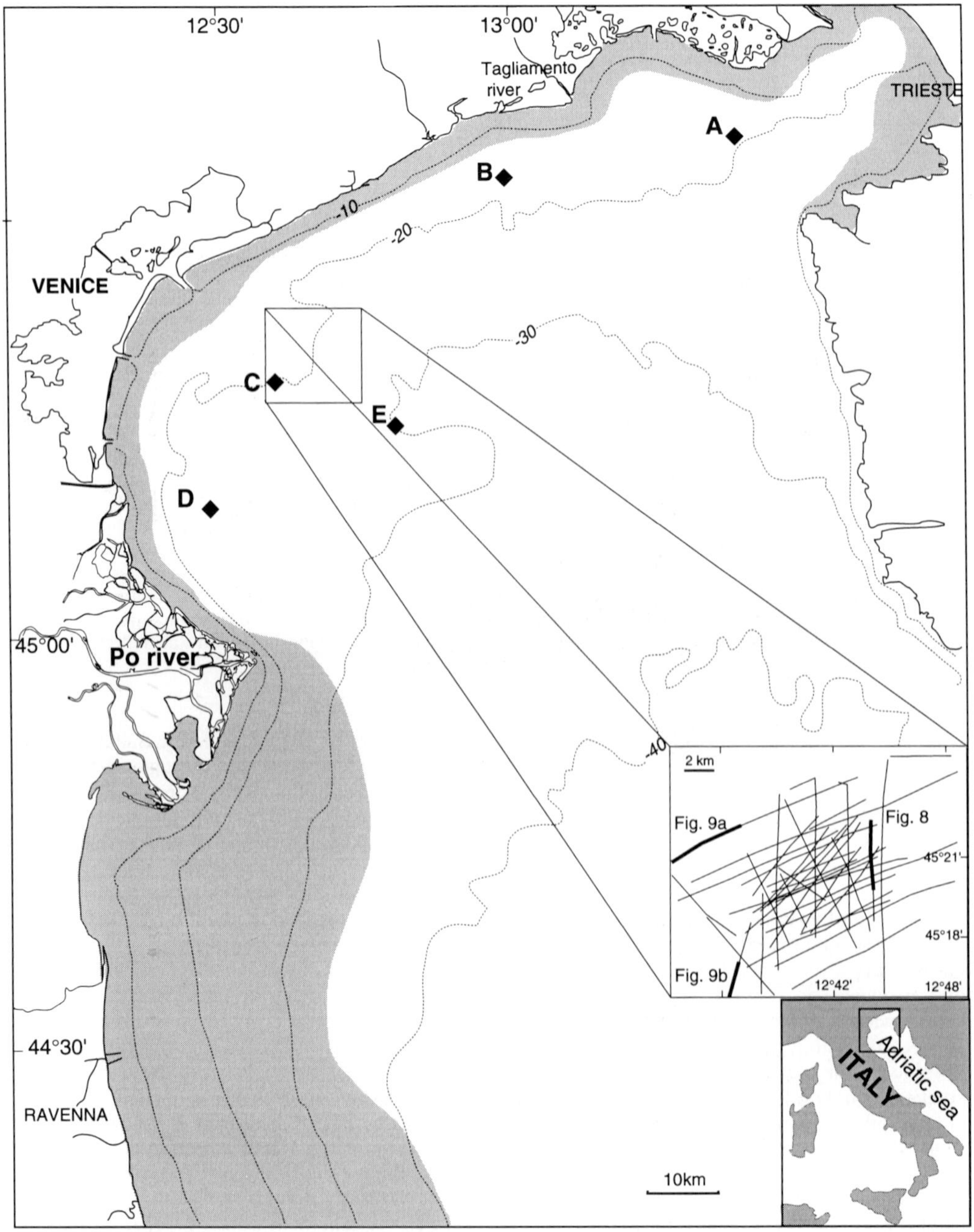

Fig. 1. Simplified bathymetry of the North Adriatic; regional contours are in metres. North of the Po delta, the highstand systems tract (HST; shaded area) consists of extensive barrier-lagoon systems, is undersupplied and shows a limited seaward extent. South of the Po delta the offshore extent of HST is greater. The inset shows the grid of seismic profiles collected in 1990 and 1991 in the study area. Capital letters denote other areas of the North Adriatic basin where the sea floor is characterized by bedforms of varied size: A, from Mosetti (1966); B, from Cavaleri & Stefanon (1980); C, bedforms in Fig. 9a,b; D, unpublished data of the Istituto per la Geologia Marina; E: sand ribbons, from Newton & Stefanon (1982). A B and C occur in water depths shallower than in the study area; D is in 20–25 m water depth; and E is in about 30 m of water.

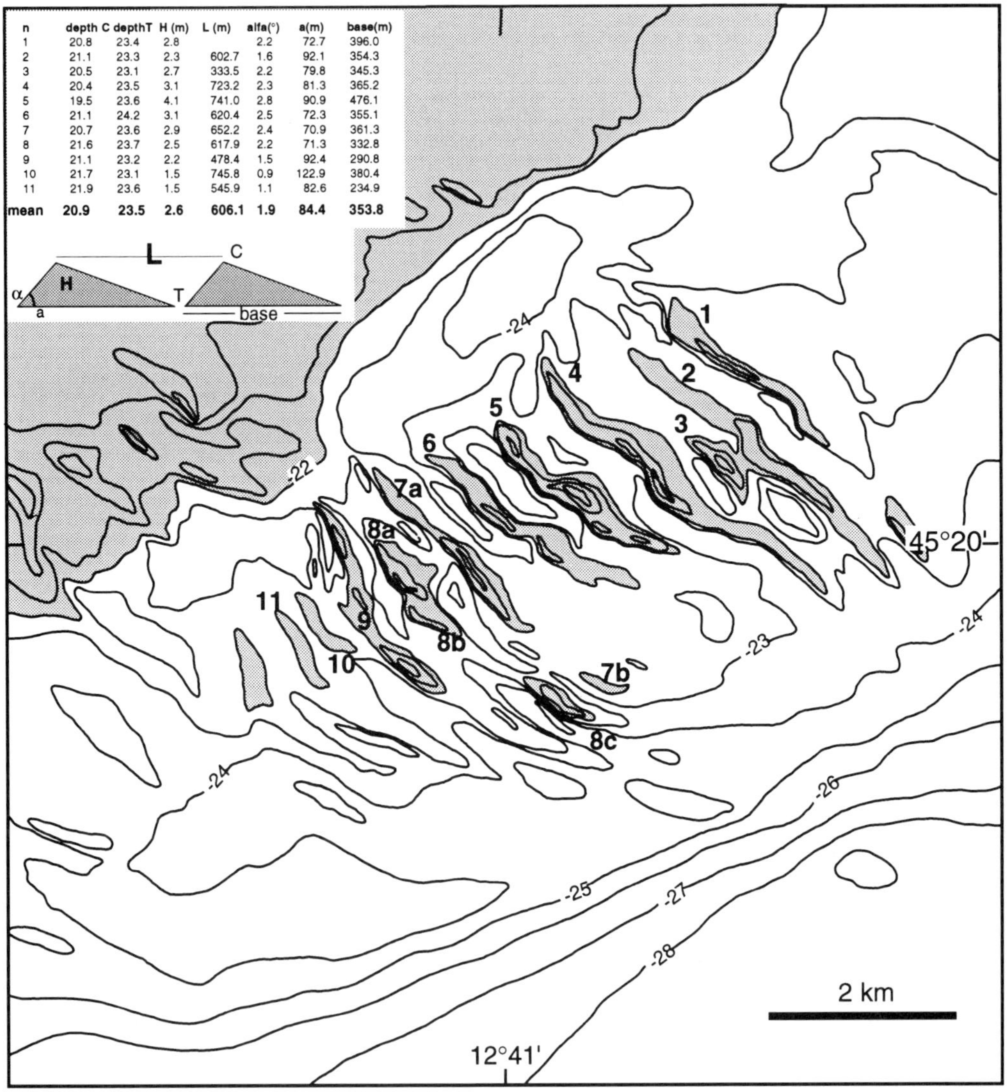

n	depth C	depthT	H (m)	L (m)	alfa(°)	a(m)	base(m)
1	20.8	23.4	2.8		2.2	72.7	396.0
2	21.1	23.3	2.3	602.7	1.6	92.1	354.3
3	20.5	23.1	2.7	333.5	2.2	79.8	345.3
4	20.4	23.5	3.1	723.2	2.3	81.3	365.2
5	19.5	23.6	4.1	741.0	2.8	90.9	476.1
6	21.1	24.2	3.1	620.4	2.5	72.3	355.1
7	20.7	23.6	2.9	652.2	2.4	70.9	361.3
8	21.6	23.7	2.5	617.9	2.2	71.3	332.8
9	21.1	23.2	2.2	478.4	1.5	92.4	290.8
10	21.7	23.1	1.5	745.8	0.9	122.9	380.4
11	21.9	23.6	1.5	545.9	1.1	82.6	234.9
mean	**20.9**	**23.5**	**2.6**	**606.1**	**1.9**	**84.4**	**353.8**

Fig. 2. Detailed bathymetry of the sand-dune field in the study area; contours are at 1 m intervals. Shading denotes areas that are shallower than 22 m. Note that the main field of bedforms belongs to an elongated shore-parallel mound that is delimited landward by an erosional trough. The inset summarizes the main geometric parameters for each of the main bedforms in the study area. Values are averaged from measurements taken from several profiles across each bedform.

genetic interpretation (Allen 1980; Belderson *et al.* 1982; Robin & Hunter 1982).

Methods

The data base on the dune field offshore the Venice Lagoon comprises approximately 480 km of 3.5 kHz high-resolution seismic-reflection profiles and 250 km of Uniboom profiles collected during two cruises in 1990 and 1991 (Fig. 1, inset). Profiles were recorded at time intervals of 0.25 s; firing and recording intervals of Uniboom profiles were 0.25 and 0.125 s respectively. Vertical resolution is on the order of 0.5 m in both kind of profiles. Navigation was based on GPS (global positioning system), and absolute positioning errors are less than 50 m. Ridges were defined through a detailed bathymetric map contoured at 0.5 m and plotted at 1 m intervals after corrections

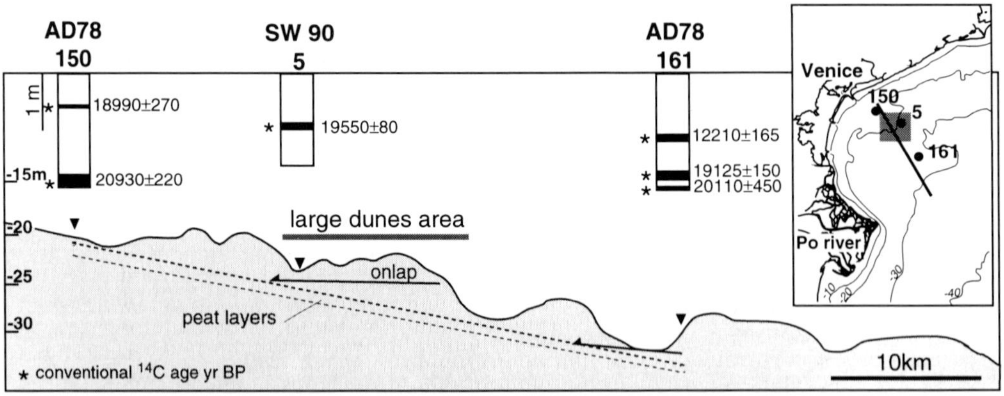

Fig. 3. Bathymetric profile showing extensive peat layers that originated during lowstand times and were drowned and transgressed during the late-Quaternary relative sea-level rise. In the North Adriatic, backstepping transgressive deposits form broad shore-parallel sediment mounds and troughs. The profile shows the location of sediment cores that contain ^{14}C-dated peat layers and the position of the sand-dune field discussed in this paper.

for velocity of sound in water and for tides. Sedimentological data on the large dunes came from four vibrocores; unfortunately, coring in the crests provided less than 1 m recovery. Gravity cores come from both the study area and the rest of the North Adriatic and allow dating and regional correlation of two peat horizons. About 30 box cores provide information on main variations in surface-sediment grain sizes on crests and troughs.

The Adriatic basin and the late Quaternary sea-level rise

Shelf morphology and surface-sediment distribution

The Adriatic Sea represents the largest (800 × 200 km) low-gradient epicontinental shelf in the Mediterranean area and corresponds to the Plio-Quaternary foreland basin of the Apennine chain. The thickness distribution of the late-Holocene highstand systems tract (HST) shows that modern deposition occurs south of the Po river delta, the main sediment entry point to the basin, and does not extend seaward more than 25 km from the modern coastline (Fig. 1) (Trincardi *et al.* 1994). Mean suspended-sediment discharge from modern rivers is negligible on the eastern side of the basin, very small in the north (where rivers deliver only 0.3×10^7 ton/year) and maximum on the western side of the basin, where the Po river and several smaller Apennine rivers provide a total of 3.9×10^7 ton/year (Frignani *et al.* 1992). The difference in sediment discharge rates north and south of the Po delta results in two contrasting regimes (Thorne & Swift 1991): river-mouth bypassing occurs where muddy sediment fed by the Po river as well as by other Apennine rivers escapes the shoreface and builds a prodelta wedge that is up to 20 m thick; on the contrary, in the north portion of the basin the shelf is sediment-starved, and modern rivers discharge into coastal-plain lagoons or build muddy prodelta wedges that are only a few metres thick (e.g. Tagliamento river; Fig. 1 and Stefanon 1979).

The modern sediment dispersal determines two distinct morphologic domains in the Adriatic epicontinental shelf: south of the Po delta and close to the western coast, the shelf floor is smooth and dips gently seaward; further offshore a rugged surface shows a variety of sediment mounds and ridges that are a few metres in relief and several kilometres in extent. This irregular sea floor extends close to the modern shoreface north of the Po delta and shows evidence of smaller-scale bedforms of varied orientation and geometry (Fig. 3).

The regional bathymetric gradient in the area north of the Po delta, if local roughness is averaged out, is less than 0.4 m/km and is the lowest in the Adriatic basin. The sea floor in this sector of the basin is mantled by a thin veneer of shelly sand between 10 and 50 cm thick (Colantoni *et al.* 1979; Trincardi *et al.* 1994). This veneer, which includes fragments of marine shells, fines upward and rests on a sharp erosional surface (ravinement surface; Trincardi *et al.* 1994). No gravel-size or pebbly lags are associated with the ravinement surface in any of the cores collected in the study area, or in the rest of the North Adriatic basin. This fact indicates that the depositional environments trans-gressed during the late-Quaternary were dominated by fine-grained sediment in a low-gradient alluvial plain.

Modern oceanography

The modern Adriatic sea is storm-dominated, microtidal and characterized by a cyclonic thermohaline circulation that forces riverine waters to flow southwards against the western side of the basin (Malanotte-Rizzoli 1994). The land-locked position, freshwater runoff and strong winter cooling of the sea surface control the formation of dense waters that flow to the south along the western side of the basin (Malanotte-Rizzoli & Bergamasco 1983; Malanotte-Rizzoli 1994).

The oceanographic setting of the North Adriatic is dominated by the following factors:

(1) tides – not generated by the direct action of gravity tidal forcing but depending on co-oscillation with Ionian Sea tides (Franco *et al.* 1982);
(2) seiches – associated with intensive winds from the SE accompanied by the passage of cyclonic areas on the Adriatic. Seiches may have diverse periods, the most important being that with 22 hour periodicity. During storm surges seiches can increase in amplitude to more than 80 cm;
(3) storm-generated currents and waves – set-ups in the north end of the basin can be in the order of 1 m (Franco *et al.* 1982). Waves up to 6 m in amplitude have periods in the order of 12 s. During November 1966 storm tide reached 1.9 m in the Venice Lagoon (Seibold & Berger 1982). During this event horizontal velocities of 112 cm/s may have developed in 22 m water depth by exceptional storm waves (Stefanon 1979).

During winter the North Adriatic is characterized by a two-gyre system driven by the vorticity input from a strong katabatic wind (named Bora) that blows offshore from the NE (Zore-Armanda & Gacic 1987). During winter the water mass in the North Adriatic is homogeneous and therefore the vertical shear is negligible so that the wind-induced transport is in the direction of the surface flow (Zore-Armanda & Gacic 1987). As a consequence, a cyclonic gyre has a shore-parallel component that affects the entire water mass in the shallow North Adriatic basin and in the study area.

Transgressive stratigraphy

Within the Adriatic basin, the late-Quaternary transgressive systems tract (TST) and related bounding surfaces show contrasting sedimentological expressions in cores and high-resolution seismic profiles taken from the central and deepest portion of the basin to the north (Trincardi *et al.* 1994). Low shelf gradients and decreasing sediment input relative to increasing basin size, favoured large landward shifts of the shoreline and resulted in the deposition of telescoped but thin transgressive parasequences. Following Thorne & Swift (1991) it is possible to split the late-Quaternary TST into its two basic components: the backbarrier and the marine components, separated by the ravinement surface. Where one of the two components is missing, the ravinement surface merges with either the underlying transgressive surface or the overlying maximum flooding surface (Trincardi *et al.* 1994; Saito 1994). The backbarrier wedge in the North Adriatic the late-Quaternary TST is typically a few metres thick and consists of a variety of backbarrier deposits truncated by the ravinement surface. In cores, the ravinement surface appears covered by a thin veneer of transgressive marine sands (Trincardi *et al.* 1994). The backbarrier wedge is thicker and more widespread in the basin sectors located south of the modern Po delta. The marine wedge above the ravinement surface corresponds either to muddy deposits under the influence of major rivers or to sandy deposits that are shaped into bedforms away from river inputs (Fig. 3).

Large dunes in the North Adriatic basin

Morphology

The bathymetric map of the area located 20 km SE of Venice contoured at 1m interval reveals a ridge-and-swale morphology in 20 to 24 m water depth (Fig. 2). The sand dunes rest on a large shore-parallel mound that is bounded landward by an elongated trough (Figs 2 and 3). Crest sinuosity is low and individual dunes are up to 2 km long and extend across the entire width of the underlying mound to the edge of the shore-parallel trough (Figs 4 and 5).

The large dunes are strikingly constant in length, shape, height, orientation and asymmetry (Figs 2 and 4). The spacing between them varies from 330 to 745 m; their height varies from 1.5 to 4.1 m. The orientation of the large dunes is NW–SE and normal to the regional contour; the dunes are relatively wide and

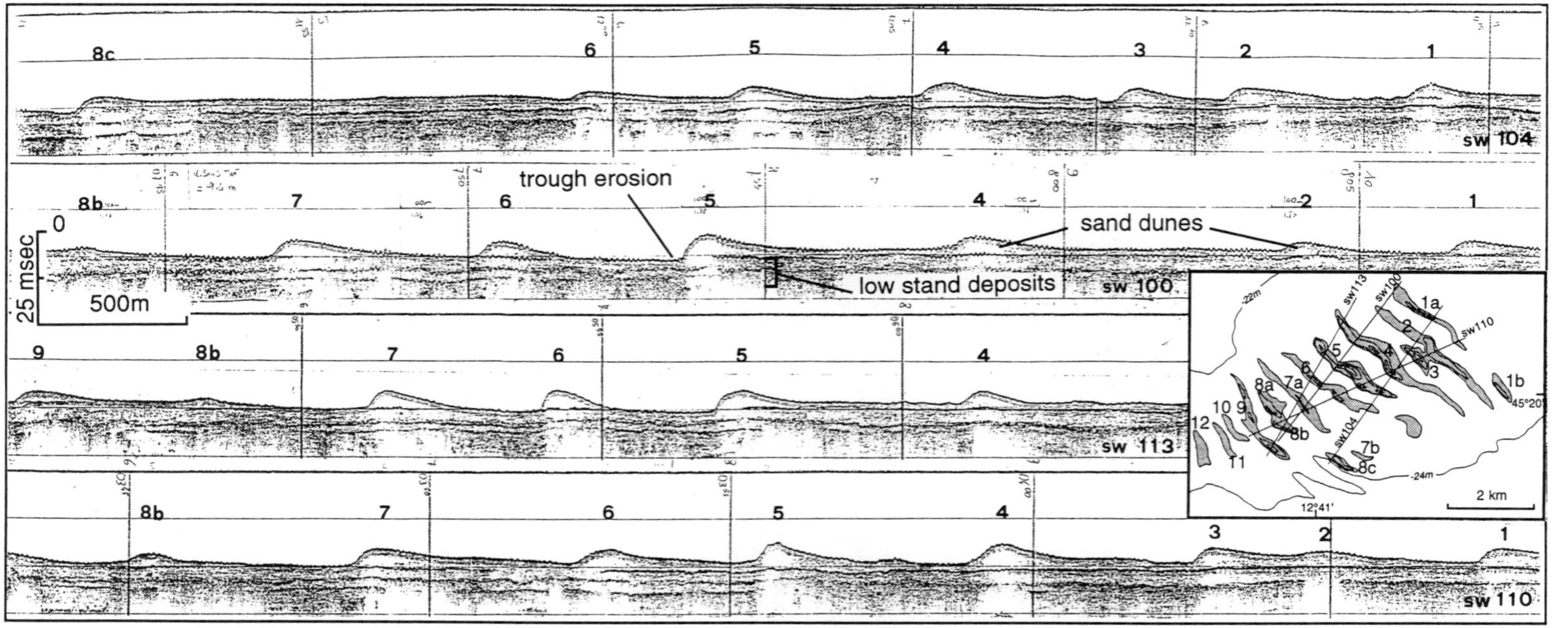

Fig. 4. High-resolution 3.5 kHz profiles across the dune crests document the lateral continuity of these features (numbers denote crests). The inset shows the tracks of profiles relative to the numbered bedforms. No internal surfaces can be detected within the dune crests; these features are undersupplied and erosion is evident in some of the troughs. Subhorizontal reflectors occur beneath the dune field and correlate throughout the study area. The shallowest reflector at the base of the dune field corresponds to a peat horizon, dated in three different sites in the study area (Fig. 6). This layer originated close to the 18-ka sea-level lowstand and occurs within fluvial deposits.

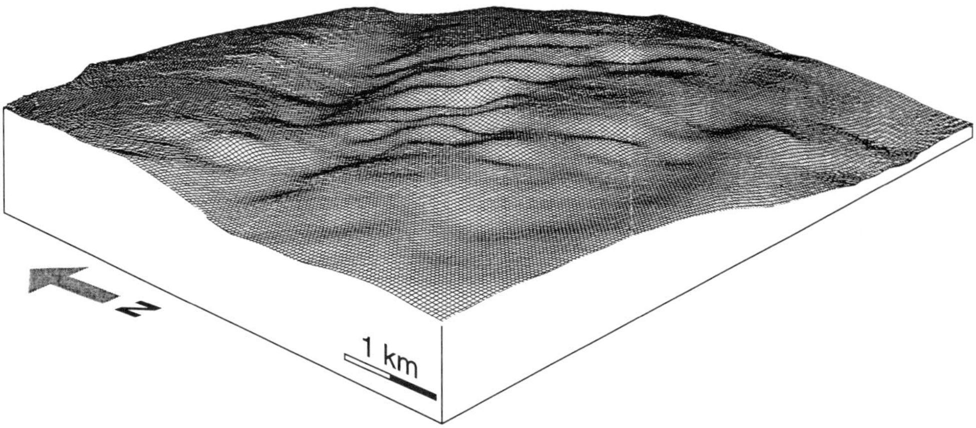

Fig. 5. Mesh diagram of the bathymetry in the study area (Fig. 1 inset), showing that the sand-dune field is superimposed on a larger-scale mound; this mound is roughly parallel to the regional contour and to the modern coastline. Note the shore-parallel trough on the landward limit of the bedform field.

markedly asymmetric with lee-side slopes of 0.9° to 2.8°, facing to the SW. The large spacing between dunes (Fig. 2) probably indicates that a reduced amount of sandy sediment was available. Furthermore, seismic profiles show clear evidence of erosion in some of the troughs; where erosion occurs, the uppermost regionally extensive peat layer is missing (Fig. 4). Trough erosion provides an additional source of sand to the dune field from the underlying fluvial sediments. A field of modern subtidal bedforms in the Bay of Bourgneuf is characterized by similar trough erosion; in this case the internal structure of the dunes is resolved and indicates climbing at a negative angle (Berné *et al.* 1991).

Sedimentology

Over two troughs and two crests we analysed 28 sediment samples that show grain sizes from fine to medium sand. Figure 6 shows schematic sedimentological logs of cores collected along one crest and the adjacent troughs; grain-size analyses were performed on both crest and trough samples. Figure 7 shows a scattergram that relates the graphic mean grain size versus sorting for all the samples: surficial sand samples from the upper 10 cm in the troughs are coarser and poorly sorted while the crest population is finer and moderately well-sorted. Within the 1 m of sand collected in the crest, however, this plot does not show major vertical trends in grain size. The coarser grain size in the trough population results from a more abundant bioclastic component while the lack of sorting is enhanced by the recycling of finer sediments from the units beneath.

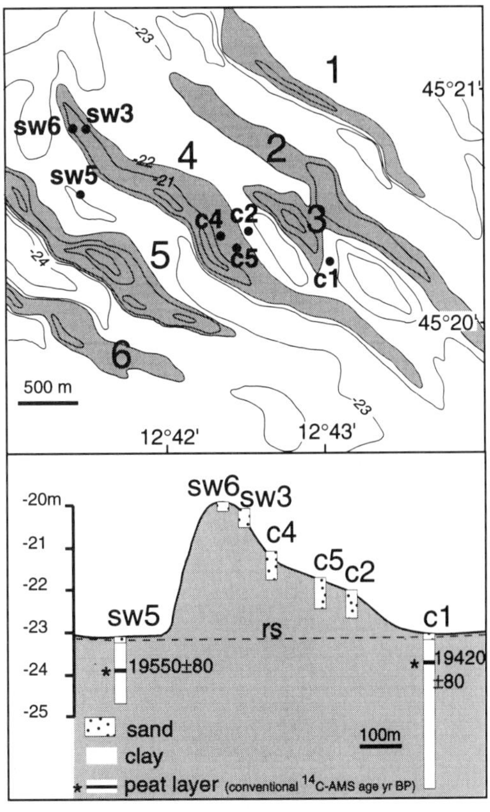

Fig. 6. Bathymetry of the central portion of the study area and location of cores recovered from the crest of dune 4 and the adjacent troughs. Sedimentological logs are projected on a schematic bathymetric profile. Note that the peat layer in cores SW 5 and C1 is about 19 ka old and belongs to the lowstand systems tract (LST). 'rs' denotes the coinciding transgressive and ravinement surfaces.

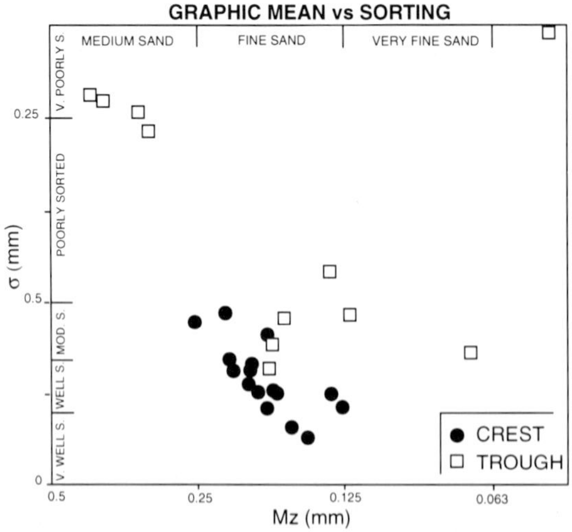

Fig. 7. Scattergram showing the grain size (graphic mean) versus sorting of surficial sediments. The crest population (filled dots) appears more sorted than the trough population (hollow squares).

The sand is lithic with quartz fragments and feldspars; the biogenic component is made of marine shell fragments, calcareous algae and serpulids; plant debris derived from *Poseidonia mattes* has been found as well. Coarse shell lags form in the trough, while the dune crests appear irregularly cemented. Foraminifera are rare (*Elphidium Crispum* and *Ammonia Beccarii*) but confirm that the sand ridges were reworked in a marine environment. The evidence of patchy cementation and the finding of *Poseidonia* roots on the ridge flanks indicate that the ridges are not active. The lack of any mud drape on top of the ridges is consistent both with the occurrence of starved conditions and with the effects of high-energy events. The occurrence of current ripples on some of the ridge crests (Colantoni & Gallignani 1980) indicates that during modern highstand conditions the ridges may be affected by exceptional storm flows, as suggested by Cavaleri & Stefanon (1980).

Stratigraphy

Laterally persistent subbottom reflectors identified in the seismic records below the sand ridge field are nearly horizontal or gently dipping to the south. Sediment cores provide some basic information on the nature and age of the units that lie immediately below the sand ridges (Figs 3, 6 and 7). These units were cored in the troughs, where the transgressive sand sheet is only about 10 cm thick; the sediment is a very fine-grained massive sand (0.09 mm); the sand is light grey, lithic and bears no evidence of biogenic content. Peat horizons are present within this massive sand and correlate throughout the entire study area; the shallowest regionally extensive peat horizon is found between a few metres and 80 cm below the sea floor (Figs 3 and 6). This correlatable peat layer is 19 ka old and was deposited in an alluvial plain environment, during the last glacial lowstand of sea level. The occurrence of this and other slightly older peat layers at shallow depths below the sea floor confirms that the study area was undersupplied during the following transgression. The peat layers that were deposited close to the last glacial maximum extend landward under the modern Venice Lagoon and are buried by coastal or paralic deposits of the HST that are younger than 5 ka (Fontes & Bertolami 1973; Trincardi *et al.* 1994). The gap in dates between 19 and 5 ka under the modern coastline deposits is derived from transgression under sediment-starved conditions and, from a sequence stratigraphical point of view, corresponds to the coinciding transgressive, ravinement and maximum flooding surfaces (Trincardi *et al.* 1994).

The shallowest peat horizon encountered in cores corresponds to a flat reflector that is weak in acoustic character and discontinuous laterally. The peat horizon is around 11.7 ka old and occurs only in the vicinity of the base of major sand ridges that helped prevent its erosion. The peat layer can also be correlated outside the study area in the North Adriatic (Colantoni *et al.* 1979); this layer, however, does not extend landward to water depths shallower than about 20 m indicating an onlap termination onto the older lowstand deposits (Fig. 3). The 11.7 ka old peat was deposited in an alluvial plain that was time-equivalent to a coastline located several tens of kilometres seaward in about 50 m of water, based on the published sea-level curves for the late-Quaternary (Fairbanks 1989).

A deep channel follows the northeast edge of the sand ridge field and can be traced for about 7 km; the thalweg is up to 10 m deep and about 100 m wide. The channel appears highly sinuous and presents a composite fill consisting of both lateral accretion (point bar?) and passive aggradation (Fig. 8). The flanks of the channel flatten under the ravinement surface indicating that this erosional feature belongs to the fluvial drainage system that originated across the Adriatic shelf close to the last glacial maximum. Smaller-scale erosional features are found below the sand ridge field; they are less than 2 m deep and filled by lateral accretion on low-angle beds. The fill consists of very fine sand and may represent an extra source of material for the ridges on this sediment-starved shelf environment.

Other fields of bedforms in the starved North Adriatic basin

Figure 1 reports all the available evidence of bedforms from the North Adriatic. Sand ridges and smaller-scale bedforms dominate the sea floor in water depths both deeper and shallower than in the study area (Fig. 9a, b). In water shallower than in the study area, bedforms of smaller size have a shore-normal orientation (A in Fig. 1 from Mosetti 1966). East of the study area, in water depths between 15 and 20 m large-scale bedforms have relief up to 3 m and spacing from 80 to 200 m (B in Fig. 1 from Newton & Stefanon 1982); their orientation is shore-normal (350°), the lee side dips westward, and the crests are capped and probably stabilized by patches of dead *Poseidonia* roots (Newton & Stefanon 1982). In deeper water, the transgressive deposit is reduced to a gravelly shell lag and is moulded into megaripples (Cavaleri & Stefanon 1980); periodic lineations spaced between 15 and 20 m and parallel to the coast are also found in 29 m water depth on hard and cohesive sea floor (E in Fig. 1 from Cavaleri & Stefanon 1980).

Starved bedforms on late-Quaternary shelves

The sand dunes present an unusual geometry given the water depth where they occur (20 to 24 m); the low height/width ratio seems to reflect the limited amount of sand available for the dunes and is consistent with the evidence of concurrent erosion in the troughs (Fig. 4); comparable trough erosion also occurs in other bedform fields outside the study area (Fig. 9a). Modern continental margins show several examples of bedforms that formed in sediment-starved environments close to the establishment of the modern high sea-level conditions. Figure 10 shows the empirical relation between bedform height and wavelength established by Flemming (1988); on this plot the sand ridges offshore Venice constitute an end member in which the wavelength is very large compared to other examples; smaller bedforms in the Adriatic basin, but in shallower waters, fall closer to the typical population. The large bedforms in the study area present a greater spacing because of the small amount of sand available. Where a limited volume of sand is available, bedforms evolve independently of the adjacent bedforms and appear isolated by the occurrence of flats (Perillo & Ludwick 1984); solitary bedforms observed in modern estuaries on low-gradient coastal plains are comparable in size and water depth to those observed in the study area (Aliotta & Perillo 1987).

In the Skagerrak basin at the entrance of the Baltic epicontinental sea, active bedforms are comparable to those observed in the study area; they are between 3 and 5 m high and show wavelengths of 250–600 m. These large-scale bedforms are attributed to non-tidal surge currents (Kuijpers *et al.* 1993), a mechanism that may have affected the sea floor in the North Adriatic. In the North Bering epicontinental shelf, large bedforms evolve under the influence of coast-parallel currents enhanced by wind forcing during storms (Field *et al.* 1981).

Discussion

The North Adriatic basin is a starved shelf setting where lowstand continental deposits are veneered by a transgressive sand lag that is only a few tens of centimetres thick. Where the thickness of the transgressive deposit increases, bedforms of various size and orientation are encountered. In contrast to the area just south of the Po delta, the North Adriatic basin does not show evidence of widespread and relatively thick backbarrier deposits in the late-Quaternary transgressive record (Colantoni *et al.* 1990; Trincardi *et al.* 1994). By studying the morphology, sediment composition and stratigraphical setting of one of the best developed fields of bedforms in the North Adriatic we have tried to answer two related questions:

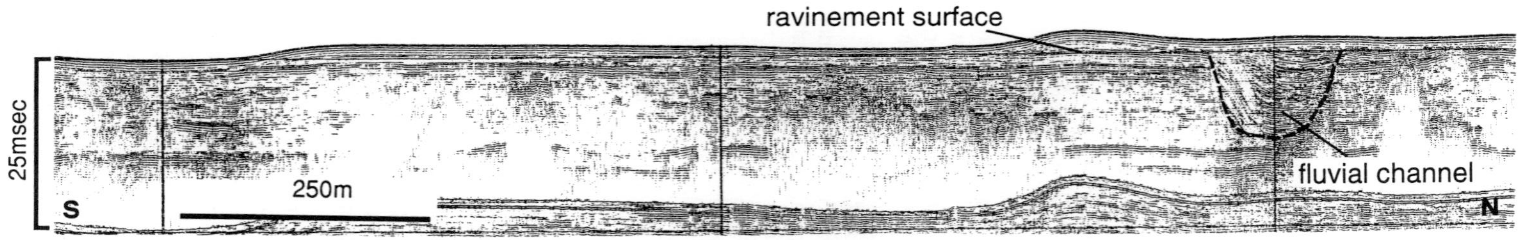

Fig. 8. High-resolution 3.5 kHz profile (located in Fig. 1 inset) showing a channel-fill deposit along the north and eastern edges of the dune field.

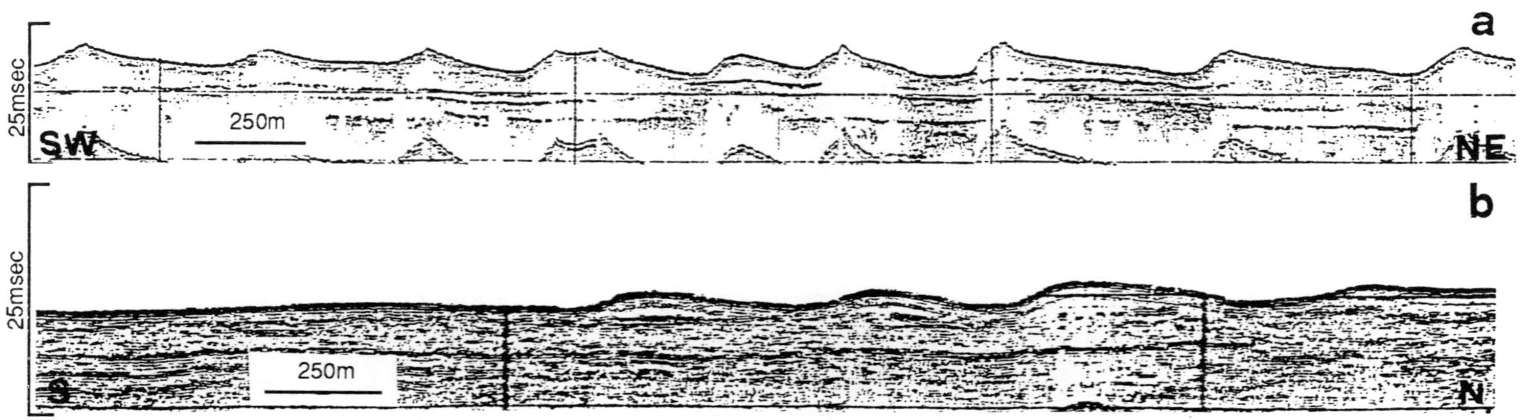

Fig. 9. Fields of asymmetric sand bedforms adjacent to the study area (see Fig. 1 for location) in shallower water (**a**), and in water depths comparable to those in the study area (**b**).

Location(reference)	spacing L(m)	height H(m)	water depth(m)
Torres Strait - Australia (Harris 1988)	80-750	2-6	16-18
Bass Strait - Australia (Malikides et al., 1989)	370	9	26-35
Adolphus Ch. - Australia (Harris, 1989)	102	3.9	20
Skagerrak Sea - Denmark (Kuijpers et al., 1993)	250-600;800-900	3.5-7	20-75
Chesapeake Bay (Perrillo and Ludwick, 1984)	200	2.1	10
Alabama Shelf - USA (Parker, 1992)	500	2	16-22
Long Island Sound - USA (Fenster et al., 1990)	118-300	4-12	35-90
Gironde Estuary- France (Berné et al., 1993)	37-182	1.5-6.7	15-25
Chesapeake Bay- USA (Ludwick, 1972)	60-245	1.5-3.4	3-18
Cook Inlet - Alaska (Bouma et al.,1980)	100-500	5-9	25-120
North Adriatic Shelf - Italy (Cavaleri and Stefanon, 1980)	80-200	1-3	20
North Adriatic Shelf - Italy **This study**	**330-745**	**1.5-4.1**	**23**

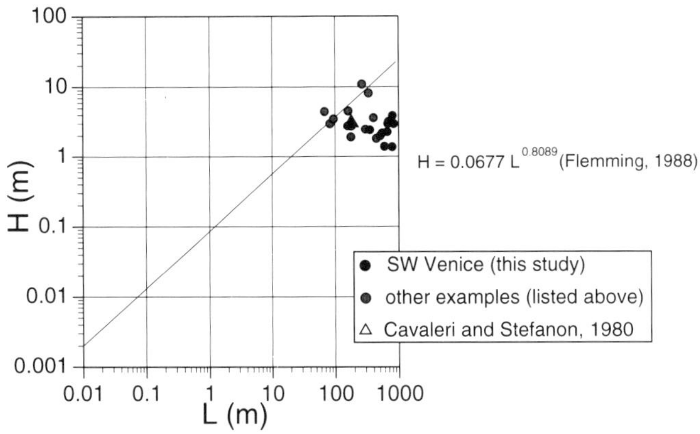

Fig. 10. Empirical relation between height (H) and wavelength (L) after Flemming (1988); on this kind of plot the sand dunes offshore Venice (black circles) clearly constitute an end member with larger spacing than in other examples because of the scarcity of sand available. Smaller bedforms in the Adriatic basin (white circles), but in shallower water depths, appear closer to the typical curve (from Cavaleri & Stefanon 1980). Examples of other large flow-transverse bedforms in shallow water on modern continental shelves are listed above, for comparison.

(1) what was the source of the sand on this sediment-starved epicontinental shelf where lowstand continental deposits are encountered within 1 m below the sea floor? and
(2) what are the mechanism and the timing of formation of these unusual bedforms relative to the late-Quaternary sea-level rise?

Two different kinds of sources may explain the accumulation of the sand that is necessary to create the observed bedforms: the breaching through lowstand alluvial plain deposits that underlie the transgressive surface by storm flows, or the drowning of a transgressive coastal sediment body. Evidence for both interpretations was found in late-Quaternary deposits on other continental margins (Stubblefield & Swift 1976; McBride & Moslow 1991). The first hypothesis requires an original ridge-and-swale morphology formed at the leading edge of the transgressive sand sheet and implies subsequent downcutting and removal of the sand from the troughs (Stubblefield & Swift 1976); downcutting stops and lateral erosion takes over when a deeper cohesive clay bed is encountered (Stubblefield & Swift 1976). In the study area this interpretation is consistent with the evidence of erosion in the troughs and would explain the unusual spacing of the large dunes for the limited water depth. However, the second hypothesis explains better why the sand dunes are superimposed on a larger shore-parallel mound (Figs 2 and 5); this large-scale relief could represent the legacy of a drowned coastal lithosome as a delta-front bar, a littoral spit or an ebb-tidal delta. In this scenario, the lithosome was drowned by the continuing transgression and since that time was progressively reworked into a ridge-and-swale topography. Shoreface attached and detached sand dunes are found along several modern coastal plain shelves either under the influence of major delta systems (Penland *et al.* 1988) or on sediment-starved barrier-island coasts (Field 1976; McBride & Moslow 1991; Siringan & Anderson 1993). In these settings significant amounts of sand are concentrated on the inner shelf by a variety of possible mechanisms that are commonly associated with the construction of river or ebb-tidal deltas (Siringan & Anderson 1993). Estuarine backfilling during sea-level rise may be accompanied by deposition in a delta-front environment; delta front deposition controls the formation of positive reliefs that rest over buried fluvial channels (Johnson *et al.* 1982). Studies on the Middle Atlantic shelf revealed the control of ancestral (lowstand) rivers on the thickness of the late-Quaternary transgressive sand sheet and thus, more indirectly, on the possible growth of bedforms (Knebel 1981).

In the North Adriatic the asymmetric dunes represent flow-transverse bedforms with the lee side facing consistently to the southwest. The orientation of these and other bedforms in the basin appears consistent with the shore-parallel circulation pattern that can be observed today; this pattern was established after the area was transgressed and drowned by the sea. The depth-average velocity required to produce sand waves like those observed in the study area is about 40 to 80 cm/s (from Rubin & McCulloch 1980). This velocity can occasionally be reached even today in the North Adriatic sea during exceptional storms or combined high tides and storm surges (Stefanon 1979; Mosetti 1985); however, storms of lower intensity but higher recurrence interval were probably capable of remoulding sand on the sea floor soon after drowning of the area but before the modern highstand was attained (Cavaleri & Stefanon 1980).

We suggest that the shelf has evolved to its present state through the succession of three discrete steps controlled by rising sea level and the onset of the modern hydraulic regime (Fig. 11): (1) the emplacement of a transgressive lithosome, (2) its drowning and marine reworking into a field of sand dunes and (3) the subsequent drowning of the reworked lithosome and its stabilization by early cementation and growth of *Poseidonia mattes*; these three steps all occurred in a short time interval close to the end of the late-Quaternary sea-level rise (Fig. 11).

The large dunes found in 20–24 m water probably formed after the modern circulation pattern was established in response to the substantial widening of the basin driven by the

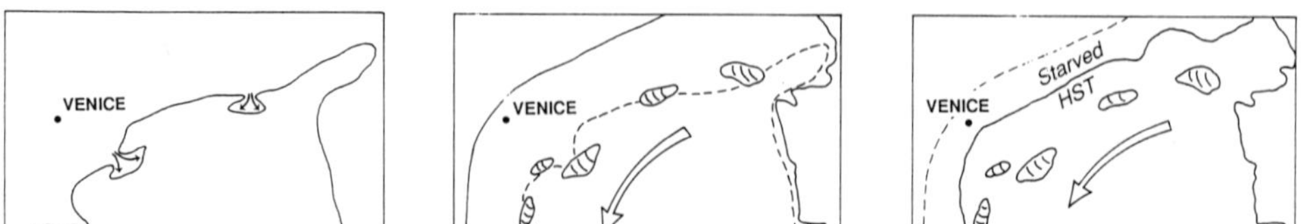

Fig. 11. Schematic reconstruction of the latest portion of the late-Quaternary sea-level rise and highstand in the North Adriatic shelf. Patches of bedforms formed wherever concentrations of sandy material were available along the shoreline. Left: ebb-tidal deltas and fluvial deltas provide concentrations of sand along the 7–8 ka shoreline. Centre: shoreline at time of maximum flooding (*c.* 3–4 ka BP): sand bodies are drowned and reworked into large dunes or smaller bedforms. Right: HST deposition reflects varied sediment input along the North Adriatic coast: where the HST is undersupplied, the sea-floor topography reflects the occurrence of reworked transgressive deposits.

late-Quaternary sea-level rise (Trincardi *et al.* 1994). However, the evidence of early cementation and the occurrence of dead roots of *Poseidonia* on their crests suggest that these features are not active during the present highstand. This observation implies that remoulding of sediment by marine processes into bedforms was most effective in the North Adriatic Sea for four concurrent factors that were maximized close to the turn-around point at the end of the transgression but before the onset of highstand progradation: (1) the decreased rate at which accommodation was created allowed more response time to physical processes (waves and currents) for re-equilibrating drowned deposits into bedforms and/or erosional features; (2) the decreased sediment discharge when river base level approached its highest position close to the end of the sea-level rise enhanced the reworking of older transgressive deposits on the sea floor; (3) the maximum widening of the semi-enclosed Adriatic epicontinental basin determined a larger fetch for the winds from the southeast and a stronger thermohaline circulation; and (4) the extension of the basin to an area that was probably more strongly affected by the very strong katabatic winds from the northeast.

The last two factors, in particular, can be taken into account when interpreting transgressive responses to high-frequency relative sea-level changes in ancient epicontinental basins. Indeed, compared to pericontinental margins, where changes in the balance between relative sea-level change and sediment input result in a relatively simple landward or seaward shift of the shoreline, epicontinental seas are characterized by larger variations in both basin size, coastline length, palaeogeography and oceanographic regime.

Conclusion

The large-scale sand dunes investigated offshore the Venice Lagoon constitute one of several fields of bedforms that originated during the latest portion of the late-Quaternary sea-level rise, in the North Adriatic Sea. We suggest that the sand dunes in the study area formed from the reworking of a drowned coastal lithosome accompanied by secondary erosion in the troughs and recycling of lowstand fluvial sand. The development of this sand dune field records an increased reworking under marine conditions, compared to other coastal lithosomes that were drowned in deeper waters during earlier phases of the late-Quaternary transgression.

Transgressive reworking is maximized in the shallow North Adriatic as a result of two sets of factors: those that characterize any continental margin close to the end of a transgression and those that are specific of a transgressed low-gradient epicontinental basin. Decreased rates of accommodation space and decreased sediment supply belong to the first set of factors; the onset of a stronger oceanographic regime in response to the widening of the semi-enclosed basin belongs to the second group and may have enhanced the potential for reworking of drowned transgressive coastal deposits close to the end of the sea-level rise.

We are indebted to Giovanni Bortoluzzi, Leonardo Langone, Stefano Miserocchi, Luca Masini and Daniela Penitenti for their help during the 1990 survey and to Marco Ligi for helping with the processing of navigation and bathymetry data. Reading and constructive criticism from Serge Berné and Dave Rubin are gratefully acknowledged. This is contribution no. 999 of the Istituto per la Geologia Marina.

References

ALIOTTA, S. & PERILLO, G. M. E. 1987. A sand wave field in the entrance to Bahia Blanca Estuary, Argentina. *Marine Geology*, **76**, 1–14.

ALLEN, J. R. L. 1980. Sand waves: a model of origin and internal structure. *Sedimentary Geology*, **26**, 281–328.

BELDERSON, R.H., JOHNSON, R.H. & KENYON, N. 1982. Bed forms. *In*: STRIDE, A. H. (ed.) *Offshore Tidal Sands*. Chapman and Hall, London, 27–57.

BERNÉ, S., CASTAING, P., LE DREZEN, E. & LERICOLAIS, G. 1993. Morphology, internal structure, and reversal asymmetry of large subtidal dunes in the entrance to Gironde Estuary (France). *Journal of Sedimentary Petrology*, **63**, 780–793.

——, DURAND, J. & WEBER, O. 1991. Architecture of modern subtidal dunes (sand waves), Bay of Bourgneuf, France. *In*: MIALL, A. D. & TYLER, N. (eds) *The Three-dimensional Facies Architecture of Terrigenous Clastic Sediments and its Implications for Hydrocarbon Discovery and Recovery*. Society of Economic Paleontologists and Mineralogists, Concepts in Sedimentology and Paleontology, **3**, 245–260.

BOUMA, A. H., RAPPERPORT, M. L., ORLANDO, R. C. & HAMPTON, M.A. 1980. Identification of bedforms in lower Cook Inlet, Alaska. *Sedimentary Geology*, **26**, 157–177.

BRAMBATI, A. & VENZO, G. 1967. Recent sedimentation in North Adriatic sea between Venice and Trieste. *Studi Trentini Scienze Naturali*, **A44–1**, 202–274.

CAVALERI, L. & STEFANON, A. 1980. Bottom features due to extreme meteorological events in the North Adriatic Sea. *Marine Geology*, **79**, 159–170.

COLANTONI, P. & GALLIGNANI, P. 1980. Ricerche sulla piattaforma continentale dell'alto Adriatico. *Progetto Finalizzato Oceanografia e Fondi Marini. Quaderno*, **2**, 1–87.

——, —— & LENAZ, R. 1979. Late Pleistocene and Holocene evolution of the North Adriatic continental shelf. *Marine Geology*, **33**, M41–M50.

——, PRETI, M. & VILLANI, B. 1990. Sistema deposizionale e linea di riva olocenica sommersi in Adriatico al largo di Ravenna. *Giornale di Geologia*, **52**, 1–18.

CORREGGIARI, A., FIELD, M. E., BORTOLUZZI, G., LIGI, M. & PENITENTI, D. 1992. *Ridge and Swale Morphology on the North Adriatic Epicontinental Shelf.* Report Commission Internationale pour l'Exploration Scientifique de la Mer Méditeranée, **33**, 125.

FAIRBANKS, R. G. 1989. A 17,000 year glacio-eustatic sea-level record: influence of glacial melting rates on the Younger Dryas event and deep-ocean circulation. *Nature*, **342**, 637–642.

FENSTER, M. S., FITZGERALD, D. M., BOHLEN, W. F., LEWIS, R. S. & BALDWIN, C. T. 1990. Stability of giant sand waves in Eastern Long Island Sound, USA. *Marine Geology*, **91**, 207–225.

FIELD, M. E. 1976. Sand bodies on coastal plain shelves: Holocene record of the U.S. Atlantic inner shelf off Maryland. *Journal of Sedimentary Petrology*, **50**, 505–528.

——, NELSON, C. H., CACCHIONE, D. A. & DRAKE, D. E. 1981. Sand waves on an epicontinental shelf: North Bering Sea. *Marine Geology*, **42**, 233–258.

FLEMMING, B. W. 1988. Zur klassifikation subaquatisher, stromungstranversaler transport-korper. *Bochumer Geologische und Geotechnische Arbeiten*, **29**, 44–47.

FONTES, J. C. & BORTOLAMI, G. 1973. Subsidence of the Venice area during the past 40,000 yr. *Nature*, **244**, 339–341.

FRANCO, P., JEFTIC, L., MALANOTTE-RIZZOLI, P., MICHELATO, A. & ORLIC, M. 1982. Descriptive model of the North Adriatic. *Oceanologica Acta*, **5**, 379–389.

FRIGNANI, M., LANGONE, L., PACELLI, M. & RAVAIOLI, M. 1992. *Input, Distribution and Accumulation of Dolomite in Sediment of the Middle Adriatic Sea.* Report Commission Internationale pour l'Exploration Scientifique de la Mer Méditeranée, **33**, 324.

HARRIS, P. T. 1988. Large-scale bedforms as indicators of mutually evasive sand transport and the sequential infilling of wide-mouthed estuaries. *Marine Geology*, **57**, 273–298.

——1989. Sandwave movement under tidal and wind-driven currents in a shallow marine environment: Adolphous Channel, northwestern Australia. *Continental Shelf Research*, **9**, 981–1002.

JOHNSON, D. P., SEARLE, D. E. & HOPLEY, D. 1982. Positive relief over buried post-glacial channels, Great Barrier Reef province, Australia. *Marine Geology*, **46**, 149–159.

KNEBEL, H. J. 1981. Processes controlling the characteristics of the surficial sand sheet, U.S. Atlantic outer continental shelf. *Marine Geology*, **42**, 349–368.

KUIJPERS, A., WERNER, F. & RUMOHR, J. 1993. Sandwaves and other large-scale bedforms as indicators of non-tidal surge currents in the Skagerrak off Northern Denmark. *Marine Geology*, **111**, 209–221.

LUDWICK, J. C. 1972. Migration of tidal sand waves in Chesapeake Bay entrance. *In*: SWIFT, D. J. P., DUAND, D. B. & PILKEY, O. H. (eds) *Shelf Sediment Transport: Processes and Pattern.* Dowden, Hutchison & Ross, Stroudsbourg, 377–410.

MALANOTTE-RIZZOLI, P. 1994. The North Adriatic Sea as a prototype of convection and water mass formation on the continental shelf. *In*: MALANOTTE-RIZZOLI, P. (ed.) *The General Circulation of the Oceans.* Istituto Veneto Scienze Lettere ed Arti, 267–288.

—— & BERGAMASCO, A. 1983. The dynamics of the coastal region of the North Adriatic Sea. *Journal of Physical Oceanography*, **13**, 1105–1130.

MALIKIDES, M., HARRIS, P. T. & TATE, P. M. 1989. Sediment transport and flow over sandwaves in a non-rectilinear tidal environment: Bass Strait, Australia. *Continental Shelf Research*, **9**, 203–221.

MCBRIDE, R. A. & MOSLOW, T. F. 1991. Origin, evolution, and distribution of shoreface sand ridges, Atlantic inner shelf, USA. *Marine Geology*, **97**, 57–85.

MOSETTI, F. 1966. Morfologia delli Adriatico Settentrionale. *Bollettino di Oceanologia Teorica e Applicata*, **XIII/30**, 138–150.

——1985. Problemi di previsione dell acqua alta nell'Adriatico Settentrionale. *Bollettino di Oceanologia Teorica e Applicata*, **III/4**, 263–282.

NEWTON, R. S. & STEFANON, A. 1982. Side-scan sonar and subbottom profiling in the North Adriatic Sea. *Marine Geology*, **46**, 279–306.

PARKER, S. J., SHULTZ, A. W. & SHROEDER, W. W. 1992. *Sediment Characteristics and Seafloor Topography of a Palimpsest Shelf, Mississippi–Alabama Continental Shelf.* Society of Economic Paleontologists and Mineralogists, Special Publication, **48**, 243–251.

PENLAND, S., BOYD, R. & SUTER, J. R. 1988. Transgressive depositional systems of the Mississippi delta plain: a model for barrier shoreline and shelf sand development. *Journal of Sedimentary Petrology*, **58**, 932–949.

PERILLO, G. M. & LUDWICK, J. C. 1984. Geomorphology of a sand wave in lower Chesapeake Bay, Virginia, USA *Geo-Marine Letters*, **4**, 105–112.

RUBIN, D. M. & HUNTER, R. E. 1982. Bedforms climbing in theory and nature. *Sedimentology*, **29**, 121–138.

—— & MCCULLOCH, D. S. 1980. Single and superimposed bedforms: a synthesis of San Francisco Bay and flume observation. *Sedimentary Geology*, **25**, 169–188.

SAITO, Y. 1994. Shelf sequence and characteristic bounding surfaces in a wave-dominated setting: Latest Pleistocene–Holocene examples from the Northeast Japan. *Marine Geology*, **120**, 105–127.

SEIBOLD, E. & BERGER, W. H. 1982. *The Sea Floor.* Springer, Berlin.

SIRINGAN, F. P. & ANDERSON, J. B. 1993. Seismic facies, architecture, and evolution of the Bolivar Roads tidal inlet/delta complex, east Texas Gulf Coast. *Journal of Sedimentary Petrology*, **63**, 794–808.

STEFANON, A. 1979. Gli affioramenti rocciosi dell'Alto Adriatico: considerazioni sulla loro distribuzione, struttura ed evoluzione, nel contesto della problematica del bacino. *Proceedings del Convegno sul P. F. Oceanografia e Fondi Marini del CNR*, Roma, 1233–1242.

STUBBLEFIELD, W. L. & SWIFT, D. J. P. 1976. Ridge development as revealed by sub-bottom profiles on the central New Jersey shelf. *Marine Geology*, **20**, 315–334.

THORNE, J. A. & SWIFT, D. J. P. 1991. Sedimentation on continental margins, VI: a regime model for depositional sequences, their component systems tracts, and bounding surfaces. *In*: SWIFT, D. J. P., OERTEL, G. F., TILLMAN, R. W. & THORNE, J. A. (eds) *Shelf Sand and Sandstone Bodies – Geometry, Facies and Sequence Stratigraphy.* International Association of Sedimentologists, Special Publication, **14**, 189–255.

TRINCARDI, F., CORREGGIARI, A. & ROVERI, M. 1994. Late Quaternary transgressive erosion and deposition in a modern epicontinental shelf: the Adriatic semi-enclosed basin. *Geo-Marine Letters*, **14**, 41–51.

ZORE-ARMANDA, M. & GACIC, M. 1987. Effects of the Bura on the circulation in the North Adriatic. *Annales Geophysicae*, **5B**, 93–102.

Seismic expression of depositional sequences associated with expansion and contraction of ice sheets on the northwestern Antarctic Peninsula continental shelf

P. J. BART & J. B. ANDERSON

Department of Geology and Geophysics, Rice University, Houston, TX 77251, USA

Abstract: We present results from seismic stratigraphical analysis of a regional grid of intermediate-resolution seismic profiles from the Antarctic Peninsula continental shelf that reveal greater complexity in the stratigraphical record than has been previously recognized using lower-resolution seismic records. Widespread unconformities and prograding strata patterns are interpreted as subglacial erosional surfaces and proglacial depositional systems respectively. Two types of units are recognized, aggrading-shelf units and prograding-slope units.

Aggrading-shelf units display minimal slope progradation (less than 2 to 3 km) beyond the pre-existing shelf edge. Units of this type are composed of either acoustically layered prograding foresets or acoustically chaotic seismic facies. Unit thicknesses vary along the shelf from less than 20 ms two-way travel time to more than 200 ms. Low-angle foresets downlap and partially fill palaeotroughs that were formed during prior glacial advances. Foresets are interpreted as proglacial depositional surfaces. The zones of active proglacial deposition were lobate in shape and ranged in width from a few kilometres to more than 30 km. Toplap truncations of foresets result from ice-sheet expansion over proglacial deposits.

Glacial erosion on the shelf was concentrated along the axes of ice streams to form deep troughs. Slope progradation occurred mainly near the mouths of these troughs. Prograding-slope units exhibit local slope progradation in excess of 2 to 3 km beyond the location of the pre-existing shelf edge.

Shifts in the position of troughs indicate lateral shifts in the positions of ice streams. Hence, sedimentation on the outer shelf and upper slope is point-sourced, not line-sourced, and individual units change from progradational to aggradational along the margin. It is not possible to interpret the ice-sheet grounding history of the shelf using changes in strata stacking patterns without constraining these lateral changes. However, detailed mapping of prograding-slope units does reveal regional trends that imply glacial periods when greater quantities of sediment were eroded from the shelf and deposited on the slope. We interpret prograding-slope units as having been formed during relatively long-duration glacial expansions. Ice-sheet retreat from the shelf probably occurred rapidly, as indicated by a lack of backstepping proglacial deposits.

The marine stratigraphical record of Antarctic glaciation is based largely on 'proxy' evidence, including eustatic curves, oxygen isotope records, down-core variations in ice-rafted debris, deep-sea hiatuses, and latitudinal shifts in microfossils with time. These methods do not reveal when and where ice sheets reached sea level and grounded on the continental shelf. Combined seismic records and drill core can provide a direct record of glaciation on the continental shelf, but there is disagreement as to how seismic data should be interpreted.

Larter & Barker (1989) and Cooper *et al.* (1991) argued that the overall stacking patterns of strata on Antarctic continental shelves are unique, relative to lower-latitude shelves. They contended that differences are due to the influence of glaciation on the shelf and presented models that related stratal patterns to ice-sheet advance and retreat. Bartek *et al.* (1991) agreed that ice sheets have profoundly influenced shelf stratal geometries, but their comparison of seismic record from the Ross Sea continental shelf and various low-latitude shelves showed broad similarities, which they attributed to eustasy. From this, Bartek *et al.* (1991) and Anderson & Bartek (1992) warned against using stratal geometries alone to reconstruct ice-sheet grounding events on the continental shelf. But the question remains, how are the observed seismic stratigraphical packages on the Antarctic continental shelf produced and how are ice-sheet grounding events manifested in these strata?

This paper presents results of a seismic stratigraphical and seismic facies analysis of this data set. One product of this research was a depositional model that relates the observed geomorphic features, seismic facies and stratal stacking patterns to ice-sheet advance and retreat on the shelf.

From De Batist, M. & Jacobs, P. (eds), 1996, *Geology of Siliciclastic Shelf Seas*, Geological Society Special Publication No. 117, pp. 171–186.

Methods

A regional grid of over 3800 km of intermediate-resolution single-channel seismic data (Fig. 1) was collected on the northwestern Antarctic Peninsula continental shelf in 1988 and 1990 using either one or two 100 in^3 waterguns (1988), or a 100 in^3 Generator-Injector airgun (1990). The survey extends from Marguerite Bay to the southern end of Bransfield Basin (Fig. 1).

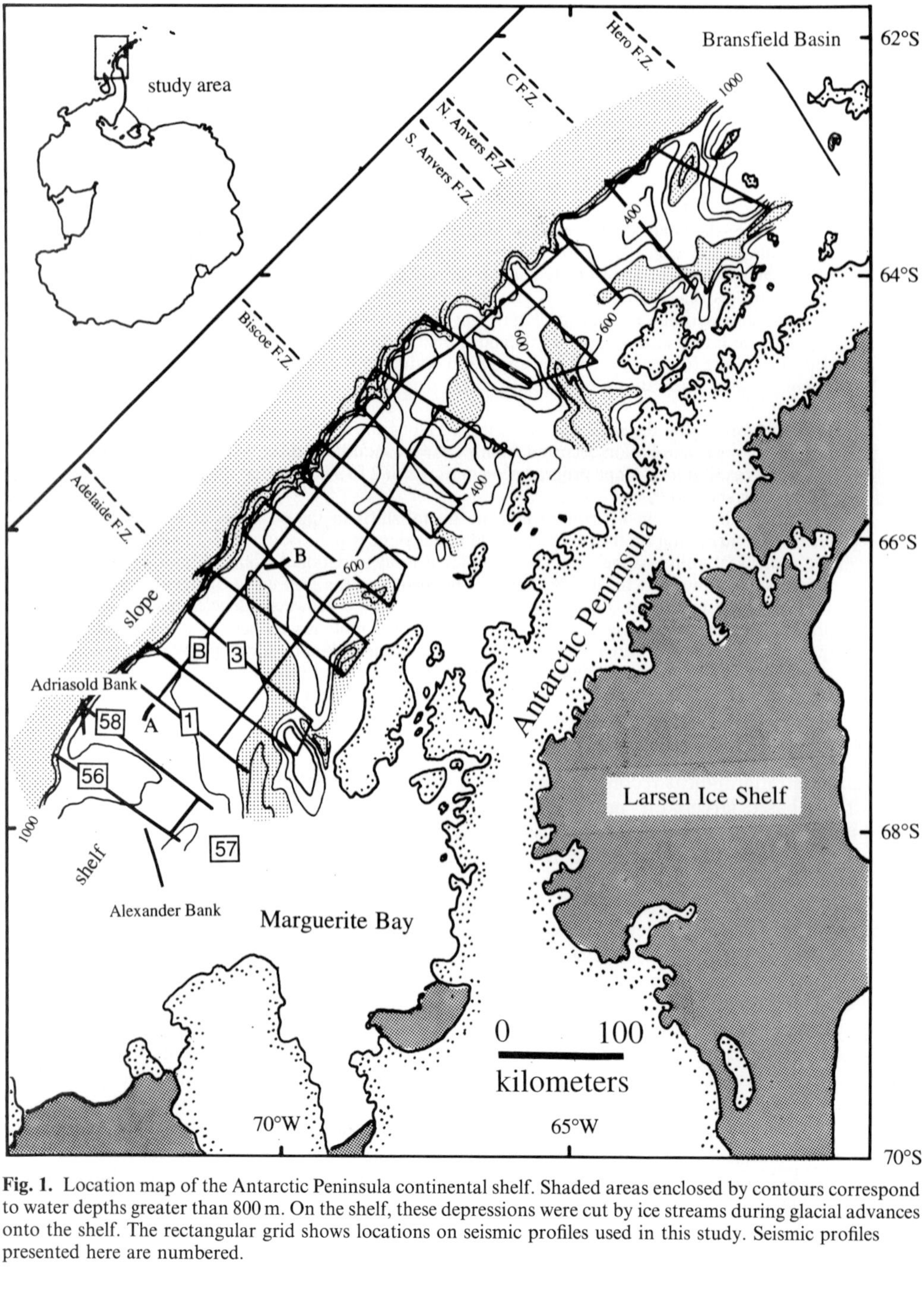

Fig. 1. Location map of the Antarctic Peninsula continental shelf. Shaded areas enclosed by contours correspond to water depths greater than 800 m. On the shelf, these depressions were cut by ice streams during glacial advances onto the shelf. The rectangular grid shows locations on seismic profiles used in this study. Seismic profiles presented here are numbered.

Seismic profiles were interpreted from vertically exaggerated plots (approximately 30:1). This scale exaggeration enhances recognition of glacial features, particularly glacial troughs.

Bathymetry

A map of the study area's major bathymetrical features was constructed using available bathymetrical charts and seismic data (Fig. 1). The average depth of the outer shelf is 400 m and the width of the shelf is approximately 150 km. Glacial troughs are the most prominent bathymetrical features on the shelf. The orientations of these troughs vary between longitudinal and transverse relative to the coast. Trough depth is greatest on the inner shelf, in excess of 1000 m water depth in some surveyed areas. In general, trough depth decreases toward the outer shelf. Trough width is variable and ranges from a few kilometres to several tens of kilometres (Vanney & Johnson 1976). The largest trough, Marguerite Trough, extends from Marguerite Bay to the outer shelf (Fig. 1). The upper slope is not well surveyed. Vanney & Johnson (1976) estimate that slope canyons in the area are as numerous as on other margins.

Description of aggrading-shelf units

Anderson *et al.* (1990) showed that the near-surface acoustic stratigraphy (upper 200 ms or so) of the Antarctic Peninsula continental shelf exhibits an overall aggradational pattern with minor slope progradation (typically less than 2 to 3 km). We consider a unit exhibiting less than 2 to 3 km of slope progradation beyond the pre-existing shelf edge to be an aggrading-shelf unit.

In our study of the Antarctic Peninsula continental shelf, we found stratigraphical relationships within the aggrading-shelf units that are quite complex along strike and dip. Acoustically structureless and acoustically layered seismic facies are bound by seismic unconformities with trough-like relief. The following descriptions of the three youngest seismic units on Profiles 56, 57 and 58 (Figs 2, 3 and 4) are intended to illustrate this complexity. These three units are bound by widespread unconformities.

Individual units are designated using numbers. The first number corresponds to a stratigraphical package and the second number corresponds to a subdivision within the package. The unconformities are numbered from the top down. For example, Unconformity 1.3 designates the third unconformity from the sea floor within package 1. Unit 1.1 is bound by unconformities labelled 1.1 and 1.2; Unit 1.2 is bound by Unconformities 1.2 and 1.3; and Unit 1.3 is bound by Unconformities 1.3 and 2.1. These number designations are the same as those used by Bart & Anderson (1995).

Strike-oriented Profile 57 (Fig. 2) was collected on the inner continental shelf. Acoustic basement occurs at approximately 800 ms below sea level and is overlain by an acoustically stratified section. Acoustic basement has irregular relief. At this location, sedimentary cover is relatively thin (between 100 and 300 ms). Continuity of internal reflectors ranges from good to poor. There are indications of downlap onto basement toward the southwest. Seismic reflectors within the stratified section are truncated at the sea floor. This section has been subdivided into three units. This subdivision was based on correlation with major seismic unconformities observed on the two adjoining dip profiles. At the location of cross-line 58, two units (Units 1.3 and 1.2) are defined. The lower unit (Unit 1.3) pinches out to the southwest and the overlying unit (Unit 1.2) is exposed at the sea floor. Toward the southwest, Unit 1.2 is overlain by Unit 1.1. At the location of cross-line 56, Unit 1.1 is exposed at the sea floor.

On dip-oriented Profile 58 (Fig. 3), the top of acoustic basement dips in an offshore direction,

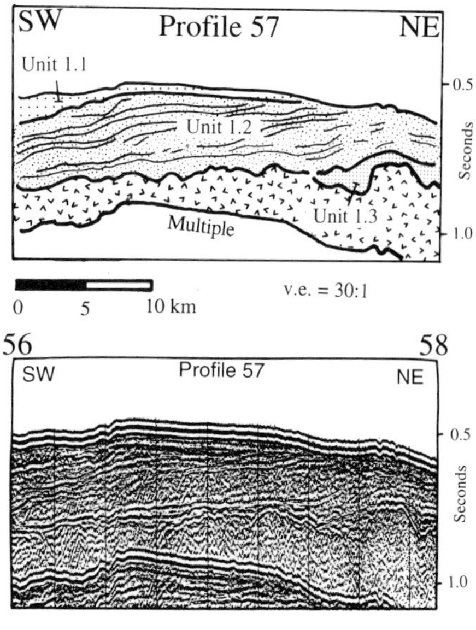

Fig. 2. Interpreted and uninterpreted inner-shelf strike-oriented Profile 57. Cross-line positions of Profiles 56 and 58 are marked. Note the truncation of strata at the sea floor and downlap onto the underlying acoustic basement.

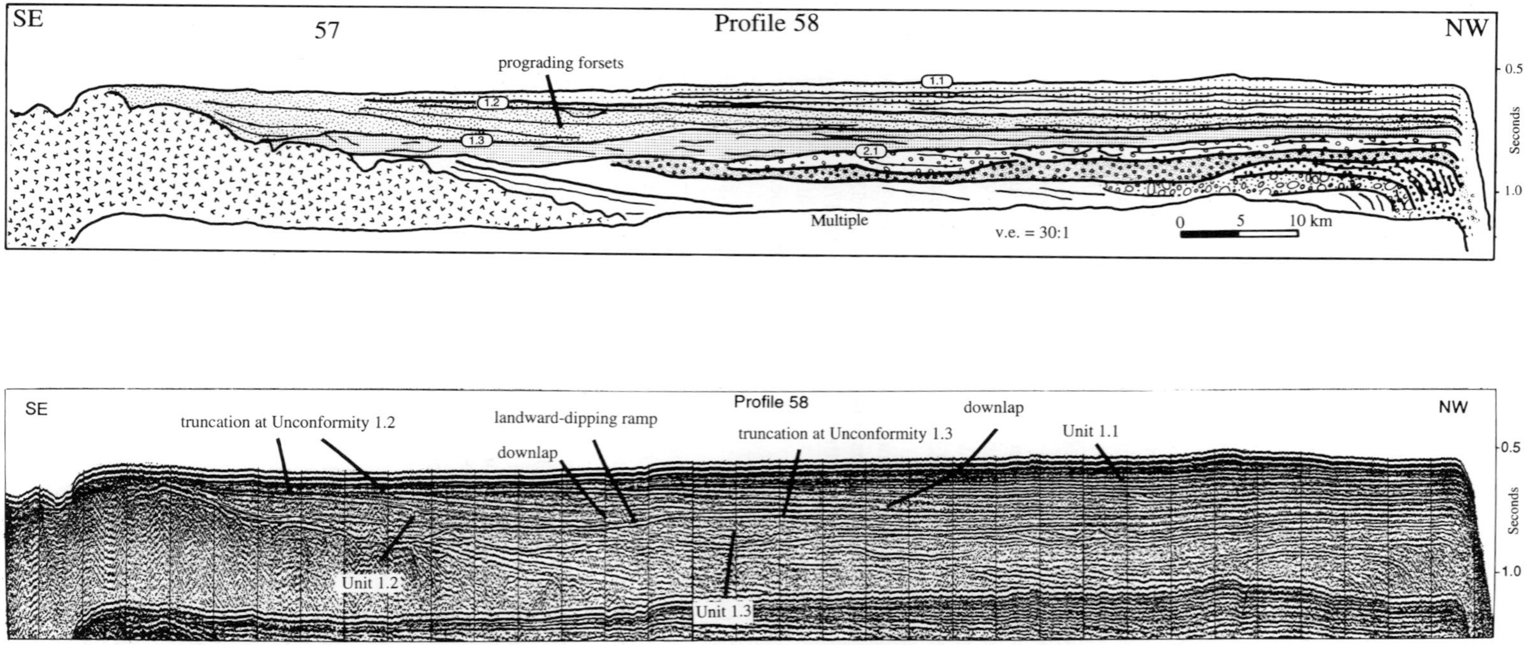

Fig. 3. Interpreted and uninterpreted dip-oriented Profile 58. The position of cross-line Profile 57 is marked. Note low-angle foresets in Unit 1.2 that downlap and prograde across the top of Unit 1.3.

eventually dipping below the water-bottom multiple. At the location of cross-line 57, Unit 1.3 downlaps acoustic basement. Seaward, Unit 1.3 rests on a locally discontinuous unconformity (Unconformity 2.1) that truncates the underlying acoustically stratified section. The top of Unit 1.3 is a very continuous subhorizontal unconformity (Unconformity 1.3). The slight dip inflection of Unconformity 1.3 forms a landward-dipping ramp with approximately 50 ms (approximately 35 m) relief over a distance of about 5 km. Unit 1.3 thins seaward, from approximately 100 ms (75 m) to about 10 ms (8 m) on the outer shelf. The internal seismic facies are variable, mostly chaotic to transparent. Unit 1.3 contains a few very low-angle, prograding foresets. One reflector can be traced a distance of over 30 km, where it downlaps directly onto the unit's basal unconformity (Unconformity 2.1). Unit 1.3 prograding foresets are truncated at the upper unconformity (Unconformity 1.3). Truncation continues seaward of the ramp. Notice the lack of a sedimentary drape above the upper and basal unconformities.

Unit 1.2 has a similar form to Unit 1.3 in dip orientation, thinning from 150 ms (approximately 110 m) on the inner shelf to less than 50 ms (approximately 35 m) on the outer shelf. The upper bounding surface of Unit 1.2 (Unconformity 1.2) does not exhibit a foredeepened ramp. Internally, a few well-defined, continuous, low-angle seismic reflectors downlap and prograde directly across Unconformity 1.3. Individual prograding reflectors extend for over 30 km. Toplap truncations of these prograding reflectors occur at Unconformity 1.2.

On dip-oriented Profile 56 (Fig. 4), acoustic basement dips offshore below the water-bottom multiple. Unconformity 1.3 has a foredeepened profile and can be traced from the inner shelf, where it is somewhat discontinuous and poorly defined, to the outer shelf, where it becomes a sea-floor unconformity that truncates the underlying seaward-dipping strata. At the shelf edge, Unconformity 1.3 forms a landward-dipping ramp with approximately 150 ms relief over a horizontal distance of about 15 km. A thin wedge of Unit 1.3 occurs on the outer shelf. There is no significant slope progradation associated with Unit 1.3. Above Unconformity 1.3, there are no appreciable draping sediments. On the inner shelf, Unconformity 1.3 is buried below Units 1.2 and 1.1. Internal reflectors within Unit 1.2 prograde at a relatively high angle across Unconformity 1.3. Within Unit 1.2, downlap appears to occur at more than one stratigraphical horizon. Individual prograding reflectors within Unit 1.2 have a horizontal length of approximately 5 km; this is shorter than the approximately 30 km horizontal length of some Unit 1.2 prograding reflectors at Profile 58 (Fig. 3). The upper unconformity (Unconformity 1.2) has a landward-sloping ramp that rises to form a platform. Prograding reflectors are truncated along the ramp and platform. Unit 1.2 pinches out on the inner shelf.

Unit 1.1 is relatively thin but has a very similar external form to Unit 1.2. It is restricted to the inner shelf, extending slightly more seaward than Unit 1.2 (Fig. 4). The top of Unit 1.1 is a sea-floor unconformity (Unconformity 1.1) and, on the innermost shelf, this surface has a landward-sloping profile with over 150 ms of relief. Unit 1.1 shows subtle evidence of progradation. The mounded, internally structureless feature at the toe of the prograding sequence has the form of a slump or debris flow.

Glacial activity related to development of aggrading-shelf units

Figure 5 is a five-stage reconstruction of the glacial deposition and erosion of Units 1.3 and 1.2 at Profiles 56 and 58. On these profiles, Units 1.3 and 1.2 are characterized by low-angle prograding foresets that downlap relatively broad and shallow-relief glacial topography. Because of the low angle of repose, the units have an overall aggradational form.

In Stage 1, the ice sheet has expanded onto the continental shelf. At the grounding zone, detritus is being added to the prograding grounding zone wedge of Unit 1.3. Foresets downlap and fill pre-existing glacial topography. By Stage 2 time, the glacial expansion reached its culmination. As the ice sheet expanded, it eroded previously deposited strata and recycled this material to the new grounding zone. At Profile 56, the ice sheet expanded to the outer shelf and at Profile 58 it reached the middle shelf. On these seismic lines, the grounding zone corresponds to the seaward-most extent of foreset truncation. Ramps may correspond to coupling zones, the zones where the ice sheet begins to be buoyant, and probably mark a transition in the thickness of the overlying ice sheet.

At the end of Stage 2 the ice sheet retreated. Stage 3 shows the distribution of Unit 1.3 after the retreat of the ice sheet. The irregular morphology of the sea-floor unconformity probably represents the basal profile of the ice sheet when it was last coupled to the sea floor at the end of Stage 2. No backstepping or draping stratal patterns occur. This suggests that deposition during and following ice-sheet retreat was minimal.

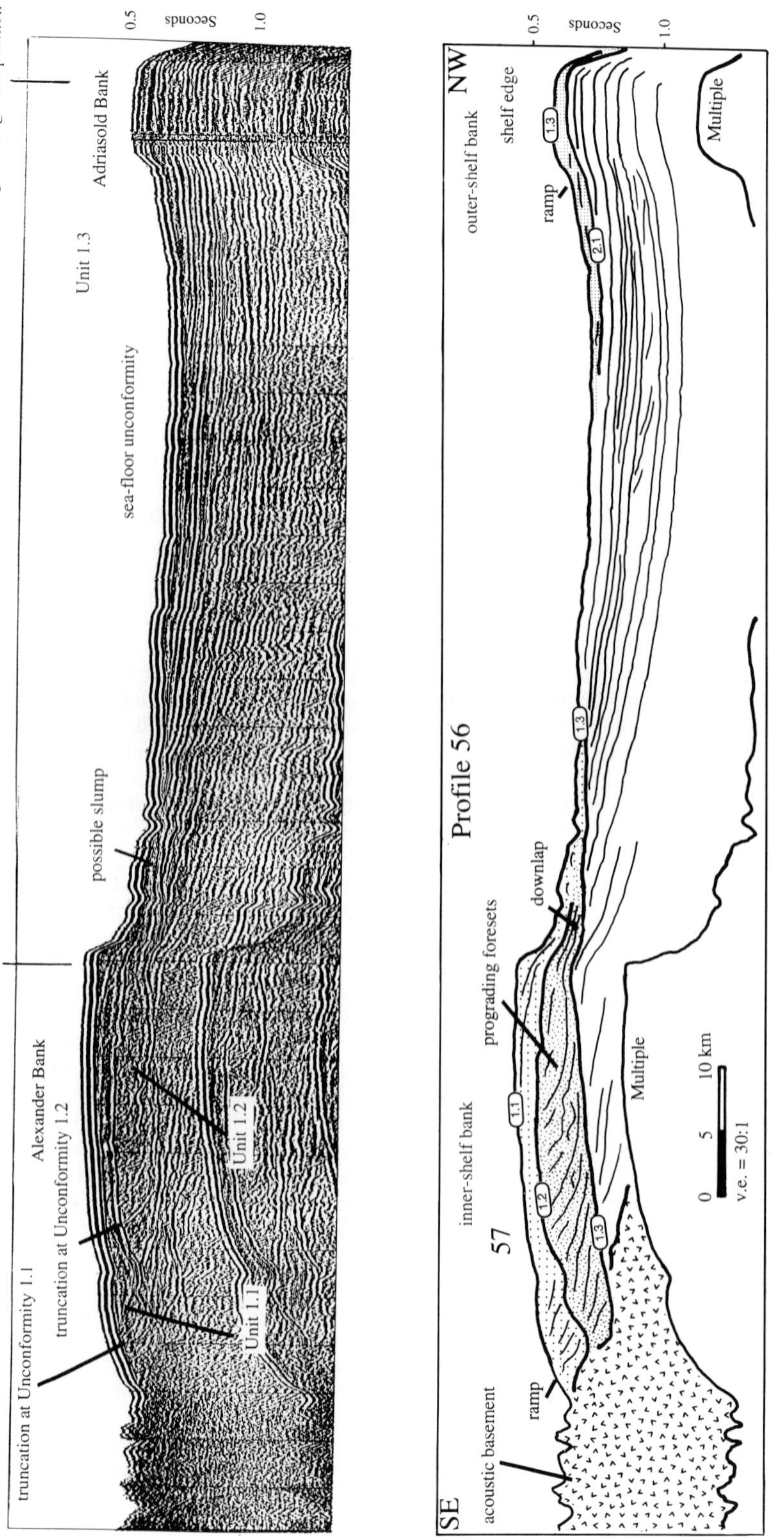

Fig. 4. Interpreted and uninterpreted dip-oriented Profile 56. Unit 1.3 is confined to the outer shelf. Unit 1.2 foresets downlap the underlying unconformity (Unconformity 1.3).

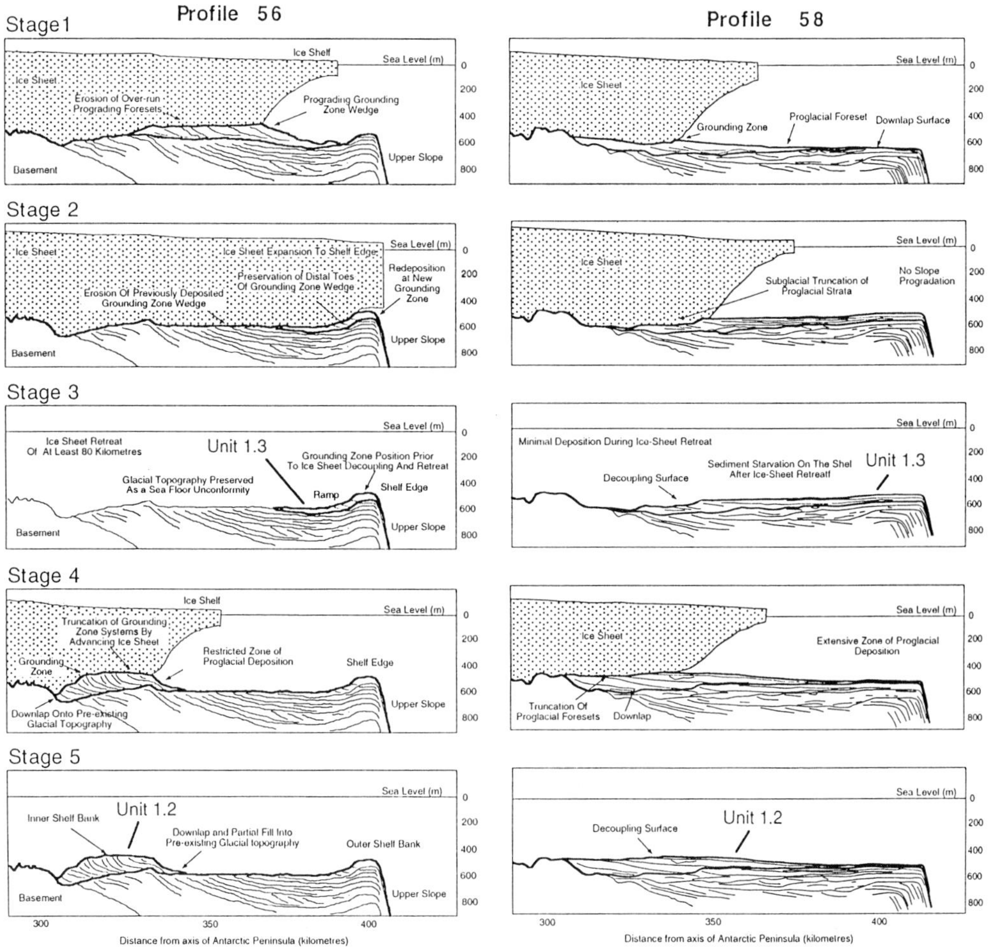

Fig. 5. Stratigraphical model illustrating the inferred progression of ice-sheet advance and retreat across the continental shelf responsible for construction of Units 1.2 and 1.3 at Profiles 56 and 58. See text for discussion.

Stage 4 of the model shows the distribution of Unit 1.2 at the culmination of another ice-sheet advance. Near the location of Profile 56, Unit 1.2 proglacial deposits are confined to the middle shelf. There, relatively high-angle foresets downlap and prograde across glacial topography created during the Stage 2 ice-sheet advance. At Profile 58, low-angle prograding foresets of Unit 1.2 extend across the outer shelf. The seaward-most truncations of Unit 1.2 foresets occur roughly on the same part of the inner shelf at Profile 56 and 58. This indicates that the ice-sheet grounding zone was roughly linear across this part of the shelf, and that the proglacial wedge of Unit 1.2 had a lobate shape. In Stage 5 the ice sheet retreated to the bays and fjords of the coast.

Figure 6 shows isopach maps of Units 1.1, 1.2 and 1.3, as determined from Profiles 56, 57 and 58. The average volume of sediment contained within Units 1.1, 1.2, and 1.3 at Profiles 57, 56 and 58 (Figs 2 3 and 4) ranges from approximately 150 to 350 km^3 (i.e. area of approximately 4900 km^2 and approximate thicknesses of 35 to 75 m). The position of the 1000 m isobath is shown as a solid bold line in Fig. 6. The units are essentially confined to the shelf. Maximum thicknesses of the units range from 100 to 200 ms (approximately 75 to 150 m). Each isopach map shows two zones, a dark-shaded zone and a light-shaded zone, separated by a dashed line. The dashed line marks the most seaward extent of truncation of prograding foresets within the underlying unit, which

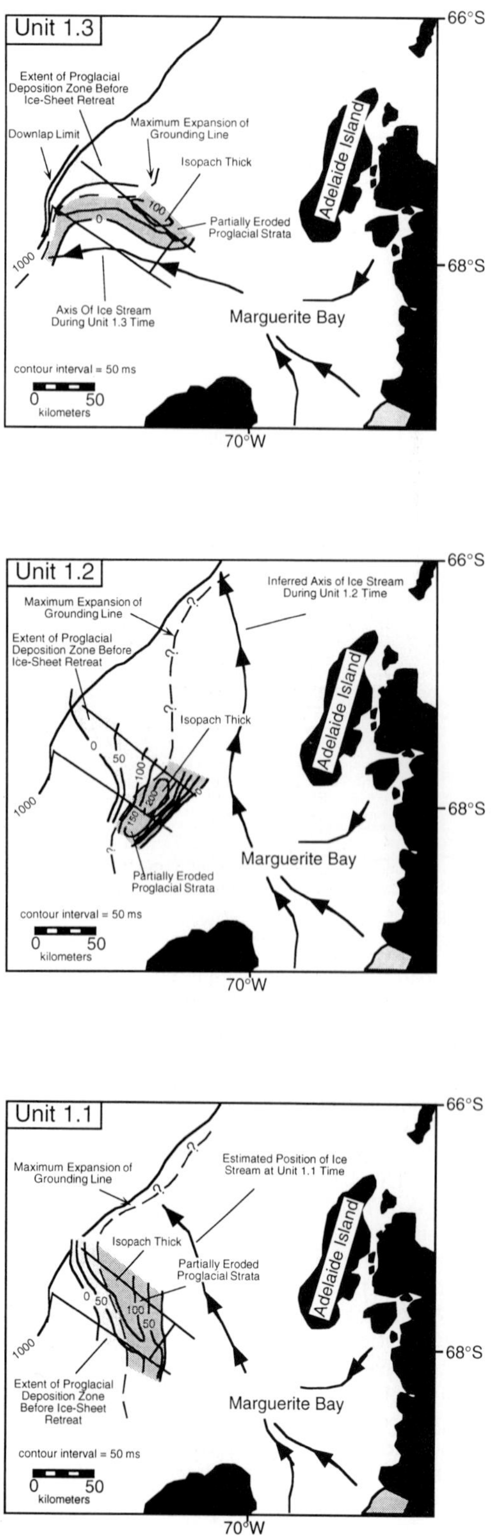

corresponds to the most seaward position of the grounding line. Thus, the dark-shaded area represents regions where grounded ice advanced over strata that were deposited proglacially. Landward of the zero-thickness line, the unit has been completely eroded. In these regions, the ice sheet cut unconformities into either older units or into acoustic basement. The light-shaded part of the isopach map corresponds to the zone of proglacial sedimentation prior to ice-sheet retreat. The seaward limit of these proglacial strata corresponds to a downlap limit. The bold lines with arrows indicate the estimated inferred positions of ice streams based on locations of palaeotroughs.

During Unit 1.3 time, the grounding line extended east–west between Profiles 56 and 58. At Unit 1.2 time, the grounding line was oriented northeast–southwest, and at Unit 1.1 time it was oriented approximately north–south. These differences in the orientation and location of the grounding line indicate changes in the locations of ice streams on the shelf.

At the location of Profile 56, Unit 1.3 proglacial deposits are restricted to the outer shelf. The downlap limit of Unit 1.3 at this location is expected to occur on the upper slope, beyond the limit of our seismic data. On Profile 58, grounded ice did not extend to the outer shelf. At the location of Profile 58, partially eroded proglacial sediments of Unit 1.3 had a maximum thickness of approximately 100 ms on the middle shelf. Based on the distribution of Unit 1.3 strata, we infer that, during their deposition, an ice stream flowed through Marguerite Bay and toward the west. Beneath the ice stream, Unit 1.3 was eroded. Rapid ice-sheet retreat left the glacial topography intact. The magnitude of ice-sheet retreat at Profile 56 was approximately 80 km. During the subsequent glacial expansions, the position of the ice stream shifted to the east. Deposits of Units 1.2 and 1.1 downlap and partially fill the glacial trough that was formed during Unit 1.3 time. At the location of Profile 56, Units 1.2 and 1.1 constitute Alexander Bank on the inner shelf (Fig. 1).

Fig. 6. Isopach and distribution of Units 1.1, 1.2 and 1.3 on the continental shelf in the vicinity of Profiles 56 and 58. The bold line marks the position of the 1000 m isobath. The dashed line represents the maximum expansion of the ice sheet. The light-shaded zone represents the distribution of proglacial strata at the end of the glacial expansion. The dark-shaded zone represents the distribution of proglacial strata that have been over-run and truncated by the ice sheet.

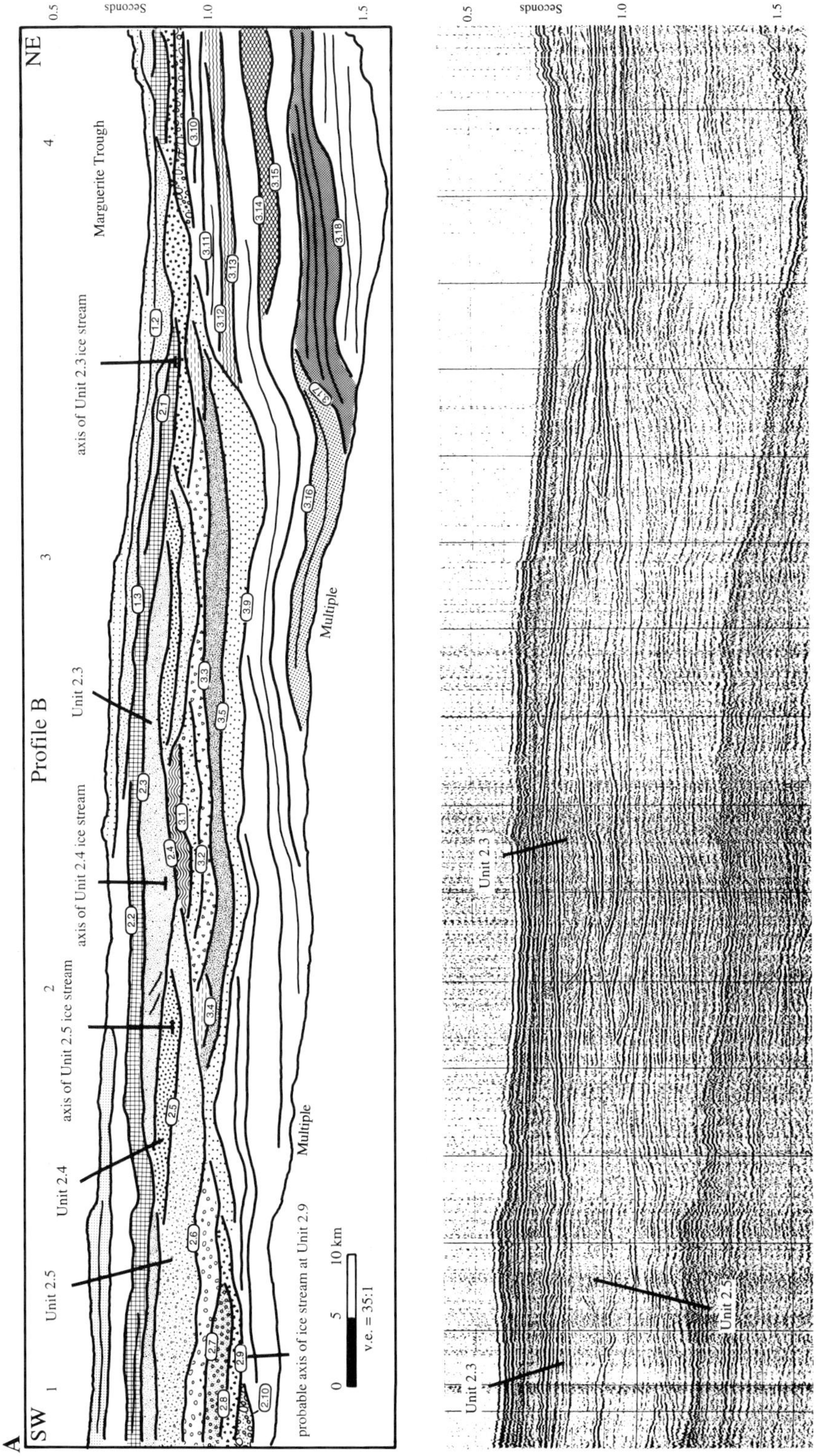

Fig. 7. Segment of strike-oriented Profile B. Positions of cross-line Profiles 1 2, 3 and 4 are marked.

The zone of active sedimentation for Unit 1.2 varied from approximately 10 km at Profile 56 to over 30 km at Profile 58. The variable extent of Unit 1.2 along strike indicates a point-source nature to the depositional system.

Seismic stratigraphical relationships of prograding-slope units and associated unconformities

We now shift focus to seismic units and unconformities associated with slope progradation and examine a larger segment of the study area so that lateral changes in stratal patterns can be demonstrated. Three seismic profiles (Figs 7, 8 and 9) are used to illustrate representative features of prograding-slope units.

Seismic Profile B (Fig. 7), a strike-oriented line collected along the outer shelf, is key to our seismic stratigraphical analysis. It provides a cross-sectional view of the outer shelf and illustrates the striking broad-scale cut-and-fill structure and poor lateral continuity of seismic units. Three things should be noted in Fig. 7. First, the palaeotroughs are of approximately the same dimensions as modern troughs in the study area. Second, the axes of troughs have shifted laterally on the shelf, which implies that deposition on the outer shelf and upper slope has also shifted. The third point is that no single dip-oriented seismic profile provides a complete record of glacial erosion and deposition on the shelf. Imagine the different glacial reconstructions that would be derived from drill sites placed along Profile 1 versus Profile 3.

Figure 8 shows a segment of dip-oriented Profile 1 which crosses a large prograding-slope package. Individual units are separated by unconformities; several of these unconformities show erosional truncation of prograding-slope foresets. Units 2.10 to 2.8 are prograding-slope units that are confined to the outer shelf. The unconformities that bound these units amalgamate on the middle shelf where they have irregular erosional relief and truncate seaward-dipping strata. The units above Unconformity 2.8 exhibit negligible slope progradation, but bounding seismic unconformities do have foredeepened profiles. Notice the lack of progradation of Unit 2.3, which is approximately 50 ms thick on the outer shelf. Using Profile B Unit 2.3 can be traced approximately 60 km to the northeast to dip Profile 3 (Fig. 9). At Profile 3 (Fig. 9), Unconformities 2.4 and 2.3 have a gentle landward-dipping profile. Unit 2.3 is roughly 50 ms thick and shows approximately 5 km of slope progradation. Similar along-strike variations in the magnitudes of slope progradation occur in Units 2.4 to 2.7.

Figure 10 shows a four-stage reconstruction of the glacial deposition and erosion associated with a single prograding-slope unit, Unit 2.5, at Profiles 1 and 3. The Stage 1 reconstruction shows very deep glacial topography. Acoustic basement and forearc basin strata are truncated on the inner shelf, and aggrading-shelf units and prograding-slope units are truncated on the outer shelf. An advancing ice sheet is situated on the middle shelf. At the grounding zone, prograding foresets downlap and partially fill the underlying glacial topography. Stage 2 shows the expansion of the ice sheet and progradation of the grounding-zone deposits. Subglacial ramps probably correspond to transitions in the thickness of the overlying ice sheet. As the ice sheet advances, it erodes into the top of the underlying unit. At Profile 1 the grounding zone is located on the middle shelf, while the ice sheet is grounded at the shelf edge in the area of Profile 3.

During Stage 3 the ice sheet covers the entire shelf. At Profile 1 no slope progradation occurs. At Profile 3 large amounts of sediments, stripped from the underlying units, are transported along the base of the ice sheet to the shelf-edge grounding zone and redeposited as prograding-slope strata. By Stage 4 the ice sheet has eroded much of the grounding zone wedge. The presence of deep palaeotroughs indicates zones of pronounced erosion, which we interpret as locations of former ice streams. Negligible slope progradation occurs at Profile 1. At Profile 3 approximately 5 km of slope progradation is indicated.

Stage 5 shows the distribution of Unit 2.5 after ice-sheet retreat from the shelf. No back-stepping or draping stratal patterns exist.

Style of outer shelf–upper slope deposition

In the previous section, we provided documentation of lateral changes in the magnitude of slope progradation along the continental margin. Does this mean that the stratal stacking patterns on the outer shelf and upper slope cannot be used to reconstruct the glacial history of the shelf? Clearly, glacial reconstructions should not be made from a single dip-oriented seismic profile, or even several profiles from the same area. In this section we will demonstrate that aggradational and progradational stacking patterns on the outer shelf and upper slope do show regional trends that indicate times of

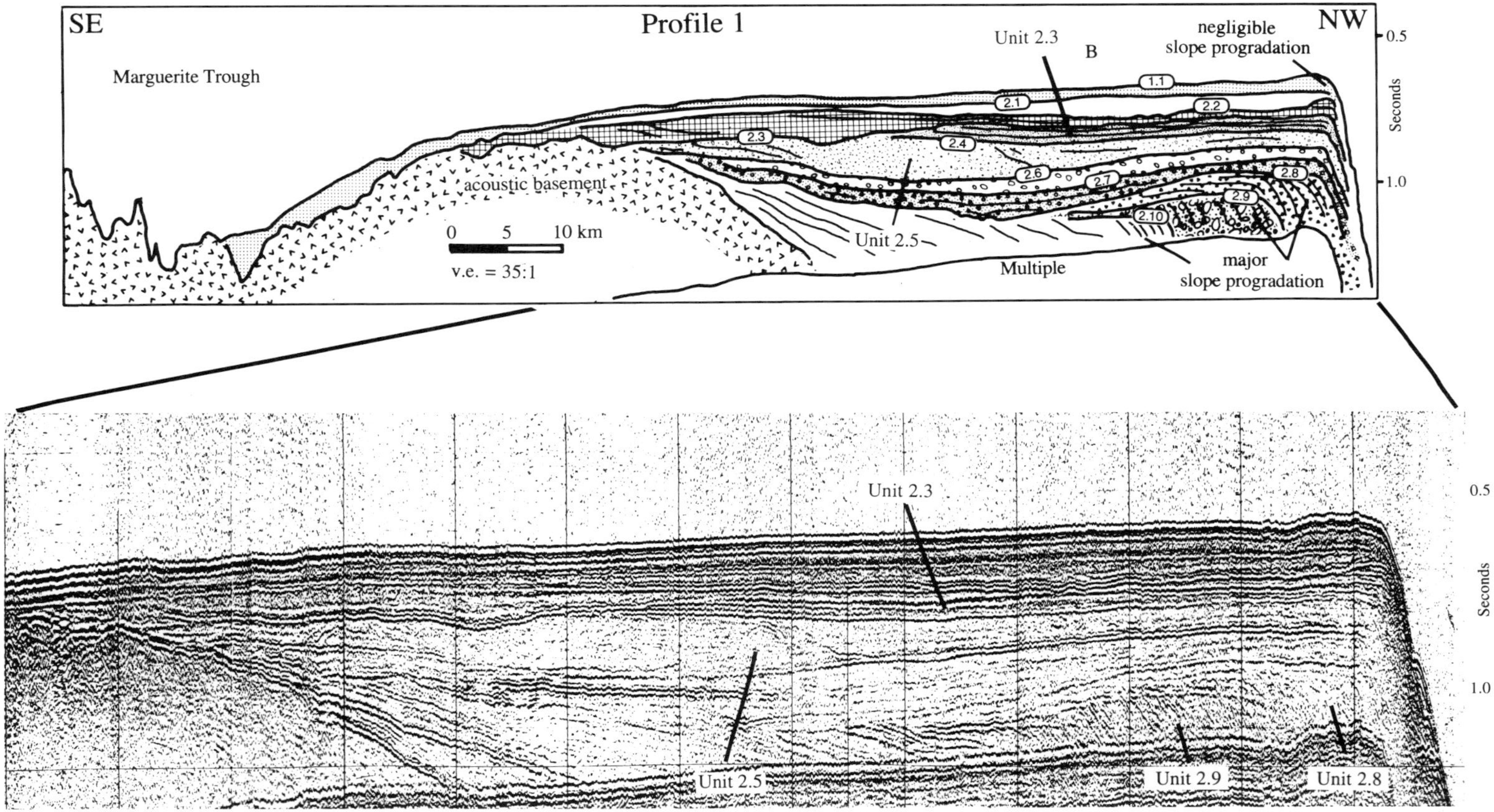

Fig. 8. Dip-oriented Profile 1. The position of cross-line Profile B is marked. Steep prograding-slope foresets are truncated at unconformities 2.10, 2.9 and 2.8. Transition from slope progradation to shelf aggradation occurs at Unconformity 2.8.

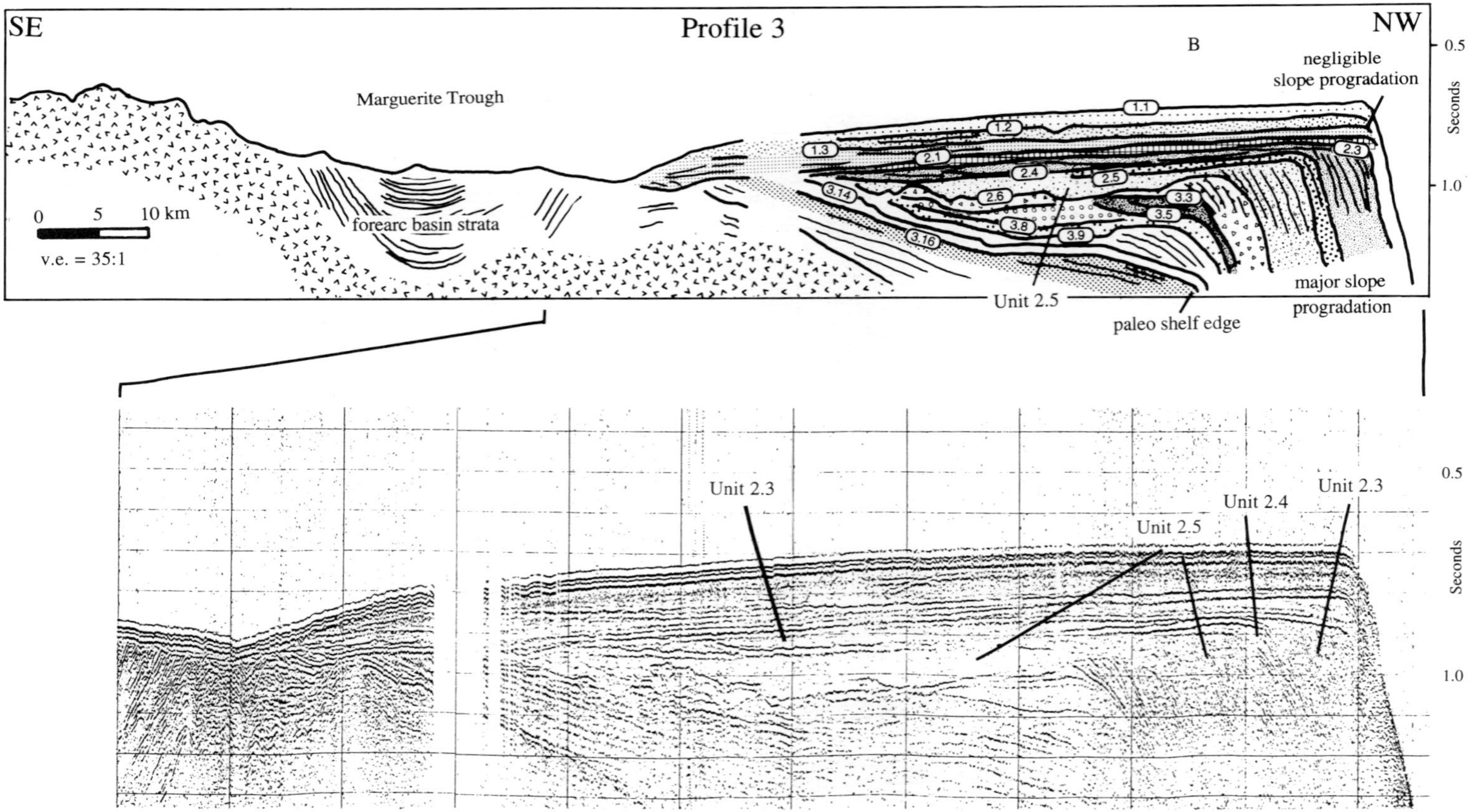

Fig. 9. Dip-oriented Profile 3. The location of cross-line Profile B is marked. Note that several kilometres of slope progradation occurs in units 2.3 and 2.5.

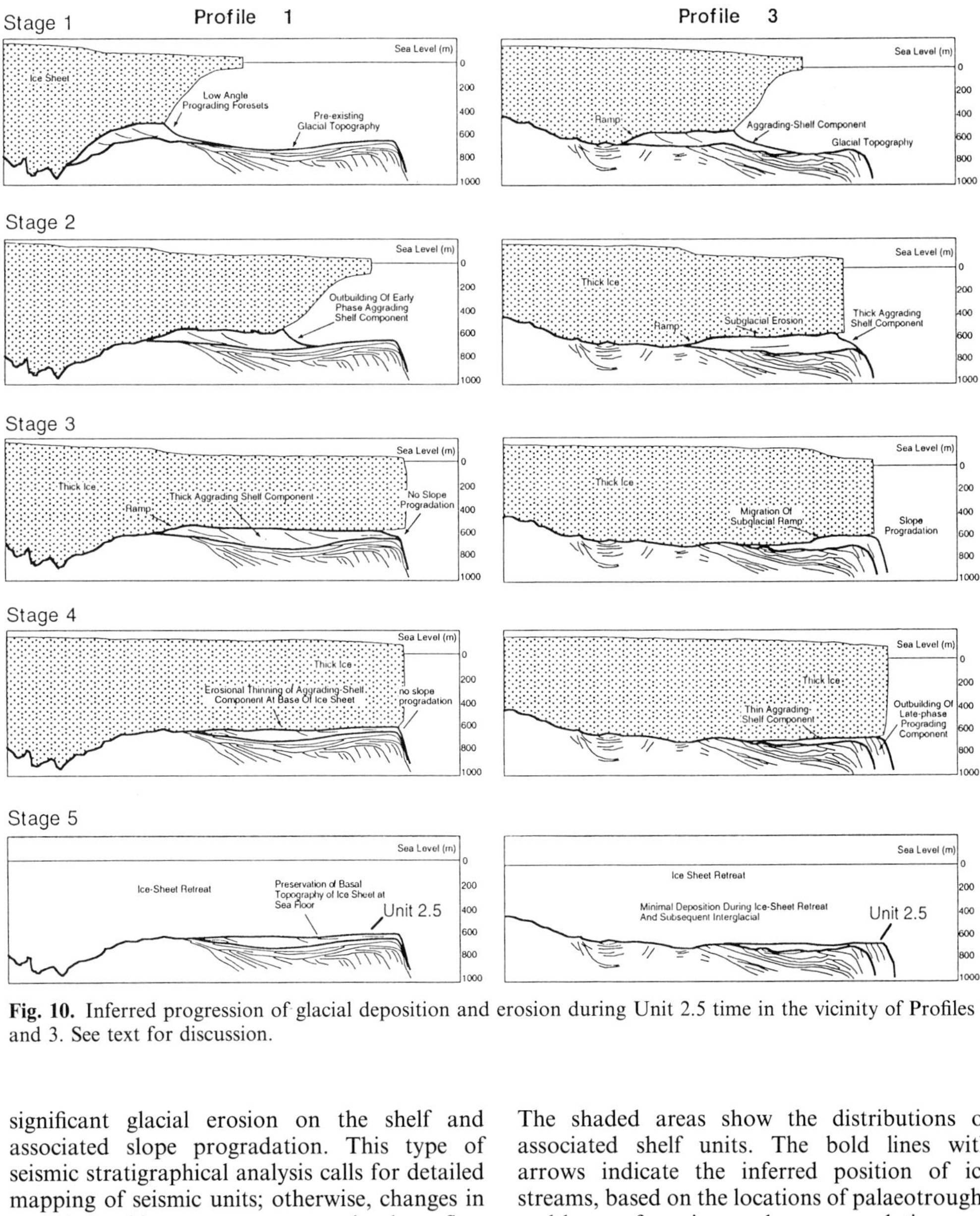

Fig. 10. Inferred progression of glacial deposition and erosion during Unit 2.5 time in the vicinity of Profiles 1 and 3. See text for discussion.

significant glacial erosion on the shelf and associated slope progradation. This type of seismic stratigraphical analysis calls for detailed mapping of seismic units; otherwise, changes in stratal stacking patterns may simply reflect shifting sediment point sources.

The distributions of eight prograding-slope units (Units 2.9 to 2.2) are shown in Fig. 11. The dashed line marks the position of the present shelf edge. The hatched pattern corresponds to areas of toplap truncation of slope foresets beyond the pre-existing shelf edge. The seaward edge of the hatched pattern marks the shelf-edge position after deposition of the prograding-slope unit. The downlap limit of prograding-slope foresets occurs beyond the limits of our data. The shaded areas show the distributions of associated shelf units. The bold lines with arrows indicate the inferred position of ice streams, based on the locations of palaeotroughs and locus of maximum slope progradation.

During the deposition of Units 2.9 and 2.8, slope progradation resulted in the construction of a bulge in the trend of the shelf edge. At the location of Profile 1 (Fig. 8), deposition of Units 2.9 and 2.8 resulted in a total of approximately 18 km of slope progradation. There was no measurable slope progradation of Units 2.9 and 2.8 at the location of Profile 3 (Fig. 9). The aggradational components of Units 2.9 and 2.8 are restricted to the outermost shelf. These are the outer shelf strata preserved after the ice sheet

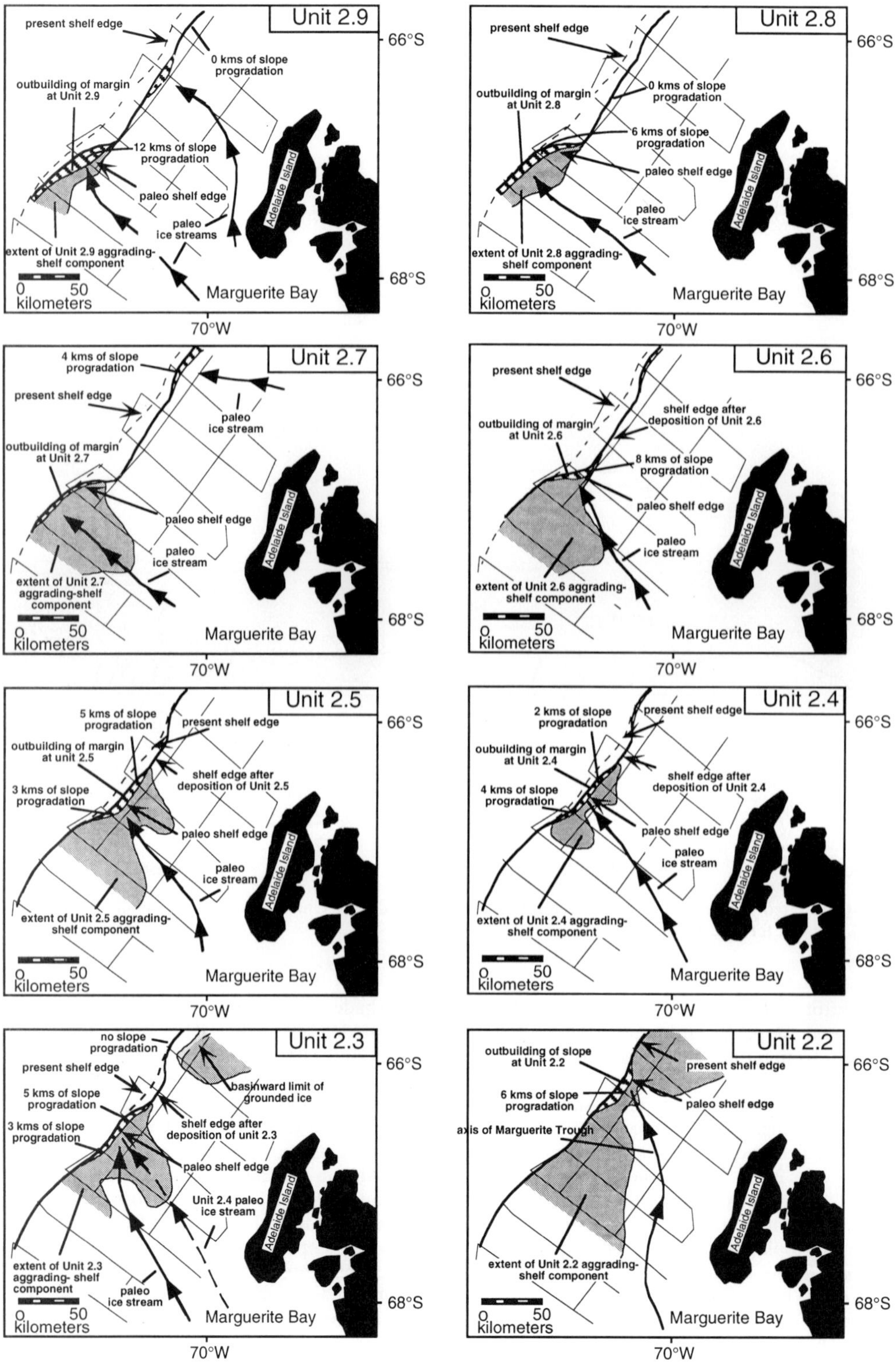
Unit 2.9
present shelf edge
0 kms of slope progradation
outbuilding of margin at Unit 2.9
12 kms of slope progradation
paleo shelf edge
paleo ice streams
extent of Unit 2.9 aggrading-shelf component
Adelaide Island
0 50 kilometers
Marguerite Bay
66°S
68°S
70°W
Unit 2.8
present shelf edge
0 kms of slope progradation
outbuilding of margin at Unit 2.8
6 kms of slope progradation
paleo shelf edge
paleo ice stream
extent of Unit 2.8 aggrading-shelf component
Adelaide Island
0 50 kilometers
Marguerite Bay
66°S
68°S
70°W
Unit 2.7
4 kms of slope progradation
present shelf edge
paleo ice stream
outbuilding of margin at Unit 2.7
paleo shelf edge
paleo ice stream
extent of Unit 2.7 aggrading-shelf component
Adelaide Island
0 50 kilometers
Marguerite Bay
66°S
68°S
70°W
Unit 2.6
present shelf edge
shelf edge after deposition of Unit 2.6
outbuilding of margin at Unit 2.6
8 kms of slope progradation
paleo shelf edge
paleo ice stream
extent of Unit 2.6 aggrading-shelf component
Adelaide Island
0 50 kilometers
Marguerite Bay
66°S
68°S
70°W
Unit 2.5
5 kms of slope progradation
present shelf edge
outbuilding of margin at unit 2.5
shelf edge after deposition of Unit 2.5
3 kms of slope progradation
paleo shelf edge
paleo ice stream
extent of Unit 2.5 aggrading-shelf component
Adelaide Island
0 50 kilometers
Marguerite Bay
66°S
68°S
70°W
Unit 2.4
2 kms of slope progradation
present shelf edge
oubuilding of margin at Unit 2.4
shelf edge after deposition of Unit 2.4
4 kms of slope progradation
paleo shelf edge
paleo ice stream
extent of Unit 2.4 aggrading-shelf component
Adelaide Island
0 50 kilometers
Marguerite Bay
66°S
68°S
70°W
Unit 2.3
no slope progradation
present shelf edge
basinward limit of grounded ice
5 kms of slope progradation
shelf edge after deposition of unit 2.3
3 kms of slope progradation
paleo shelf edge
Unit 2.4 paleo ice stream
extent of Unit 2.3 aggrading- shelf component
paleo ice stream
Adelaide Island
0 50 kilometers
Marguerite Bay
66°S
68°S
70°W
Unit 2.2
outbuilding of slope at Unit 2.2
present shelf edge
6 kms of slope progradation
paleo shelf edge
axis of Marguerite Trough
extent of Unit 2.2 aggrading-shelf component
Adelaide Island
0 50 kilometers
Marguerite Bay
66°S
68°S
70°W

expanded over its proglacial deposits. Landward of the shaded patterns, shelf strata were completely eroded.

During the ice-sheet expansion that deposited Unit 2.7, the magnitude of slope progradation was relatively small, but the extent of this unit on the continental shelf is widespread, being confined mainly to the glacial trough that was formed during the previous glacial advances. In the northern sector of the study area, approximately 4 km of slope progradation occurred.

Following deposition of Unit 2.7, the locus of slope progradation shifted toward the northeastern flank of the shelf-edge bulge. At the location of Profile 1 Unit 2.6 strata show dominant shelf aggradation with negligible slope progradation. By this time, the shelf edge in the southern part of the study area had reached its present position. At the location of Profile 2 there was 8 km of Unit 2.6 slope progradation.

Three kilometres of slope progradation within Unit 2.5 is seen on Profile 2 and 5 km on Profile 3. The zone of maximum slope progradation is situated at the mouth of a glacial trough incised into the top of Unit 2.5 (Fig. 7). The distribution of Unit 2.5 on the shelf and the relief of the overlying Unconformity 2.5 indicate that the axis of the glacial trough was situated along the position of Profile 2.

Slope progradation during Unit 2.4 time was similar to that of Unit 2.5. However, Unit 2.4 is only preserved on the outer shelf and upper slope. There was up to 4 km of slope progradation between Profiles 2 and 3. The axis of the glacial trough during Unit 2.4 time shifted by approximately 10 km to the northeast (Fig. 7).

The distribution of Unit 2.3 on the shelf and slope indicates that an ice stream was located between Profiles 1 and 2 on the middle shelf. Unit 2.3 fills the glacial trough formed during the previous (2.4) glacial advance (shown in Fig. 11 as a dashed bold line with arrows). The location of slope progradation remained essentially unchanged from the previous two (2.5 and 2.4) glacial advances. Up to 5 km of slope progradation occurred during the 2.3 ice-sheet advance.

Unit 2.2 is very widespread on the shelf. Erosional truncation of the unit suggests that the axis of the ice stream shifted to the northeast. This essentially is the location of Marguerite Trough today (Fig. 7). The maximum extent of slope progradation associated with Unit 2.3 was approximately 6 km.

Conclusions

1. Two types of stratigraphical units occur on the outer shelf and upper slope of the Pacific sector of the Antarctic Peninsula continental margin, aggrading-shelf units and prograding-slope units. Within aggrading-shelf units, widespread unconformities with trough-like relief and low-angle foresets are interpreted as glacial unconformities and truncated proglacial grounding zone depositional systems, respectively. The zone of proglacial deposition was very broad (from 5 to over 30 km) and had a lobate shape.
2. Prograding-slope units exhibit more than 2 to 3 km of slope progradation beyond the pre-existing shelf edge and represent glacial deposition during times when the ice sheet advanced to the edge of the continental shelf and remained grounded long enough to excavate shelf deposits and transport them across the shelf. Palaeotroughs are interpreted as marking the locations of former ice streams. Maximum slope progradation appears to have occurred near the mouths of these large glacial troughs. The magnitudes of slope progradation beyond the pre-existing shelf edge vary across the margin and range from 0 to 12 km. At no single stratigraphical level did we find a consistent magnitude of slope progradation for any appreciable distance along the slope.
3. Preservation of toplap-truncation surfaces suggests that transition from glaciated shelf to a non-glaciated shelf occurs rapidly (i.e. no backstepping proglacial strata were observed). The glacial unconformities may represent ice-sheet decoupling events. Glacial unconformities are directly overlain by low-angle prograding foresets. Absence of draping strata above glacial unconformities suggests that interglacials were characterized by sediment starvation.

Fig. 11. Distribution of prograding-slope units 2.9 to 2.2. The pre-existing shelf edge is shown as a bold line. The locus of slope progradation shifted through time, generally shifting toward the northeast. The dashed line represents the present position of the shelf edge. The hatched pattern indicates the locus of slope progradation. The shaded area shows the distribution of the truncated proglacial strata on the continental shelf. The bold lines with arrows mark the inferred position of ice streams, which is based on the locations of palaeotroughs (Fig. 7) and/or prograding-slope strata.

4. Our models for sedimentation on the continental shelf call for sedimentation being concentrated seaward of the grounding zone, which implies that shelf strata consist mainly of glacial marine and sediment gravity flow deposits. This is a significant difference between our model and that of Larter & Barker (1989); their model implies that shelf deposits are mostly subglacial in origin. Our model for sedimentation on the continental shelf also differs from that of Larter & Cunningham (1993), who argue that sedimentation on the continental slope is line-sourced. We have demonstrated that sedimentation on the outer shelf and upper slope is point-sourced and that the locus of sedimentation changes with time. These are important differences. Our results indicate that the overall stacking patterns of strata on the outer shelf and upper slope cannot be used to reconstruct the ice-sheet grounding history on the shelf until lateral changes in stratal geometries have been constrained by detailed mapping of seismic units.

References

ANDERSON, J. B., POPE, P. G. & THOMAS, M. A. 1990. Sequence stratigraphy and hydrocarbon potential of the northern Antarctic Peninsula continental shelf. *In*: ST. JOHN, B. (ed.) *Antarctica as an Exploration Frontier*. American Association of Petroleum Geologists, Studies in Geology, **33**, 1–12.

—— & BARTEK, L. R. 1992. Cenozoic glacial history of the Ross Sea revealed by intermediate resolution seismic reflection data combined with drill site information. *In*: KENNETT, J. P. & WARNKE, D. A. (eds) *The Antarctic Paleoenvironment: A Perspective on Global Change*. American Geophysical Union, Antarctic Research Series, **56**, 231–263.

BART, P. J. & ANDERSON, J. B. 1995. Seismic record of glacial events affecting the Pacific margin of the Northwestern Antarctic Peninsula. *In*: COOPER, A. K., BARKER, P. F. & BRANCOLINI, G. (eds) *Geology and Seismic Stratigraphy of the Antarctic Margin*. American Geophysical Union, Antarctic Research Series, **68**, 74–95.

BARTEK, L. R., VAIL, P. R., ANDERSON, J. B., EMMET, P. A. & WU, S. 1991. Effect of Cenozoic ice sheet fluctuations in Antarctica on the stratigraphic signature of the Neogene. *Journal of Geophysical Research*, **96**(B4), 6753–6778.

COOPER, A. K., BARRETT, P. J., HINZ, K., TRAUBE, V., LEITCHENKOV, G. & STAGG, H. M. J. 1991. Cenozoic prograding sequences of the Antarctic continental margin: a record of glacio-eustatic and tectonic events. *Marine Geology*, **102**, 175–213.

LARTER, R. D. & BARKER, P. F. 1989. Seismic stratigraphy of the Antarctic Peninsula Pacific margin: a record of Pliocene–Pleistocene ice volume and paleoclimate. *Geology*, **17**, 731–734.

—— & CUNNINGHAM, A. P. 1993. The depositional pattern and distribution of glacial interglacial sequences on the Antarctic Peninsula Pacific margin. *Marine Geology*, **109**, 203–219.

VANNEY, J. R. & JOHNSON, G. L. 1976. Geomorphology of the Pacific continental margin of the Antarctic Peninsula. *In*: HOLLISTER, C. D. & CRADDOCK, C. *et al.* (eds) *Initial Reports of the Deep Sea Drilling Project*. US Government Printing Office, Washington, **35**, 279–289.

Quaternary erosion and depositional processes: western Norwegian fjords, Norwegian Channel and North Sea Fan

H. P. SEJRUP[1], E. L. KING[1], I. AARSETH[1], H. HAFLIDASON[1] & A. ELVERHØI[2]

[1] *Department of Geology, University of Bergen, Allégaten 41, N-5007 Bergen, Norway*
[2] *Geological Institute, University of Oslo, PO Box., 1047 Blindern, N-0136 Oslo, Norway*

Abstract: The southern Norwegian fjords, the northern North Sea and North Sea margin represent major pathways of sediment flux from the southern part of Fennoscandia to the deep sea basin. Volumes, genesis and chronology of the temporary and semi-permanent Quaternary sediment repositories are reviewed. The fjord infill is dominated by sediments from the last deglaciation (150 km^3). The Norwegian Channel is occupied by tills and glaciomarine/marine sediments (5000 km^3). The North Sea Fan comprises glacially fed debris-flow fingers/lobes (constituting 80% of deposits in the mid- to late Pleistocene) together with hemipelagic and mass-transport slide sediments, totalling 15 000 km^3. The Norwegian Channel represents a dynamic transport conduit whereby during peak glaciation a fast-flowing ice stream has drained large parts of the southern Fennoscandian ice sheet several times. The sediment flux to the deep sea is temporarily interrupted (between glacials) when stored in the channel but under glacial maxima the channel is probably largely bypassed, with material transported as subglacial load and deposited as debris flows on the North Sea Fan. Sediments on the fan have a much longer storage time (numerous glaciation cycles) but are further transported to the deep sea in large slide events. The volume of Quaternary sediments found along this sediment pathway, plus that which slid to the deep sea, total at least 25 000 km^3, corresponding to an averaged 120 m of erosion across southern Norway.

Quaternary sedimentation in the basin of the Norwegian Sea and on the adjacent continental margin occurred predominantly under glacial stages during which up to 90% of deposition occurred (e.g. Kellogg 1976; Sejrup *et al.* 1991). A wide range of sedimentary processes including marine, gravity-induced and glacial processes, have acted at very different intensities during the glacial stages relative to interglacial intervals. Little is known about the relative importance in time and space of the sedimentary processes and fluxes of sediments from the adjacent continents to the deep Norwegian Sea through a glacial–interglacial cycle. Aiming towards a better understanding of the processes along a sediment pathway from the mainland, across the shelf and to the deep sea, the present paper provides a review and synthesis of data from the south-eastern Norwegian Sea margin.

The western Norwegian fjords, the Norwegian Channel and the North Sea Fan (Fig. 1) are important temporary sediment repositories of glacial erosion products from southern Fennoscandia en route to the deeper part of the Norwegian Sea. Investigations of these sedimentary bodies during the last decade, mainly through seismic and borehole studies, have contributed to our understanding of the marine and glacial processes acting on this margin through the Quaternary. In addition, new information concerning the climatic history of the region has been obtained (Sejrup *et al.* 1995; King *et al.* 1996). The present study reviews and addresses the geometry and lithology of the Quaternary sediment bodies encountered along a transect from the western Norwegian fjords via the Norwegian Channel to the North Sea Fan. Based on interpretation of the genesis and the chronology of the sediments, the role of the different repositories through a glacial–interglacial cycle will be discussed.

Western Norwegian fjord sediments

The glacial basins of western Norwegian fjords act as effective sediments traps during deglaciation as well as during interglacial and interstadial phases. The fjords are the result of glacial erosion along former fluvial valleys and along bedrock fracture zones (e.g. Holtedahl 1975; Nesje & Whillans 1994). All western Norwegian fjords have outer bedrock or morainic sills in the order of 200–300 m deep, and water depths of 500–700 m are common in the fjord basins. The two deepest, Hardangerfjord and Sognefjord (Fig. 2), are 900 and 1300 m deep respectively.

Acoustical and sedimentological investigations have been used in estimating quantities

From De Batist, M. & Jacobs, P. (eds), 1996, *Geology of Siliciclastic Shelf Seas*, Geological Society Special Publication No. 117, pp. 187–202.

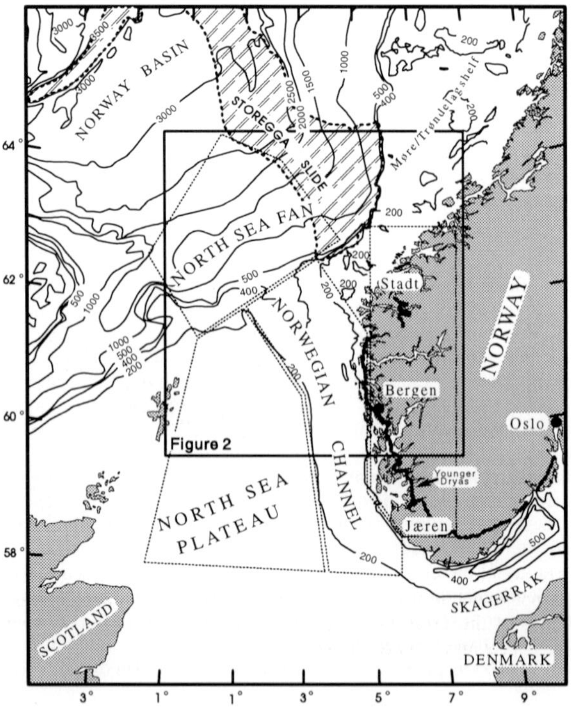

Fig. 1. Location map of the investigated area with geographic names mentioned in text. The area affected by erosion/deposition of the Storegga slide event (Bugge *et al.* 1987) is shown. Also shown is the extent of the Fennoscandian ice sheet during Younger Dryas. The four stippled polygons show the areas (North Sea Fan, North Sea Plateau, Norwegian Channel and Norwegian fjords) over which volumes of the Quaternary sediments have been estimated (Fig. 11). The area in the box is shown in more detail in Fig. 2.

and composition of sediments present in the fjords today. Correlations to raised (rebound) marine sediments along the fjords as well as chrono- and biostratigraphical analyses of sediment cores have made determinations of ages and sedimentation rates possible. Interglacial and interstadial sediments older than Late Weichselian are found on land at several localities along the outer part of the western Norwegian coast (Mangerud *et al.* 1981; Andersen *et al.* 1983; Sejrup 1987; Larsen *et al.* 1987; Larsen & Sejrup 1990). Sedimentary infill in some coast-parallel (ice-transverse) fjords which, based on seismostratigraphy, clearly predates the main deglaciation, can be equivalent to these older deposits. Such deposits older than Late Weichselian may account for *c.* 10% of the fjord sediments in southern Norway.

During the last deglaciation of the coastal areas the glaciers were subject to several oscillations (Mangerud 1970; Aarseth & Mangerud 1974) of which the Allerød/Younger Dryas can be traced in terrestrial as well as in fjord sediments (Fig. 1). In many fjords the moraines deposited during Younger Dryas comprise large accumulations with glaciofluvial foreset beds up to 100 ms two-way travel time (TWT) (75 m) thick resting on even thicker bottomset beds (Fig. 3). The distal glaciomarine sediments may amount to 400 ms TWT in thickness in some fjords (*c.* 320 m), most of which was deposited during the Allerød deglaciation. These sediments typically consists of 25–55% clay, 70–45% silt and <2–3% fine sand except for occasional ice-rafted gravels (Aarseth *et al.* 1989).

The final deglaciation of the fjords took place shortly after the retreat from the Younger Dryas moraines. Most glaciers then grounded at the head of the fjords and deposited large ice-marginal deltas with distal glaciomarine silty clays in the fjord basins (e.g. Anundsen & Simonsen 1968; Vorren 1973; Mangerud *et al.* 1979). High sedimentation rates during the deglaciation created sediment slopes subject to gravity failure (Aarseth *et al.* 1989). The largest slides took place on slopes connecting tributary fjords to the much deeper trunk fjords (hanging

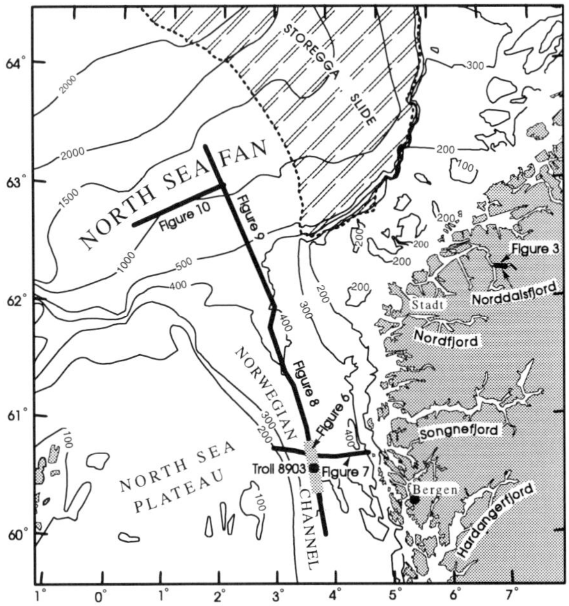

Fig. 2. Location map of the northern North Sea also showing the larger of the Norwegian west coast fjords which enter the Norwegian Channel, which in turn empties onto the North Sea Fan. The fan is bounded to the NE by the Storegga slide. Locations of the Troll borehole 8903 and the various geologic sections and illustrations of seismic profiles (Figs 3, 6, 7, 8, 9 and 10) are shown.

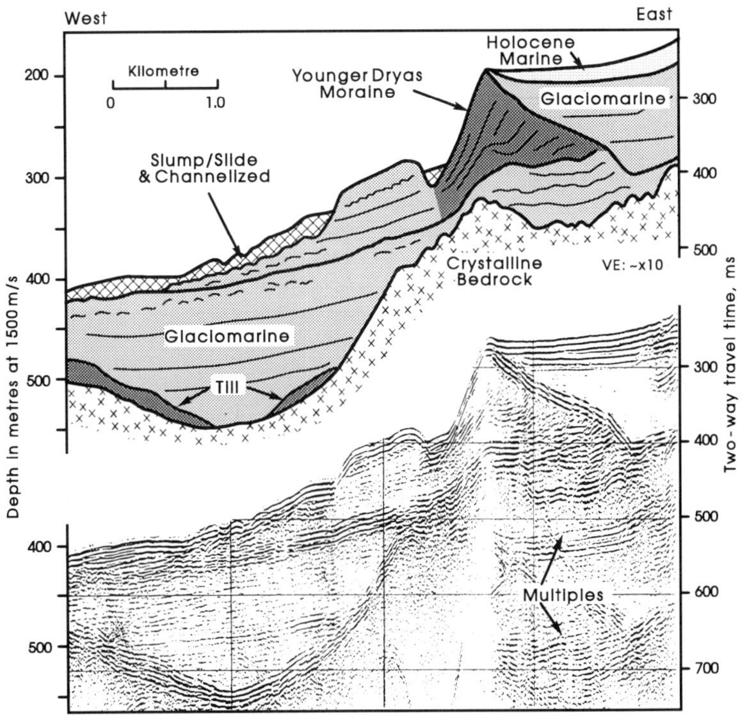

Fig. 3. Example of a sparker reflection profile across a Younger Dryas subaqueous ice contact moraine in Norddalsfjord (for location see Fig. 2) showing possible till overlain by glaciomarine sediments resting on bedrock. The lower glaciomarine unit was deposited during Allerød before the Younger Dryas glacial readvance. The distal moraine equivalents have been eroded by slumps/slides and turbidity currents.

valleys). Core samples in the slide scars have revealed sediments containing silty clays with numerous laminae of fine-grained sand a few milli-metres thick. Such laminae can be potential slide surfaces. Most slide activity happened during, and just after the deglaciation, but slides of up to $15 \times 10^6\,m^3$ took place just three decades ago (Aarseth *et al.* 1989). Turbidites from old as well as young mass movements are frequent in fjord sediments (Holtedahl 1965, 1975; Aarseth *et al.* 1989), but in areas of no sliding, sedimentation rates are commonly in the order of only 0.3 mm/year except for a few fjord heads which receive an ample sediment input derived from present glacial rivers. Figure 4 is an idealized longitudinal fjord profile illustrating some of the aspects presented above.

Seabed erosion is observed both in sill areas and at the bottom of fjord basins, immediately inside the sills, where it is thought to be caused by episodic inflow of heavier Atlantic water masses (Fig. 4). In these areas, Holocene sediments are absent while their thickness elsewhere may vary from a few metres up to 10 m. The Holocene sediments are better sorted than the glaciomarine sediments below; they have clay values of 20–45% and may have a higher sand content especially in the surface samples (Aarseth *et al.* 1975).

Sediment volume estimates based on tight seismic grids in most fjords and prorated for uninvestigated fjords in the area between 58°45′ N (Jæren area) and 63° N (area shown in Fig. 1) give a value of $150\,km^3$. Subglacial processes effectively remobilize the fjord deposits for further transport out of the fjord during subsequent glaciations.

The Norwegian Channel

The Norwegian Channel (also denoted the Norwegian Trench, the Rinne and in Norwegian, Norskerenna) is a morphological trough feature originating in the Oslofjord and continuing via the Skagerrak around the south and southwest coast of Norway and north to reach the continental margin west of Stadt (Fig. 1). The Norwegian Channel is deepest in Skagerrak (*c.* 700 m) with a shallower threshold west of Jæren (*c.* 270 m) from where it slopes gradually northward to the mouth at the shelf edge in about 400 m water depth. Early investigations (Sellevold & Sundvor 1974) and later, systematic seismostratigraphic studies (Aarseth & Godvik 1984) and mapping of part of the channel (Rise *et al.* 1984) identified numerous extensive units lying above an angular unconformity found at between 100 and 200 m below seabed, extending across and along the Norwegian Channel from Jæren to the shelf edge. The age and origin of this feature has been discussed for several decades (Floden & Sellevoll 1972; Sellevoll & Sundvor 1974; Rise *et al.* 1984) based on morphology and extensive sets of seismic data, and there is general agreement that the feature bears a strong imprint of glacial erosion and deposition. Sejrup *et al.* (1995) published stratigraphic data from the Troll petroleum field which made it possible to date and establish the genesis of the extensive seismostratigraphic units. This stratigraphy has further been tied into the seismostratigraphy of the North Sea Fan located beyond the mouth of the Channel (King *et al.* 1996). This will be discussed in the next section.

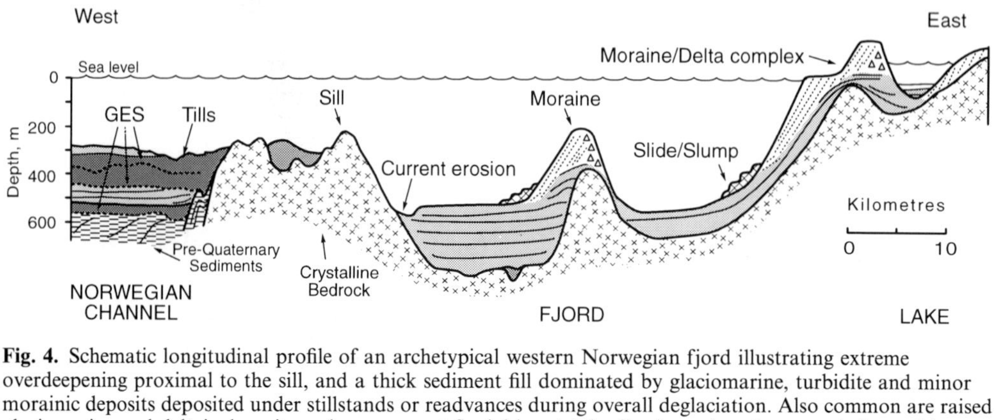

Fig. 4. Schematic longitudinal profile of an archetypical western Norwegian fjord illustrating extreme overdeepening proximal to the sill, and a thick sediment fill dominated by glaciomarine, turbidite and minor morainic deposits deposited under stillstands or readvances during overall deglaciation. Also common are raised glaciomarine and deltaic deposits and more recent fluvial input. The fjords are largely scraped clean with subsequent glaciations. GES; glacial erosion surface (stipples and symbols largely as in Fig. 8).

The seismic framework has been tied together with lithostratigraphical, geochronological and biostratigraphical data from Borehole 8903 (219 m long) from the Troll Field (Fig. 2). The borehole penetrated the upper, flat-lying reflectors (*c*. 200 m) above the angular unconformity and core studies have allowed an interpretation of the depositional history of the Norwegian Channel sediments and a chronology based on palaeomagnetism, amino acid diagenesis, strontium isotopes, AMS dates and biostratigraphy (Sejrup *et al.* 1995). A synthesis of the stratigraphy from the Troll region is presented in Fig. 5 while a correlation between the borehole stratigraphy and seismostratigraphy is provided in Figs 6, 7 and 8.

The Quaternary sediments of the Norwegian Channel comprise sequences which (when they are fully developed) are composed of:

(1) an extensive basal erosional unconformity formed through glacial sheet erosion, followed by
(2) one or more thick (commonly several metres to 30–40 m) till units separated by glacial erosion surfaces (GES); and
(3) glaciomarine/marine sediments conformably overlying the tills.

The till units are diamictons commonly containing more than 20% material >125 μm, including far-transported chalks, a mixture of reworked fossils, and medium to high shear-strength values (Sejrup *et al.* 1995). The marine units have less coarse component, higher water content and *in situ* fossil assemblages.

At Troll three such sequences have been distinguished (Fig. 5). The oldest of these includes a till (Unit L6) from 1.1 Ma which forms a nearly continuous blanket up to 70 m thick covering the eastern channel floor and indicating that the ice extended to near the palaeo-shelf edge. A *c*. 50 m thick marine/glaciomarine unit spanning *c*. 500 ka overlies the till (Unit L5). In this glaciomarine/marine unit two interglacial events (Norwegian Trench and Radøy Interglacials,

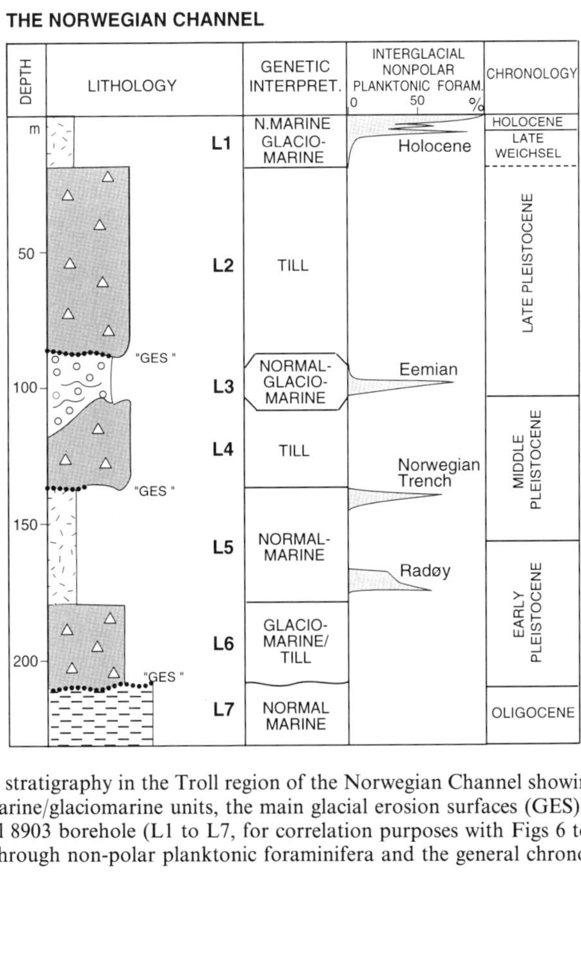

Fig. 5. Synthesis of the stratigraphy in the Troll region of the Norwegian Channel showing differentiation of the Quaternary tills and marine/glaciomarine units, the main glacial erosion surfaces (GES), the lithostratigraphic subdivision at the Troll 8903 borehole (L1 to L7, for correlation purposes with Figs 6 to 8), the Interglacial intervals as identified through non-polar planktonic foraminifera and the general chronology (modified from Sejrup *et al.* 1995).

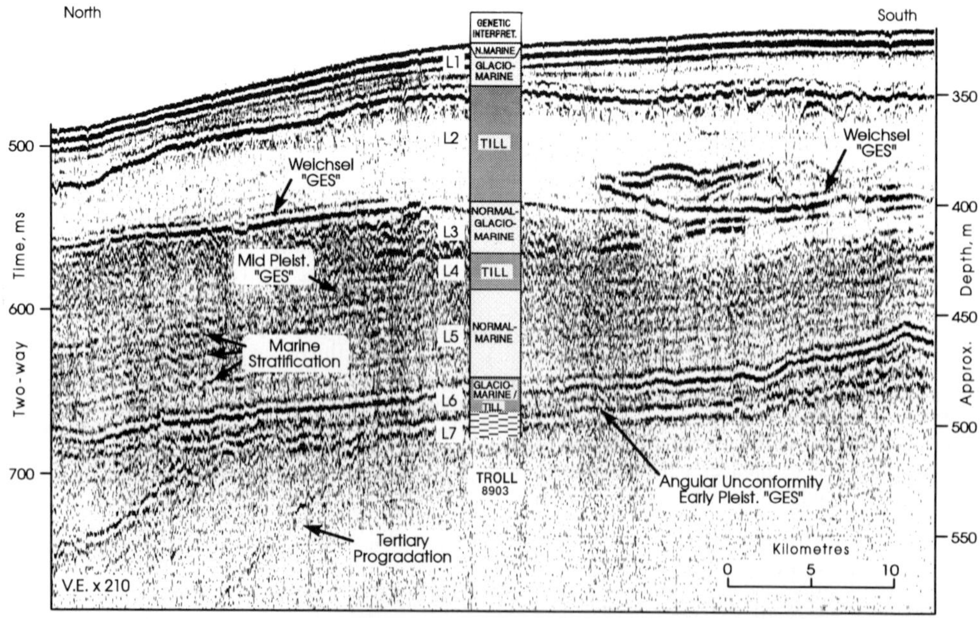

Fig. 6. North–south 1 kJ sparker profile through the Troll borehole location showing correlation of the major chrono-lithostratigraphic units and bounding surfaces with the seismostratigraphic units. Most of these units extend partially or fully across and along the Norwegian Channel as depicted in Figs 7 and 8. The window in Fig. 8 corresponds to this horizontally compressed seismic profile (for location see Fig. 2).

Fig. 5) are recorded. The next sequence is composed of a till of Mid-Pleistocene age (L4) which is followed (disconformably) by a pocket of more sorted sediments (L3) which also include sediments of Last Interglacial age. The last sequence is composed of at least two till units (L2) and a subsequent late glacial/Holocene marine sequence (L1) (Sejrup *et al.* 1994, 1995). Thus while the first and last sequences are largely preserved, with their till and subsequent glaciomarine package, the middle sequence is much more fragmentary with large hiatuses both within and bounding it.

As seen in Fig. 8 the number of glacial erosion surfaces and seismostratigraphic units increases

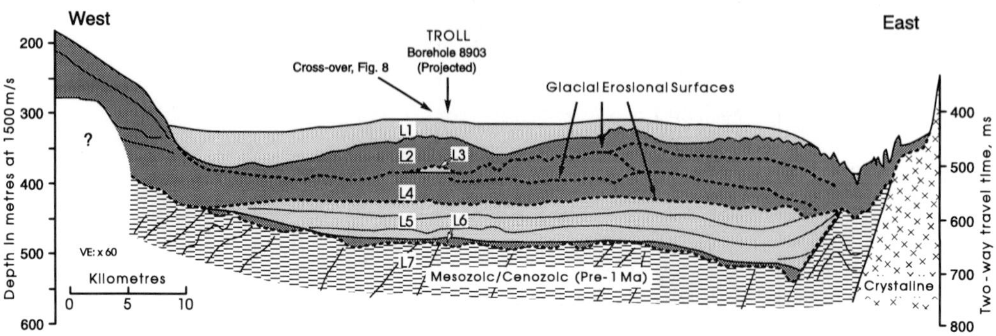

Fig. 7. Interpretation of an east–west seismic profile across the Norwegian Channel immediately north of the Troll borehole (for location see Fig. 2). The L1 to L7 designations correspond to the lithostratigraphic breakdown in the Troll 8903 borehole (Fig. 5) while L1 and L2 correlate with the Kleppe Senior and the Norwegian Trench Formations respectively of Rise *et al.* (1984). The eastward slope of the angular unconformity as well as the trough along the crystalline 'fall line' apparently result from glacial erosion. The lower till is thin here but locally thickens up to 70 m. The thick, overlying early to mid-Pleistocene marine unit pinches out westward. Numerous glacial sheet erosion surfaces separate mid- and late Pleistocene tills. Stipples and symbols as in Fig. 8. Modified from Sejrup *et al.* (1995).

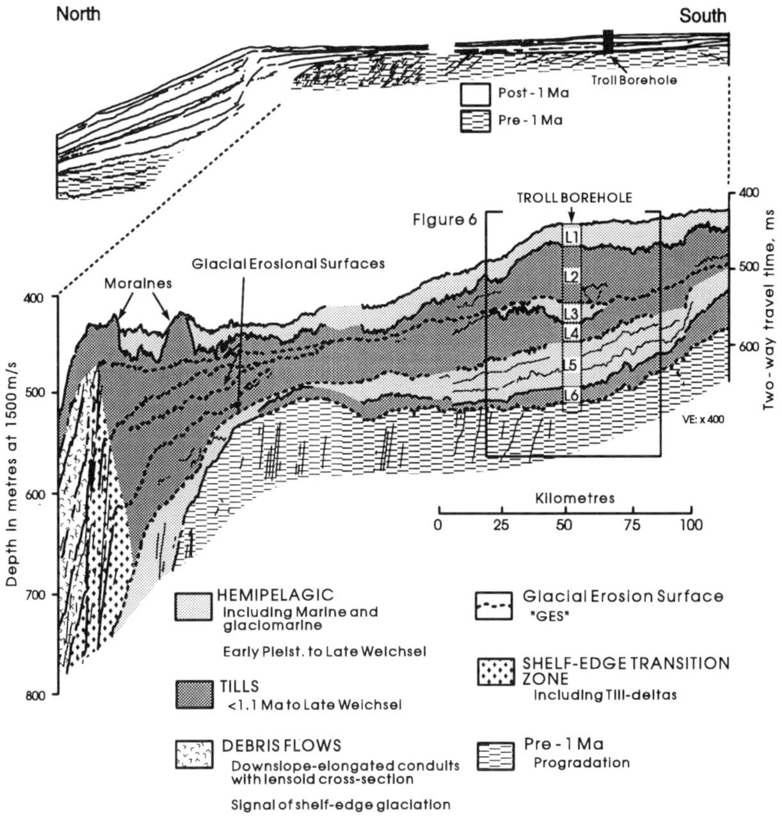

Fig. 8. Interpretation of a composite of north–south seismic profiles through the Troll borehole (showing stratigraphic unit breakdown) and along the Norwegian Channel (for location see Fig. 2). The window represents position of the Fig. 6 seismic profile. Preservation of expansive and thick early and early mid-Pleistocene deposits changed to a more sporadic preservation following the onset of multiple mid-Pleistocene glaciations (Troll unit L4 and bounding unconformities). The near-palaeoshelf extent of the *c.* 1.1 Ma lower till helps establish a chronostratigraphic horizon in the North Sea Fan as does the lower mid-Pleistocene glacial erosion surface (GES). The progressive steepening with age of the GES in the outer channel is interpreted to reflect tectonic subsidence which allowed preservation of some of the multiple glaciation shelf/slope record.

towards the shelf edge. On the other hand, very little of the complexity of the Quaternary package remains intact in the southern (ice upstream) end of the channel, south of Jæren. In the deeper part of the Skagerrak the sediments other than those from the last glacial cycle are rare above the pre-Quaternary strata (von Haugwitz & Wong 1993). General North Sea subsidence (Clarke 1973; Sejrup *et al.* 1987) combined with enhanced subsidence in the outer Norwegian Channel associated with massive sediment loading (note progressively steeper dip with depth of originally flat-lying GES, Fig. 8), allowed preservation of several glaciation cycles. These are represented by sheetform, erosion-bounded tills and laterally associated subglacial, progradational delta-like units. Termed 'till deltas', these are identified with glaciation to the shelf edge (glacier provides the base level analagous to sea level in a normal delta) and they evolve distally into more chaotic, gently sloping flow bodies. Preservation of much of the shelf/slope transition on a long-term basis allows a preliminary chronology of the fan sequences through correlation with the Troll borehole sediments in the Norwegian Channel.

The North Sea Fan and adjacent shelf margins

The North Sea Fan has acted as the major depocentre for Quaternary sediment shed from mainland Norway and part of the North Sea and then transported along the Norwegian Channel. The shelf margin has prograded into the SE Norwegian Sea during the Quaternary by at least 30 km by sediments fed through the Norwegian Channel.

Approximately 1000 km of recently collected high-resolution sleeve gun array transects (five dip and three strike lines) are the basis for dividing the fan body into numerous seismic sequences based either on unique seismic character and bounding prominent reflectors or the slope equivalents of unconformities in the outer shelf region (King *et al.* 1996). A minimum Quaternary thickness has been determined on the fan by projection of the basal glacial erosion horizon associated with the 1.1 Ma till (Figs 8 and 9) found at Troll. However this horizon has been truncated by a large slide; thus Quaternary sediments on the fan referred to in this study are clearly less than 1 Ma old.

The North Sea Fan contrasts with most deep-sea fans by the absence of canyons, turbidity channels and turbidites. Three dominant styles of sedimentation within the thick (>900 m) Quaternary sediment wedge include:

(1) thick (commonly 50 to 120 m) blankets of terrigenous hemipelagic sedimentation,
(2) large-scale masswasting to the ocean basin as enormous translational slides (mega-slides); and
(3) aprons of glaciogenic debris-flow deposits contributing up to 80% of fan construction since shelf-edge glaciations began (Fig. 9).

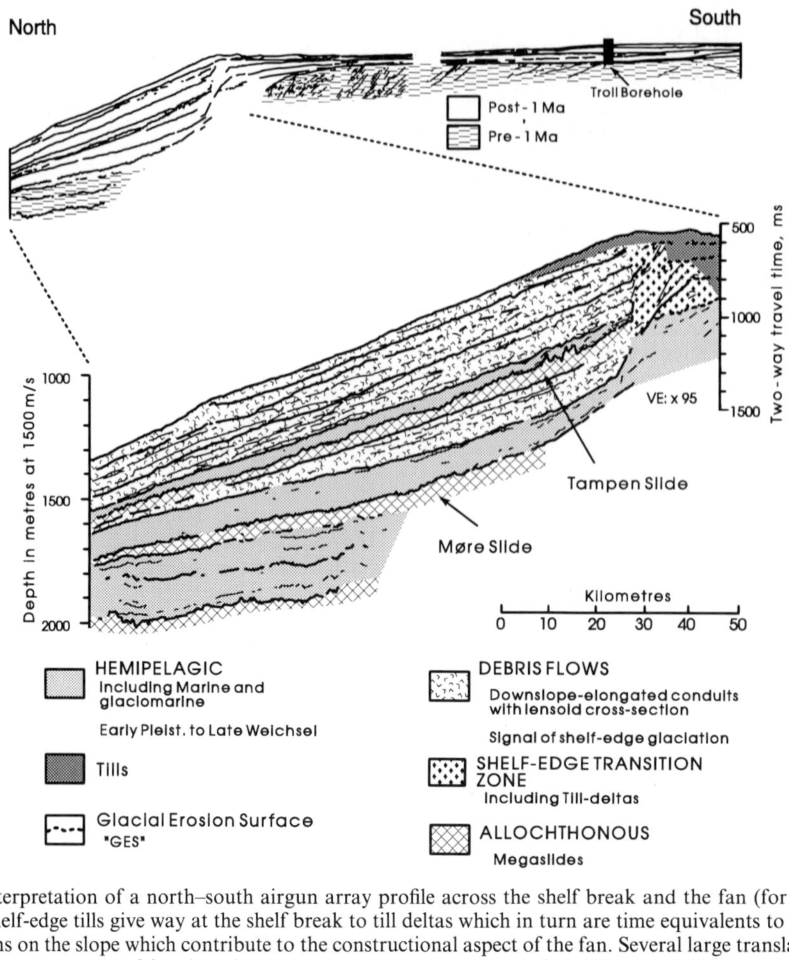

Fig. 9. Interpretation of a north–south airgun array profile across the shelf break and the fan (for location see Fig. 2). Shelf-edge tills give way at the shelf break to till deltas which in turn are time equivalents to thick debris-flow aprons on the slope which contribute to the constructional aspect of the fan. Several large translational slides removed large amounts of fan deposits to the deep sea and complicated shelf–fan correlations.

Hemipelagites

The hemipelagic fan sequences display a moderately transparent, often structureless to weakly stratified internal character which, together with the overall geometry, indicates ponding and draping. They become successively thinner up through the fan succession, the upper one being 50 to 60 m thick over much of the fan. Two hemipelagic sequences of Quaternary age are mainly deposited within the scar formed by preceding slides which may have created or at least enhanced a sediment trap effect for material input from the channel mouth and possibly via contour currents. Even if the sedimentation process for these sequences resembled the continuous and presumably high rates of glaciomarine input associated with the last deglaciation on the fan, which rarely left more than 3 m of sediment, then time periods exceeding 100, perhaps 200 ka are required for these thicknesses.

Translational slides

At least three major, buried slide events have been identified within the North Sea Fan from the recently collected seismic data. The earliest is buried at least 600 m below seabed and involved large (10 km × 100 m) block sliding. Two subsequent slide events (King *et al.* 1996), designated Møre and Tampen slides (Fig. 9), exhibit high and steep head- and sidewalls with sharp lower inflections which develop into smooth, gently dipping (0.5°), expansive décollement or translational glide-plane horizons upon which thick, partially or wholly disturbed sediments

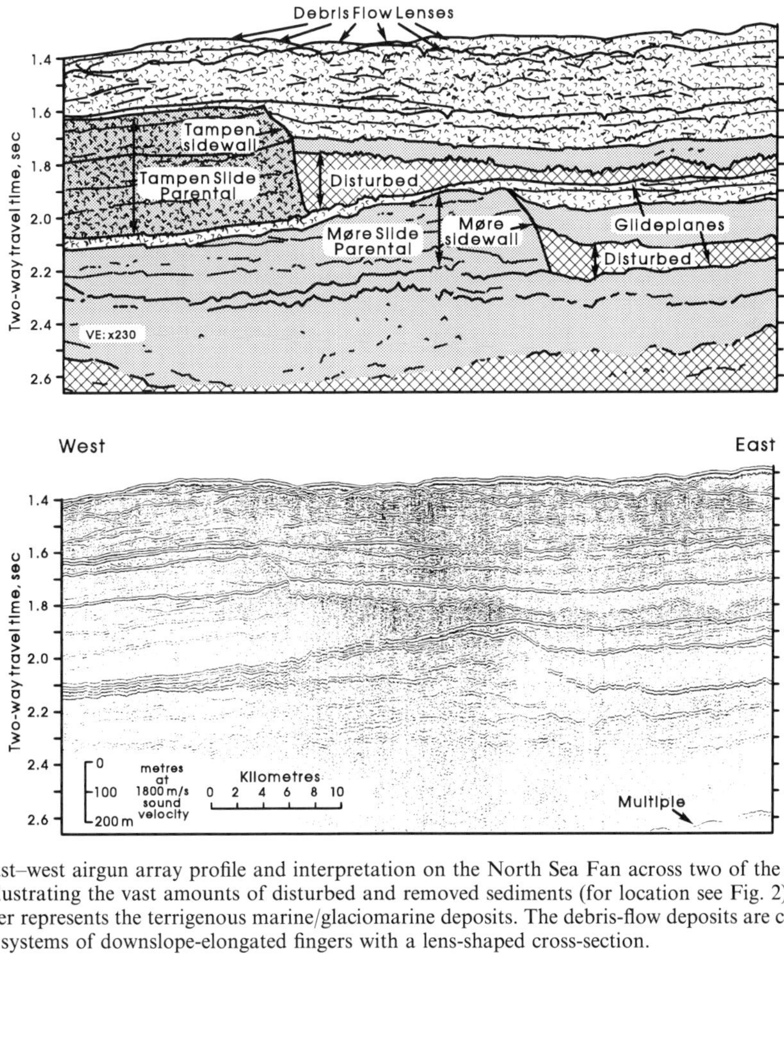

Fig. 10. East–west airgun array profile and interpretation on the North Sea Fan across two of the major slide sidewalls illustrating the vast amounts of disturbed and removed sediments (for location see Fig. 2). A more draped cover represents the terrigenous marine/glaciomarine deposits. The debris-flow deposits are characterized by stacked systems of downslope-elongated fingers with a lens-shaped cross-section.

have been conveyed. The Møre slide appears to be an early Pleistocene event while the Tampen slide post-dates the lower mid-Pleistocene glacial equivalents on the shelf. Volume measurements of disturbed, partially translated sediment remaining within the fan body are in the order of 2500 km^3 for the Møre and Tampen slides, but estimates based on assumed pre-slide seabed reconstructions suggest that up to twice this volume may have been involved, the difference being removed to the ocean basin. The better investigated Storegga slide complex (Bugge *et al.* 1987; Jansen *et al.* 1987), located northeast of the main fan body, is exposed at the seabed (Fig. 1) and is dated at 30 to 50 ka (Bugge *et al.* 1987). This and succeeding slides during the Holocene removed about 4200 km^3 to the Norway Basin while 1300 km^3 remain within the slope scar (Jansen *et al.* 1987). Subsequent, but smaller slides have also occurred in post-glacial times (Bugge *et al.* 1987; Evans *et al.* 1996).

Debris flows

The fan equivalents to the shelf-edge tills and till deltas at the channel mouth are numerous thick (100 m or greater) sequences of debris flows. The till to debris flow transition at the shelf edge is complex and is usually via a coarsely prograding till delta (Alley *et al.* 1989) whose detailed relationship to glacial advance, erosion and subsequent retreat events remains unclear. The relatively steep delta strata give way to chaotic, irregular to lenticular bodies apparently associated with failure and flow in an overall more shallow-dipping (0.5°) setting. Farther down the fan each sequence is made up of one or several aprons comprising complex systems of stacked, interwoven, downslope-elongated constructional, lenticular and/or lobate debris flows which range from 2 to 40 km in width and 15 to 60 m in thickness. The lenses are usually stacked in such a manner that succeeding lenses are situated at the lowest topographic position between two pre-existing lenses, indicative of an origin as flowing bodies. Shallow cores (up to 6 m) yield a homogeneous diamict with 38% clay, 34% silt, and 28% sand and exhibiting almost no lateral variation. Similar forms have been recognized on other glacial trough mouth fans (Laberg & Vorren 1995; Stoker 1995; Aksu & Hiscott 1992) and they are interpreted as debris flows. The earliest examples of these aprons are recognized at about 600 m below the seabed in the proximal reaches of the fan and they are the approximate stratigraphic equivalents of the earliest of the Mid-Pleistocene tills in the channel.

The debris flows form the building block of the fan and are attributed to periodic input from several phases of a Norwegian Channel ice stream reaching the shelf edge (King *et al.* 1996). At least five sequences of these debris flows have been preserved on the fan representing well over 6000 km^3 of material. The large slides almost certainly removed other full sequences though it is not clear what proportion of these removed deposits consisted of typical stacked debris flow as opposed to the more normal and glaciomarine hemipelagites. Ironically, following the rapid and voluminous sedimentation from the glacial phases, the North Sea Fan may be more stable with respect to large, regional slide events. Evans *et al.* (1996) showed some evidence for an inherent stability in the debris-flow deposits in contrast to the more normal, glaciomarine sediments on the slope outside the debris-flow regime.

The North Sea Fan bears many similarities to other glaciated trough mouth fans in terms of debris flows, glaciogenic hemipelagites and slides (Laberg & Vorren 1993, 1995; Stoker, 1995; Aksu & Hiscott 1992).

Margin adjacent to the fan

The North Sea Fan is volumetrically the most important depocentre for sediments derived from southern Norway via the Norwegian Channel (south of Stadt) but the mainland-derived ice sheet both east and west of this also reached the shelf edge in multiple events, i.e. across the Møre/Trøndelag shelf (Bugge 1983; Holtedahl 1993; Haflidason *et al.* 1991) and north and west of Shetland (Johnson *et al.* 1993; Holmes *et al.* 1993). On the margin beyond the North Sea Fan proportionately larger volumes of till and glaciomarine sediment are preserved on the shelf while slope equivalents are dominated by seismically laminated deposits which thin significantly in deeper water depths. These are low-energy distal glaciomarine deposits derived through ice rafting, plumes, contourites, occasional upper-slope debris flows interbedded with the laminated deposits (Evans *et al.* 1996; Stoker 1995), and thick, lower to mid-slope slide deposits (e.g. Miller Slide; Stevenson 1991; Long & Bone 1990). As seen within the fan itself, on the immediate eastern fan margin there is evidence for large-scale mass-wasting in the form of large translational slides (Bugge *et al.* 1987).

The fan and adjacent margins contrast in terms of the great volume, constructional geometry and near-complete dominance of gravity flow features where the greatest ice flux built the fan and less voluminous flow and hemipelagic aprons elsewhere. Analogous, though much smaller-scale fan/interfan contrasts are recognized on the West Shetland and Hebrides shelf/slope (Stoker 1995).

Discussion and conclusions

Controls on erosion, deposition and transport processes on the shelf: a fast-flowing ice stream

A minimum number of five glaciations have apparently reached the shelf edge as indicated by shelf tills and bounding glacial erosion surfaces (GES) and by associated debris-flow complexes on the slope. The first recorded glacial advance to the shelf edge was in the Early Pleistocene but no corresponding debris-flow apron on the fan has been recognized. Several mid-Pleistocene events were erosion-dominated in all but the outer shelf, and are responsible for a minimum of half the Quaternary fan buildup, possibly considerably more. One GES which shows evidence of particularly strong erosion, can be traced over the whole Norwegian Channel (corresponding to Level B of Rokoengen & Rønningsland 1983). During this phase the U-shaped configuration of the outer Norwegian Channel was enhanced through cutting of thick Quaternary deposits now preserved only at and beyond the east and west channel margins. At least one phase of Weichselian-age glacial erosion cut down to this level but it remains unclear if this major erosional phase was purely a Weichselian phenomenon. Weichselian deposition was clearly multi-phased, at least locally, and in general deposits are thick in comparison to those preserved from earlier glaciations.

Rokoengen & Rønningsland (1983) proposed an ice-flow pattern for the upper part of the Quaternary package in the Norwegian Channel, based on pattern of erosion and presence of tills, whereby ice flowed west, out from the fjords, then swung north, feeding the main northward flow along the channel and to the shelf edge. There are parallels between this ice-flow pattern and erosion and deposition associated with ice streams which, through multiple glaciations, build a thick succession. Furthermore, the close lateral association between thick tills and coarsely prograding packages at the palaeoshelf edge can be likened to 'till deltas' deposited below an ice stream as envisioned by Alley *et al.* (1989). They reasoned that progradation occurs in a zone proximal to the grounding line below an Antarctic ice stream concurrent with proximal and 'topset' sheets of laterally and possibly vertically aggrading 'active' or 'deforming till'. In this model the deltas are continuously fed with water-saturated, subglacial, deforming till in a conveyor-belt fashion beneath the ice stream. Analogous, though much older sheet erosion-bounded till sequences with occasional 'subglacial deltas' and ice-stream erosional troughs are recognized beyond present-day ice streams in the Ross Sea (Anderson & Bartek 1992; Alonso *et al.* 1992). Subglacial fluvial processes in such a situation would be much reduced owing to lack of communicability in the subglacial hydraulic system in a clay-rich, constantly mixed deforming till layer. This is consistent with the complete lack of fluvial-type channelling in these deposits. During shelf-edge ice maxima in the Norwegian Channel this water-saturated till would directly source the debris-flow aprons on the fan. The foresets, being the steepest surface recognized in this system, are probably the site of debris-flow initiation. During most of the glaciation the subglacial environment over most of the Norwegian Channel as well as the delta fronts would be primarily sediment bypass zones. It was under these conditions that the large sediment volumes in the slope-situated debris-flow aprons would have accumulated. The last fan-situated sequence alone comprises over 700 km^3 of debris-flow material. However there could have been some subglacial deposition in the outer channel for much of the duration of the glaciation given the plentiful accommodation space at the shelf break.

Depositional conditions on the shelf can only be met and maintained during successive glaciations given adequate accommodation space and sediment supply. Sediment supply is abundant given the broad catchment area and efficient transport mechanism beneath a fast-flowing ice stream. Accommodation space is plentiful on the slope but on the shelf is dependent in the long term on tectonic subsidence and frequency/amount of previous glaciogenic input and the erosional vigour of the following glaciation. The fact that the ice rides out across deformable beds probably governs its development into a fast-flowing ice stream but its capability to erode and transport and the ensuing channel shape and gradient may be, in addition to basal temperature/hydraulics, governed more by factors analogous to the hydraulic geometry of a fluvial system. As in any subglacial system there must have been a temporal and spatial balance between deposition and erosion throughout the evolution of the ice stream. The fact that the base of each till unit in the Norwegian Channel is a regional GES while the top often has preserved features such as moraines and iceberg scours, and the fact that the GES projects across till deltas whose time equivalents comprise the bulk of the fan/slope deposits of a sequence, suggests that the major till depositional phase follows the most vigorous erosional phase, i.e. that shelf deposition is mainly a post-glacial-maximum phenomenon.

Within the framework of a fast-flowing ice stream in the initial deglaciation stage there is large potential to deposit very thick subglacial deposits on the shelf simply through a sudden increase in accommodation space, much as Alley *et al.* (1989) and Bentley (1994) envisioned. The combination of a presumably warming glacier together with tectonic subsidence (with sediment loading), and a rising relative sea level (and the positive feedback flow and sediment supply factors that these can entail) can all contribute to increased accommodation space and deposition within the channel rather than bypass to the fan. Other factors appear to dominate in the Skagerrak which, apart from its much larger scale, behaves more like the fjords than the rest of the Norwegian Channel. Overdeepening, sill development, and purging of Quaternary sediments with each glaciation occur despite an apparently ample sediment supply and accommodation space. Perhaps in this more up-ice position lower velocities and consequently less friction melting inhibit development of a basal water-saturated till layer.

In summary, thick till/glaciomarine sequences have accumulated in the Norwegian Channel, largely under deglacial phases, but their preservation has been varied. The first sequence appears relatively complete, possibly because there was ample subsidence time (*c.* 0.5 Ma) before the ensuing glaciation. A much-fragmented record of numerous Mid-Pleistocene glaciations remains in the Norwegian Channel apparently reflecting erosion and sediment bypass due to lack of accommodation space as input outpaced subsidence. Evidence for most of these phases is only present at the palaeoshelf break and on the fan. Accommodation space was essentially unlimited under glacial retreat resulting in thick till/glaciomarine deposition in the final (Weichsel) sequence. However, if the pattern of most previous glaciations continues, then without significant subsidence before another glaciation the marine/glaciomarine blanket will be removed, the depositional features of the till (moraines, etc.) will be destroyed and provided a warm-based ice stream develops, everything will be incorporated into a subglacial till and re-transported to the slope to be deposited as debris flows.

The fjord, channel, fan sediment pathway

The following discussion focuses on depositional and transport processes and their relative importance through the various stages in the sediment path from the southern part of Fennoscandia to the Norwegian Sea. It is evident that there are several unknown variables in this sedimentation system and we present a generalized first approximation. Firstly, it is important to note that during interglacial periods very little sediment is contributed in this region. Transport from the fjords to the shelf at times when ice caps did not fill the fjords and in interglacial times is negligible as the fjords act as effective sediment traps (Eisma 1981). Holocene sedimentation in the northern North Sea is mostly a redistribution whereby wave and current winnowing of glacial sediments in shallow areas is transported to more protected and basinal areas. This process dominates during sea-level lowstand immediately following deglaciation. Holocene sediment thicknesses in the investigated areas are usually less than 0.5 m and rarely more than a few metres (e.g. Sejrup *et al.* 1981; Jansen *et al.* 1983; Nagy & Ofstad 1980; Sejrup *et al.* 1994). This pattern is also reflected in the thickness of interglacial sediments recorded in

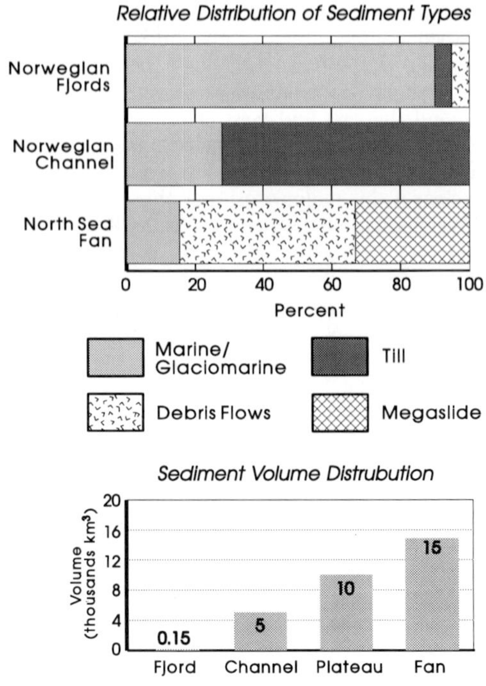

Fig. 11. Relative distribution of Quaternary sediment types (upper diagram) and total volumes in the fjord, channel, plateau and fan system (lower diagram).

the region (Sejrup *et al.* 1987, 1989, 1995; Sejrup & Knudsen 1993). Thus the major Quaternary deposition/transport system is linked with an arctic climate, low sea level and glacial processes.

Coarse volume estimates of the different Quaternary sediment packages in the fjords, channel and fan are presented in Fig. 11. Sediment volume calculations apply for those areas outlined in Fig. 1. Estimates for each area are based on several unpublished and published seismic profile interpretations. The North Sea Plateau volume was calculated from the Quaternary sediment isopach of Johnson *et al.* (1993, Fig. 72). Owing to the complexity of the Quaternary stratigraphy of the North Sea Plateau area, the Quaternary sediments are not broken down into genetic groups. The Norwegian Channel estimates are based on seven transverse profiles and a longitudinal profile. Most sequences on the North Sea Fan are of sufficient uniformity to allow 'first-order' volume estimates and, where applicable, extrapolations outside survey coverage are based on projections to the topographically defined fan margins. Ongoing work in the channel and on the fan will constrain these figures better.

An overview of the sedimentary deposition and transportation processes among the different sediment repositories between southern Norway and the Norwegian Sea is shown in Fig. 12. This figure illustrates some of the major trends in the flux of sediments from Fennoscandia to the deep sea. A first approximation Quaternary (in this case post-1 Ma) sediment budget calculation for mainland Norway erosion is about 120 m assuming a catchment area the size of the mainland shown in Fig. 1 (south of Stadt), and a total sediment volume of 25 000 km^3 (including slide sediments removed to the deep sea). Clearly, large parts of this erosion are related to the glacial fjords and valleys observed to be cut into the mature landscape of the pre-Quaternary, uplifted palaeosurface across the mainland. However, a very significant component may also have been removed from coastal areas during strandflat evolution (Larsen & Holtedahl 1985; Holtedahl, pers. comm. 1994).

The following points together with Figs 11 and 12 summarize the fjord, channel, slope margin and fan system in terms of dominant processes and volumes.

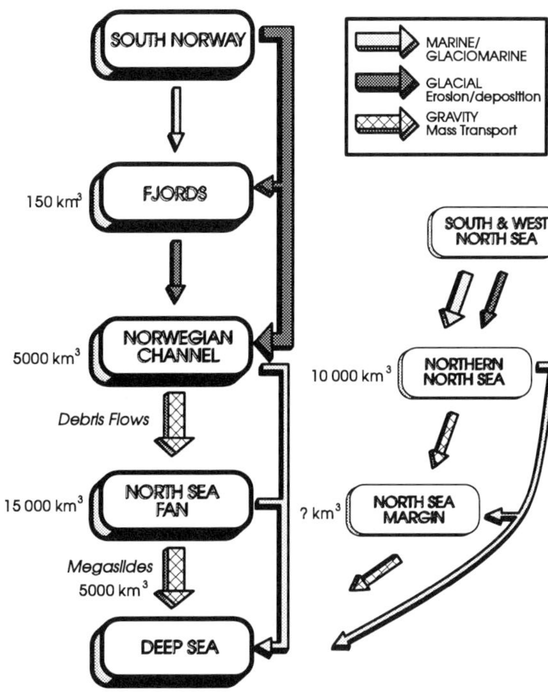

Fig. 12. Sediment transport processes in the fjord, channel, fan system. Arrow thickness roughly indicates relative importance in terms of volume. At least 5000 km^3 are thought to have been transported from the North Sea Fan to the deep sea via the megaslides.

1. Sedimentation from the mainland into the fjords is dominated by glaciomarine sediments deposited very rapidly during deglaciations. Long-lived land-based glaciations could conceivably fill the fjords to sea level to be subsequently emptied under larger advances. Despite small sediment volumes (Fig. 11) a more frequent fill and flush cycle enhances the significance of the fjord sediment budget.
2. The only (significant) transport mechanism from the fjords to the Norwegian Channel is by glacial ice. Together with sediments derived by glacial erosion on land, the fjord sediments are largely removed during subsequent glaciations and sediment bypass prevails in the fjords under ice maximum conditions while erosion/overdeepening is accentuated. Tills dominate volumetrically in the channel and much of the deposition is thought to be during the late glacial stages from an ice stream flowing along the channel. Transport is thought to occur via a thin, water-saturated subglacial sediment layer in a conveyor-belt fashion. During peak glaciation the entire transport path from mainland to shelf edge is thought to experience mainly erosion and bypass, with erosion more important in up-ice locations. When and where accommodation space becomes available this layer can be deposited, resulting locally in thick tills. In common with the fjords, most of the marine sediments are deposited during deglaciations in the Norwegian Channel, but an additional contribution is through increased winnowing of the North Sea Plateau during low sea level.
3. During shelf-edge maximum glaciation, sediment which was temporarily deposited in the channel is transported subglacially to the shelf edge where gravity processes then take over on the slope. The debris-flow process can distribute diamicton over 100 km down the fan while numerous, presumably successive conduits bring about a broad across-fan distribution.
4. The North Sea Fan is volumetrically the largest repository (15 000 km^3), and the debris flows make up 80% of the sediments still remaining on the fan that were deposited since the initial mid-Pleistocene shelf-edge glaciations. The fact that the North Sea Fan (and other glacial-fed fans) exist as positive forms of the glaciated margin, indicates that these parts of the Norwegian Sea margins are relatively stable. However, the identification of Pleistocene slide scars involving such large sediment volumes within the fan sequence, show that slide events are the major transport mechanism of these sediments to the deep sea.
5. With our present knowledge it is difficult to evaluate the flux of sediment from the margin not occupied by a fast-flowing ice stream. A much lower input to the slope via minor debris flows and plume sedimentation occurs during the entire shelf-edge glaciation and the stacked debris-flow aprons characteristic of the fan are absent. Stoker (1995) observed a similar contrast in smaller fan and 'between fan' areas on the NW British margin.

The consideration above suggests that the major transport from southern Fennoscandia to the deep sea occurs while the ice margin is situated on the shelf and the major process is subglacial transport through the Norwegian Channel by a fast-flowing ice stream extending to the shelf edge beyond which gravity-controlled processes dominate.

This work was funded by the European Commission through the MAST II programme under the auspices of the ENAM (European North Atlantic Margin; sediment pathways processes and fluxes) project, and in part by the Research Council of Norway (NFR). The Geological Survey of Denmark (DGU) and The Netherlands Institute of Sea Research provided a processed seismic section used in the Fig. 9 interpretation. The manuscript was critically reviewed by M. Stoker and T. Vorren. The support of these people and institutions is gratefully acknowledged.

References

AARSETH, I., & MANGERUD, J. 1974. Younger Dryas end moraines between Hardangerfjorden and Sognefjorden, Western Norway. *Boreas*, **3**, 3–22.

—— & GODVIK, J. 1984. Seismostratigraphy in the Norwegian Channel west of Hordaland and Sogn. *In*: AARSETH, I. & SEJRUP, H. P. (eds) *Quaternary Stratigraphy of the North Sea*, Abstract Volume, **79**, University of Bergen.

——, BJERKLI, K., BJØRKLUND, K., BØE, *et al.* 1975. Late Quaternary sediments from Korsfjorden, western Norway. *Sarsia*, **58**, 43–66.

——, LØNNE, Ø. & GISKEØDEGAARD, O. 1989. Submarine slides in glaciomarine sediments from some western Norwegian fjords. *Marine Geology*, **88**, 1–21.

AKSU, A. E. & HISCOTT, R. N. 1992. Shingled Quaternary debris flow lenses on the northeast Newfoundland Slope. *Sedimentology*, **39**, 193–206.

ALLEY, R. B., BLANKENSHIP, D. D., ROONEY, S. T. & BENTLEY, C. R. 1989. Sedimentation beneath ice shelves. The view from Ice Stream B. *Marine Geology*, **85**, 101–120.

ALONSO, B., ANDERSON, J. B., DÍAZ, J. I. & BARTEK, L. R. 1992. Pliocene–Pleistocene seismic stratigraphy of the Ross Sea: evidence for multiple ice sheet grounding episodes. *In*: ELLIOT, D. H. (ed.) *Contributions to Antarctic Research III*. American Geophysical Union, Antarctic Research Series, **57**, 93–103.

ANDERSEN, B. G., SEJRUP, H. P. & KIRKHUS, Ø. 1983. Eemian and Weichselian deposits at Bø on Karmøy. *Norges Geologiske Undersøkelse*, **380**, 189–201.

ANDERSON, J. B. & BARTEK, L. R. 1992. Cenozoic glacial history of the Ross Sea revealed by intermediate resolution seismic reflection data combined with drill site information. *In*: KENNETT, J. P. & WARNKE, D. A. (eds) *The Antarctic Paleoenvironment: A Perspective on Global Change*. American Geophysical Union, Antarctic Research Series, **56**, 231–263.

ANUNDSEN, K. & SIMONSEN, A. 1968. *Et preborealt breframstøt på Hardangervidda og i området mellom Bergensbanen og Jotunheimen*. Universitetet i Bergen Årbok, Matematisk-Naturvitenskapelig Serie 1967, **7**.

BENTLEY, E. R. 1994. Sedimentation by the Antarctic ice sheet from a glaciological perspective. *Terra Antarctica*, **1**, 249–250.

BUGGE, T. 1983. *Submarine Slides on the Norwegian Continental Margin, with Special Emphasis on the Storegga Area*. IKU Publication, **110**, 1–152.

——, BEFRING, S., BELDERSON, R. H., EDVIN, T., JANSEN, E., KENYON, N. H., HOLTEDAHL, H. & SEJRUP, H. P. 1987. A giant three-stage submarine slide off Norway. *Geo-Marine Letters*, **7**, 191–198.

CLARKE, R. H. 1973. Cainozoic subsidence in the North sea. *Earth and Planetary Science Letters*, **18**, 329–332.

EISMA, D. 1981. Supply and deposition of suspended matter in the North Sea. *In*: NIO, S. D., SCHÜTTENHELM, R. T. E. & VAN WEERING, TJ. E. (eds) *Holocene Marine Sedimentation in the North Sea Basin*. International Association of Sedimentologists, Special Publication, **5**, 415–428.

EVANS, D., KING, E. L., KENYON, N. H., BRETT, C. & WALLIS, D. 1996. Evidence for long-term instability in the Storegga slide region off Western Norway. *Marine Geology*, **130**, 281–292.

FLODEN, T. & SELLEVOLD, M. A. 1972. Two seismic profiles across the Norwegian Channel west of Bergen. *Stockholm Contr. Geol.*, **24**(2), 25–32.

HAFLIDASON, H., AARSETH, I., HAUGEN, J. E., SEJRUP, H. P., LØVLIE, R. & REITHER, E. 1991. Quaternary stratigraphy of the Draugen area, Mid Norwegian shelf. *Marine Geology*, **101**, 125–146.

HOLMES, R., JEFFERY, D. H., RUCKLEY, N. A. & WINGFIELD, R. T. R. 1993. *Quaternary Geology around the United Kingdom (North Sheet)*. 1 : 1 000 000. British Geological Survey, Edinburgh.

HOLTEDAHL, H. 1965. Recent turbidites in the Hardangerfjord, Norway. *In*: WHITTARD, W. F. & BRADSHAW, R. (eds) *Submarine Geology and Geophysics*. Colston Papers, **17**, 107–141.

——1975. The geology of the Hardangerfjord, West Norway. *Norges Geologiske Undersøkelse*, **323**, 1–87.

——1993. *Marine Geology of the Norwegian Continental Margin*. Norges Geologiske Undersøkelse, Special Publication, **6**, 11–50.

JANSEN, E., SEJRUP, H. P., FJÆRAN, T., HALD, M., HOLTEDAHL, H. & SKARBØ, O. 1983. Late Weichselian paleoceanography of the southeastern Norwegian Sea. *Norsk Geologisk Tidsskrift*, **63**, 117–146.

——, BEFRING, S., BUGGE, T., EDVIN, T., HOLTEDAHL, H. & SEJRUP, H. P. 1987. Large submarine slides on the Norwegian Continental Margin: Sediments, transport and timing. *Marine Geology*, **78**, 77–107.

JOHNSON, H., RICHARDS, P. C., LONG, D. & GRAHAM, C. C. 1993. *United Kingdom Offshore Regional Report: The Geology of the Northern North Sea*. HMSO for the British Geological Survey, London.

KELLOGG, T. B. 1976. Late Quaternary climatic changes: evidence from cores of the Norwegian and Greenland Seas. *Geological Society of America, Memoir*, **145**, 77–110.

KING, E. L., SEJRUP, H. P., HAFLIDASON, H., ELVERHØI, A. & AARSETH, I. 1996. Quaternary seismic stratigraphy of the North Sea Fan: Glacially-fed gravity flow aprons, hemipelagic sediments, and large submarine slides. *Marine Geology*, **130**, 293–316.

LABERG, S. L. & VORREN, T. O. 1993. A Late Pleistocene submarine slide on the Bear Island Trough Mouth Fan. *Geo-Marine Letters*, **13**, 227–234.

—— & —— 1995. Late Weichselian submarine debris flow deposits on the Bear Island Trough Mouth Fan. *Marine Geology*, **127**, 45–72.

LARSEN, E. & HOLTEDAHL, H. 1985. The Norwegian strandflat: A reconsideration of its age and origin. *Norsk Geologisk Tidsskrift*, **65**, 247–254.

—— & SEJRUP, H. P. 1990. Weichselian land–sea interactions: western Norway–Norwegian Sea. *Quaternary Science Reviews*, **9**, 85–97.

——, GULLIKSEN, S., LAURTIZEN, S. E., LIE, R., LØVLIE, R. & MANGERUD, J. 1987. Cave stratigraphy in western Norway: multiple Weichselian glaciations and interstadial vertebrate fauna. *Boreas*, **16**, 267–292.

LONG, D. & BONE, B. D. 1990. Sediment instability on the continental slope of northwest Europe. Proceedings of *Oceanology International*, **90**, Brighton.

MANGERUD, J. 1970. Late Weichselian vegetation and icefront oscillations in the Bergen district, Western Norway. *Norsk geografisk Tidsskrift*, **24**, 121–148.

——, LARSEN, E., LONGVA, O. & SØNSTEGAARD, E. 1979. Glacial history of western Norway 15,000–10,000 B.P. *Boreas*, **8**, 179–187.

——, SØNSTERGAARD, E., SEJRUP, H. P. & HALDORSEN, S. 1981. A continuous Eemian–Early Weichselian sequence containing pollen and marine fossils at Fjøsanger, western Norway. *Boreas*, **10**, 137–208.

NAGY, J. & OFSTAD, K. 1980. Quaternary foraminifera and sediments in the Norwegian Channel. *Boreas*, **9**, 39–52.

NESJE, A. & WHILLANS, I. M. 1994. Erosion of Sognefjord, Norway. *Geomorphology*, **9**, 33–45.

RISE, L., ROKOENGEN, K., SKINNER, A. C. & LONG, D. 1984. *Nordlige Nordsjø. Kvartærgeologisk kart mellom 60°30′N og 62°N, og øst for 1°Ø.* (Northern North Sea, Quaternary geology map between 60°30′N and 62°N, and east of 1°E). M 1 : 500 000. Institutt for kontinentalsokkelundersøkelser (IKU).

ROKOENGEN, K. & RØNNINGSLAND, T. M. 1983. Shallow bedrock geology and Quaternary thickness in the Norwegian sector of the North Sea. *Norsk Geologisk Tidsskrift*, **63**, 83–102.

SEJRUP, H. P. 1987. Molluscan and foraminiferal biostratigraphy of an Eemian–Early Weichselian section on Karmøy, southwestern Norway. *Boreas*, **16**, 27–42.

—— & KNUDSEN, K. L. 1993. Correlation and paleoenvirnoments of interglacial sediments in the North Sea. *Boreas*, **22**, 223–225.

——, AARSETH, I., ELLINGSEN, K. L., *et al.* 1987. Quaternary stratigraphy of the Fladen area, central North Sea: a multidisciplinary study. *Journal of Quaternary Science*, **2**, 35–58.

——, —— & HAFLIDASON, H. 1991. The Quaternary succession in the northern North Sea. *Marine Geology*, **101**, 103–111.

——, ——, ——, LØVLIE, R., BRATTEN, Å., TJØSTHEIM, G., FORSBERG, C. F. & ELLINGSEN, K. L. 1995. Quaternary of the Norwegian Channel: palaeoceanography and glaciation history. *Norsk Geologisk Tidsskrift*, **75**, 65–87.

——, FJÆRAN, T., HALD, M. *et al.* 1981. Benthic foraminifera in surface samples from the Norwegian continental margin between 62°N and 65°N. *Journal of Foraminiferal Research*, **11**, 277–295.

——, HAFLIDASON, H., AARSETH, I., FORSBERG, C. F., KING, E., LONG, D. & ROKOENGEN, K. 1994. Late Weichselian glaciation history of the northern North Sea. *Boreas*, **23**, 1–13.

——, NAGY, J. & BRIGHAM-GRETTE, J. 1989. Foraminiferal stratigraphy and amino acid geochronology of Quaternary sediments in the Norwegian Channel, northern North Sea. *Norsk Geologisk Tidsskrift*, **69**, 111–124.

SELLEVOLL, M. & SUNDVOR, E. 1974. The origin of the Norwegian Channel – A discussion based on seismic measurements. *Canadian Journal of Earth Science*, **11**, 224–231.

STEVENSON, A. G. 1991. *Miller, Sheet 61°N 02°W, Quaternary Geology*. 1 : 250 000 map series. British Geological Survey.

STOKER, M. S. 1995. The influence of glacigenic sedimentation on slope-apron development on the continental margin off Northwest Britain. *In*: SCRUTTON, R. A., STOKER, M. S., SHIMMIELD, G. B. & TUDHOPE, A. W. (eds) *The Tectonics, Sedimentation and Palaeoceanography of the North Atlantic Region*. Geological Society of London, Special Publication, **90**, 159–177.

VON HAUGWITZ, W. R. & WONG, H. K. 1993. Multiple Pleistocene ice advances into the Skagerrak: A detailed seismic stratigraphy from high resolution seismic profiles. *Marine Geology*, **111**, 189–207.

VORREN, T. 1973. Glacial geology of the area between Jostedalsbreen and Jotunheimen, south Norway. *Norges Geologiske Undersøkelse*, **291**, 1–46.

Detailed survey of the western end of the Hurd Deep (English Channel): new facts for a tectonic origin

G. LERICOLAIS[1], P. GUENNOC[2], J.-P. AUFFRET[3], J.-F. BOURILLET[1] & S. BERNE[1]

[1] *Département Géosciences Marines, IFREMER, BP 70, F-29280 Plouzané, France*
[2] *Département Infrastructure Géologique et Géophysique, BRGM, BP 6009, F-45060 Orléans, France*
[3] *Université de Caen, Esplanade de la Paix, F-14032 Caen, France*

Abstract: During the 'Sédimanche 2' survey, a study of the western end of the Hurd Deep was carried out using new high-resolution geophysical techniques. The obtained results, i.e. bathymetry, acoustical imaging and seismic profiling, determine a tectonic control on the first excavation of Hurd Deep, a complex infilling and present-day sediment dynamics.

In size and geometry, the Hurd Deep is a unique feature in the English Channel, terminating abruptly at its western end. It is incised into Jurassic and Late Cretaceous sediments. Except for the trough itself, the sea floor is extensively levelled at a depth between 70 and 90 m.

The Hurd Deep infill, revealed by high-resolution seismic data, is characterized by a set of five main sequences. The thickness of the infill averages 80 m along the axis of the trough and reaches a maximum of 140 m. The side-scan sonar mosaic shows active sediment dynamics characterized by the presence of dunes on the southern part of the present trough.

A Neogene tectonic origin for the Hurd Deep and the interplay of erosion and sedimentation during the last glacial and interglacial periods are presented in terms of geomorphology and sequence stratigraphy.

Even though several surveys have been carried out along the Hurd Deep axis (Boillot 1963; Dangeard 1929; Gibbard 1988; Hamilton & Smith 1972; Quesney 1983; Smith 1985; Wingfield 1990), the lack of drilling information is a significant constraint restricting an adequate understanding of its origin, its formation and the nature and the age of its sedimentary infilling. One aim of the 'Sédimanche 2' cruise on board 'R/V Le Suroit' was to obtain very high-resolution geophysical data using new equipment on one of the intriguing features of the Hurd Deep: its western ending.

The Hurd Deep is a unique feature in the English Channel (Fig. 1). It is an elongate depression about 150 km long and between 2 and 5 km wide with a maximum water depth reaching 170 m and an undulating long axis. Outside the Deep, the sea floor is very flat, at between 70 and 90 m water depth. Incised into deformed Jurassic and Upper Cretaceous strata, the Hurd Deep is located along the major fault zone of the Western Channel, orientated ENE–WSW (Hamilton & Smith 1972; Evans 1990). The Deep terminates southwestwards at about 4° W.

During the 'Sédimanche 2' survey this particular western ending was investigated using the Simrad EM 1000 shallow-water swath-bathymetry and acoustic imaging system, and a digital high-resolution seismic reflection data-acquisition system. In order to extend our understanding of the formation of such a deep in the middle of the English Channel, and to correlate its infill with the up-stream tributary system described by Auffret & Larsonneur (1971) and Auffret *et al.* (1975), we describe here the bathymetrical map and the imaging mosaic, obtained from Simrad EM 1000 system, the very high-resolution (VHR) seismic profiles, and the three-dimensional (3D) representation of the interpreted seismic horizons obtained by gridding and smooth-surface fitting of the seismic lines.

The Hurd Deep infilling is characterized by a set of five main sequences (not yet dated). The thickness of sediment infilling averages 80 m along the axis of the trough and reaches a maximum of 140 m while the bottom of the incision reaches a depth of about −240 m below present sea level. The imaging mosaic shows an active sedimentary dynamic characterized by some dunes located along the southern part of the present trough. The dune amplitude is about 10 m and the presence of numerous sand ribbons confirms present transport to the west. The 3D representation of the Hurd Deep five main sedimentary horizons allows us to determine

From De Batist, M. & Jacobs, P. (eds), 1996, *Geology of Siliciclastic Shelf Seas*, Geological Society Special Publication No. 117, pp. 203–215.

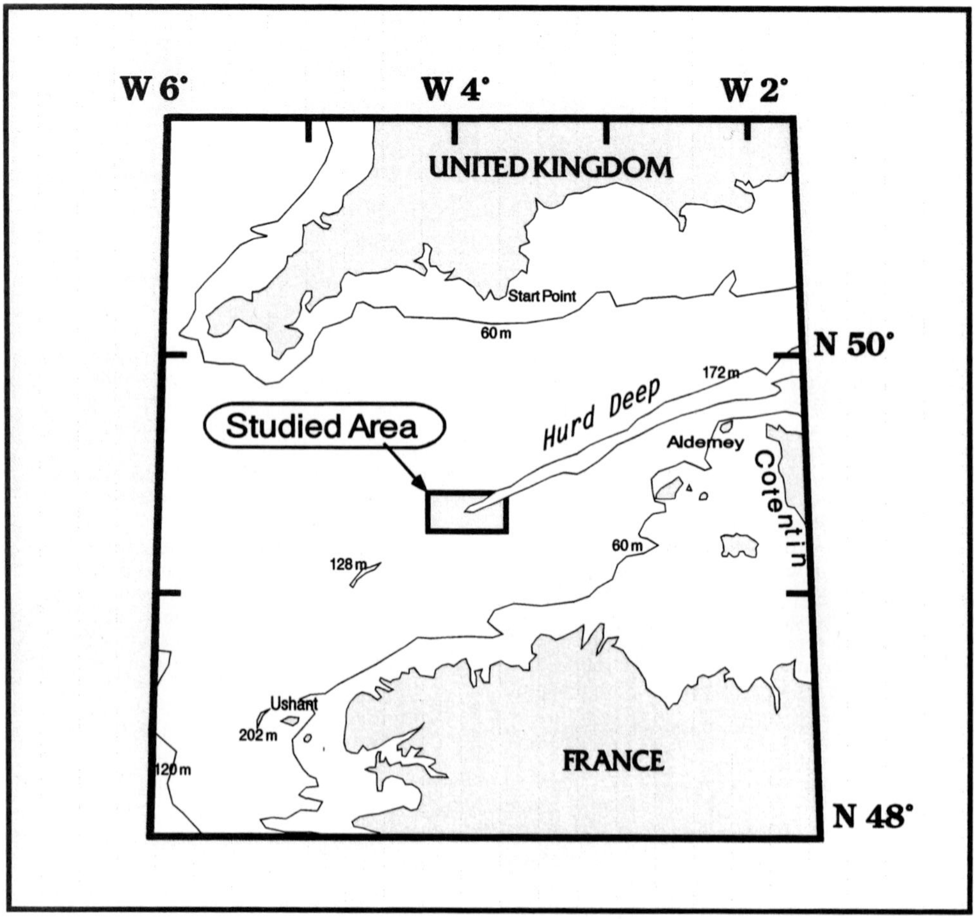

Fig. 1. Location of the Hurd Deep in the English Channel.

the infilling of the Deep and in particular to point out the tectonic control at depth on the western end and to enhance the role of faulting in the origin of the Hurd Deep.

Geological setting

The trend of the Hurd Deep is parallel to the substratum major faults of the western English Channel (Hamilton & Smith 1972; Evans 1990). Different sedimentary infills with an average thickness of 80 m and a maximum thickness of 140 m have been recognised (Hamilton & Smith 1972). Dredging carried out in the deep itself has revealed bioclastic material and boulders derived from Hercynian rocks (Andreieff *et al.* 1972). The Armorican Massif was the main source for these blocks, although the larger blocks may have been brought by ice-rafting ('radeaux de glaces', Dangeard 1929) or river action (Boillot in questions to Hamilton & Smith 1972).

The western English Channel is characterized by dominant ENE–WSW structures in the shallow substratum composed of various Mesozoic and Cenozoic series more or less deformed or faulted. The deep structure is dominated by a wide Permo-Triassic and Jurassic basin, the Western Approaches Basin (Ziegler 1987). This basin terminates northeastwards along the Start–Cotentin ridge and is intersected by the Central Channel E–W fault system, and is subdivided into different sub-basins. To the north, the base of the very deep Permo-Triassic Plymouth Bay Basin shows a wide synclinal form, more than 10 km deep (Evans 1990; Smith 1985). No large-scale basement block faulting is detected on the deep seismic reflection lines (Bois *et al.* 1991*a*, *b*). The Permo-Triassic series outcrop along the English coast and can be more than 5000 m thick in the basin. Further offshore they are conformably overlain at depth by Jurassic strata up to 3000 m thick in the Southwest Channel Basin and Brittany Trough (Ziegler 1987; Evans 1990).

By the end of the Jurassic and during the early Cretaceous, these basins were strongly reactivated under tensional forces linked to the opening of the North Atlantic and of the Bay of Biscay. Thick detrital series were deposited along the main axis of the basins. From the Albian and during the Upper Cretaceous and lower Tertiary, major tectonic movements resumed and under high sea-level conditions chalk was deposited over large areas (Smith & Curry 1975). Starting during the Eocene, with peak activity during the Oligocene and continuing in the Miocene (Ziegler 1987; Evans 1990), compressive movements linked to the Alpine orogeny led to strong inversion of the Mesozoic basins. According to Ziegler (1987) relief up to 3000 m was created along an anticlinorium zone located along the southern part of the Western Approaches Basin. On the southern side of the Channel, Mesozoic series are much reduced or absent south of the Cotentin–Ushant line. This is often depicted as the Alderney–Ushant Fault system bounding the Mesozoic basin to the south (Smith & Curry 1975). The importance of this trend is reinforced by the presence of elongated high magnetic anomalies just to the north of this fault system. The Channel magnetic anomaly is interpreted as an old Precambrian suture (Lefort & Segoufin 1978) or as Permo-Triassic intrusions (Ziegler 1987).

From SWAT-ECORS profiles 7 to 11, it can be clearly seen (in Bois *et al.* 1991*b*, p. 191) that the southern border of the Mesozoic Western Channel Basin and the contact with the Armorican basement is quite variable. The boundary between the North Armorican basement and the Channel Basin series is clearly faulted offshore northwestern Brittany and the Channel Islands, whereas between these two areas and westwards, in the Western Approaches Trough (Ziegler 1987; Bois *et al.* 1991*a*, *b*), it appears as a more progressive slope or steep flank. These variations can be explained by the different role played by N60°-trending faults, the Chan nel Median Fault or the Alderney–Ushant Fault that are quite discontinuous along strike, during the Mesozoic extension and the Tertiary inversion.

Boillot (1963, 1964) was one of the first to describe the Hurd Deep. He proposed a karstic origin for the Deep and related its origin to westward and smaller deeps such as the Ushant Trough and the Vierge Trough (Boillot & Hom 1966). Hamilton & Smith (1972) proposed that

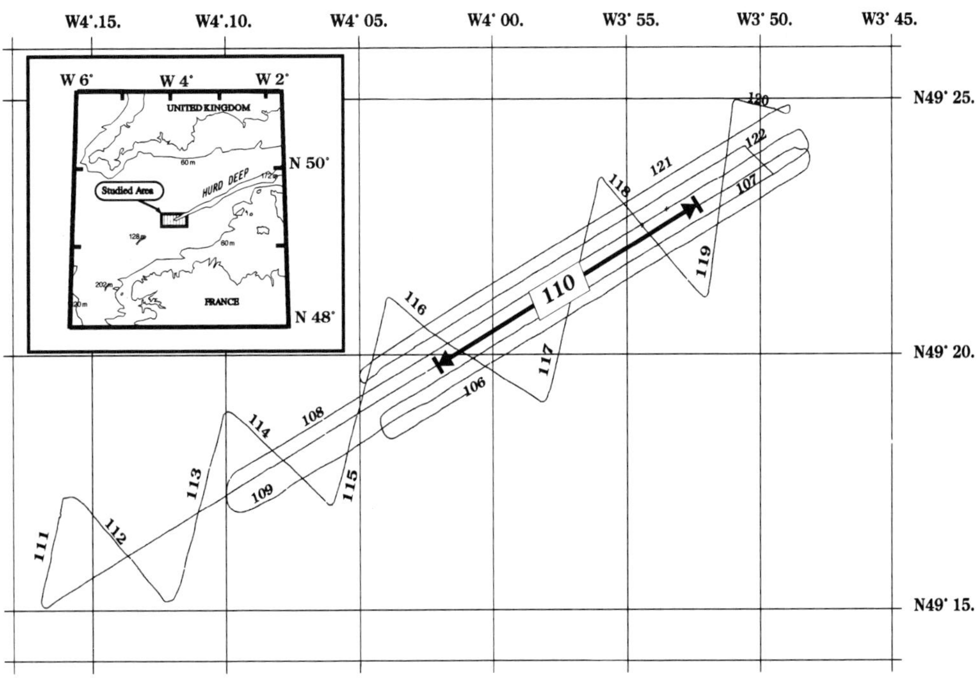

Fig. 2. Positioning chart of the 'Sedimanche 2' lines.

the Deep origin was due to incision resulting in strong tidal scouring during a glacial lowstand period. The Deep would have been partially infilled during the subsequent highstand. However, the deep incision seems difficult to explain only by tidal scouring. An additional catastrophic origin is invoked by Smith (1985), who imagines a breakthrough of the chalk barrier across the Dover Strait by melting waters from the North Sea with sufficient energy to incise deeply enough the Dover Strait valleys and to create the eastern anastomosing valley system described by Auffret & Larsonneur (1971). In the Central Channel, the flood would have reconcentrated its energy to carve the Hurd Deep; this flood reconcentration is explained by the topographic focusing north of the Cotentin Peninsula.

Wingfield (1990) compared the Hurd Deep to the British continental shelf deeps. The origin of these deeps is linked to the discharge of glacial lakes during deglaciation. The one drawback of such an assumption is that the caning of the Hurd Deep should have been from a strong stream coming from west to east. However, it is difficult to explain the anastomosing river system of the Eastern Channel with such a flood.

Methods

Acquisition and processing of data

The Hurd Deep was surveyed during the 'Sédimanche 2' survey on board the 'R/V Le Suroit' in March 1993 (Fig. 2). Precise positioning was given by a DGPS (differential global positioning system). Real-time processing of the navigation with an accuracy of a few metres made it possible to follow parallel track lines spaced 400 m apart, at a speed of 5 knots. The vessel was equipped with the Simrad EM 1000 swath-bathymetry system and very high-resolution seismic lines were shot simultaneously using a surface-towed sparker. All data acquisition was synchronized and digitally recorded.

The Simrad EM 1000 multi-beam echosounder system is a shallow-water precision sea-floor mapping system, providing both topographic maps and acoustic sonar imaging of the sea floor in the same operation (Fig. 3). Navigation, bathymetry and image data were processed and synthesized, in order to plot positions automatically, to produce bathymetrical maps (Trismus software; Bourillet 1996) and to produce an image mosaic (Trias software; Augustin 1985).

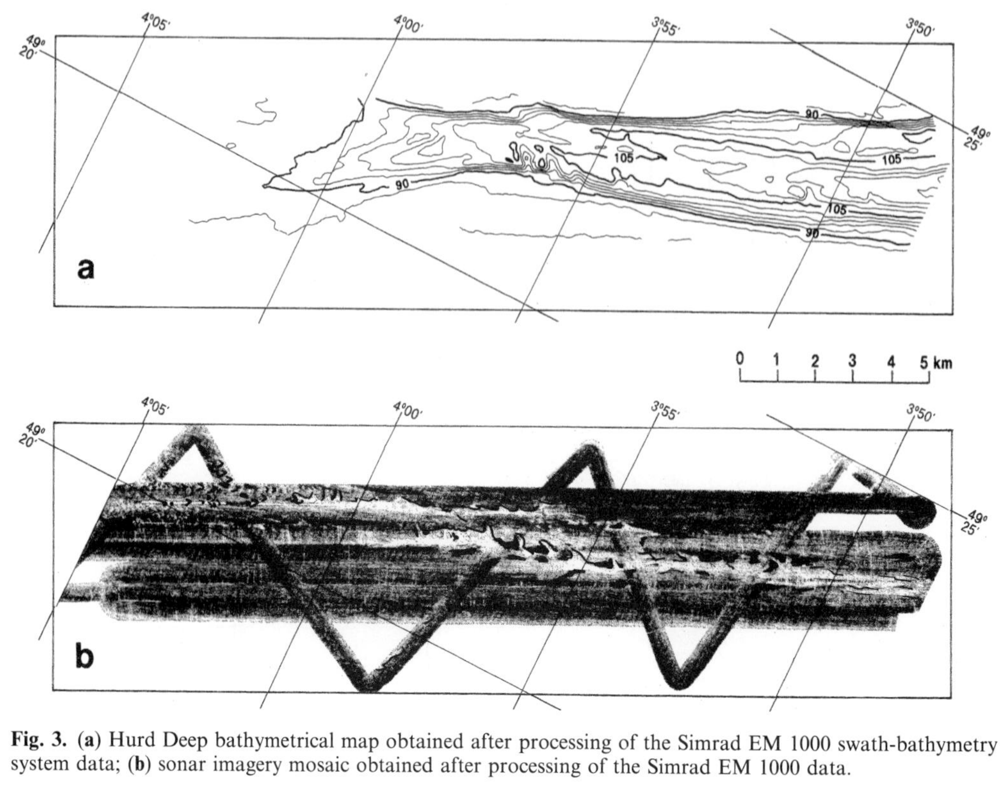

Fig. 3. (**a**) Hurd Deep bathymetrical map obtained after processing of the Simrad EM 1000 swath-bathymetry system data; (**b**) sonar imagery mosaic obtained after processing of the Simrad EM 1000 data.

The very high-resolution seismic reflection source was a single-channel sparker, with a central frequency around 800 Hz. The digital data acquisition was done in real-time on the PC-based Elics Delph system. The data were processed with Sithere processing software (Lericolais *et al.* 1990) with swell removal, deconvolution, and the main horizon digitized on computer monitor and interpolated to obtain a data terrain model.

Bathymetrical and mosaic imaging interpretation

The Hurd Deep is limited on its northwestern side by a 'bank' striking N65° E with a 5% gradient (Fig. 3a). This side lies on a branch of the major Alderney–Ushant Fault system, which limits the Deep all along its length. From east to west, the southeastern limit first strikes N75° with a 2% gradient and then strikes towards N65° with a 5% gradient. This southern bank is also installed on a branch of the Alderney–Ushant Fault system. These two branches delineate a Jurassic horst (Bouysse *et al.* 1975; Taylor *et al.* 1980).

Inside the Deep, on the southwestern steep side, stand four discrete sand dunes (Fig. 3b). They lie against the side and are observed only inside the Deep where they present a strong asymmetry with the lee side towards the southwest. Their height ranges from 9 to 12 m, with a length of about 200 m. The sand dunes area is limited to the east by the deeper water. The westernmost part of the area studied, outside the

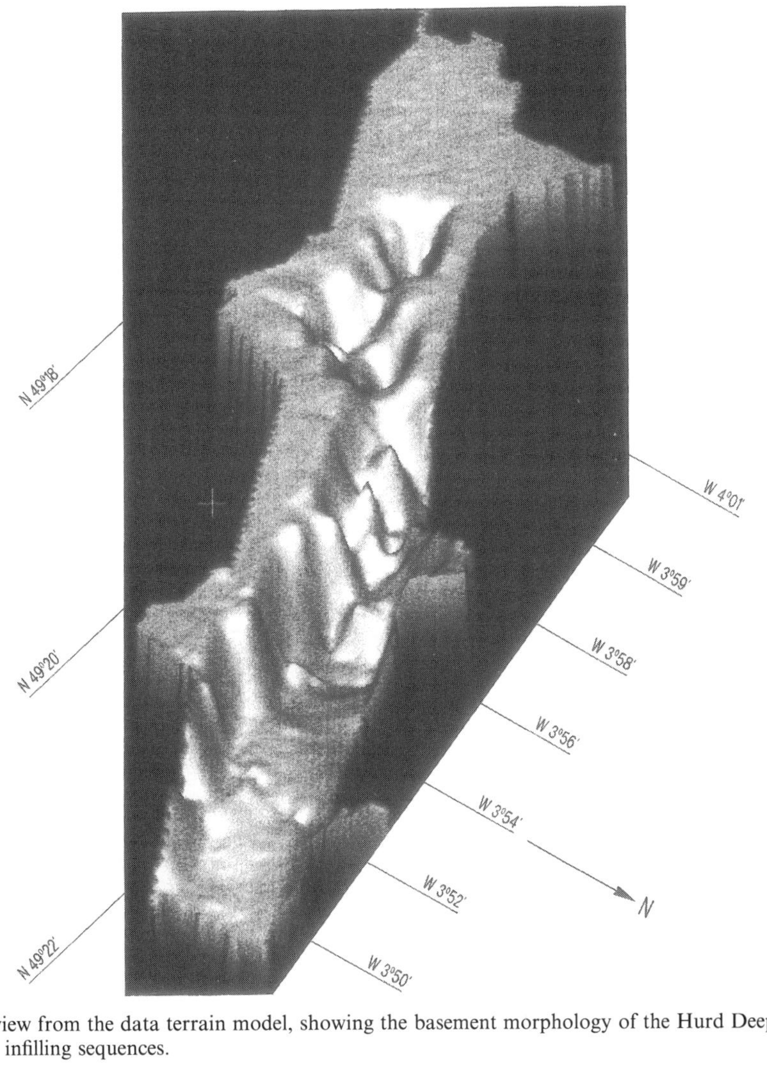

Fig. 4. 3D view from the data terrain model, showing the basement morphology of the Hurd Deep without the sedimentary infilling sequences.

Deep, presents some other sand dunes that are not controlled by the bathymetry. All these asymmetrical dunes have their steeper face towards the present direction (235°) of sediment transport in the Channel (Hamilton 1979; Pingree & Griffiths 1979).

Seismic interpretation

The Hurd Deep has the shape of a fault trough (Dangeard 1929). The seismic profiles reveal that this depression is the partly infilled remnant of much more deeply eroded features. The deep is infilled by a set of fill-in-fill terraces (Quesney 1983), separated by main erosional surfaces. Five main sequences were recognized on the seismic profiles. A selection of reflector points was used to construct a 3D data terrain model. A geometrical model has been built by mapping the interpreted reflector boundaries and an isochron map of each boundary has been compiled (Fig. 4).

The five seismic units are divided into four depositional sequences (IS1 to IS4 on Fig. 5), the fifth and deepest one (ES1 on Fig. 5) being interpreted as the weathered uppermost part of the bedrock. From the 1/250 000 geological map (Taylor *et al.* 1980), the western end of the Hurd

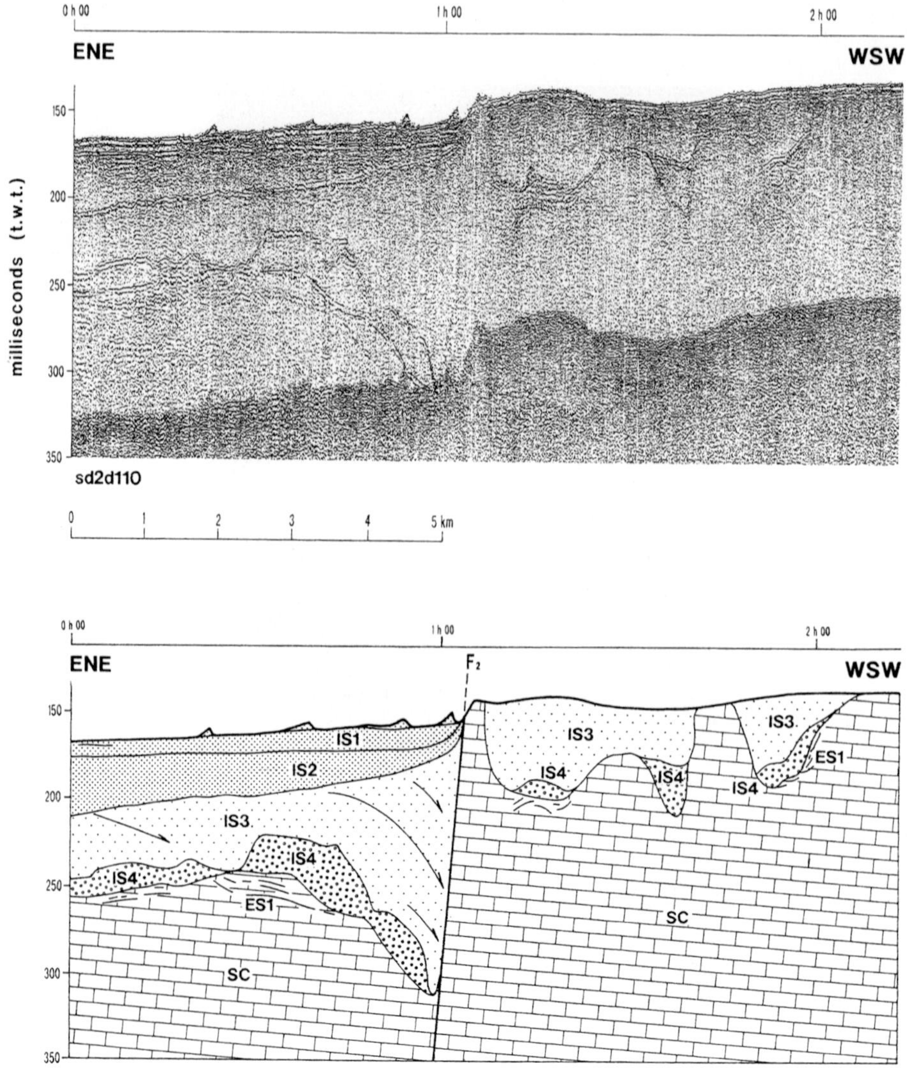

Fig. 5. Seismic line along the Hurd Deep and its interpretation. IS*i* = infilling sequence (*i* = 1 to 4); ES1 = bedrock erosional sequence; SC = Cretaceous bedrock.

Deep is incised into Santonian chalks easily recognizable on seismic lines by their subparallel, strong and regularly spaced reflectors. Located at the northwestern part of the Deep, the Cretaceous–Eocene contact is characterized by cross-bedding, onlap reflectors and then probably faulted reflectors.

The deeper sequence (IS4) presents a chaotic aspect and is strongly eroded by the boundary surface IS3. Its width ranges from 0 to 30 m. This sequence is present in the Deep itself and in the two satellite pits. Its seismic facies, more chaotic due to coarser sediments, is different from the other units and could correspond to the first channel lag deposit of the initial Channel river discharge.

The next sequence (IS3), which is the thickest sequence of the Hurd Deep infill, presents westward prograding foresets downlapping on the underlying formations, characterizing a filling from east to west. The IS3 sequence terminates on fault relief and on bedrock highs but is not affected by any tectonics. The reflector characteristics and their foresets prograding westwards give this sequence a deltaic configuration. This could have happened when the Hurd Deep was the path of the Channel river. The strong reflector separating IS3 and IS2 outcrops in two pits located westward and offset southwestwards from the present deep axis. This erosional boundary surface could correspond to one of the last planation surfaces. A mono- or polycyclic origin is still undetermined.

The sequence IS2 presents at its base a strong and continuous reflector (Fig. 5) and thins over the bedrock relief in the Deep and also at the Deep banks. This sequence is interpreted as an infilling phase during a Pleistocene highstand period. These two last sequences (IS1 and IS2) are located inside the present deep, whereas IS3 sequence is filling the present deep but also the two pits. The top of IS2 sequence is eroded by the subsequent one (IS1).

The upper sequence (IS1), characterizing the last infilling phase of the Hurd Deep, is limited at its base by a strong continuous reflector. It presents in some parts a transparent seismic facies that matches the EM 1000 mapped dune area (Fig. 5). The seismic facies suggests this sequence is essentially composed of medium to coarse sands; the thinnest part of this sequence is thought to have been reworked to form the observed sand bodies. Because its extension follows the −100 m isobath, this sequence seems to be controlled by the bathymetry. The base of IS1 is interpreted as a transgressive surface and could correspond to the beginning of the Flandrian transgression. Nevertheless, the bottomset of the sand dunes is well marked by a reflector. The question is whether this reflector corresponds to a present reactivation surface in the contemporaneous sedimentary deposit (Berné *et al.* 1988) or to a ravinement surface itself. In the latter case, the age of the IS1 lower sequence boundary is to be related to an earlier transgression than Flandrian.

Discussion

Tectonic pre-control and origin of the western Hurd Deep

In the western Hurd Deep zone, the deep structure, as shown by the SWAT 9 profile (Bois *et al.* 1991*b*) or on multi-channel deep seismic lines (Fig. 6), is controlled by the Median Channel Fault limiting a basement block to the north and an important half graben to the south which could be 7 km deep (3 s two-way travel time (TWT)) (Fig. 6). Mesozoic series are dipping to the NW and are offset by steep south-facing normal faults. They are overlain by subhorizontal Upper Cretaceous and Tertiary strata about 700 m thick (0.5–0.6 s TWT). Reactivation of these median Channel faults is evidenced by short upright folds and by faults within the Mesozoic series (Fig. 6).

Only mild undulations are seen within the Upper Cretaceous and Tertiary strata on the deep seismic profiles. The Hurd Deep zone has been surveyed for oil investigations and a dense grid of multi-channel profiles has been obtained in the area with a 4 km spacing between transverse NNW–SSE lines. Through the courtesy of Coparex, a general structural scheme has been established from these lines for the Mesozoic substratum. Although it may appear as a roughly linear and rather homogeneous structure, this structural scheme shows at a smaller scale the lateral discontinuity of N60°-striking faults and the existence of multiple relay zones between them (Fig. 7a).

The faults detected within the Mesozoic series display flower-type structure and small anticline forms typical of tectonic inversion (Cooper & Williams 1989). A closer examination of the deep structure (Fig. 7a) also shows that the Hurd Deep is located over a syncline form and between two normal faults in the Mesozoic series which is narrowing westwards and is replaced by a faulted zone where Mesozoic strata are more strongly folded and tilted (Fig. 6). Thus, it is evident that the western termination of Hurd Deep is located above a relay zone between two important faults of the Mesozoic Channel Basin.

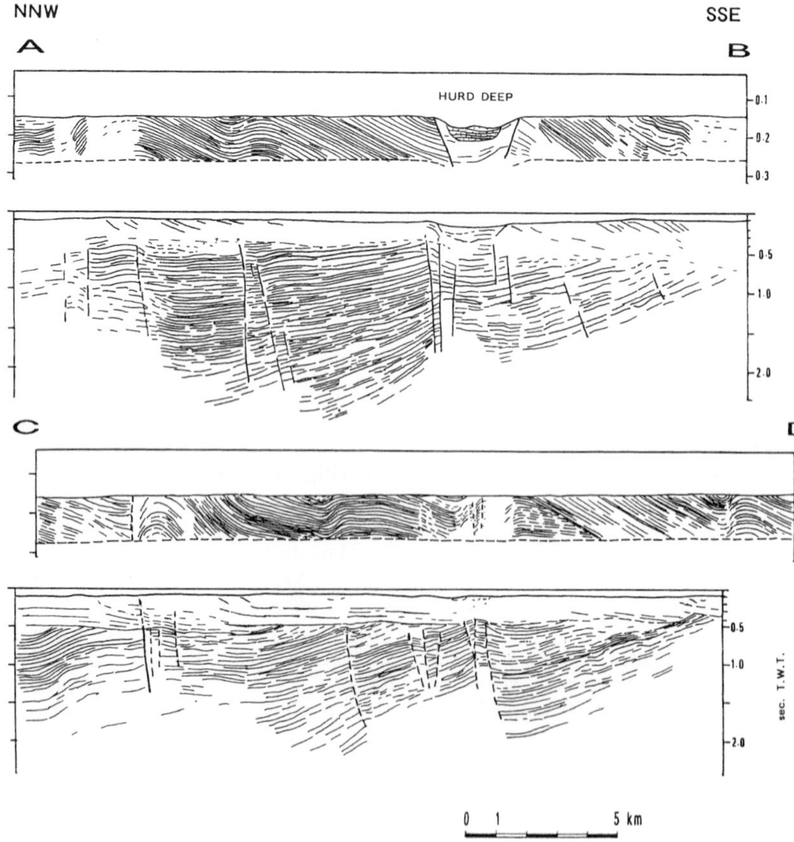

Fig. 6. Line-drawings of deep and shallow seismic profiles along two cross-sections (A–B, C–D) in the western area of the Hurd Deep (see location on Fig. 7a). Top: sparker profiles showing slightly deformed Upper Cretaceous and Tertiary series. Bottom: multi-channel seismic profiles (courtesy of Coparex) showing the deep half-graben structure in the Triassic–Jurassic series.

Because of relatively limited thicknesses and shallow water depths inducing strong multiples, the shallow structures are not clearly seen on the multi-channel seismic reflection profiles. However these structures have been investigated by numerous sparker profiles obtained by French and British teams for the geological mapping of the sea floor of the English Channel (Bouysse *et al.* 1975; Smith & Curry 1975; Evans 1990). The 1/250 000 Guernsey solid geological map (Taylor *et al.* 1980) (Fig. 7b) shows a rather continuous Upper Cretaceous, and Lower Tertiary cover except along the Hurd Deep east of 3°45′ W where Jurassic series subcrop below the Quaternary infill or outcrop on the flanks. On the Guernsey sheet, the Hurd Deep appears to be limited by two subparallel faults converging westwards. Upper Cretaceous and Lower Tertiary series are folded and faulted along the N60°-striking faults but some E–W trends are also depicted.

Comparison between deep structures revealed by the multi-channel seismic lines and shallow structures revealed by the 'Sédimanche 2' high-resolution seismic lines shows some good correlation between fault reactivation within Triassic and Jurassic series and anticlines in the Upper Cretaceous strata. The latter are also faulted mainly in the Hurd Deep area. This is further evidence, together with the strong deformation observed in the Jurassic series in the same area, that this was the site of the most important tectonic movements during Upper Tertiary inversion.

Most of the seismic profiles obtained during the 'Sédimanche 2' cruise are subparallel to the structures and show limited penetration within the substratum. But the pseudo-3D technique

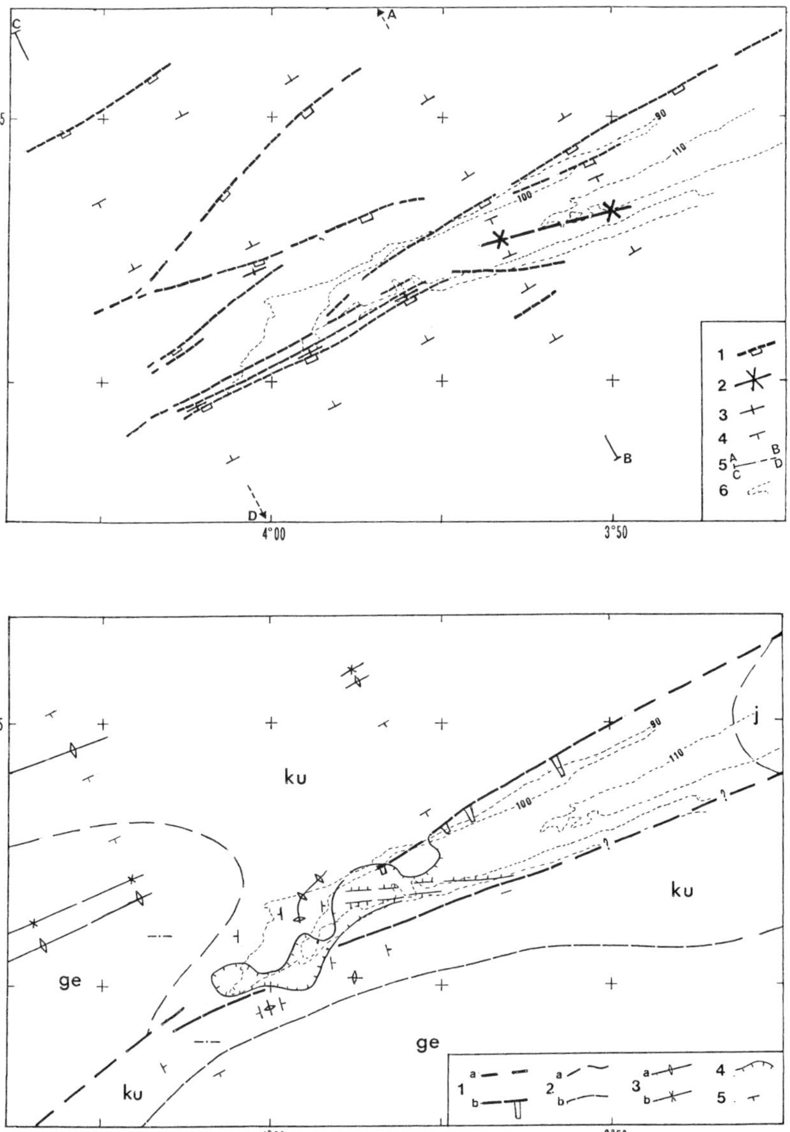

Fig. 7. (**a**) Structural scheme of the Triassic–Jurassic series established from Coparex industrial lines (deep structures). Note the general normal faulting in the Mesozoic series and the termination of the Hurd Deep above a relay zone between two major faults. 1, normal faults; 2, syncline axis; 3, fold axis; 4, general dip; 5, location of cross-sections (shown on Fig. 6); 6, main bathymetrical contours of the present deep. (**b**) Upper Cretaceous and Lower Tertiary structures redrawn from the 1/250 000 Guernsey sheet and 'Sédimanche 2' survey (surficial structures). 1, faults, and 2, geological contours, drawn from Guernsey sheet (a) or 'Sédimanche 2' survey (b); 3a, anticline axis; 3b, syncline axis; 4, steep slopes observed on shallow seismics; 5, dips in Upper Cretaceous and Lower Tertiary series.

allows construction of a substratum isochron map (Fig. 4) which can be discussed together with the Hurd Deep western area structural scheme. Just outside the Deep, Upper Cretaceous or Eocene strata are only slightly deformed and the dips within the layered series are well observed on the profiles. The western Hurd Deep ending area is characterized by a linear northern flank. The present slope corresponds to a substratum fault so that a structural control can be inferred; a slight bend is also observed to the west. The southern flank of the Deep has been less well imaged by the 'Sédimanche 2' profiles. As this flank has an opposite slope to that of the Mesozoic substratum faults observed below it (Fig. 7a), no direct structural control may be inferred there. West of 3°54′ W, the Hurd Deep rapidly narrows and the substratum morphology becomes quite complex as shown on the isochron diagram (Fig. 4). Because of this complexity and the spacing of the seismic lines, the exact structure of the substratum cannot be precisely described. If the termination of the present deep is clearly controlled by highs (horsts ?) (Fig. 5), it is not clear whether it corresponds to a linear E–W horst structure as is drawn on the structural scheme (Fig. 7a) or if a succession of highs and lows exists as indicated on the calculated isochron map (Fig. 4). Indeed, exact mapping of structures with such limited extent appears to be beyond the limits of two-dimensional profile investigation despite their close spacing.

If the detailed structure below the ending of the present trough is more or less uncertain, just west of it, clear anticline forms are observed in the Upper Cretaceous series. Thanks to the good positioning, it is evident that this anticline structure is not linear and that its direction changes from N60° to N–S.

One of the main results of the 'Sédimanche 2' cruise is the discovery and mapping of other infilled small troughs to the west of 3°58′ W (Fig. 4). The two western troughs are clearly offset to the south of those located below the western end of the present deep. Again, because of their limited extent, the profile spacing and the strong deformation within the substratum, precise mapping of the structures was difficult to establish there. The existence of small faults and their orientation are not clear. To the north, substrate highs could correspond to the discontinuous axis of fold structures within Upper Cretaceous strata. Indeed, from the analysis and comparison between the deep and shallow seismic profiles, it appears that these westernmost ancient troughs are mainly controlled by complex folding of the Upper Cretaceous series with 'en échelon' discontinuous anticline and syncline forms instead of small short faults. These fold structures are, however, clearly located above a major fault zone of the Permo-Triassic basin showing tectonic inversion in the Jurassic series (Fig. 6). At the Hurd Deep western end, the fold axes become closely spaced and, as a result of Mid-Tertiary compressive movements, more upright faults in Mesozoic series and faults reaching the Upper Cretaceous and Lower Tertiary were formed. The fault trough was created.

Relationship with the Channel drainage system and the Hurd Deep sedimentary infill

The foundations of the Channel river drainage pattern should have appeared at the end of Oligocene, or early Miocene (Gibbard 1988), when earth movements associated with both the Alpine orogeny and the opening of the North Atlantic Ocean were at their maximum (Ziegler 1987). The Channel maximum inversion phase at the early Miocene (Curry *et al.* 1978; Curry & Smith 1975) gave birth to a regional sea-level drop at the end of the Serravalian (Ziegler 1987; Gibbard 1988). This regression marked the beginning of the development of the London and Paris drainage basins. These became the drainage basins of the Channel river system set up on the present eastern English Channel shelf. The question is determine where this river ended at this time. The western English Channel was functioning through Neogene time as a gulf where sea-level fluctuations occurred. The Channel river was flowing into the Hurd Deep acting as a lake (Quesney 1983). The river could have passed the Deep during an important sea-level drop such as the Messinian regression and would have been flowing through the Shamrock and Black-Mud canyons (Reynaud *et al.* 1995).

The onset of true cold climates in the Praetiglian stage caused a marked change in depositional style, with the input of coarse sediments such as the IS4 sequence (Fig. 5). The return to temperate conditions in the Tiglian caused a readopting of the meandering river form for the Channel river and a renewed deposition of fine sediments, as for the Seine, the Rhine and Meuse rivers. The return to cold climates in the Eburonian–Menapian period established periglacial weathering and consequently the loading of rivers with weathered detrital material (Boillot 1964). The river incised its course and deposited coarse sediments. This increase in sediment deposits is well marked by the vast expansion of deltas in The Netherlands (Zagwijn 1974). Such a delta could be imagined in the Hurd Deep (IS3, presenting westward progradation clinoforms).

Throughout the successive glacial–interglacial stages of the Pleistocene the planation surface should be the result of additional factors such as pre-erosional permafrost with cryoturbation of the soil and the following landward-stepping erosional surface accompanied by tidal erosions. The first records of the prevalence of true permafrost have been dated as Tiglian age at the Dutch–Belgian border (Vandenberghe & Kasse 1989) and Eburonian age at La Londe in the Seine region (Lautridou 1982). The pattern of repeated erosive and depositional phenomena related to glacio-eustasy is the origin of the modification of the fluvial system of the central English Channel (new incisions, sedimentary deposits and erosion). These variations determined the river course in the central and eastern English Channel and the complexity of the drainage system that is characterized by the stacking of several sequences, abandonment and resumption of its valleys.

The emersion phases in the western Channel were less numerous than in the eastern Channel and thus the sedimentation was less important, the Hurd Deep being used nearly every time as the exsurgence for the Channel river. The Channel river extension to the shelf break (−180 m) should have happened only during strong glacial maxima. There is no evidence of incision with sediment infills westwards of the Hurd Deep, except under the Celtic Sea banks. Only these banks can witness the existence of a delta or a river mouth. As for the Hurd Deep, the karstic origin is difficult to explain, firstly because of the elongated morphology and secondly, because of the evidence for tectonic transpressive faulting initiating the trough. Nevertheless, during the 'Sédimanche 2' survey, the eastern anastomosing valley system was also surveyed, and it seems that a thermokarstic origin combined with tidal scouring of a small circular depression, prevealed from first interpretation of the Simrad EM 1000 data, could certainly be envisaged (Lericolais *et al.* 1995). The smaller westward deeps such as the 'Ile Vierge' Trough (Boillot 1961) do not present any sediment infill and their origin, which was not studied here, could be related to combined erosion: pre-erosion during lowstand in periglacial conditions giving birth to thermokarsts, completed by excavation by the tidal currents and the tidal waves during the following sea-level rise. This removal to the west of the river incision can be explained by strong marine transgression erosion following the previous lowstand. The results of the last strong variations in the Saalian–Eemian and Weichselian–Holocene should have been preservation of the tidal reworked sediments in the Deep (IS2 and IS1). The river mouth should have acted as an estuary during sea-level rise. The last phase (Holocene) is characterized inside the Deep by the present tidal dynamic (sand dunes) clearly observed on the geophysical data.

Conclusion

The previous description indicates that the structure below the western ending of the Hurd Deep is due to fault inversion movements in the Jurassic substratum inducing some faults and mainly folds within the Upper Cretaceous and Lower Tertiary series. This decoupling between 'basement' and 'sedimentary cover' tectonics agrees well with the geological history of the Channel. It illustrates the differences between the Permo-Triassic and Jurassic basin structure acquired during the Lower Cretaceous rifting phase and the unconformable, poorly deformed Upper Cretaceous and Lower Tertiary sediments. The deep structural scheme clearly shows that the Hurd Deep terminates to the west in a relay zone between important faults of the Permo-Triassic–Jurassic substratum that were reactivated by Tertiary compressive movements. By Late Eocene, when these movements were initiated, the deep half graben in mid-Channel was covered by subhorizontal Upper Cretaceous and Eocene strata. The globally N–S compressive movements induced reactivation of the major faults within the Mesozoic substratum and created some secondary faults with the same direction. In order to understand the general structure of that area one must take into account the following observations or hypotheses:

- because the pre-existing structures were oblique to the main compressive axis, transpressive movements must have been generated along the N60°-trending faults;
- in the relay zone between two faults, complex stress conditions must have induced changes in the direction of the compressive main axis (s1 direction);
- when the space between major faults was large enough, Mesozoic strata were not strongly deformed;
- with closer spacing between faults, synclinal form was initiated as under the Hurd Deep;
- in the area with very closely spaced faults, i.e. the western end of the Hurd Deep, more upright faults in Mesozoic series and faults reaching the Upper Cretaceous and Lower Tertiary were formed. At that time an elongated trough was formed.

Further northeastwards, the large syncline form observed below the present Hurd Deep can be explained by a gentle folding of the Upper Jurassic series between two normal faults that were inverted under Tertiary positive movements. The creation of such a low in the Mesozoic substratum must have induced folding and some small downfaulting in the overlying thin Upper Cretaceous strata. Thus if the Hurd Deep structure appears to be limited by two faults in the surficial substratum, the deep structure appears to be quite different.

Although no precise age can be ascertained for the lower sedimentary units filling the Hurd Deep at its western end, we assume that the troughs generated by Middle Tertiary movements may have been filled first by Upper Tertiary series. Indeed, the more ancient infilling deposits could be as old as Pliocene. Once the westernmost troughs were filled, the Western Channel abrasion surface was shaped by active subaerial processes (mostly periglacial) and submarine processes during the numerous Upper Pliocene and Quaternary regressive and transgressive phases (Curry 1989). Northeastwards, the deeper main trough, i.e. the present Hurd Deep, was incompletely filled by lower sedimentary units; the presence of progradational foresets in the main infilling sequence indicates that the Deep could have functioned as an exsurgence lake. Then various processes, such as very strong tidal and submarine currents, water mass movements linked to glacial/interglacial phases, must have been active and must have reworked the sedimentary infill. Such active processes and strong currents are evidenced by the present sea-floor features and those observed in the upper sedimentary infilling units. The sedimentary sequence thicknesses and the topography of the substratum as revealed by the 'Sédimanche 2' survey also indicate that sediment infilling of the Hurd Deep was controlled by Tertiary compressive structures. The Hurd Deep then appears as a long-lived and multi-phased feature. It is a prioity candidate area for a shallow borehole that would help to decipher the Upper Tertiary to Quaternary history of the western English Channel.

References

ANDREIEFF, P., BOUYSSE, P., HORN, R. & MONCIARDINI, C. 1972. Contribution à l'étude géologique des approches occidentales de la Manche. *Mémoires du Bureau de Recherches Géologiques et Minières*, **79**, 30–48.

AUFFRET, J. P. & LARSONNEUR, C. 1971. Paléovallées et bancs sableux entre l'estuaire de la Seine et le Nord Cotentin. *Bulletin de la Société Géologique de Normandie*, **64**, 21–34.

——, BIGNOT, G. & BLONDEAU, A. 1975. Géologie du bassin tertiaire de la Manche orientale au large du Pays de Caux. *Philosophical Transactions of the Royal Society of London*, **279**, 169–176.

AUGUSTIN, J. M. 1985. *Logiciel de Traitement des Images Acoustiques des Sonars Latéraux.* IFREMER, Internal Report, DIT/DI/DLS.

BERNÉ, S., AUFFRET, J. P. & WALKER, P. 1988. Internal structure of subtidal sand waves revealed by high-resolution seismic reflection. *Sedimentology*, **35**, 5–20.

BOILLOT, G. 1961. Sur une nouvelle fosse de la Manche Occidentale, la 'fosse de l'île Vierge'. *Comptes Rendus de l'Académie des Sciences. Paris*, **252**, 156–158.

——1963. Sur la fosse centrale de la Manche. *Comptes Rendus de l'Académie des Sciences. Paris*, **257**, 4199–4202.

——1964. *Géologie de la Manche Occidentale: Fonds Rocheux, Dépôts Quaternaires. Sédiments Actuels.* Annales de l'Institut Océanographique, **17**.

—— & HORN, R. 1966. Prospection sismique de la fosse d'Ouessant (Manche occidentale) par la méthode 'sparker'. *Comptes Rendus de l'Académie des Sciences, Paris*, **263**, 1677–1680.

BOIS, C., CAZES, M., GARIEL, O., LEFORT, J. P., LE GALL, B., PINET, B. & SIBUET, J. C. 1991*a*. Principaux apports scientifiques des campagnes SWAT et WAM à la géologie de la mer Celtique, de la Manche et de la marge atlantique. *Mémoires de la Société Géologique de France*, **n.s.**, 185–217.

——, GARIEL, O., MASCLE, A. & SCHROEDER, I. 1991*b*. Les bassins sédimentaires de mer Celtique et de Manche: apport des profils sismiques SWAT. *Mémoires de la Société Géologique de France*, 25–66.

BOURILLET, J. F., EDY, C., RAMBERT, F., SATRA, C. & LOUBRIEU, B. 1996. Swath mapping system processing: bathymetry and cartography. *Marine Geophysical Researches*, **18**, 487–506.

BOUYSSE, P., HORN, F., LEFORT, J. P. & LE LANN, F. 1975. Tectonique et structures post- paléozoiques en Manche Occidentale. *Philosophical Transactions of the Royal Society of London*, **279**, 41–54.

COOPER, M. A. & WILLIAMS, G. D. 1989. *Inversion Tectonics.* Geological Society of London, Special Publication, **44**.

CURRY, D. 1989. The rock floor of the English Channel and its significance for the interpretation of marine unconformities. *Proceedings of the Geologists' Association*, **100**, 339–352.

—— & SMITH, A. J. 1975. New discoveries concerning the geology in the central and eastern parts of the English Channel. *Philosophical Transactions of the Royal Society of London*, **279**, 155–167.

ADAMS, C. G., BOULTER, M. C., DILLEY, F. C., EARNES, F. E., FUNNELL, B. M. & WELLS, M. K. 1978. *A Correlation of Tertiary Rocks in the British Isles.* Geological Society of London, Special Report, **12**.

DANGEARD, L. 1929. Observations en géologie sousmarine et d'océanographie relatives à la Manche. *Annales de l'Institut Océanographique*, **6**, 1–295.

EVANS, C. D. R. 1990. *United Kingdom Offshore Regional Report: the Geology of the Western English Channel and its Western Approaches.* HMSO for the British Geological Survey, London.

GIBBARD, P. L. 1988. The history of the great northwest European rivers during the past three million years. *Philosophical Transactions of the Royal Society of London*, **318**, 559–602.

HAMILTON, D. 1979. The geology of the English Channel, South Celtic Sea and continental margin, South Western Approaches. *In*: BANNER, F. T.,

COLLINS, M. B. & MASSE, K. S. (eds) *The North-West European Shelf Seas: the Sea Bed and the Sea in Motion – I. Geology and Sedimentology*. Elsevier, Amsterdam, 61–87.

—— & SMITH, A. J. 1972. The origin and sedimentary history of the Hurd Deep, English Channel, with additional notes on other deeps in the western English Channel. *Mémoires du Bureau de Recherches Géologiques et Minières*, **79**, 59–78.

LAUTRIDOU, J. P. 1982. The Quaternary of Normandy. *Bulletin du Centre de Géomorphologie, Caen*, **26**.

LEFORT, J. P. & SEGOUFIN, J. 1978. Etude comparée des structures profondes et des anomalies magnitiques allongées reconnues en Baie d'Audierne: existence possible d'une suture cryptique au nord-ouest du Masif armoricain (France). *Tectonophysics*, **46**, 65–76.

LERICOLAIS, G., ALLENOU, J. P., BERNÉ, S. & MORVAN, P. 1990. A new system for acquisition and processing of very high-resolution seismic reflection data. *Geophysics*, **55**, 1036–1046.

——, AUFFRET, J. P. & BOURILLET, J. F. 1995. *A Periglacial Landform in the Central Channel: A Pingo Revealed by High Resolution Geophysical Data*. Publication ASF, **22**, 93.

PINGREE, R. D. & GRIFFITHS, D. K. 1979. Sand transport paths around the British Isles resulting from M2 and M4 tidal interactions. *Journal of the Marine Biology Association of the UK*, **59**, 497–513.

QUESNEY, A. 1983. *Manche Occidentale et Mer Celtique. Etude des Paléovallees. des Fosses et des Formations Superficielles*. PhD Thesis, Université de Caen.

REYNAUD, J. Y., TESSIER, B., PROUST, J. N., LERICOLAIS, G., MARSSET, T., BERNÉ, S. & CHAMLEY, H. 1995. Apport de la sismique très haute résolution à l'interprétation génétique d'un banc sableux de la mer Celtique. *Comptes Rendus de l'Académie des Sciences, Paris*, **320**, 125–132.

SMITH, A. J. 1985. A catastrophic origin for the paleovalley system of the eastern English Channel. *Marine Geology*, **64**, 65–75.

—— & CURRY, D. 1975. The structure and geological evolution of the English Channel. *Philosophical Transactions of the Royal Society of London*, **279**, 3–20.

TAYLOR, R. T., BOUYSSE, P., FLETCHER, B. N. & LEFORT, J. P. 1980. *Guernsey, Sheet 49N 04W. Scale 1 : 250 000*. Crown Edit.

VANDENBERGHE, J. & KASSE, C. 1989. Periglacial environments during the early Pleistocene in the Southern Netherlands and Northern Belgium. *Palaeogeography. Paleoclimatology. Palaeoecology*, **72**, 133–139.

WINGFIELD, R. T. R. 1990. The origin of major incisions within the Pleistocene deposits of the North Sea. *Marine Geology*, **91**, 31–52.

ZAGWIJN, W. H. 1974. Paleogeographic evolution of the Netherlands during the Quaternary. *Geologie en Mijnbouw*, **53**, 369–385.

ZIEGLER, P. A. 1987. Evolution of Western Approaches Trough. *Tectonophysics*, **137**, 141–146.

The São Francisco strandplain: a paradigm for wave-dominated deltas?

J. M. L. DOMINGUEZ

Curso de Pós-Graduação em Geologia – PPPG/IGEO/UFBA, Rua Caetano Moura 123, Federaão, 40210–340, Salvador, Bahia, Brazil

Abstract: The São Francisco 'delta' (total area 800 km^2) has been considered for more than two decades as a paradigm for wave-dominated deltas. Although it is frequently cited in the literature, there is a paucity of published data about this sedimentary feature. At the time the modern models of deltaic sedimentation were proposed, many of those deltas receiving modifiers such as 'wave-dominated' and 'tide-dominated' were very poorly understood. Circularity of reasoning did the rest. Inevitably phenomena described for well-studied areas such as the Mississippi River (i.e. dip-feeding and autocyclicity) were uncritically extrapolated to other areas and assumed to be valid for all other members of the so-called delta category.

There are three major aspects that make sedimentation in the São Francisco strandplain very different from the classical deltas.

(i) Sea-level history for the east coast of Brazil is characterized by a maximum around 5100 years ago when sea level reached 5 m above the present level. This sea-level history has exerted a major control on the evolution of the São Francisco Quaternary plain. During the 5.1 ka highstand the coastal plain was drowned and a barrier island/lagoonal system formed. When sea level dropped afterwards the shoreline prograded and Holocene beach ridges were deposited. As a result, sand body distribution is intimately controlled by sea-level history.

(ii) In wave-dominated environments, the fluvial effluent behaves effectively as a groyne, retaining in the updrift side sediments transported shorewise by the longshore currents. This phenomenon creates a clear asymmetry in facies distribution at the Quaternary plain. Whereas the updrift side is characterized by a continuous sand sheet, progradation on the downdrift side occurs by incorporation of sandy islands (reworked river-mouth bar sediments) that extend coastwise, protecting small lagoons rapidly filled by suspended sediments. This manner of sediment dispersal is also reflected along the entire shoreline and inner shelf in the vicinity of the river mouth. Therefore, the São Francisco 'delta' is not simply a dip-fed system. In the updrift side progradation of the shoreline is accomplished with sediment contribution from outside sources.

(iii) Autocyclic processes such as changes in the lower river course (the delta cycle concept) have not been observed for the São Francisco strandplain.

Since first used by Herodotus to describe the alluvial land between the two distributaries of the Nile River, the definition of the term 'delta' has been continuously modified to accommodate each new description of coastal accumulations associated with river mouths (see for example the historical review in LeBlanc 1975). This trend culminated with the classical work of Fisher (1969) and Fisher *et al.* (1969) according to whom the term 'delta' was simply defined as a river-fed depositional system that results in an irregular progradation of the shoreline. Wright (1978) defined 'delta' as a 'coastal accumulation of river-derived sediments adjacent to or in close proximity to the stream source, including the deposits secondarily moulded by various marine agents'.

Coleman and Wright published a series of papers in the early to mid-1970s (Coleman & Wright 1971, 1975; Wright & Coleman 1972, 1973) which in subsequent years have strongly influenced geologic thinking concerning deltaic sedimentation. These authors, elaborating on the earlier work of Fisher *et al.* (1969), have developed a well-known classification system (see also Galloway 1975), according to which the morphology and three-dimensional structure of deltaic accumulations are controlled, among other factors, by river discharge, tidal range and wave energy. Unfortunately, these classification schemes drew heavily on Holocene examples whose evolution by that time (and even now) were not well understood (e.g. São Francisco, Senegal, Nile rivers), particularly those receiving modifiers such as 'wave-dominated' and 'tide-dominated'.

Although during the 1980s and '90s several authors have questioned these classification schemes, either modifying them or adding new elements (Boyd *et al.* 1992; Walker 1992; Orton

From De Batist, M. & Jacobs, P. (eds), 1996, *Geology of Siliciclastic Shelf Seas*, Geological Society Special Publication No. 117, pp. 217–231.

& Reading 1993), very broad definitions of delta systems are still in use today. The inclusion, in one unique category, of coastal accumulation forms whose only common aspect is the geographical association with a river mouth usually leads to circularity of reasoning. Inevitably, phenomena described for well-studied areas such as the Mississippi River (e.g. dip-feeding, autocyclicity) are uncritically extrapolated and assumed to be valid for all other members of the 'delta' category. The São Francisco River 'delta' in particular, has been considered by several authors as the paradigm of the wave-dominated category. However, as discussed below, a significant volume of sediments for progradation of the shoreline was provided by the wave-generated longshore drift of sediments (strike-feeding). Also, autocyclic processes have not been observed for the São Francisco strandplain. Instead, the river has maintained its course during the last 5000 years, experiencing only limited lateral migration. For these reasons we will avoid using the term 'delta' to refer to the São Francisco Quaternary plain.

The purpose of this paper is to present the major features of the São Francisco strandplain. These features show that deltaic sedimentation in wave-dominated environments, if we take as a paradigm the so-called São Francisco 'delta', does not result exclusively from the interaction between a decelerating river effluent and wave reworking and lateral dispersion of river-borne sediments (Dominguez 1986, 1990; Barbosa & Dominguez 1994*a*). Other factors such as sea-level history and strike-feeding play important roles in the evolution of the plain.

Regional setting

The São Francisco River drainage basin

Geology. The São Francisco River is one of the lengthiest rivers (2700 km) draining the South American platform. The entire drainage basin, with a total area of 608 770 km^2, is contained within the São Francisco craton and its marginal foldbelts, an important geotectonic feature of the South American platform stabilized at the end of the Braziliano cycle (Upper Proterozoic) (Almeida 1977; Dominguez & Misi 1993) (Fig. 1). Lithologies in the drainage basin comprise essentially Precambrian sedimentary rocks, mostly limestones and quartzites with limited exposures of the crystalline basement.

Geomorphology. The portion of the drainage basin of the São Francisco River that is located within the São Francisco craton has a mean altitude of 400 m. After exiting the craton,

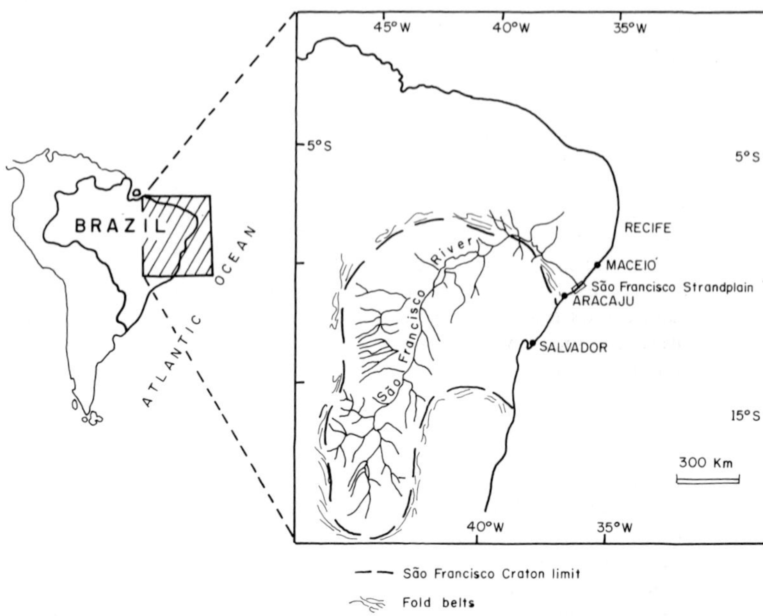

Fig. 1. Location map of the São Francisco strandplain. Also shown are the limits of the São Francisco craton and its marginal foldbelts and the São Francisco River drainage basin.

altitudes drop very rapidly, and the river incises the crystalline basement, forming a narrow, deep canyon, before reaching the coastal plain. During most of the Cenozoic, the São Francisco craton has experienced epeirogenic uplift which has resulted in the development of several denudation surfaces (Valadâo & Dominguez 1994).

Climate. A semi-arid climate, with six or more dry months during the year, characterizes most of the drainage basin. Rainfall varies from a maximum of 1250 mm/year at the river's headwaters and coastal zone to a minimum of 750–500 mm/year in the central portion of the drainage basin (Nimer 1989).

River discharge. Average discharge as measured 200 km from the river mouth is about 2980 m^3/s (minimum of 660 m^3/s and maximum of 15 000 m^3/s). Maximum discharges occur during the summer (January to April) whereas minimum discharges take place during the winter months (June to September).

Existing dams. Practically all the hydroelectric power of the São Francisco River has been tapped. A total of nine major dams have been built along the São Francisco River course, for the purpose of electricity generation and flood control.

The coastal zone in the vicinity of the São Francisco strandplain

Climate. The coastal zone in the vicinity of the São Francisco strandplain is characterized by a semi-humid climate with four consecutive dry months during the year. Away from the strandplain, the climate gets progressively more humid, with just one or two dry months during the year. Annual rainfall varies from 1750 to 1250 mm/year along the coastal zone (Nimer 1989). The rainy season occurs from March to August (autumn–winter). Dominant winds blow from E and SE with average speeds of 4 to 6 m/s.

Geology–geomorphology. The hinterland is reasonably flat, reaching altitudes of 100 m within 100 km of the shoreline. Late Pliocene alluvial fan deposits (the Barreiras Formation) fringe the entire coastal zone. A line of Pleistocene sea cliffs carved into the sediments of this formation marks the landward limit of the Quaternary plains. In some sectors of the coastal zone the Barreiras Formation reaches the shoreline, forming active sea cliffs. Other lithological types occurring in the coastal zone include sedimentary rocks of Cretaceous–Tertiary age of the Sergipe–Alagoas basin, whose origin is related to the break-up of Gondwana. Sediments in the Quaternary plain comprise beach-ridge terraces, lagoonal, freshwater marsh, mangrove swamp and aeolian deposits. Also present are coral–algal and sandstone reefs (Barbosa *et al.* 1986*a*; Bittencourt *et al.* 1982, 1983*a*).

Tides and wave regime. The coastal zone in the vicinity of the São Francisco strandplain is located in the lower mesotidal range (2.6 m spring tide range). No long-term wave measurements are available for this section of the coast. Data published in Hogben & Lumb (1967) show that waves approach predominantly from the east in this region, followed in importance by southeast and northeast waves. These waves have periods of 5 to 7 s and heights of 1.0 to 1.5 m.

Longshore transport. Although no long-term measurements of wave-generated longshore transport rates are available for this stretch of the coastal zone, the preferential direction of longshore drift can be estimated from the direction of propagation of wave trains relative to the initial shoreline orientation. These estimates can then be checked against geomorphic indicators (e.g. direction of migration of sand spits, general shoreline configuration etc.). The initial orientation of the shoreline is considered here to be the line of fossil sea cliffs carved into the Late Tertiary Barreiras Formation, sometime during the Pleistocene (Barbosa *et al.* 1986*a*; Bittencourt *et al.* 1983*a*). Assuming this initial shoreline and the dominance of eastern waves, the expected preferential longshore transport of sands is NE–SW. Longshore transport associated with SE waves is probably minor because these waves approach almost parallel to the shoreline. Geomorphic indicators along this stretch of the shoreline as observed on various published maps (Bittencourt *et al.* 1983*a*, *b*; Barbosa *et al.* 1986*a*, *b*) and aerial photographs, corroborate these conclusions; support is also provided by Viana (1972), who determined longshore transport rates of 658 000 m^3/year, directed from NE to SW, and 132 000 m^3/year, directed from SW to NE, for the shorezone in front of Aracaju city (Fig. 1).

The shoreface–inner shelf: bathymetry and sedimentation. Steep offshore profiles characterize the shoreface-inner shelf in front of the coastal zone in the vicinity of the São Francisco strandplain (Fig. 2). The width of the continental shelf varies from 20 to 40 km with gradients ranging from 1 : 700 to 1 : 100. The shelf break is located at depths varying from 30 to 50 m. Submarine canyons such as the São Francisco characterize the slope.

Different styles of sedimentation characterize the continental shelf northeastward (updrift) and southwestward (downdrift) of the São Francisco strandplain. Updrift of the São Francisco strandplain, except for a narrow band of siliciclastic sediments that occurs close to the shoreline, sedimentation is essentially carbonatic (Barbosa *et al.* 1986*a*). The major sediment constituents are fragments of calcareous algae (*Halimeda* and *Coralline* algae). Additionally the coastal zone is dotted by numerous coral–algal reefs.

Downdrift of the São Francisco strandplain, carbonates are restricted to the outer shelf. Siliciclastic sands characterize sedimentation in the inner shelf (Bittencourt *et al.* 1983*a*). Muds are presently accumulating in front of the São Francisco strandplain and around the heads of submarine canyons (Tintelnot *et al.* 1994).

Regional sea-level history. Two important transgressive episodes affected the east-northeastern coast of Brazil during the Quaternary, including the coastal zone in the vicinity of the São Francisco strandplain. The older one, named the Penultimate Transgression, reached a maximum at 120 ka when sea level was positioned 8 ± 2 m above present sea level (Martin *et al.* 1982). The younger episode, named the Last Transgression, reached a maximum at 5.1 ka when sea level rose 5 m above present sea level (Martin *et al.* 1980*a*, *b*; Suguio *et al.* 1980).

Relative sea-level curves for the last 7000 years have been constructed for several sections of the Brazilian coast (Martin *et al.* 1979, 1980*a*, *b*; Suguio *et al.* 1980; Dominguez *et al.* 1990) (Fig. 3). Radiocarbon dating of samples of vermetid encrustations, calcareous algae, and corals collected from above the modern life zone of these organisms were used to reconstruct ancient positions of sea level. Radiocarbon dating of shells and wood collected in beach and lagoonal deposits were also used. In this case, sedimentological studies were necessary to determine the approximate position of mean sea

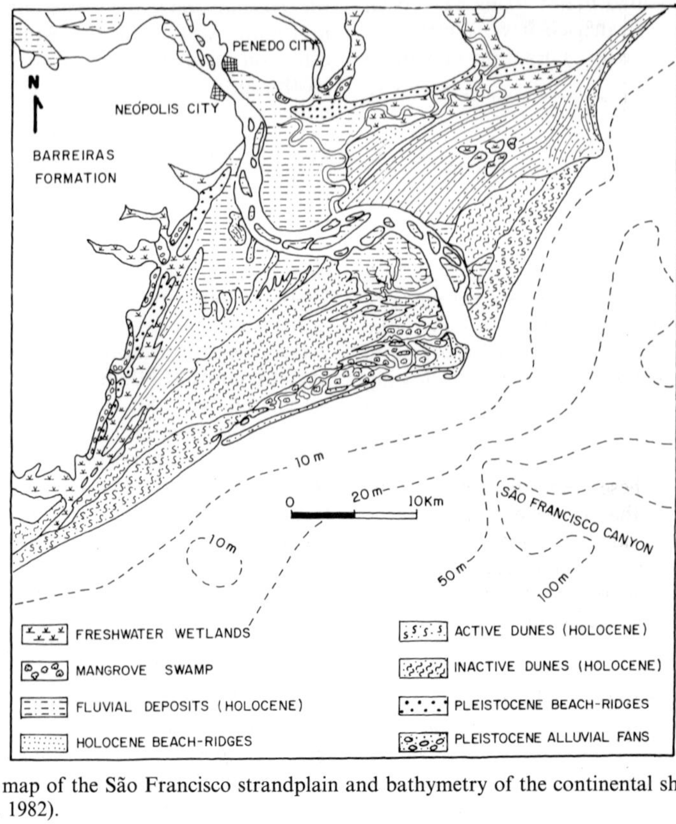

Fig. 2. Geologic map of the São Francisco strandplain and bathymetry of the continental shelf (modified from Bittencourt *et al.* 1982).

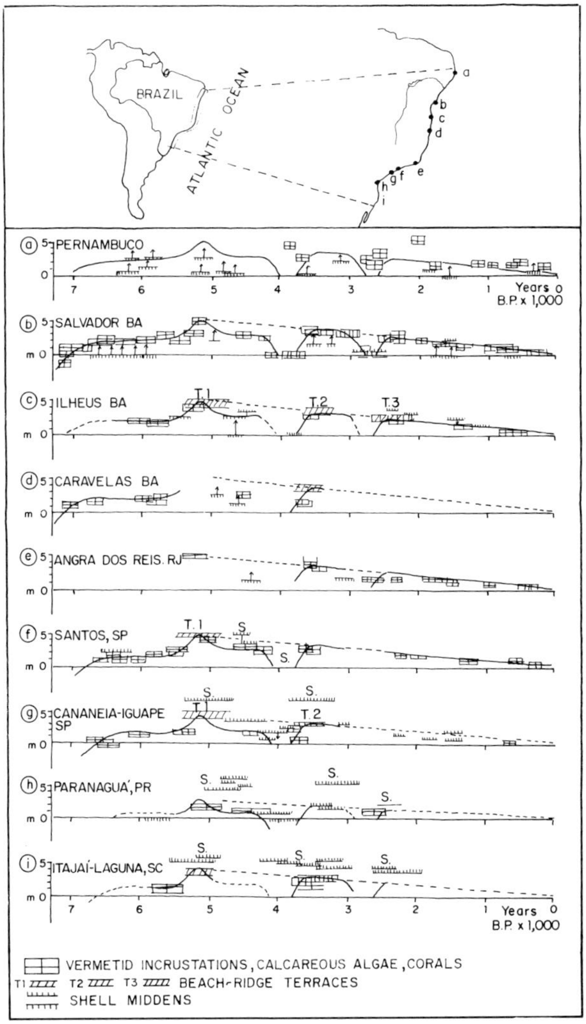

Fig. 3. Sea-level curves constructed for several sectors of the east coast of Brazil (modified from Suguio *et al.* 1985; Dominguez *et al.* 1990). Types of data are: sea-level reconstructions using indicators such as vermetid encrustations, calcareous algae and corals; beach-ridge terraces; and shell middens.

level at the time these deposits were formed. These curves show that sea-level lowering from the 5.1 ka highstand was not regular but interrupted by high-frequency relative sea-level oscillations as small as 2–3 m acting on time scales of no more than 200–300 years (Fig. 3).

The São Francisco strandplain Quaternary deposits

Quaternary sedimentary deposits on the São Francisco strandplain comprise beach-ridge terraces and lagoonal, freshwater marsh, mangrove swamp, alluvial fan, fluvial and aeolian deposits (Fig. 2). The strandplain has a total area of 800 km^2.

Pleistocene alluvial fans

These deposits comprise moderately sorted massive white-leached medium to coarse sands resting at the foot of fossil sea cliffs carved into the Barreiras Formation. The upper surface dips noticeably seaward. Its interpretation as alluvial fans is based on physical correlation with similar deposits occurring south of the São Francisco strandplain and which has been unquestionably interpreted as such (Vilas Boas *et al.* 1979).

Pleistocene beach-ridge terraces

The Pleistocene beach-ridge terraces occur in the internal portion of the strandplain; they form terraces 8 to 10 m high, almost continuously bordering the Pleistocene alluvial fan deposits and the line of fossil sea cliffs of the Barreiras Formation. These terraces are made up of massive white-leached medium to coarse sand with local intercalations of pebble beds. Aerial photographs show remnants of beach ridges on the surface. No radiometric ages are available for these sediments. We attribute an age of 120 ka to them based on physical correlation with similar terraces occurring along the east coast of Brazil, which have been dated using the Io/U method (Martin *et al.* 1982).

Lagoonal deposits

The lagoonal deposits are essentially restricted to the low-lying area separating the Pleistocene from the Holocene beach-ridge terraces (described below). These sediments comprise dark grey muds rich in wood fragments and other detrital organic matter. Lagoonal sediments outcrop only along river banks and are presently capped by light grey muds deposited in freshwater marshes that flourish in this area today. A radiocarbon date of a piece of wood collected in these lagoonal sediments provided an age of 5730 ± 200 years BP.

Holocene beach-ridge terraces

The Holocene beach-ridge terraces have altitudes varying from 4 to 0.5 m with heights decreasing towards the present-day shoreline. The surfaces of these terraces are covered with beach ridges readily visible in the field. Outcrops of these terraces along the São Francisco River banks show that the dominant sedimentary structure is a beach-face stratification dipping gently towards the sea. No radiocarbon dates for these terraces are available. A Holocene age is inferred by the following:

(i) the fact that lagoonal sediments infilling the low-lying area separating these terraces from the Pleistocene ones have an age older than 5.1 ka, as mentioned before, and
(ii) physical correlation with other terraces, showing the same characteristics, that occur along the coastal zone and that have been dated using the radiocarbon method (Bittencourt *et al.* 1983*a*, *b*; Barbosa *et al.* 1986*a*, *b*).

Dune deposits

Bordering the present shoreline there are two dune generations: an inactive dune field of Holocene age located landward of and being transgressed by active dunes.

Integration of detailed mapping with overflights, field work, and sediment sampling along the shoreline and at the dune field has shown that dune development and dune morphologies are intimately controlled by available grain size (Barbosa & Dominguez 1992, 1994*b*).

Dune migration takes place exclusively during the dry season (October to March). During the rainy season the interdune areas are flooded. The surface of the active dunes is also wet, inhibiting sand movement. In the flooded interdune areas, wind-generated wavelets give origin to miniature beaches in the stoss side of the dunes. When dune migration resumes in the next dry season, these miniature beaches are left behind as arcuate ridges marking the former position of the dune. Systematic measurement of distances between two adjacent ridges and comparison with aerial photographs taken in different periods, show that the average rate of dune migration for a 2 m high dune is about 18–20 m/year (Barbosa & Dominguez 1994*b*). Therefore it takes about 100–200 years for an isolated dune to migrate from the place where it initially forms, near the shoreline, to the furthest inland position. These measurements suggest that dune formation is a relatively new phenomenon at the study area, suggesting an increase in aridity. Other phenomena in the Quaternary plain suggesting an increase in aridity during the Holocene include the following (Fig. 2):

(i) early sedimentation in the coastal plain during the Holocene was characterized by the accumulation of beach ridges without dune development; and
(ii) some time after 5.1 ka, dune development on both sides of the river mouth was characterized by parabolic dunes, which normally form in the presence of vegetation, thus suggesting a wetter climate. These parabolic dunes are presently inactive and stabilized by vegetation.

Fluvial deposits

The São Francisco River has its main channel choked with numerous islands, which gives it the appearance of a braided river (Fig. 2). Well-defined natural levees are not observed along the margins of the river. The fluvial deposits observed along the river banks, comprising interbedded muds and fine sands, result from lateral migration of the river channel. Distribution of fluvial sediments indicates that during the last 5000 years the river has maintained its present course, with limited lateral migration (Fig. 2).

Freshwater and mangrove swamps

Freshwater marshes and swamps occupy low-lying areas of the strandplain, whereas mangrove swamps are restricted to the downdrift side of the São Francisco River mouth. These deposits are made up essentially of light to dark grey organic-rich muds.

Quaternary coastal evolution

The integration of the regional sea-level history and the previous knowledge of coastal evolution for eastern Brazil (Dominguez *et al.* 1987, 1992; Dominguez & Wanless 1991), with mapping and the few radiocarbon dates of the sedimentary deposits occurring in the São Francisco strandplain, has allowed the establishment of three major evolutionary stages for the São Francisco strandplain (Fig. 4).

Stage I – Pleistocene beach-ridge plain

This stage corresponds to the regressive event which followed the maximum of the Penultimate Transgression (120 ka). During this period, a drop in sea level favoured progradation of the coastline. Remnants of these strandplains are preserved today as the Pleistocene beach-ridge terraces (Fig. 4A).

Stage II – Maximum of the last transgression

The rise in sea level that followed the last glacial period partly eroded and drowned the Pleistocene plains. By the time the maximum of the Last Transgression was reached (5.1 ka) a barrier island/lagoonal system existed at the São Francisco strandplain (Fig. 4B).

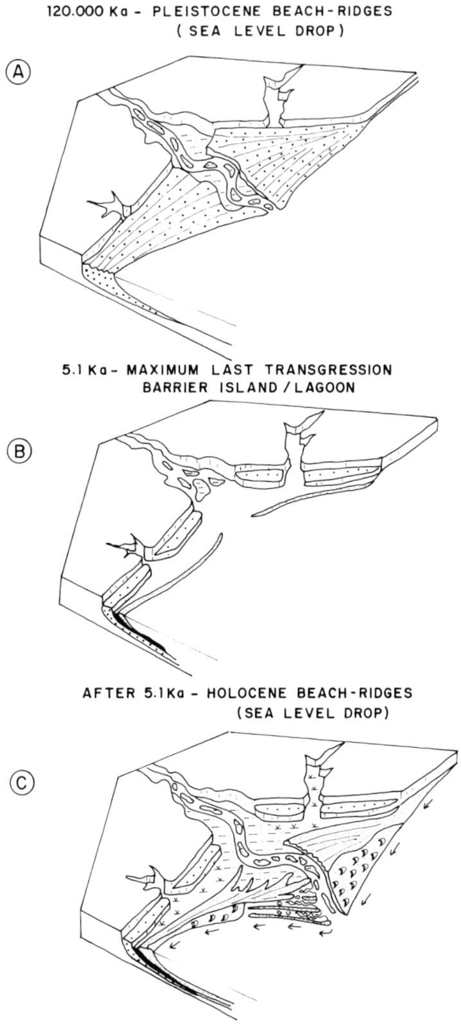

Fig. 4. Evolutionary model for the São Francisco strandplain (see text for details).

Stage III – Holocene beach-ridge plain

The drop in sea level that followed the maximum of the Last Transgression favoured the seaward progradation of the coastline which resulted in the deposition of the Holocene beach-ridge terraces (Fig. 4C). At a more recent time, possibly as a result of an increase in aridity, dune development took place at the external fringe of the strandplain.

River mouth dynamics

Figure 5C shows a clear asymmetry in facies distribution between updrift and downdrift sides

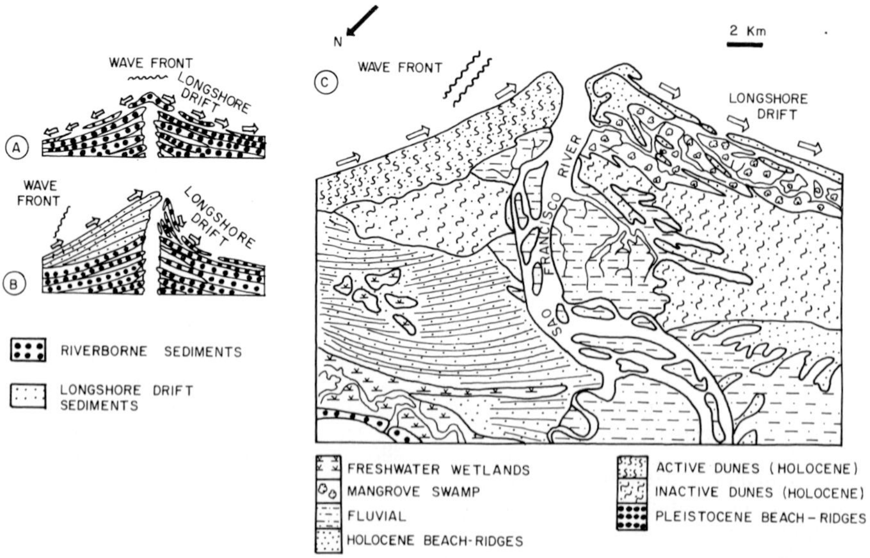

Fig. 5. Groyne effect associated with river mouths in wave-dominated environments. (**A**) When the wave fronts approach parallel to the shoreline, the longshore drift will redistribute fluvial sediments to both sides of the river mouth. (**B**) When the wave fronts approach the shoreline forming an angle, the fluvial discharge will act as a groyne retaining updrift of the river mouth those sediments transported by the wave-induced longshore drift. River-borne sediments will accumulate downdrift of the river mouth. (**C**) A geologic map of part of the São Francisco strandplain illustrating the asymmetry in facies distribution between updrift and downdrift sides of the São Francisco river mouth. This asymmetry results from the groyne effect (modified from Dominguez 1990). See text for details.

of the São Francisco River mouth. The updrift side is characterized by active migrating dunes, whereas the downdrift side is characterized by mangrove swamps.

When the wave-front approach is parallel to the shoreline, in the presence of a point source of sediment such as a river mouth, the longshore drift produced by the waves will redistribute sediments to both sides of the river mouth (Fig. 5A). A theoretical cuspate wave-dominated delta will thus form. This is probably an extremely rare case in the geologic record. More commonly the wave front will form an angle with the shoreline (Fig. 5B). In this situation the offshore build-up of the fluvial effluent will act as a groyne, trapping sediment transported by the wave-generated longshore drift in the updrift side of the river mouth (the 'Groyne effect', Dominguez *et al.* 1983; Dominguez 1990). The downdrift side will continue to be nourished by river-borne sediments. Zenkovitch (1967) reported this phenomenon for river mouths along the Caucasian Black Sea coast. Komar (1973), using computer simulation models, has also concluded that rivers act as a barrier to the longshore drift of sands, effectively behaving as a jetty or groyne. Because of the groyne effect, different sediment grain sizes and modes of progradation for the updrift and downdrift sides of the river mouth will result (Fig. 5C). On the updrift side, nourished by the wave-generated longshore drift, sediments are continuously incorporated into the beach prism causing the shoreline to advance seaward. The result is a continuous sheet of beach sands capped by beach-ridge and/or dune deposits. On the downdrift side, progradation takes place through the reworking of river-mouth bars into sandy islands or, alternatively, by the coastwise progradation of sand spits rooted in the downdrift side of the river mouth. These sandy islands and spits protect small lagoons, filled with fine-grained sediments. The final result is then a succession of linear sand bodies separated from each other by a low-lying wetland area. This difference in modes of progradation between the updrift and downdrift sides of the river mouth has been observed for the major rivers emptying onto the east-northeastern coast of Brazil (Dominguez *et al.* 1983; Dominguez 1990). Also, because of these different modes of progradation, mean grain size for beach sediments will differ for updrift and downdrift sides. For the São Francisco plain, mean grain size on the updrift side (longshore-drift fed) averages 2.90 phi whereas in the downdrift side (river-fed) sands are usually coarser, with mean grain size averaging 2.27 phi (Barbosa & Dominguez 1994*b*).

Radiocarbon dates of beach ridges, and other geomorphic features at river mouths located along the east-northeastern coast of Brazil, indicate that progradation rates in the updrift side (longshore-drift fed) were observed to be equal to or even greater than those on the downdrift side (river-fed) (Dominguez *et al.* 1983; Dominguez 1990). Thus, the volume of sediment provided by the longshore drift (strike-feeding), may be equivalent to or even greater than the volume provided by the river itself (dip-feeding) in promoting progradation of the shoreline.

Figure 6 shows the bathymetry of the river-mouth bar and immediate river channel. The river mouth bar has a clear asymmetrical shape in plan view with a well-marked subaqueous levee in its updrift side. This asymmetry is also a consequence of the groyne effect described above, according to which the wave-generated longshore drift impedes the fluvial effluent to expand updrift.

According to Wright (1977), wave refraction by the fluvial effluent, in wave-dominated environments, causes local concentration of power around the effluent. The waves breaking promote abrupt mixing and deceleration of the effluent and the wave-induced set-up inhibits outflow. This process results in rapid seaward deceleration of the effluent flow, as well as subdued effects of buoyancy due to destruction of stratification. Wave reworking causes the shoreward return of sediment over subaqueous levees as swash bars. Swift *et al.* (1991) point out that the seaward face of the river-mouth bar can be viewed as a shoreface wrapped around the river mouth. Shoreface processes dominate on the flanks of the river-mouth bar, whereas channel processes dominate along the axis. Therefore, lithofacies successions on the mouth-bar flanks will resemble those of prograding shorefaces formed by storm wave and current processes. In such successions, a laminated shale lithofacies is overlain by an interstratified sand and shale lithofacies, which in turn is overlain by an amalgamated sandstone lithofacies (Swift *et al.* 1991).

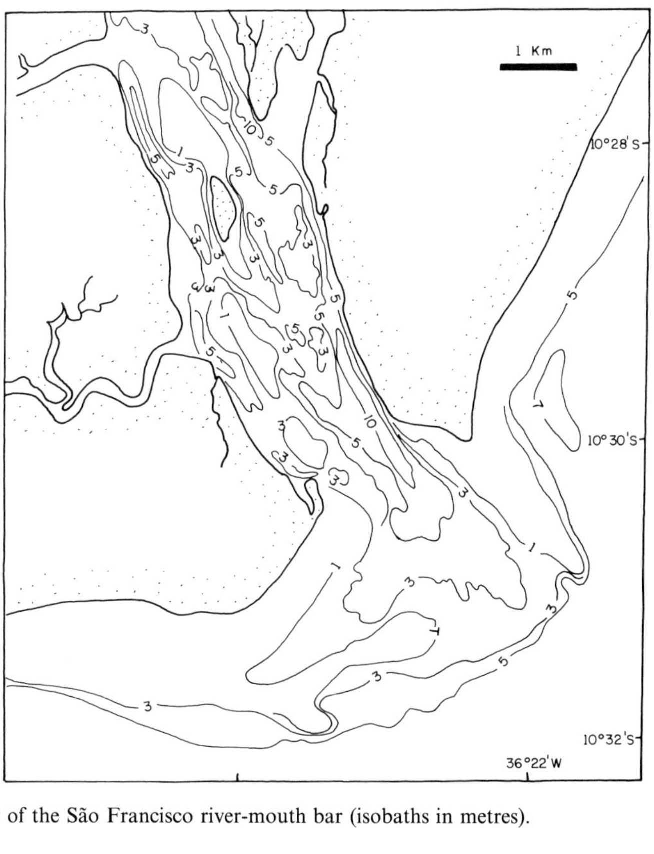

Fig. 6. Bathymetry of the São Francisco river-mouth bar (isobaths in metres).

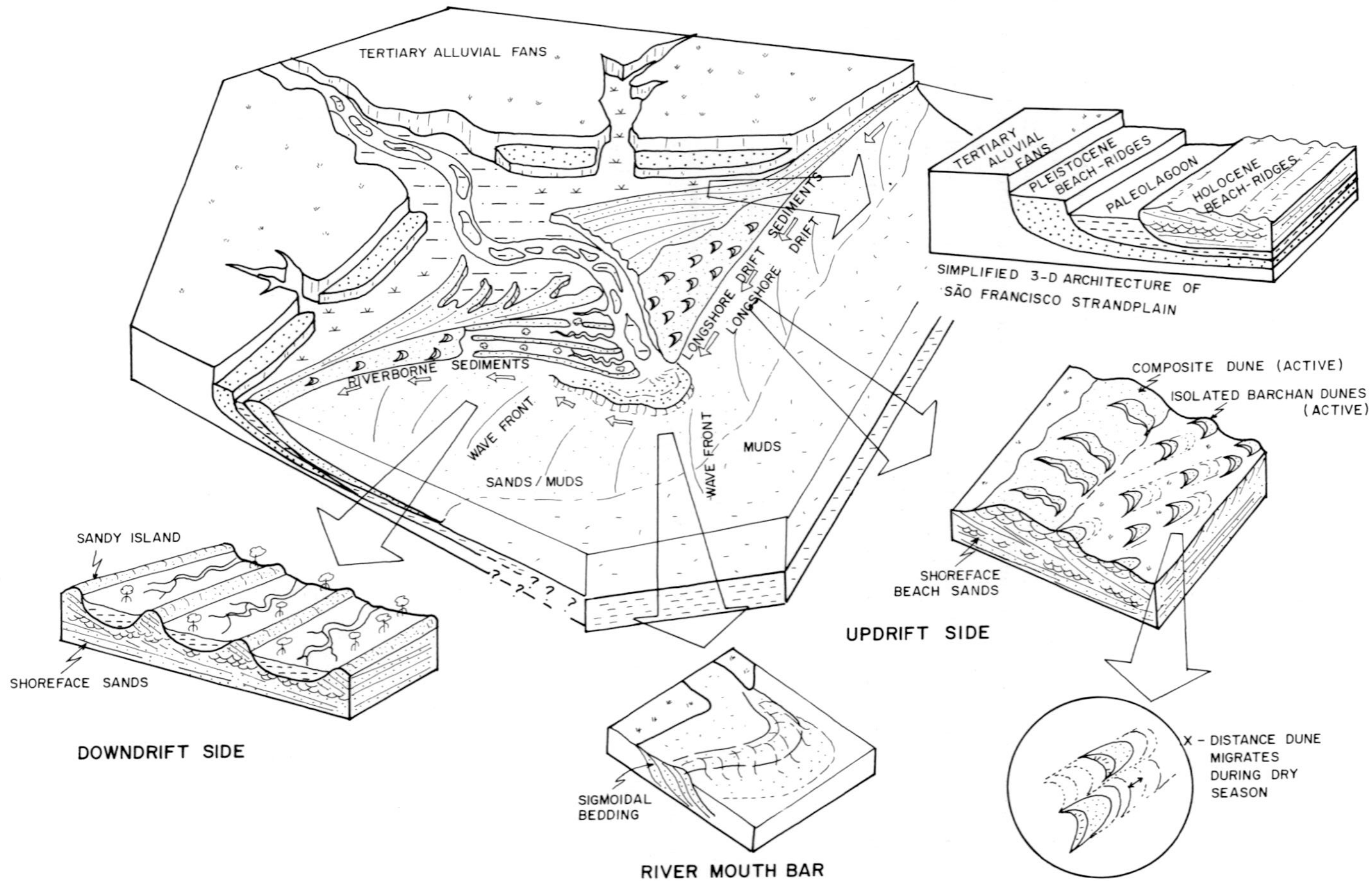

Fig. 7. Conceptual model for the São Francisco strandplain. Please note that in order to improve readability, the inactive dune fields are not represented.

Discussion

Figure 7 presents a simplified conceptual model summarizing the major features of the São Francisco strandplain. Except for the river-mouth bar, which is probably the only feature where the river leaves its imprint, the rest of the plain is essentially similar to any prograding wave-dominated shoreface sequence. At this point, trying to apply the existing delta definitions is a very tricky business. Attempts to do so usually result in circular reasoning. Bhattacharya & Walker (1992) commenting on wave-dominated deltas call attention to the fact that 'in the geologic record, one vertical facies succession of this type indicates only a prograding wave-storm-dominated shoreface. Good three dimensional control is necessary before such a shoreface can be positively associated with a delta'. Thus, the so-called wave-dominated deltas have no distinctive features – except for very localized fluvial sediments and the river-mouth bar complex – that help in distinguishing them from normal wave-dominated shoreface sequences. Concerning this aspect it is interesting to note that the Nayarit plain on the Pacific coast of Mexico (Curray *et al.* 1969) has never been considered as the wave-dominated delta of the Rio Grande de Santiago. This plain has the same characteristics of the São Francisco River plain. Instead, it is considered as a classical example of a strandplain, although every author agrees that sand for progradation is supplied by the Rio Grande de Santiago and a number of smaller rivers.

The results presented here also show the importance of strike-feeding in wave-dominated settings. The fluvial effluent effectively acts as a groyne retaining sediment transported by the wave-generated longshore drift in the updrift side of the river mouth. The downdrift side will continue to be nourished by river-borne sediments.

A strong longshore drift may also force an intermittent downdrift migration of the river mouth. Although this phenomenon is not observed at the São Francisco strandplain, it is

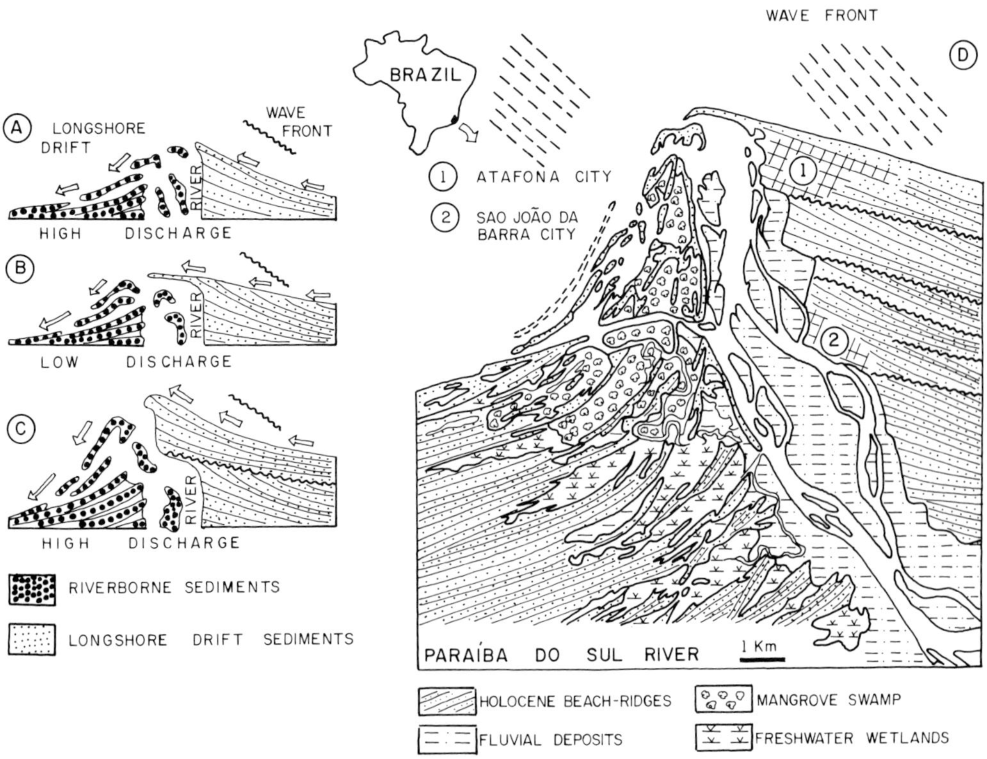

Fig. 8. Scheme showing mechanism of intermittent downdrift migration of a river mouth (**A**, **B** and **C**). This migration results from the interaction between the longshore drift and fluctuations in river discharge. (**D**) Geologic map of the Paraíba do Sul river strandplain. The map clearly shows the erosive truncations of beach-ridge sets updrift of the river mouth (see text for details) (after Dominguez 1990).

described here because it characterizes other river mouths along the wave-dominated east-northeastern coast of Brazil. Figure 8 illustrates the mechanism involved in this migration. During periods of high fluvial discharge the river acts effectively as a groyne stopping the longshore drift of sediments (Fig. 8A). The shoreline in the updrift side of the river mouth then progrades. During periods of low fluvial discharge the groyne effect of the river mouth is less efficient. The shoreline in the updrift side will then experience erosional retreat and reorient itself. This reorientation causes a truncation of the beach-ridge alignment (Fig. 8B). The sediments eroded are moulded into a spit that partly obstructs the river mouth forcing it to a new position slightly downdrift. In the next period of high discharge, the river will again effectively behave as a groyne, stopping the longshore drift of sediments (Fig. 8C). Progradation of the shoreline will resume on the updrift side of the downdrifted migrated river mouth. As a result of repetition of this process the updrift side of the river mouth will be characterized by an echelon pattern of beach-ridge sets separated by erosional truncations. This pattern is beautifully illustrated by the Paraíba do Sul river mouth (Fig. 8D).

An extreme case of the mechanisms described above is exemplified by the Senegal River (Fig. 9). In this region, a strong unidirectional wave-generated longshore drift forces the river mouth to continuously migrate in the downdrift direction. River-borne sediments in a situation like this will be deposited far downdrift of the river mouth. Note that the sand barrier and beach deposits that are used to classify this area as a wave-dominated delta, do not have a genetic relationship with the river itself. Instead, the sediments which make them up were provided by the longshore drift of sands eroded from unconsolidated pre-Holocene aeolian deposits located updrift of the river plain (Michel 1968). This situation is extremely common in the wave-dominated east-northeastern coast of Brazil where rivers with smaller discharges exhibit behaviour similar to the Senegal river.

The 'delta cycle' concept was first introduced by Scruton (1960) and later expanded by Coleman & Gagliano (1964). According to this concept, deltas rarely build indefinitely in one direction. Continued progradation of a delta at approximately sea level results in a progressive decrease in the overall gradient of its distributary system. The river then shifts to a shorter, steeper-gradient route to the sea. This

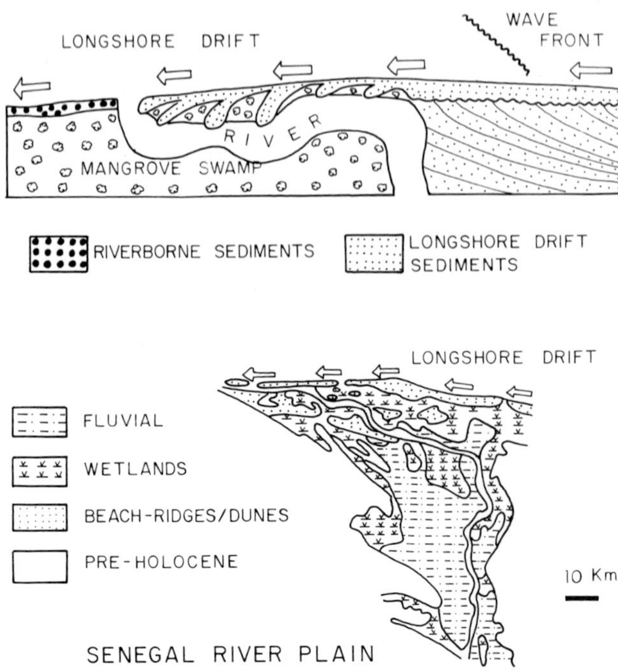

Fig. 9. Continuous downdrift migration of a river mouth as a result of a strong longshore drift as exemplified by the Senegal River (geologic map modified from Michel 1968).

shift in the point source of sediment supply is responsible for the abandonment of an active delta and the initiation of a second cycle related to the new point source. The abandoned delta, deprived of nourishment, undergoes coastal retreat and inundation due to compaction/subsidence coupled with marine wave and current action. The mechanism described above is known as autocyclic because it is self-regulating and self-perpetuating and excludes the necessity of external forcing (e.g. pulses of basinal subsidence, sea-level changes, changes in sediment supply etc.) to explain cyclicity in deltaic systems. This model has been frequently invoked to interpret cyclicity in coastal systems associated with a river mouth. Autocyclic processes have not been observed along the east-northeastern coast of Brazil. Documented cases of change in the lower river course during the Holocene have been demonstrated to be not a result of autocyclic processes but rather a response to external forcing (i.e. sea-level changes) (Dominguez 1987; Dominguez *et al.* 1987). For the São Francisco strandplain, as mentioned before, no changes in its lower river course have occurred during the last 5000 years.

Finally, another important aspect to be taken into consideration is the fundamental role played by the Late Quaternary sea-level changes in the evolution of the Brazilian Quaternary strandplains such as the São Francisco. This role has been ignored or not properly incorporated in the formulation of deltaic sedimentation models, probably because of the general belief, which persisted until very recently, that sea level has been continuously rising or stable everywhere during the last 5000 years. It is virtually impossible to understand the evolution of those areas without a previous knowledge of the sea-level history.

Although the São Francisco continues to be widely used in the literature as a paradigm for wave-dominated deltas, the information presented here indicates that additional discussion or even elaboration of a new major synthesis for sedimentation in wave-dominated settings is now needed, in order to reconcile existing conflicts of interpretation, some of which have been presented here.

The author thanks Drs Marc De Batist and Patric Jacobs for the invitation to participate in this symposium volume. The comments presented by E. Kosters greatly enhanced the paper. Financial support for this research was provided by CNPq (Conselho Nacional de Desenvolvimento Científico e Tecnológico).

References

ALMEIDA, F. F. M. 1977. O Cráton do São Francisco. *Revista Brasileira de Geociências*, **7** 349–364.

BARBOSA, L. M. & DOMINGUEZ, J. M. L. 1992. Dunas costeiras associadas à desembocadura do Rio São Francisco: morfologia e implicações paleoclimáticas. *Proceedings of the 37th Congres Brasilian Geology*. Sociedade Brasileira Geologia, São Paulo, Brazil, **1**, 101–102.

—— & 1994*a*. The São Francisco delta: a paradigm for wave-dominated deltas? *14th International Sedimentological Congres*, Recife, Brazil, D10–D11.

—— & 1994*b*. Climate changes and grain size control on coastal dune development at the São Francisco Quaternary plain – Northeastern Brazil. *14th International Sedimentological Congres*, Recife, Brazil, D12–D13.

——, BITTENCOURT, A. C. S. P., DOMINGUEZ, J. M. L. & MARTIN, L. 1986*a*. *Mapa Geológico do Quaternário Costeiro do Estado de Alagoas* (included in the Mapa Geológico do Estado de Alagoas, scale 1:250.000). Recife, DGM/DPM, **2**, 73–76.

——, ——, —— & —— 1986*b*. The Quaternary deposits of the State of Alagoas: influence of the relative sea level changes. *In*: RABASSA, J. O. (ed.) *Quaternary of South America and Antarctica Peninsula*. Balkema, Rotterdam, 269–290.

BHATTACHARYA, J. P. & WALKER, R. G. 1992. Deltas. *In*: WALKER, R. G. & JAMES, N. P. (eds) *Facies Models – Response to Sea Level Change*. Geological Association of Canada. 157–178.

BITTENCOURT, A. C. S. P., MARTIN, L. & DOMINGUEZ, J. M. L. 1983*a*. Mapa Geológico do Quaternário Costeiro do Estado de Sergipe. *In*: BRUNI, M. A. L. & SILVA, H. P. (eds) *Mapa Geológico do Estado de Sergipe, 1:250.000*. Min. das Minas e Energia/DNPM, Governo do Estado de Sergipe, Brazil.

——, —— & —— 1983*b*. Evolução paleogeográfica quaternária da costa do Estado de Sergipe e costa sul do Estado de Alagoas. *Revista Brasileira de Geociências*, **13**, 93–97.

——, —— —— & FERREIRA, Y. DE A. 1982. Dados preliminares sobre a evolução paleogeográfica do delta do Rio São Francisco – SE/AL, durante o Quaternário: influência das variações do nível do mar. *Proceedings of the 4th Symposium Quaternário no Brasil*. ABEQUA, Rio de Janeiro, 49–68.

BOYD, R., DALRYMPLE, R. & ZAITLIN, B. A. 1992. Classification of clastic coastal depositional environments. *Sedimentary Geology*, **80**, 139–150.

COLEMAN, J.M. & GAGLIANO, S.M. 1964. Cyclic sedimentation in the Mississipi river deltaic plain. *Gulf Coast Association of Geological Societies, Transactions*, **14**, 67–80.

—— & WRIGHT, L. D. 1971. *Analysis of Major River Systems and their Deltas: Procedures and Rationale, with Two Examples*. Louisiana State University, Coastal Studies Institute Technical Report, **95**.

—— & —— 1975. Moder river deltas: variability of processes and sand bodies. *In*: BROUSSARD, M. L. (ed.) *Deltas, Models for Exploration*. Houston Geological Society, Texas. 99–149.

CURRAY, J. R., EMMEL, F. J. & CRAMPTON, P. J. S. 1969. Holocene history of a strandplain, lagoonal coast, Nayarit, Mexico. *In*: AYALA-CASTANARES, A. & PHLEGER, F. B. (eds) *Coastal Lagoons, a Symposium*. Universidad Nacional Autónoma de Mexico. 63–100.

DOMINGUEZ, J. M. L. 1986. Reevaluation of deltaic sedimentation models in wavedominated settings, evidences from eastern Brazil. *Geological Society of America 99th Annual Meeting and Exposition*, San Antonio, Texas, 586–587.

——1987. *Quaternary Sea Level Changes and the Depositional Architecture of Beach-ridge Strandplains along the Coast of Brazil*. PhD Dissertation, University of Miami, Florida.

——1990. Deltas dominados por ondas: críticas ás idéias atuais com referência particular ao modelo de Coleman & Wright. *Revista Brasileira de Geociências*, **20**, 352–361.

—— & MISI, A. (eds) 1993. *O Cráton do São Francisco*. Sociedade Brasileira de Geologia, Bahia-Sergipe Section.

—— & WANLESS, H. R. 1991. Facies architecture of a falling sea level strandplain, Doce river coast, Brazil. *In*: SWIFT, D. J. P., OERTEL, G. F., TILLMAN, R. W. & THORNE, J. A. (eds) *Shelf Sand and Sandstone Bodies – Geometry, Facies and Sequence Stratigraphy*. International Association of Sedimentologists, Special Publication, **14**, 259–289.

——, BITTENCOURT, A. C. S. P., LEÃO, Z. M. A. N. & AZEVEDO, A. E. G. 1990. Geologia do Quaternário Costeiro do Estado de Pernambuco. *Revista Brasileira de Geociências*, **20**, 208–215.

——, —— & MARTIN, L. 1983. O papel da deriva litorânea de sedimentos arenosos na construção das planícies costeiras associadas ás desembocaduras dos rios São Francisco (SE/AL), Jequitinhonha (BA), Doce (ES) e Paraíba do Sul (RJ). *Revista Brasileira de Geociências*, **13**, 98–105.

——, —— & —— 1992. Controls on Quaternary coastal evolution of the east-northeastern coast of Brazil: roles of sea level history, trade winds and climate. *Sedimentary Geology*, **80**, 213–232.

——, MARTIN, L. & BITTENCOURT, A. C. S. P. 1987. Sealevel history and Quaternary evolution of rivermouthassociated beachridge plains along the eastern/southeastern Brazilian coast: a summary. *In*: NUMMEDAL, D., PILKEY, O. H. & HOWARD, J. D. (eds) *SeaLevel Fluctuation and Coastal Evolution*. Society of Economic Paleontologists and Mineralogists, Special Publication, **41**, 115–127.

FISHER, W. L. 1969. Facies characterization of Gulf coast basin delta systems, with some Holocene analogues. *Gulf Coast Association of Geological Societies, Transactions*, **19**, 239–261.

——, BROWN, L. F. JR., SCOTT, A. J. & MCGOWEN, J. H. 1969. *Delta Systems in the Exploration of Oil and Gas*. Bureau of Economic Geology, University of Texas.

GALLOWAY, W. E. 1975. Process framework for describing the morphologic and stratigraphic evolution of deltaic depositional systems. *In*: BROUSSARD, M. L. (ed.) *Deltas, Models for Exploration*. Houston Geological Society, Texas, 87–98.

HOGBEN, N. & LUMB, F. E. 1967. *Ocean Wave Statistics*. National Physical Laboratory, Ministry of Technology, London.

KOMAR, P. D. 1973. Computer models of delta growth due to sediment input from the rivers and longshore transport. *Geological Society of America Bulletin*, **84**, 2217–2226.

LEBLANC, R. J. 1975. Significant studies of modern and ancient deltaic sedimentation. *In*: BROUSSARD, M. L. (ed.) *Deltas, Models for Exploration*. Houston Geological Society, Texas, 13–85.

MARTIN, L., BITTENCOURT, A. C. S. P. & VILAS BOAS, G. S. 1982. Primeira ocorrência de corais pleistocônicos na costa brasileira; datação do máximo da Penúltima Transgressão. *Revista Ciências da Terra*, **1**, 16–17.

——, ——, —— & FLEXOR, J. M. 1980*b*. *Mapa Geológico do Quaternário Costeiro do Estado da Bahia. Scale 1:250,000*. Text. Coordenação da Produção Mineral, Secretaria das Minas e Energia do Estado da Bahia, Brazil.

——, FLEXOR, J-M., VILAS BOAS, G. S., BITTENCOURT, A. C. S. P. & GUIMARÃES, M. M. M. 1979. Courbe de variations du niveau relatif de la mer au cours des 7000 dernières années sur un secteur homogène du littoral brésilien (nord de Salvador, Bahia). *In*: SUGUIO, K., FAIRCHILD, T. R., MARTIN, L. & FLEXOR, J-M. (eds) *Proceedings of the International Symposium on Coastal Evolution in the Quaternary*, São Paulo, Brazil, 264–274.

——, SUGUIO, K. , FLEXOR, J. M., BITTENCOURT, A. C. S. P. & VILAS BOAS, G. S. 1980*a*. Le Quaternaire marin brésilien (littoral pauliste, sudfluminense and bahianais). *Cahiers Office de la Recherche Scientifique et Technique d Outre-Mer, série Géologie*, **10**, 95–124.

MICHEL, P. 1968. Genèse et évolution de la vallée du Senegal, de Bakel a l'embouchure (Afrique Occidentale). *Zeitschrift Für Geomorphologie*, **Band 12**, 318–349.

NIMER, E. 1989. *Climatologia do Brasil*. Instituto Brasileiro de Geografia e Estatística, Rio de Janeiro.

ORTON, G. J. & READING, H. G. 1993. Variability of deltaic processes in terms of sediment supply, with particular emphasis on grain size. *Sedimentology*, **40**, 475–512.

SCRUTON, P. C. 1960. Delta building and the deltaic sequence. *In*: SHEPARD, F. P., PHLEGER, F. B. & ANDEL, T. H. (eds) *Recent Sediments, Northwest Gulf of Mexico: a Symposium*. American Association of Petroleum Geologists, 82–102.

SUGUIO, K., MARTIN, L., BITTENCOURT, A. C. S. P., DOMINGUEZ, J. M. L. & FLEXOR, J. M. 1985. Flutuações do nivel relativo do mar durante o Quaternário superior ao longo do litoral brasileiro e suas implicações na sedimentação costeira. *Revista Brasileira de Geociências*, **15**, 273–286.

——, —— & FLEXOR, J-M. 1980. Sea-level fluctuations during the past 6000 years along the coast of the State of São Paulo (Brazil). *In*: MORNER, N. A. (ed.) *Earth Rheology, Isostasy and Eustasy*. Wiley, New York, 471–486.

SWIFT, D. J. P., PHILLIPS, S. & THORNE, J. A. 1991. Sedimentation on continental margins, IV: lithofacies and depositional systems. *In*: SWIFT, D. J. P., OERTEL, G. F., TILLMAN, R. W. & THORNE, J. A. (eds) *Shelf Sand and Sandstone Bodies – Geometry, Facies and Sequence Stratigraphy*. International Association of Sedimentologists, Special Publication, **14**, 89–152.

TINTELNOT, M., PASENAU, H., BRICHTA, A. & IRION, G. 1994. Sedimentological and geophysical investigations on the continental margin off the São Francisco river. *14th International Sedimentological Congres*, Recife, Brazil, D76.

VALADÃO, R. C. & DOMINGUEZ, J. M. L. 1994. Opening of the South Atlantic Ocean and Denudation of the São Francisco Craton, Brazil. *14th International Sedimentological Congres*, Recife, Brazil, S9/10-S9/11.

VIANA, J. B. 1972. *Estimativa do Transporte Litorâneo em Torno da Embocadura do Rio Sergipe*. Instituto de Pesquisas Hidroviárias, Belo Horizonte, Brazil.

VILAS BOAS, G., MARTIN, L., BITTENCOURT, A. C. S. P. & FLEXOR, J-M. 1979. Paleogeographic and paleoclimatic evolution during the Quaternary in the northern half of the coast of the State of Bahia, Brazil. *In*: SUGUIO, K., FAIRCHILD, T. R., MARTIN, L. & FLEXOR, J-M. (eds) *Proceedings of the International Symposium on Coastal Evolution in the Quaternary*, São Paulo, Brazil, 254–263.

WALKER, R. G. 1992. Facies, facies models and modern stratigraphic concepts. *In*: WALKER, R. G. & JAMES, N. P. (eds) *Facies Models – Response to Sea Level Change*. Geological Association of Canada. 1–14.

WRIGHT, L. D. 1977. Sediment transport and deposition at river mouths: a synthesis. *Geological Society of America Bulletin*, **88**, 857–868.

——1978. River Deltas. *In*: Davis, R. A. JR. (ed.) *Coastal Sedimentary Environments*. Springer, Berlin, 5–68.

—— & COLEMAN, J. M. 1972. River delta morphology, wave climate and the role of the subaqueous profile. *Science*, **176**, 282–284.

—— & —— 1973. Variations in morphology of major river deltas as functions of ocean waves and river discharge regimes. *AAPG Bulletin*, **57**, 370–398.

ZENKOVITCH, V. P. 1967. *Processes of Coastal Development*. Wiley, New York.

Evolution of a nearshore and coastal macrotidal sand transport system, Queen Charlotte Islands, Canada

J. V. BARRIE & K. W. CONWAY

Pacific Geoscience Centre, Geological Survey of Canada, PO Box 6000, Sidney, British Columbia, V8L 4B2, Canada

Abstract: An extensive (40 km) coastal plain is prograding on the north coast of the Queen Charlotte Islands off the Pacific margin of Canada while the entire 120 km eastern coast of Graham Island is actively eroding. At the junction of these two systems is Rose Spit, which extends northeastward at a point of sediment transport convergence. Historical bathymetrical data and aerial photographs indicate that a spit platform extends 10 km to the northeast and at least four portions (up to 4 km in length) of the platform bank have become emergent since 1911. The area is seismically active and subject to storm-related sediment transport events superimposed on strong semi-diurnal tidal flows. The rapid emergence of the offshore banks and the net coastal changes are primarily controlled by: (1) convergence of sediment transport between the waters of Hecate Strait to the south and Dixon Entrance to the north, particularly during winter storms; and (2) localized tectonic uplift.

Sediment is transported north from the shallow waters of Hecate Strait by southeasterly winter storms and east, by longshore transport, along the accreting northern coast of the Queen Charlotte Islands to converge at Rose Spit. A 'hydraulic fence' has formed resulting in the creation of the elongate spit platform with emergent sand bars, east of the spit. Transpressional folding has occurred in this area east of the Queen Charlotte Fault, which separates the North American and Pacific Plates. This has resulted in localized uplift since Pliocene time, complementing the accretion process.

A wide (up to 2 km) and long (40 km) beach plain is actively prograding on the north coast of Graham Island, the northern island of the Queen Charlotte Islands in the northwestern Pacific (Fig. 1), while up to 120 km of the adjacent eastern coast are actively eroding (Fig. 2A). An intervening spit, Rose Spit, extends northeast where Hecate Strait meets Dixon Entrance (Fig. 1). Seaward of the spit is a shallow shelf extending 10 km to the east on which occur five bars which are emergent during all but the most extreme spring high tides.

Progradation rates of 0.3 to 0.4 m/year have been calculated by Harper (1980) along the north beach of McIntyre Bay (Fig. 1), while coastal bluff retreat on the east beach has been preliminarily estimated to be 1 to 2 m/year (Conway & Barrie 1994*a*). Initial observation would suggest that the volume of sediment eroded from the Quaternary bluffs on the eastern coast of Graham Island should account for the development of the extensive bank or platform off Rose Spit. What is not so clear is how a sand bar or ridge can develop so quickly and become emergent; and does sediment bypass a region of sediment transport convergence and supply the accretion on the north beaches of Graham Island?

A study was initiated by the Geological Survey of Canada to determine the processes driving this large-scale coastal change in a macrotidal and tectonically active area. Here, we present a scenario to explain this phenomenon based on observational data of coastal changes and nearshore surficial geology.

Sea-level history

The Queen Charlotte Islands coastline has been changing constantly since glacial times. During the late Wisconsian glaciation the Queen Charlotte Islands supported local mountain ice caps and piedmont glaciers which, at the glacial maximum, coalesced with ice lobes of Cordilleran ice on eastern and northern Graham Island (Blaise *et al.* 1990). Recent stratigraphical and palaeoenvironmental analysis of sediments recovered from Dogfish Bank (Fig. 1), east of Graham Island, indicates that this late Wisconsian ice was thin and limited in extent and the area supported terrestrial environments (Barrie *et al.* 1993). Sea level was at least 35 m below present levels at that time. By 10400 years BP relative sea level was more that 100 m below that of today and large areas of the shelf were exposed (Luternauer *et al.* 1989; Barrie *et al.* 1991). Eustatic sea level rise, coupled with subsidence of a glacioisostatic forebulge, allowed sea-levels to rise very rapidly (7 to 10 cm/year) and reach the present shoreline on the Queen

From De Batist, M. & Jacobs, P. (eds), 1996, *Geology of Siliciclastic Shelf Seas*, Geological Society Special Publication No. 117, pp. 233–247.

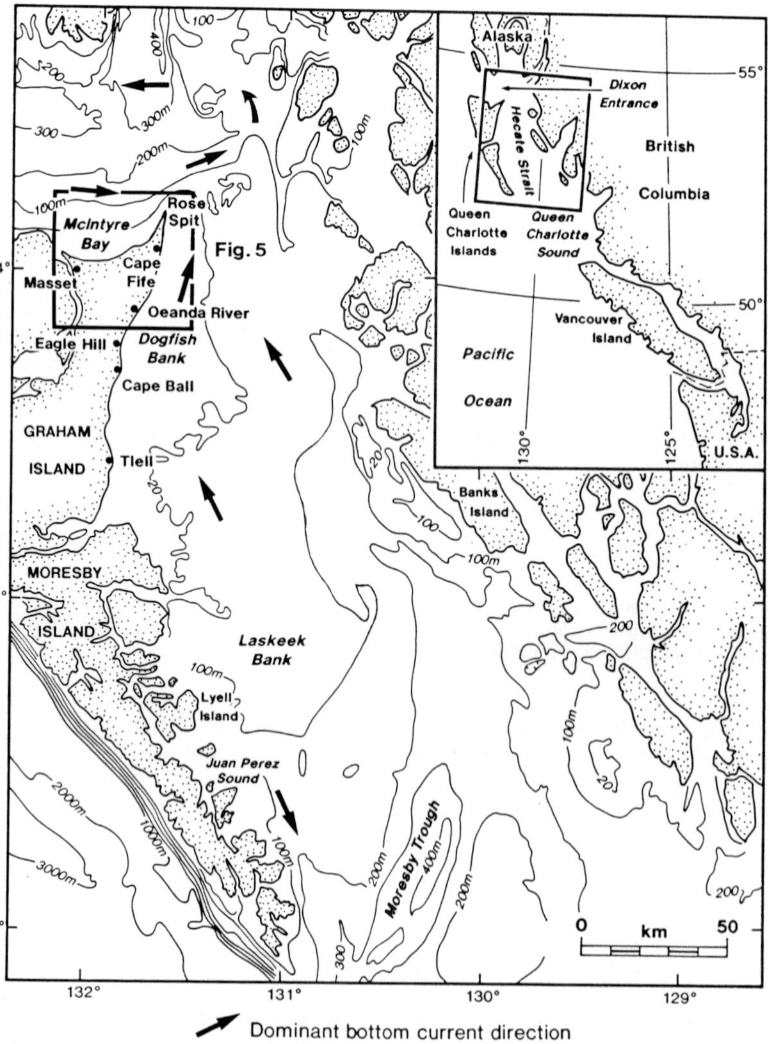

Fig. 1. The Pacific margin of northwestern Canada showing the study area on and adjacent to the northeastern Queen Charlotte Islands. Arrows represent the dominant bottom current directions.

Charlotte Islands by about 9100 years BP (Clague *et al.* 1982). This rise continued until between approximately 9000 and 7500 years when sea level was 15 m above present levels. Sea levels then fell until about 5000 to 6000 years ago after which the rate of fall was greatly reduced (Clague *et al.* 1982). Presently sea levels are still falling by 1–2 mm/year.

Regional setting

This area of the western Canadian continental margin is adjacent to the transform fault boundary between the Pacific and America Plates just along the west coast of the Queen Charlotte Islands. Relative movement along the fault has been calculated at 50–60 mm/year (Riddihough 1988). Consequently, this is a region of elevated seismicity with Canada's largest earthquake (magnitude 8.1) recorded off the northwestern coast of the Queen Charlotte Islands in 1949.

The region is underlain by a Tertiary basin (the Queen Charlotte Basin) consisting of marine and non-marine sandstones and mudstones. The Basin extends across most of the northwestern Canadian continental shelf and northern Graham Island (Sutherland-Brown 1968; Shouldice 1973;

Fig. 2. Photographs of: (**A**) coastal erosion at Cape Fife on the eastern beach of Graham Island (Fig. 1); (**B**) rafted beach log accumulation south of Cape Fife; and (**C**) emergent bars/islets on the Rose Spit platform.

Higgs 1991). During deglaciation, outwash sediments were distributed over the broad lowlands of the Basin; these are laterally and vertically extensive. Warner *et al.* (1982) suggest the outwash sands are of local origin, having been deposited by glaciers and meltwater streams issuing from the mountains west and southwest of Cape Ball (Fig. 1). Blaise *et al.* (1990) suggest that the sands were also deposited, at least in part, from meltwater streams discharging from the advancing Hecate lobe to the east.

The nearshore shelves bordering the north and east coasts of Graham Island have subsequently been reworked with a resultant thin covering of early Holocene transgressive sand and gravel deposits that occur to below 100 m water depth (Barrie & Bornhold 1989; Barrie *et al.* 1991). With the exception of three small areas of bedrock, the entire coastal zone for a distance of over 160 km, from Masset on the north coast to south of Tlell on the east coast, consists of sand and gravel beaches (Fig. 1).

Oceanographic setting

Processes of erosion and sediment transport are active in areas of the present-day western Canadian shelf to depths greater than 100 m based on observations of sedimentary bedforms (Luternauer 1986; Barrie *et al.* 1988; Barrie & Bornhold 1989). The pattern of bottom currents is a prevailing northward current along the eastern side of Hecate Strait, and a southward flow along the southwest portion of the Strait into Queen Charlotte Sound (Fig. 1) (Crawford *et al.* 1988; Crawford & Thomson 1991). In northern Hecate Strait, during summer conditions the current is primarily tidal (semi-diurnal) with a tidal range of 4 to 5 m. Winter circulation is both tidal and wind-driven with a net northward component (Fig. 1) resulting from southeasterly storm winds (Crawford & Thomson 1991). Current velocities exceeding 0.80 m/s, measured 15 m above the sea floor, occur during winter storms and spring tides. In Dixon Entrance, an eddy, termed the Rose Spit Eddy, rotates counter-clockwise throughout the entire eastern end of Dixon Entrance (Crawford & Greisman 1987; Crawford & Thomson 1991) resulting in an eastward current along the north coast of Graham Island (Fig. 1).

Wave conditions over the area vary in response to localized fetch restrictions and the sheltering effect of land on oceanic swell. Mean winter significant wave height for Hecate Strait is approximately 2 m corresponding with mean peak wave periods of 9 s. Wave propagation also shows alignment with the Strait, the main direction being from the southeast and less frequently from the northwest (Thomson 1981).

Methods

A total of five beach profile sites, two on the north coast and three on the east, have been monitored on an annual basis since 1990 (Fig. 3). Standard survey techniques were used to establish local benchmarks against which erosion could be measured and beach profiles run. The sites were selected to provide an even distribution of observation sites within the study area as well as being representative of the beach/shoreline conditions. Vertical air photos (1937, 1980 and 1984), oblique air photos (February 1994) and a SPOT satellite image (May 1992) were used to obtain estimates of erosion for the east coast of Graham Island south of Rose Spit to Tlell. The lack of good reference points on the ground inhibits continuous measurement of erosion by photogrammetry along the entire east coast of Graham Island. Selected areas with good morphologic reference features (lakes, meanders in streams, scarps) were chosen to provide a relative reference point for measurements.

Offshore bathymetrical changes were computed from Canadian Hydrographic Service surveys undertaken in 1911 and 1984 for northeastern Hecate Strait. These were supplemented with high-resolution side-scan sonar and sub-bottom profile survey lines run during a scientific cruise of the 'CSS Tully' in June 1991 and 1995 on northern Dogfish Bank.

Some 82 sediment samples were collected from McIntyre Bay at a 2 km grid in 1989 using the 'MV Sea Prince III' (Fig. 3). Intertidal beach samples were collected for the east coast between Cape Fife and Rose Spit in 1990 (26 samples and four trenches) and for the north coast in 1992 (eight samples; Fig. 3). A core was taken in Kumara Lake in 1988 soon after it had been drained and another core was taken at the same location in 1993 (Fig. 3). Two cores were also obtained at the leading edge of the spit platform during the June 1991 cruise.

The cores were split, described and photographed in the laboratory. Textural analyses were completed using a settling tube for the sand fraction (<2 mm to >63 μm) and a Sedigraph for the mud fraction (<63 μm).

Coastal and nearshore surficial geology

The northwest coast of Graham Island is dominated by rocky headlands and arcuate

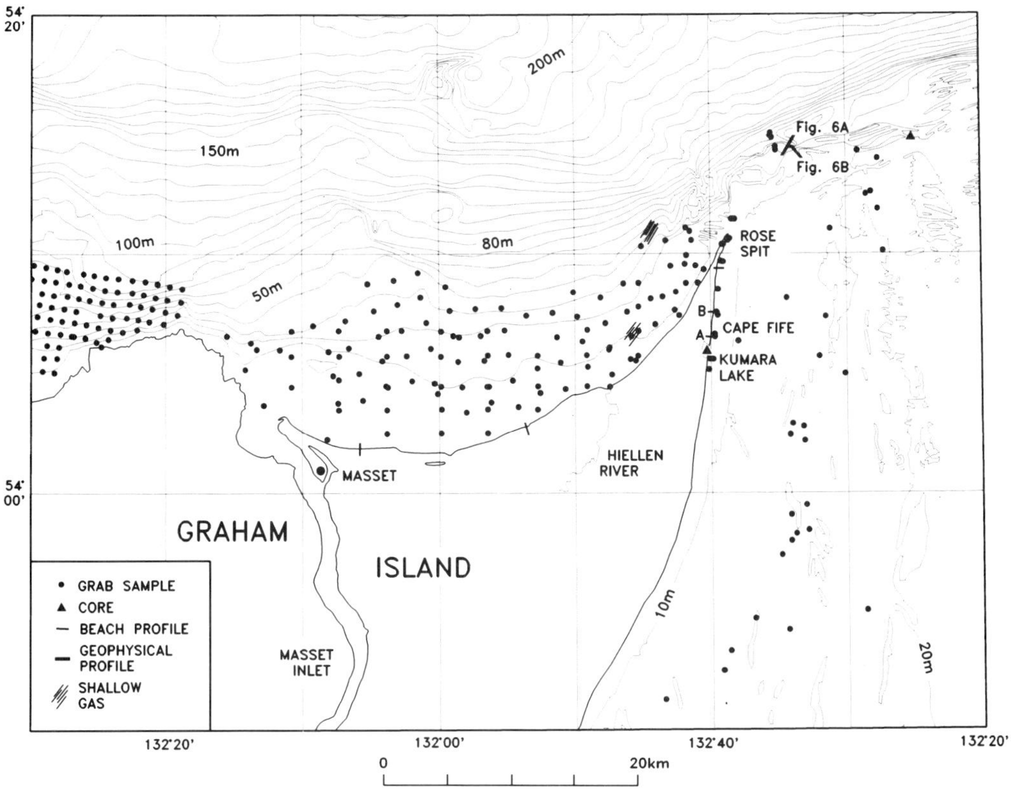

Fig. 3. Location of sediment grab samples, cores and beach profiles collected on and offshore of northern and eastern Graham Island. Note the location of gas pockmarks in McIntyre Bay.

embayments. In McIntyre Bay, which makes up the eastern portion of northern Graham Island, rapid accretion has resulted in a wide intertidal zone (greater than 200 m). The backshore consists of a series of Holocene beach ridges up to 10 m above present sea level, developed as a result of rapid beach progradation and the above-mentioned Holocene sea-level changes (Clague & Bornhold 1980). Offshore, there are no bars and an eastward thickening wedge of medium to fine sand occurs, culminating in Rose Spit and adjacent offshore platform. Bedrock is exposed on the sea floor along the north coast of Graham Island and Tertiary sandstones are covered by a drape of sand as much as 20 m thick over most of central McIntyre Bay (Conway & Barrie 1994*b*) to a depth of up to 80 m. Where the sediments thicken in eastern McIntyre Bay extensive areas of shell accumulations and shallow gas with seabed pockmarks occur (Fig. 3).

Rose Spit and the associated platform extend east-northeastward into northern Hecate Strait and southern Dixon Entrance for 10 km with a depth of less than 5 m. Meistrell (1972) and Nielsen *et al.* (1988) define a spit platform as a large-scale primary sedimentary structure formed by sediment transport along the coast and is a rise over the shelf but below mean low tide. However, the Rose Spit platform also has bars that have become subaerial islets under all but the most extreme high tides; these are not connected to the spit itself.

A comparison of hydrographic surveys undertaken in 1911 and 1984 is shown in Figs 4A and 4B. Although the precision of the 1911 survey in such a dynamic environment is not comparable to modern standards, the overall trend is readily apparent. Between the two surveys the spit platform has extended farther to the north and northeast and has shoaled to less than 5 m water depth. In 1911 only one emergent bar or islet was subaerial under the most extreme high tides whereas in 1989 five islets had emerged along the extending spit (Fig. 4B). The islets have increased in size markedly from 1984, based on SPOT satellite

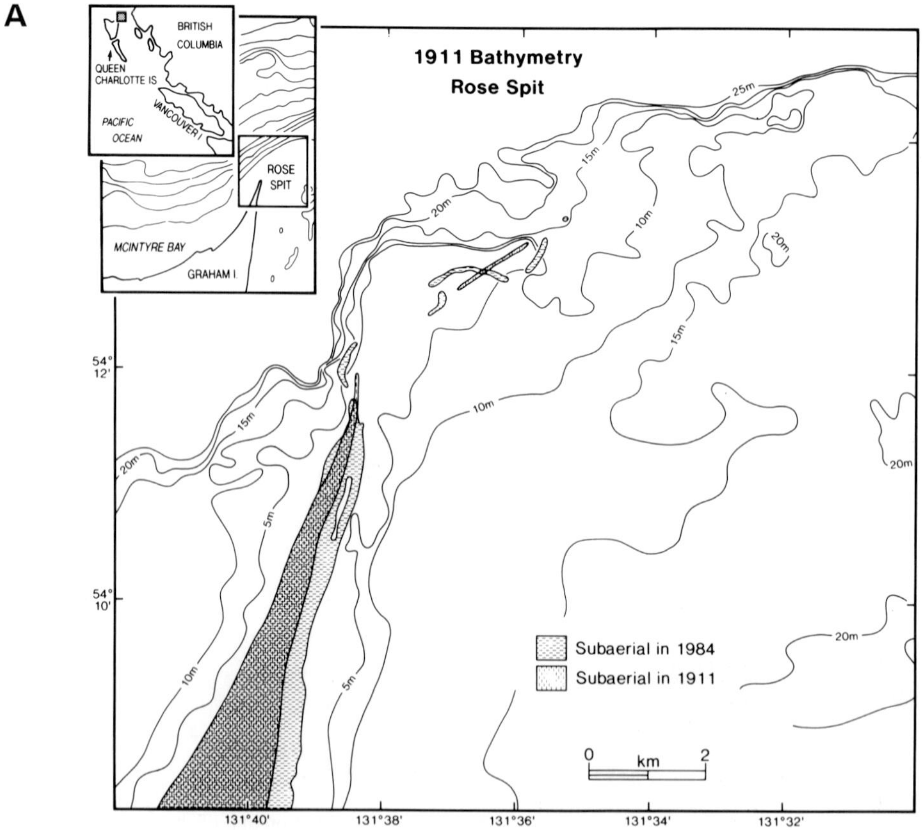

Fig. 4. Interpretation of hydrographic surveys of the area surrounding Rose Spit in (**A**) 1911 and (**B**) 1984. Subaerial portions of the spit and spit platform are shown for the two different bathymetrical surveys.

imagery taken in May 1992 (Fig. 5). Three of the islets mapped in the 1984 survey (Fig. 4B) can be seen to have coalesced to form one 4 km island by 1992. Aerial observations in 1993 confirmed this development (Fig. 2C).

Evidence for the progradation of the entire spit platform can be seen seaward of the emerging islands. Subbottom profiles collected in 1991 and 1995 (Figs 6A and 6B) show evidence of prograding sand foreset cross-beds developing in a northerly to northeasterly direction. Meistrell (1972) demonstrated in his wave tank experiments that at the end of a headland beach a spit platform forms as sediment is deposited into deeper water in a successive series of foreset beds. The cross-bedded reflectors seen on the high-resolution seismic records of the Rose Spit platform resemble a similar late Pleistocene sequence in Denmark, where the coarse-grained cross-bedded foreset unit corresponded to the prograding subaqueous spit platform (Nielsen *et al.* 1988). Nielsen *et al.* (1988) attribute the large cross-bedded unit in Denmark to an avalanche process down the steep front of the subaqueous spit platform. Whether a similar process controls the rapid build-up of the Rose Spit platform cannot be determined from the data available, but such a process is probable considering the seismic structure of the cross-beds (Figs 6A and 6B), the very coarse-grained sands that occur on the platform front, the rapid growth in one dominant direction and the potential sediment supply to the area.

Nielsen *et al.* (1988) suggest, from observations of the Pleistocene spit platform sequence, that these systems are rarely or never developed under macrotidal conditions, as is the case for the Rose Spit platform. However, the construction of the platform off Rose Spit is enhanced by the migration of subaqueous dunes and hence an abundant sediment supply. Dune migration direction is northeastwards along the axis of the spit platform and southward from Dixon Entrance up the slope leading to the spit platform (Fig. 6B).

B

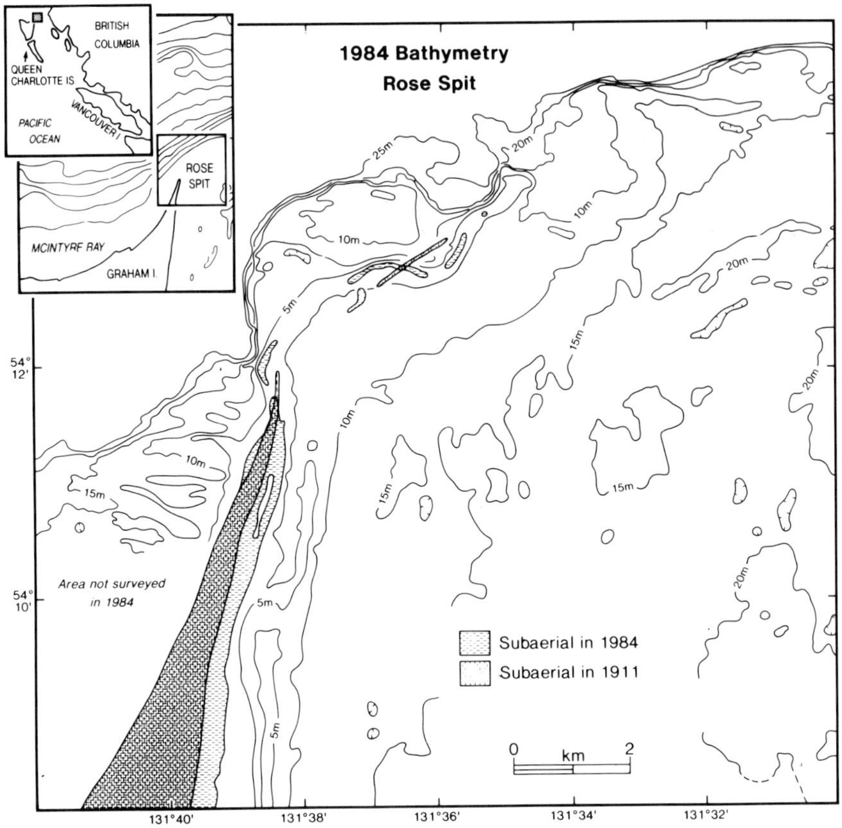

South of Rose Spit on the eastern shore of Graham Island the coast is characterized by a steep beach profile and a narrow (<70 m) intertidal zone, except where an accumulation of beached logs occurs (Fig. 2B). The sources of sediment for the beaches are the large unconsolidated Pleistocene bluffs and, toward the northern end of the east coast, raised dune and relict beach material (Harper 1980).

Surficial shoreface sands of northeastern Graham Island grade from coarse (1 mm) sand in the lower intertidal zone to fine (0.12–0.25 mm) sand within the supratidal zone. Heavy mineral concentration increases landward, in association with the fining grain sizes, to make up 90% of the sediment composition (Barrie & Emory-Moore 1994). There does not appear to be any systematic alongshore variation in texture or heavy mineral abundance within the surficial beachface sediments. Three sediment facies are evident within the beach zone: a laminated sand facies, a sand and gravel facies and a matrix-supported gravel and cobble facies (Barrie & Emory-Moore 1994).

Shore-attached bars, evident on air photos and in the SPOT image (Fig. 5), occur from Cape Ball to Rose Spit along the entire length of the coastline. The bars are attached at their southern end and indicate that strong northward longshore drift is prevalent. There are at least nine bars, the longest of which is more than 8 km, along the Cape Ball–Rose Spit coast. Where the bars attach to the coast the beach may be more than 1 km wide.

Usually these areas coincide with areas of accumulated beach logs (Fig 2B). The logs, derived primarily from modern logging practices, have arrived on the beach by the dominant northerly flowing current and southeasterly storms, escaping log booms to the south. They settle in the trough between the shore-attached bars and form a framework that traps sand in transport by longshore and wind transport. These areas then form protrusions on the normally very straight coastline.

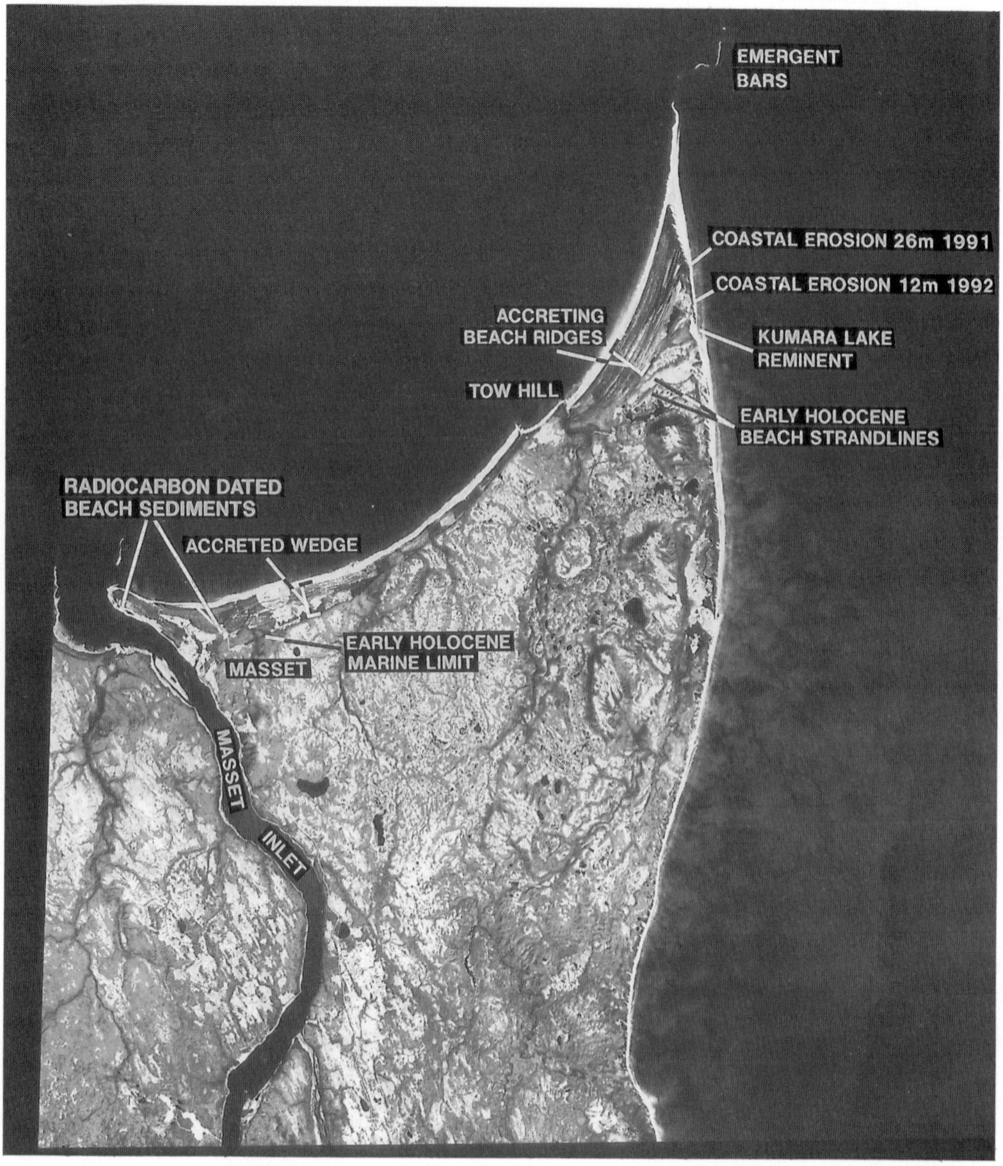

Fig. 5. SPOT satellite image of northeastern Graham Island taken on 19 May 1992 showing the accreted wedge on the north coast and the emergent bars on the Rose Spit platform.

Coastal stability

Accretion

The SPOT satellite image taken in May 1992 gives clear evidence of the accretion that has taken place since the early Holocene (Fig. 5). The amount of accretion has been notably greater both to the northeast and southwest ends of McIntyre Bay and the least where rocky headlands occur. The zone of accretion ends just to the west of Masset Inlet.

Harper (1980) calculated progradation rates (0.3 to 0.4 m/year) for the central part of the McIntyre Bay beach, based on dates from an old beach ridge. A similar, but slightly higher rate of progradation (0.4 to 0.5 m/year) has been established from four radiocarbon ages derived

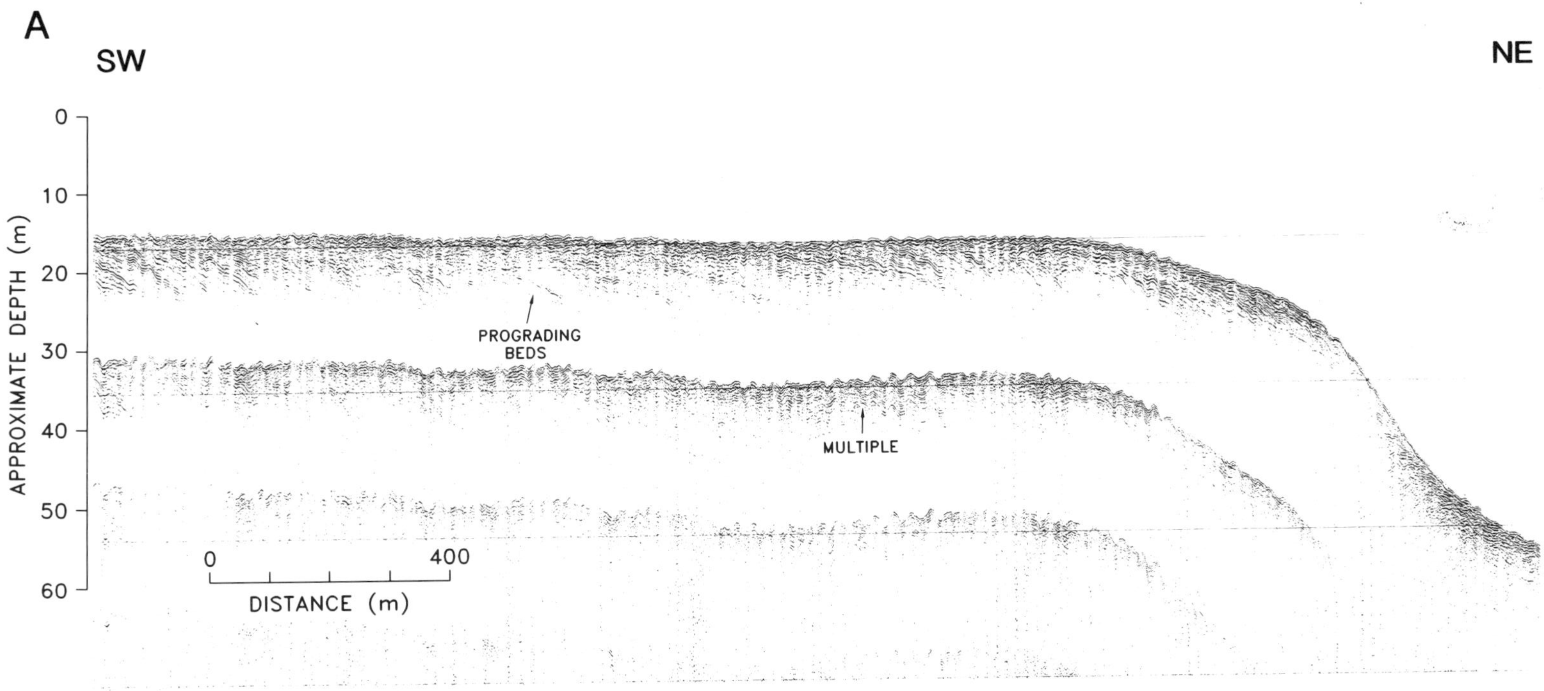

Fig. 6. (**A**) High-resolution Seistec DTS subbottom profile across the easternmost portion of the Rose Spit platform, showing the prograding beds of the leading edge of the spit platform. Location of the profile is shown in Fig. 3.

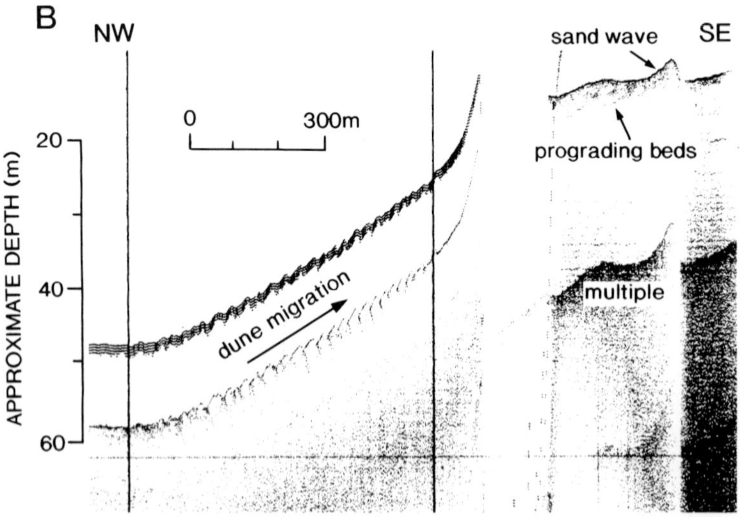

Fig. 6. (**B**) High-resolution Huntec DTS subbottom profile across the easternmost portion of the Rose Spit platform, showing the northern portion of the spit platform and the migration of subaqueous dunes upslope from Dixon Entrance. Location of the profile is shown in Fig. 3.

from samples collected within raised beach ridges on the southwestern end of McIntyre Bay (Fig. 5 and Table 1). It appears that the rate of progradation has been reasonably consistent over time for the region and only varies in relation to the coastal morphology and the sediment dynamics acting on this morphology.

Erosion

Erosion of the coastline of the eastern shore of Graham Island (Fig. 2A) has been observed by the Haida, Rose Spit being near the mythic site of creation of the Haida people, and reported on by the earliest Europeans that came to the Queen Charlotte Islands. This erosion has exposed gold-bearing heavy mineral deposits that have been exploited on a small scale since the late 1800s (Mandy 1934; Barrie & Emory-Moore 1994).

Table 1. *Radiocarbon dates from the two raised beach ridge locations north of Masset on Graham Island (Fig. 5)*

Sample	Radiocarbon date (years BP)	Laboratory no.
Marine shell	1790 ± 70	AECV-1792Cc
Marine shell	2130 ± 70	AECV-1793Cc
Marine shell	1900 ± 70	AECV-1794Cc
Marine shell	1760 ± 60	TO-4883

Measured amounts of erosion have been documented since 1990 on the east coast of Graham Island (Fig. 7). It is noteworthy that the erosion occurring in a one-day period in February 1994 at the Cape Fife site is greater than the measured erosion during all of 1993 at that location. At Kumara Lake, south of Cape Fife (Fig. 3), comparison of 1980 vertical air photos and 1994 oblique air photos coupled with ground-truthing traverses indicate that about 60 m have been lost in this area between 1980 and 1994 (Fig. 8). The seaward shore of Kumara Lake was breached by the ocean in 1988 and since that time has become a lagoon or embayment. Coarse marine sediments (sands and gravels), which are storm washover deposits, are rapidly infilling the seaward side of the former lake basin. Just north of Eagle Hill (Fig. 1) a small unnamed lake has drained in a similar fashion to Kumara Lake; erosion at this site appears to be roughly 80 m over the period 1937–1980.

Erosion at certain locations is estimated at between 10 and 70 m over the period from 1937 to 1980 by comparing vertical air photos for different sites in the area between Eagle Hill and the Oeanda River mouth (Fig. 1). The mean erosion was 36 m or slightly less than a metre per year. Between Cape Ball and Eagle Hill estimates of erosion between 1937 and 1980 varied from 0 to 70 m. The mean erosion over the period here was 48 m. These mean values are included for discussion only, as erosion may occur in short time intervals separated by relatively inactive periods.

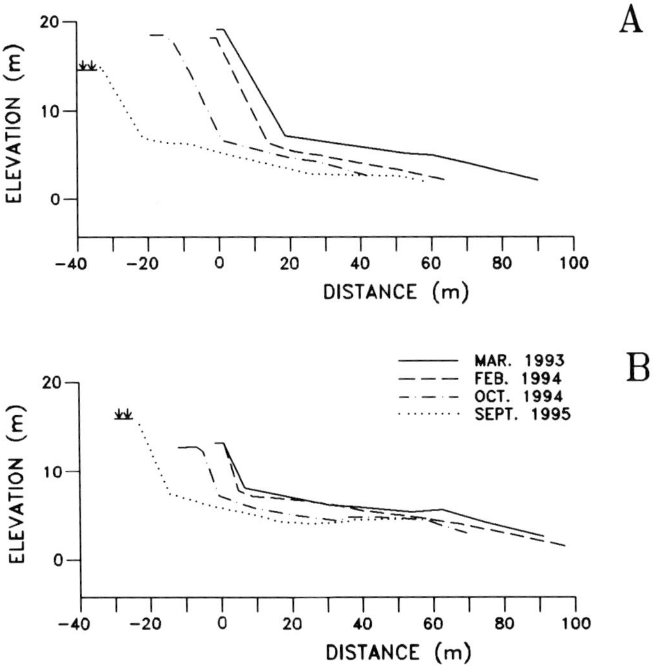

Fig. 7. Successive cliff erosion at two locations of the east beach of Graham Island as illustrated by beach profiles between March 1993 and September 1995. Location of beach profiles is shown in Fig. 3.

During a storm on 8 February 1994, 1.5 m of erosion was measured at the Cape Fife profile within 24 h. Winds at Rose Spit gusted at 83 to 96 km/h from the southeast during the peak of the storm. Winds were from the southeast and over 55 km/h for an 8 h period, then dropped off completely within 1 h. Wave height normally approaches 4 m under these wind conditions in Hecate Strait. The time during which the storm occurred covered one high water in a period of spring tides. Wind conditions similar to or more extreme than these are expected to occur several times each year.

While the erosion does affect the entire coastline as seen from visual examination during the February storm, the pattern appears to reflect a punctuated assault on different sections of the coast which varies from year to year. The net retreat of the coast may be moderate but the effect on any given segment of the coast in one year (or one storm) may be very severe.

Observations made during and after the February storm indicate that the mechanism by which the shoreline retreats within the study area is by failure of oversteepened cliffs as retrogressive, small land slips or slumps (Fig. 9). These slumps vary in size from 20 to 1000 m^3 of material. The oversteepening of the cliff face probably occurs during the storm as wave runup removes material from the foot of the slope. Following slope failure the material is then reworked and washed away in the surf zone.

A series of linear cracks was found at the top of the scarp at the Cape Fife site on the day following the storm (9 February 1994). Extensive cracks up to 5 mm in width and 3–5 m long, subparallel to the crest of the cliff to about 5 m back of the present scarp were noted. These cracks, which are probably crown cracks of incipient failures, suggest more slumping will occur at the Cape Fife site before the slope temporarily stabilizes. At Cape Fife the material underlying the dune deposits consists of Holocene lacustrine muds, which are in turn underlain by early Holocene indurated gravels. These sediments appear to be broken up during storms by wave impact on the beachface and cliff sections, and excavated as large (20–50 cm) clasts.

Discussion

From observations of coastal dynamics derived from airborne and satellite imagery, sedimentological characteristics and high-resolution

Fig. 8. Oblique air photo of Kumara Lake bed taken in February 1994 (bottom) compared to a Government of British Columbia vertical air photo taken in 1980. Note that most of the seaward forested margin of the lake has been removed. Erosion is estimated to be about 60 m in 14 years.

marine geophysical surveys, a pattern of sediment transport and deposition is apparent. The erosion of the eastern coastline and net northward longshore transport provides the requisite sediment for the development of the spit platform. Migrating dunes from Dixon Entrance show a net migration direction upslope towards the spit platform in a southerly direction. In McIntyre Bay, to the west, longshore drift is to the east-northeast terminating at Rose Spit (Clague & Bornhold 1980). Consequently the extending spit platform is a zone of sediment transport convergence with no evidence of spillover into the deeper waters of Dixon Entrance. Amos *et al.* (1995) have defined this condition as a 'hydraulic fence'. This hydraulic fence accounts for the development of the spit platform and the ultimate emergence of bars through the intertidal zone to form islets.

A sediment transport model was used for the same region to predict the sediment movement during a severe storm that occurred in February 1984 (Amos *et al.* 1995). The resultant patterns suggest that storm sediment transport is the dominant process in the evolution of the coastal and nearshore surficial geology deposits regardless of the strong macrotidal conditions of the region. This would appear to support the model for spit platform development as suggested by Nielsen *et al.* (1988). During this particular storm, sand was transported alongshore and to the north until it reached the Rose Spit platform.

Fig. 9. Retrogressive failures in unconsolidated dunes in the Cape Fife area (Fig. 3). The cliff is about 12 m high.

The model did not predict any spillover of sand into Dixon Entrance as indicated from the geophysical data.

The model also shows a portion of the sediment being funnelled into McIntyre Bay through the passage just seaward of the present spit (Fig. 5). This could provide much of the sediment required for the north coast accretion of approximately 0.3 to 0.5 m/year. Harper (1980) suggests that the immediate offshore area is the source of the accretion. Geophysical and sample data confirm that eastern McIntyre Bay is sediment-starved over some areas and the transport direction derived from sand ribbons and linear bands of sand is in the shoreward direction (Barrie & Tucker 1992). Sediment, upon entering the bay past Rose Spit, must be quickly transported to the shoreface zone.

Finally, the accretion process could be complemented by localized uplift due to transpressional folding. In northern Hecate Strait and Dixon Entrance transpressional folding and uplift has occurred inboard of the fault-bounded Queen Charlotte Islands crustal blocks since Pliocene time (Rohr & Dietrich 1992). For example, a magnitude 5.1 earthquake in north-central Hecate Strait in January 1990 corresponds with a fault within an inverted basin that comes to or near the seabed just to the southeast of the Rose Spit platform. In June 1995 a further three earthquakes, with magnitudes of between 3.0 and 4.3, were recorded in the same location.

Conclusions

The coast of the Queen Charlotte Islands off the Pacific margin of Canada has been changing under the influence of isostasy, eustasy, tectonics and sediment transport. Progradation rates of 0.3 to 0.5 m/year have been calculated along the north beach while coastal bluff retreat on the east beach, determined from beach profiles and repetitive aerial photography, is estimated to be between 1 and 3 m/year, with up to 1.5 m of coastline observed to have retreated in 24 h during one winter storm. The area is subject to storm-related sediment transport events superimposed on strong semi-diurnal tidal flows.

The development of a spit platform, the emergence of offshore bars and the net coastal changes are primarily controlled by: (1) convergence of sediment transport between the waters of Hecate Strait to the south and Dixon Entrance to the north to form a 'hydraulic fence', and (2) localized tectonic uplift due to transpressional folding in this area east of the Queen Charlotte Fault. During southeasterly storms, which coincide with ebbing tides, sediment is funnelled into the north coast through a passage directly seaward of Rose Spit providing sediment for north coast accretion. These dramatic sediment transport and sea-level conditions result in a coastline that is in constant change.

The field data could not have been obtained without the help of Carl Amos, Kathy Tucker, Bob MacDonald, Fredrick Okoth, Bill Hill and David Phillips. Graphics were created by Brian Sawyer and Richard Franklin and early versions of the manuscript were improved by Brian Bornhold. The final manuscript was critically reviewed and improved by John Luternauer. This is the Geological Survey of Canada contribution no. 31295.

References

AMOS, C. L., BARRIE, J. V. & JUDGE, T. J. 1995. Storm-enhanced sand transport in macrotidal setting, Queen Charlotte Islands, British Columbia, Canada. *In*: FLEMMING, B. W. & BARTOLOMÄ, A. (eds) *Tidal Signatures in Modern and Ancient Sediments*. Special Publication of the International Association of Sedimentologists, **24**, 53–68.

BARRIE, J. V. & BORNHOLD, B. D. 1989. Surficial geology of Hecate Strait, British Columbia continental shelf. *Canadian Journal of Earth Sciences*, **26**, 1241–1254.

—— & EMORY-MOORE, M. 1994. Development of marine placers, northeastern Queen Charlotte Islands, British Columbia, Canada. *Marine Georesources & Geotechnology*, **12**, 143–158.

—— & TUCKER, K. 1992. *Nearshore surficial geology of the Queen Charlotte Islands – northwestern Graham Island*. Geological Survey of Canada, Open File Report **2523**.

——, BORNHOLD, B. D., CONWAY, K. W. & LUTERNAUER, J. L. 1991. Surficial geology of the northwestern Canadian continental shelf. *Continental Shelf Research*, **11**, 701–715.

——, CONWAY, K. W., MATHEWES, R. W., JOSENHANS, H. W. & JOHNS, M. J. 1993. Submerged late Quaternary terrestrial deposits and paleoenvironment of northern Hecate Strait, British Columbia continental shelf, Canada. *Quaternary International*, **20**, 123–129.

——, EMORY–MOORE, M., LUTERNAUER, J. L., & BORNHOLD, B. D. 1988. Origin of modern heavy mineral deposits, northern British Columbia continental shelf. *Marine Geology*, **84**, 43–51.

BLAISE, B., CLAGUE, J. J. & MATHEWES, R. W. 1990. Time of maximum Late Wisconsin glaciation, west coast of Canada. *Quaternary Research*, **34**, 282–295.

CLAGUE, J. & BORNHOLD, B. 1980. Morphology and littoral processes of the Pacific coast of Canada. *In*: McCann, S.B. (ed.) *The Coastline of Canada*. Geological Survey of Canada, Paper **80–10**, 339–380.

——, HARPER, J. R., HEBDA, R. J. & HOWES, D. E. 1982. Late Quaternary sea level and crustal movements, coastal British Columbia. *Canadian Journal of Earth Sciences*, **19**, 597–618.

CONWAY, K. W. & BARRIE, J. V. 1994*a*. *Coastal Erosion on the East Coast of Graham Island, Queen Charlotte Islands, British Columbia*. Geological Survey of Canada, Current Research, Paper **1994-E**, 53–58.

—— & —— 1994*b*. *Late Quaternary Stratigraphy of Dixon Entrance, British Columbia Continental Shelf*. Geological Survey of Canada, Open File Report **3003**.

CRAWFORD, W. R. & GREISMAN, P. 1987. Investigation of permanent eddies in Dixon Entrance, British Columbia. *Continental Shelf Research*, **7**, 851–870.

—— & THOMSON, R. E. 1991. Physical oceanography of the western Canadian continental shelf. *Continental Shelf Research*, **11**, 669–683.

——, HUGGETT, W. S. & WOODWARD, M. J. 1988. Water transport through Hecate Strait, British Columbia. *Atmosphere-Ocean*, **26**, 301–320.

HARPER, J. 1980. *Coastal Processes on Graham Island, Queen Charlotte Islands, British Columbia*. Geological Survey of Canada, Current Research, Paper **80–1A**, 13–18.

HIGGS, R. 1991. Sedimentology, basin-fill architecture and petroleum geology of the Tertiary Queen Charlotte Basin, British Columbia. *In*: WOODSWORTH, G. J. (ed.) *Evolution and Hydrocarbon Potential of the Queen Charlotte Basin, British Columbia*. Geological Survey of Canada, Paper **90–10**, 337–371.

LUTERNAUER, J. L. 1986. Character and setting of sand and gravel bedforms on the open continental shelf off western Canada. *In*: KNIGHT, R. L. & MCLEAN, J. R. (eds) *Shelf Sands and Sandstones*. Canadian Society of Petroleum Geologists Memoir, **II**, 45–55.

——, CLAGUE, J. J., CONWAY, K. W., BARRIE, J. V., BLAISE, B. & MATHEWES, R. W. 1989. Late Pleistocene terrestrial deposits on the continental shelf of western Canada: evidence for rapid sea-level change at the end of the last glaciation. *Geology*, **17**, 357–360.

MANDY, J. T. 1934. Gold-bearing black-sand deposits of Graham Island, Queen Charlotte Islands. *Bulletin of the Canadian Institute of Mining and Metallurgy*, **37**, 563–572.

MEISTRELL, F. J. 1972. The spit-platform concept: laboratory observation of spit development. *In*: SCHWARTZ, M. L. (ed.) *Spits and Bars*. Dowden, Hutchinson & Ross, Stourdsberg, PA, 225–283.

NIELSEN, L. H., JOHANNESSEN, P. N. & SURLYK, F. 1988. A late Pleistocene coarse-grained spit-platform sequence in northern Jylland, Denmark. *Sedimentology*, **35**, 915–937.

RIDDIHOUGH, R. P. 1988. The northeast Pacific Ocean and margin. *In*: NAIRN, A. E. M., STEHLI, F. W. UYEDA, S. (eds) *The Ocean Basins and Margins. Vol.7B: The Pacific Ocean*. Plenum, New York, 85–118.

ROHR, K. M. M. & DIETRICH, J. R. 1992. Strike-slip tectonics and development of the Tertiary Queen Charlotte Basin, offshore western Canada: evidence from seismic reflection data. *Basin Research*, **4**, 1–19.

SHOULDICE, D. H. 1973. Western Canadian continental shelf. *In*: MCCROSSAN, R. D. (ed.) *Future Petroleum Provinces of Canada: Their Geology and Potential*. Canadian Society of Petroleum Geologists Memoir, **1**, 7–35

SUTHERLAND-BROWN, A. 1968. *Geology of the Queen Charlotte Islands, British Columbia*. British Columbia Department of Mines and Petroleum Resources Bulletin **54**.

THOMSON, R. E. 1981. *Oceanography of the British Columbia Coast*. Canada Special Publication of Fisheries and Aquatic Sciences, **56**.

WARNER, B. G., MATHEWES, R. W. & CLAGUE, J. J. 1982. Ice-Free conditions on the Queen Charlotte Islands, British Columbia, at the height of Late Wisconsin glaciation. *Science*, **218**, 675–678.

The influence of inherited geological framework upon a hardbottom-dominated shoreface on a high-energy shelf: Onslow Bay, North Carolina, USA

W. J. CLEARY[1], S. R. RIGGS[2], D. C. MARCY[1] & S. W. SNYDER[3]

[1] *Department of Earth Sciences, University of North Carolina at Wilmington, Wilmington, NC 28403, USA*

[2] *Department of Geology, East Carolina University, Greenville, NC 27658, USA*

[3] *Department of Marine, Earth and Atmospheric Sciences, North Carolina State University, Raleigh, NC 27695, USA*

Abstract: The southeastern coast of North Carolina is a major tourist destination that has experienced rapid population growth resulting in increased revenues. Most development is sited on transgressive barriers and headland beaches located along chronic erosion zones. Beach replenishment is viewed as the only viable option for erosion mitigation. However, a detailed sand resource assessment is non-existent. Furthermore, few data exist on the inter-relationships between the underlying geological framework and the morphology, sediments and evolution of the coastal system.

The shoreface of the two constraining headlands of this sand-starved system in Onslow Bay was studied utilizing cores, remotely sensed data, and diver surveys. The (coquina-dominated) subaerial headland at Fort Fisher and the (limestone) submarine headland at New River clearly have influenced the coastwise evolution of the intra-headland perched beaches. These shorefaces, as well as those of adjacent barriers, are characterized by hardbottoms of varying relief, morphology, and lithology. Prominent submarine scarps and karstic topographic highs that extend above the ravinement surface have resulted in major changes in barrier orientation and development of distinct subcompartments with respect to coastal processes.

Holocene sea-level rise has produced a thin (<70 cm) variable sequence, deposited unconformably over bioeroded hardbottoms. Bioerosion represents a major source and supply of new sediment. The thin sediment veneer is a mosaic of remobilized graded palimpsest and residual shell-rich sand and gravels. Data reinforce the concept that a common equilibrium profile for all shorefaces is neither realistic nor adequate for an understanding of coastal processes in the study area.

The coastal communities along Onslow Bay, North Carolina are major tourist destinations and represent prime examples of coastal areas experiencing massive population growth with rapidly increasing revenues. The primary driving force behind this rapid economic development is the presence of extensive sandy beaches. Thus, management and preservation of this dynamic coastal shoreline is of best concern to local, state, and federal governments. A variety of erosion abatement projects have been attempted along various segments of the coastline over the past five decades with minimal success. Today, beach nourishment is considered to be the primary option towards 'stabilizing the beach', yet very little has been accomplished with respect to sand resource assessment or understanding the geologic constraints.

The 100 km segment of shoreline from Onslow Beach to Cape Fear, which includes the study areas, has an average shoreline erosion rate of 1.0 m/year. (NC DEHNR 1993). However, the average annual erosion rates for selected segments of recently developed coastlines within this area often exceed 4 m/year and occasionally exceed 7 m/year. Developed areas with such high erosion rates are under severe political and economic pressure to initiate shoreline nourishment programmes. However, before mitigation projects can be undertaken, it is necessary to develop a basic understanding of the shoreface geologic framework in order to both delineate adequate sand borrow sites and to maximize the engineering design of the project without impacting environmentally sensitive areas.

On a larger scale, perhaps the most important applied problem in coastal geology is determining the response of the shoreline to the predicted global sea-level rise (Pilkey & Davis 1987; Komar *et al.* 1991). Prediction of shoreline retreat, land loss rates, and the cost of management alternatives is critical to planning of

From De Batist, M. & Jacobs, P. (eds), 1996, *Geology of Siliciclastic Shelf Seas*, Geological Society Special Publication No. 117, pp. 249–266.

coastal zone management strategies. The potential loss to society due to sea-level rise could be immense (Barth & Titus 1984). The results of this study could be a crucial component in dealing with the long-term consequences of continuing or accelerating sea-level rise.

Background

Coastal scientists generally agree that the shoreface is the least understood portion of the inner continental margin. This region between the surf zone and the inner shelf acts as a source and conduit for sediment transport both to the beaches and the deeper continental shelf. The wave and current dynamics of the shoreface determine how the adjacent shoreline and the recreational beach will respond to storms, and ultimately to the effects of rising sea level. Understanding the processes and three-dimensional geologic framework that govern the shoreface characteristics is vital to determining the behaviour of beaches, particularly those that have been replenished artificially.

Artificial nourishment of beaches provides erosion mitigation, storm protection, and, at the same time, increases the recreation and tourism potential. Almost all developed shorelines in the vicinity of the study area either have a history of replenishment, have replenishment permits pending, or are situated in chronic erosion zones and have requested assistance to stem rapid land loss.

Pilkey *et al.* (1993) contend that there are many 'shoreface profiles of equilibrium' and the Bruun (1962) model commonly used in the coastal engineering community is too simplistic. According to them, the profile of equilibrium is based on a number of erroneous assumptions. These include:

(1) all sediment movement is only driven by incoming wave orbitals acting on a sandy shoreface;
(2) closure depth with no net cross-shore transport of sediment to and from the shoreface exists and can be quantified;
(3) the shoreface is sand-rich and the underlying or offshore geology does not influence the profile shape; and
(4) if a shoreface is sand-rich, the smoothed profile described by the equilibrium profile equation must provide a useful approximation of the real shoreface shape.

The present paper focuses on assumption (3) concerning the role that the inherited geologic framework plays in determining barrier morphology, coastal cell evolution, and shoreface dynamics. The conclusions of this paper seem self-evident, but it has become obvious in our interactions with non-coastal geologists that the concept of shoreface profile of equilibrium has been over-simplified, is poorly understood, and is somewhat controversial.

Along many beach systems, profiles of equilibrium are never achieved due to eustatic sea-level fluctuations, lack of adequate sediment supplies, or the variable influence of the inherited geologic framework upon which the barrier beaches are superimposed. Obvious examples of the latter are the active Pacific continental margin coastlines of the US west coast which are dominated by wave-cut platforms and associated perched beaches. These shoreface profiles are unquestionably dictated by the lithology and physical characteristics of eroding headlands. In a similar but less dramatic way, passive margin coastlines with limited sand supplies are also significantly influenced by the geological framework occurring underneath and seaward of the shoreface. For example, many US east coast barrier islands are perched on pre-modern sediments. The stratigraphic section underlying these perched barriers commonly controls the three-dimensional morphology of the shoreface and strongly influences modern beach dynamics, as well as sediment composition and sediment fluxes.

Perched barriers will not develop a profile of equilibrium, as previously defined by Bruun (1962), for several reasons. First, perched barriers consist of thin and variable layers of surficial beach sands on top of older, eroding, stratigraphic units with highly variable compositions and geometries. Depending upon composition, the underlying platforms can act as a submarine headland forcing different responses to shoreface dynamics that will dictate the nature of the shoreface profile. Stratigraphically controlled shorefaces are often composed of compact muds, limestones, or sandstones. Such lithologies exhibit a greater effect upon both the planform of barriers and morphology of the shoreface than those composed of unconsolidated materials. Second, along many parts of the inner shelf, bathymetric features that occur modify incoming energy regimes, affecting the patterns of erosion, transport, and deposition on the adjacent shorelines.

This paper considers the shoreface off one subaerial headland-dominated coastal system (Fort Fisher to Carolina Beach) and one submarine headland-dominated coastal system (Topsail Island to Onslow Beach) in Onslow Bay, NC (Fig. 1).

The shoreface geologic framework defines:

(i) the compositional character of the sediment blanket;

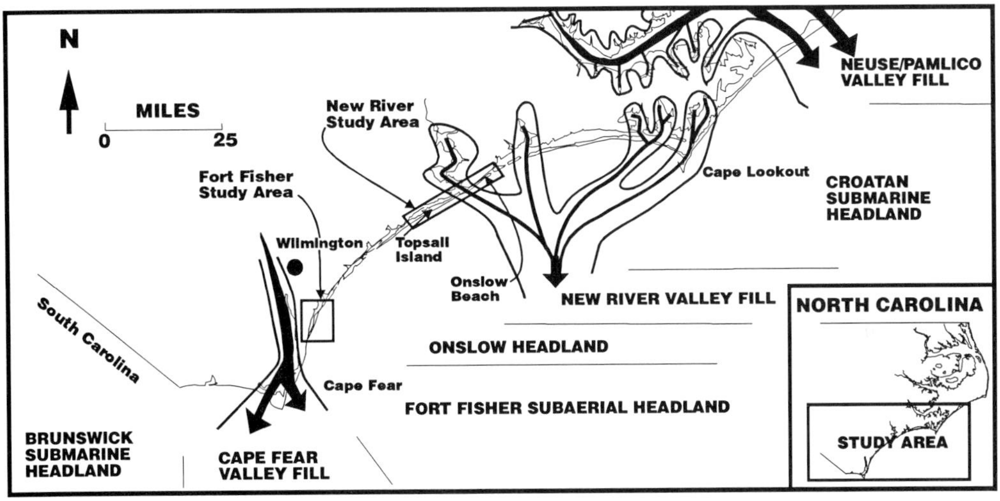

Fig. 1. Map of the southern North Carolina coastal zone showing major palaeofluvial valleys, associated interfluve headland features and the study sites.

(ii) the role of hardbottoms in shaping the shoreface profile and determining the patterns of erosion on adjacent beaches;
(iii) the rates and processes of degradation of hardbottom habitats by natural physical and biological processes that contribute a significant amount of new sediment; and
(iv) the stability and longevity of beach renourishment projects on headland-dominated coastal systems.

Study area and previous work

Onslow Bay (Fig. 1) represents one large coastal compartment located on the SE North Carolina continental margin, between Cape Lookout and Cape Fear and the associated cape-shoal retreat massifs. Onslow Bay is a broad, shallow, high-energy continental shelf system flanked on the landward side by a string of barrier islands, inlet systems, and associated estuaries. The New River study site is situated in the middle of the coastal compartment on a major bathymetric high in the Oligocene limestone that forms a major submarine headland. The study site includes the Topsail Island and Onslow Beach barriers (Fig. 1). Fort Fisher, the second study site, is located at the southwestern extremity of this compartment, where a subaerial headland forms a prominent coastal feature.

Holocene sediment accumulation in Onslow Bay is negligible due to low fluvial input, entrapment of sediments in extensive estuarine systems, and minimal sediment exchange between adjacent shelf embayments (Luternauer & Pilkey 1967; Cleary & Pilkey 1968; Cleary & Thayer 1973; Mixon & Pilkey 1976; Blackwelder *et al.* 1982). Consequently, Onslow Bay is generally a sediment-starved shelf system dominated by hardbottoms (Mearns *et al.* 1988).

Holocene sands are scattered throughout Onslow Bay as a discontinuous veneer. Except for the carbonate component produced by the benthic biota, most Holocene sediment is derived from the erosion and reworking of outcrops of older stratigraphic units (Luternauer & Pilkey 1967; Cleary & Pilkey 1968; Macintyre & Pilkey 1969; Cleary & Thayer 1973; Mixon & Pilkey 1976; Riggs *et al.* 1985). Composition and distribution of Holocene sediments are largely controlled by the outcrop pattern of Tertiary and Quaternary depositional sequences (Fig. 2) (Crowson 1980; Blackwelder *et al.* 1982; Snyder *et al.* 1982; Hine & Snyder 1985).

South of Cape Lookout, the inner shelf is dominated by Tertiary and Cretaceous units that have been highly dissected and backfilled by various types of Pleistocene sediments (Riggs *et al.* 1985). Older and more lithified, offlapping stratigraphic sequences wrap around the Carolina Platform High, a major basement structural feature that occurs between Cape Fear and Cape Romain, and crops out across much of the continental shelf in Onslow and Long Bays (Riggs *et al.* 1990). These Tertiary and Cretaceous stratigraphic units, along with local, remnant Quaternary units, form a basal platform with variable topography upon which some of the modern barriers are 'perched'. Consequently, lithofacies

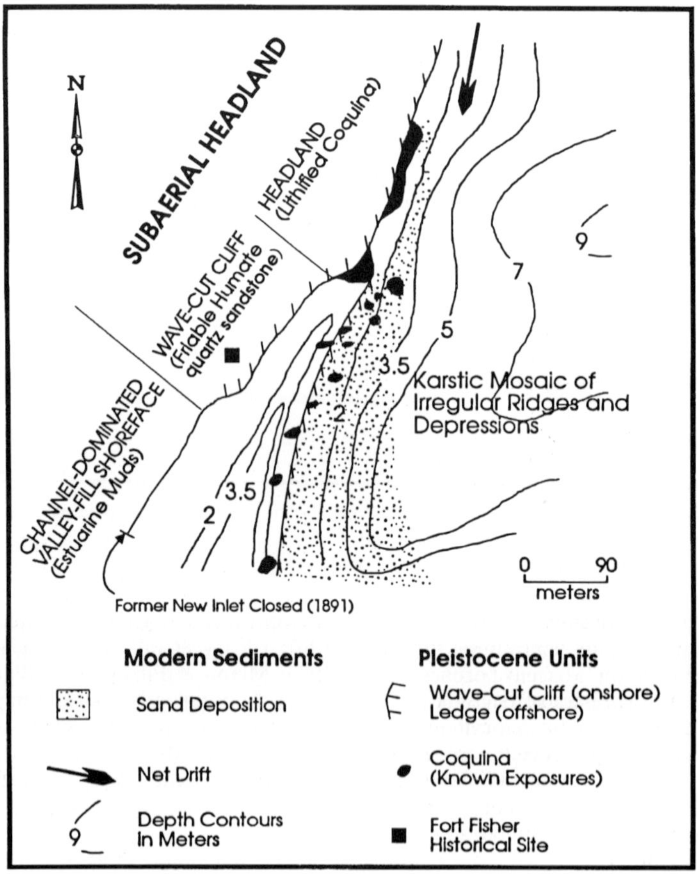

Fig. 2. Map of the subaerial headland in the Fort Fisher area showing the following: three different coastal compartments each with a different underlying geologic framework and resulting shoreline morphology; the nearshore bathymetry (in feet below MSL); location of Pleistocene coquina/sandstone outcrops on the shoreface and southward across the inner shelf; and the surface sediment distribution on the inner shelf (modified from Moorefield 1978).

differences within the palaeotopographic platform result in varying responses to the erosional forces and thus the actual morphology of both the shoreface and the shoreline. The present shape of many of the Carolina barriers is controlled by this underlying geologic framework.

Fisher (1967) described the mid-Atlantic coastal system as a series of coastal compartments. Each compartment consisted of an eroding mainland beach at the northern end with a barrier spit extending southward and grading into a series of barrier islands fronting a major estuarine system. Fisher (in Swift, 1969) interpreted the entire North Carolina barrier island system as a 'southern spit' that formed off a single eroding headland at Cape Henry, Virginia.

However, when the submarine geologic framework is considered, there is actually a series of eroding headlands that occur along the North Carolina coast (Cleary & Hosier 1979; 1986; Pilkey *et al.* 1993). The more important geologic headlands and associated valley-fills for SE North Carolina are summarized in Fig. 1. Headlands represent palaeotopographic highs of Pleistocene or older units with varying composition that generally occur in the subsurface. Palaeotopographic highs composed of sands and soft clays have little obvious effect upon the three-dimensional geometry. However, those composed of compact muds, limestones, or sandstones will have dramatic effects upon both the aerial shape of the barriers and the morphology of the shoreface and inner shelf. These underlying framework units occur in the shallow subsurface and may be either subaerially exposed or submerged below the shallow coastal waters.

Ancient sediment deposits have been vibracored under many shorefaces along the Atlantic coast (Kraft & John 1979; Rampino & Sanders 1981). Marsh peats, tidal flat muds, fluvial sands and gravels, bay-fill sands and muds, flood-tide delta sands, and inlet-fill sands and gravels have been cored below a thin veneer of modern shoreface sands that are generally less than a metre thick. Such coastal areas are characterized by a seaward thinning and fining veneer of modern shoreface sands resting disconformably on Pleistocene or older strata. The modern sand veneer is ephemeral and easily removed from the shoreface during storms, exposing the older underlying strata on the shoreface to erosion (Niedoroda *et al.* 1985; Cleary *et al.* 1992). Thus, the erosional response and post-storm shape of the shoreface profile is at least partially controlled by degree of consolidation of the underlying sediments.

During storms, ancient strata cropping out on the shoreface also provide an immediate source of 'new' sediment to the modern beach system. This process of older units supplying sediment to the shoreface of barrier islands was termed 'shoreface bypassing' by Swift (1976). In North Carolina, the general grain size characteristics of barrier island beach sands provide solid evidence that relict sediments are being eroded from the shoreface (Moorefield 1978; Crowson 1980; Cleary & Hosier 1987). Support for the conclusion that relict and residual sediments are actively being eroded from the shoreface and deposited on the subaerial beach in North Carolina include the following.

1. The extinct fossil oyster *Ostrea gigantisima* and associated Oligocene rock lithoclasts occur in great abundance on Onslow Beach and Topsail Island after storms (Crowson 1980; Cleary & Hosier 1987). The eroded gravels are derived from the bioerosion of Oligocene hardbottom scarps that crop out on the inner shelf. These gravels are subsequently transported up the beachface during high-energy storms and left on the subaerial beach in much the same fashion as heavy minerals are at the top of the swash zone on the storm beach.
2. Overwash terraces on Masonboro Island contain abundant cobble-size coquina clasts and mollusc shells derived from hardbottoms exposed on the adjacent shoreface. Also, much of the coarse-grained component of the beach sediment can be attributed to the onshore transport of reworked palimpsest sediments that mantle these hardbottoms. Storm reworking of the thin shoreface sediment cover and the degraded character of underlying rock units appear to contribute significant amounts of coarse material to the adjacent beaches (Cleary *et al.* 1992).
3. Black-stained oysters and other estuarine fossils are the dominant shell on many North Carolina beaches. These shells always produce pre-modern Holocene ages when dated by carbon-14 techniques (Pilkey *et al.* 1969; Wehmiller *et al.* 1995).
4. Mixed assemblages of Pleistocene age marine shells occur in great abundance on many of the southern Onslow Bay shoreline beaches analysed by amino-acid racemization dating techniques (Wehmiller *et al.* 1995).

Influence of inherited geological framework

The two study areas consist of portions of the shoreface in the southern and central region of Onslow Bay, NC (Fig. 1). The Fort Fisher site is a 40 km^2 area that lies seaward of the perched beach cut into Pleistocene units of the mainland peninsula between Fort Fisher and Carolina Beach. The New River site consists of two 25 km^2 areas off Topsail Island and Onslow Beach, which represent a set of barrier islands that are perched on top of Oligocene Limestone on either side of New River Inlet (Fig. 1). Our analysis of inherited geologic framework is based on data from vibracores, side-scan, seismic, and diver observations obtained during the field seasons between 1992 and 1995.

Fort Fisher subaerial headland

In the Fort Fisher area, an extensive eroding subaerial headland intersects the coastal zone without a barrier island–estuarine system (Fig. 1). The coastal system consists of a wave-cut platform incised into a series of Pleistocene sediment units of the mainland peninsula with a thin beach perched on top of the irregular geometry of the Pleistocene units (DuBar *et al.* 1974; Moorefield 1978; Meisburger 1979; Cleary & Hosier 1986; Snyder *et al.* 1994).

Figures 3A and 3B show the dramatic relationship between three different geologic framework situations in the Fort Fisher area and the geometry of the shoreline and upper shoreface. Erosion-resistant, lithified and cross-bedded coquina sandstone forms a mini-headland in the shoreline north of Fort Fisher. Friable humate and iron-cemented Pleistocene sandstone (Fig. 3A) forms a 2 m high wave-cut cliff and terrace that fronts the shoreline immediately

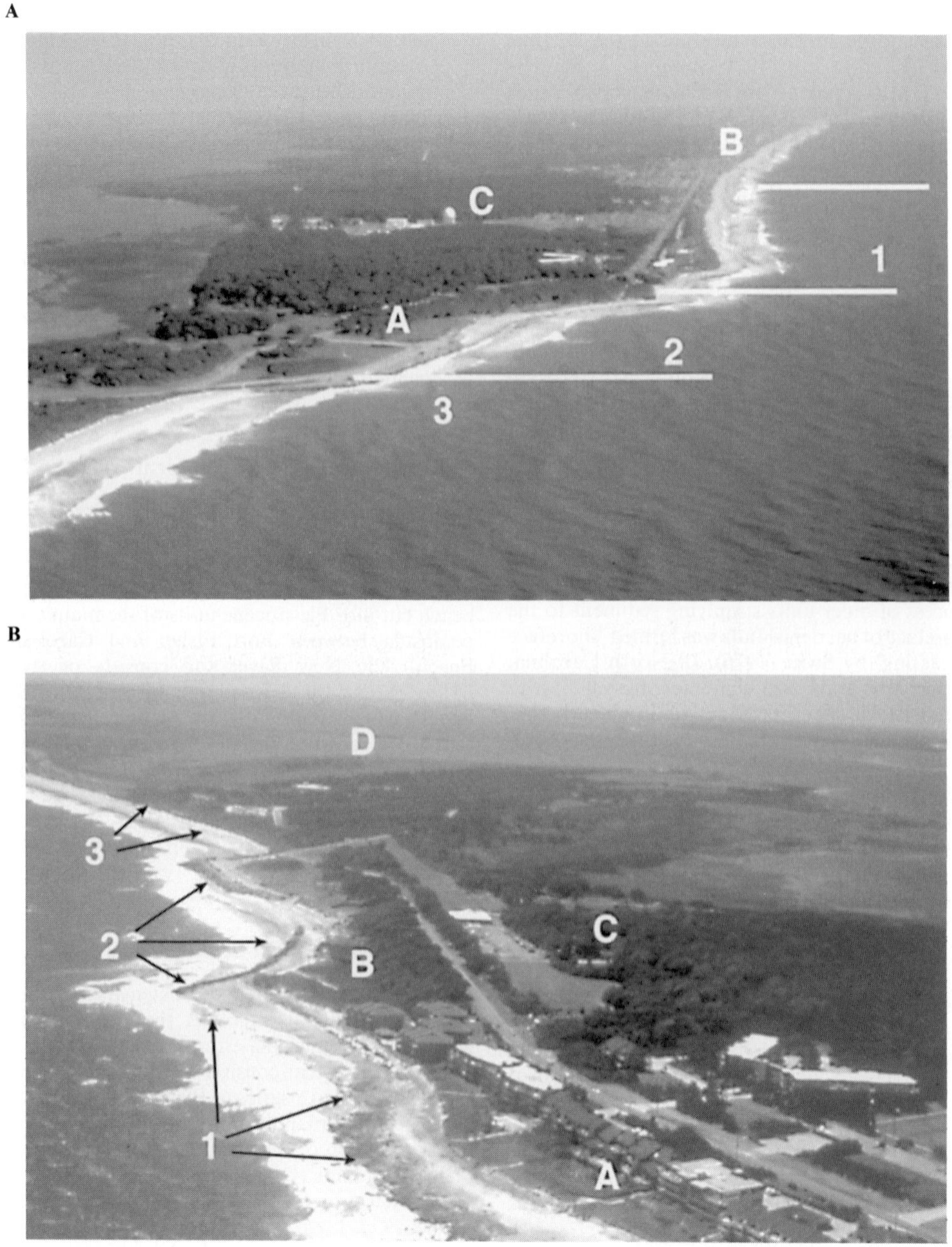

Fig. 3. (**A**) Oblique aerial photograph looking north from Fort Fisher (A) to Kure Beach (B) where the Fort Fisher subaerial headland (C) intersects the Atlantic Ocean. This photo also shows three coastal segments: (1) mini-headland formed by the outcropping Pleistocene coquina sandstone; (2) eroding wave-cut cliff of Pleistocene friable humate quartzose sands; and (3) the rapidly retreating shoreline associated with the channel-dominated valley-fill shoreface. (**B**) Oblique aerial photograph looking southwest from Kure Beach (A) to Fort Fisher (B), across the subaerial headland (C), and to the Cape Fear River estuary (D). This photo also shows the following features: (1) Pleistocene coquina sandstone outcrop in the surf zone; (2) man-made rock revetments to slow the rates of shoreline recession along the wave-cut cliff of Pleistocene friable humate quartzose sands; and (3) rapidly retreating shoreline associated with the channel-dominated valley-fill shoreface.

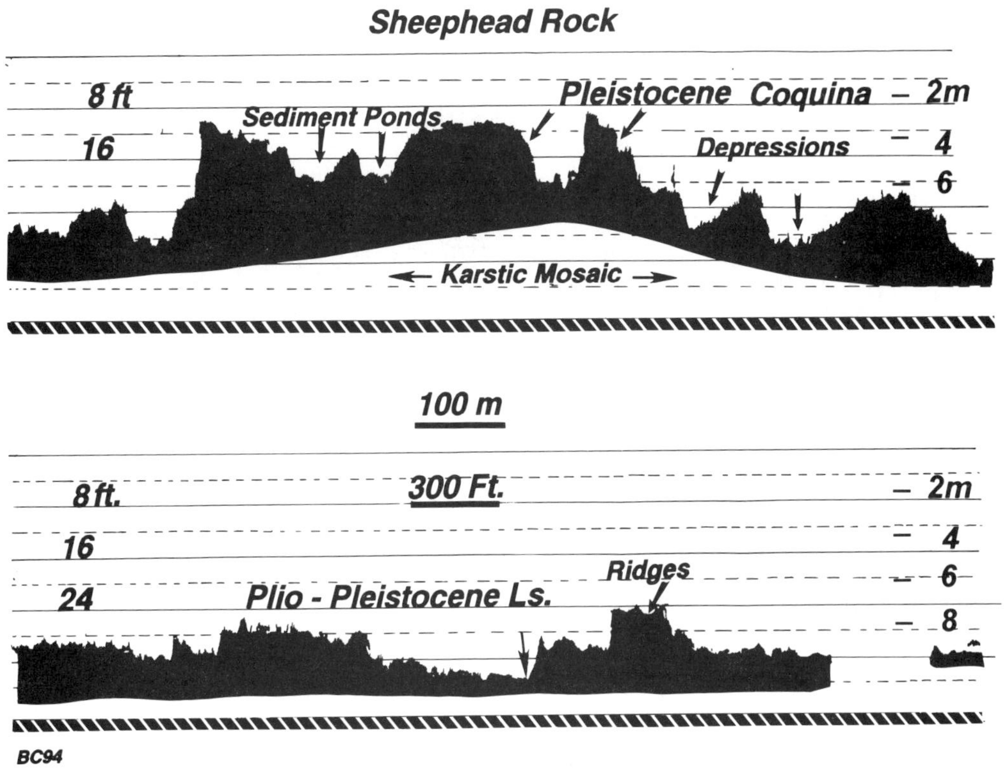

Fig. 4. Sheephead Rock fathometer traces. Representative bathymetric profiles across area of highest relief and on flank. Coquina exposed on sea floor is composed of coarse shell-hash and 10 cm pelecypod valves. Extremely well indurated Pliocene barnacle-rich hardbottom surfaces underlie calcarenite sequence. Depressions at various depths are partially sediment-filled. See Figs 7 and 8 for location.

south of the mini-headland and seaward of Fort Fisher (Figs 3A and 3B). South of Fort Fisher is a coastal segment that is a non-headland-dominated system characterized by an inlet channel-dominated shoreface underlain by 10 m of muddy estuarine sediment. The shape and evolution of the three different coastal segments around Fort Fisher is clearly related to the presence and lithology of the outcropping and underlying Pleistocene geologic framework.

Moorefield (1978) mapped the outcrops of Pleistocene coquina that occur on the beach north of Fort Fisher and the seaward extensions in the nearshore area (Fig. 2). Core data, in conjunction with groundtruthing of numerous side-scan profiles, clearly indicate that the coquina and its associated lithologies form a series of widespread, irregular, bathymetric high hardbottom features with local relief greater than 3 m (Figs 4 and 5D). The karstic mosaic includes one extensive hardbottom area locally known as Sheephead Rock that lies in 9 m of water with pedestal-like features rising to within 2.5 m of the ocean surface (Fig. 4). This hardbottom feature is blanketed with a variety of epifauna, while infauna have extensively bored all exposed surfaces.

Side-scan images reveal that the coquina sequence occurs as a series of variable relief shore-subparallel ridges, some with pronounced overhangs (Figs 5A to 5D). Scarps of 1–2 m relief are often fronted by rubble mounds composed of rotational slump blocks and broken segments of overhangs (Figs 5C and 5D). Portions of the rubble fields are partially covered by large-scale (15 m) bedforms that abut the scarps (Fig. 5B). Commonly the larger-scale bedforms imaged on side-scan and mapped by divers are composed of coarse sand and gravels. Thin megarippled sands extend across the low-relief limestone and coquina hardbottoms and are imaged as a fish-scale pattern. Granules, shell gravel, and coarse sand make up the wave troughs, while orange brown to grey, fine quartz sand forms a 10–20 cm thick veneer over the gravel on the wave crests (Figs 5A and 5B). Deposition of the fine sands presumably

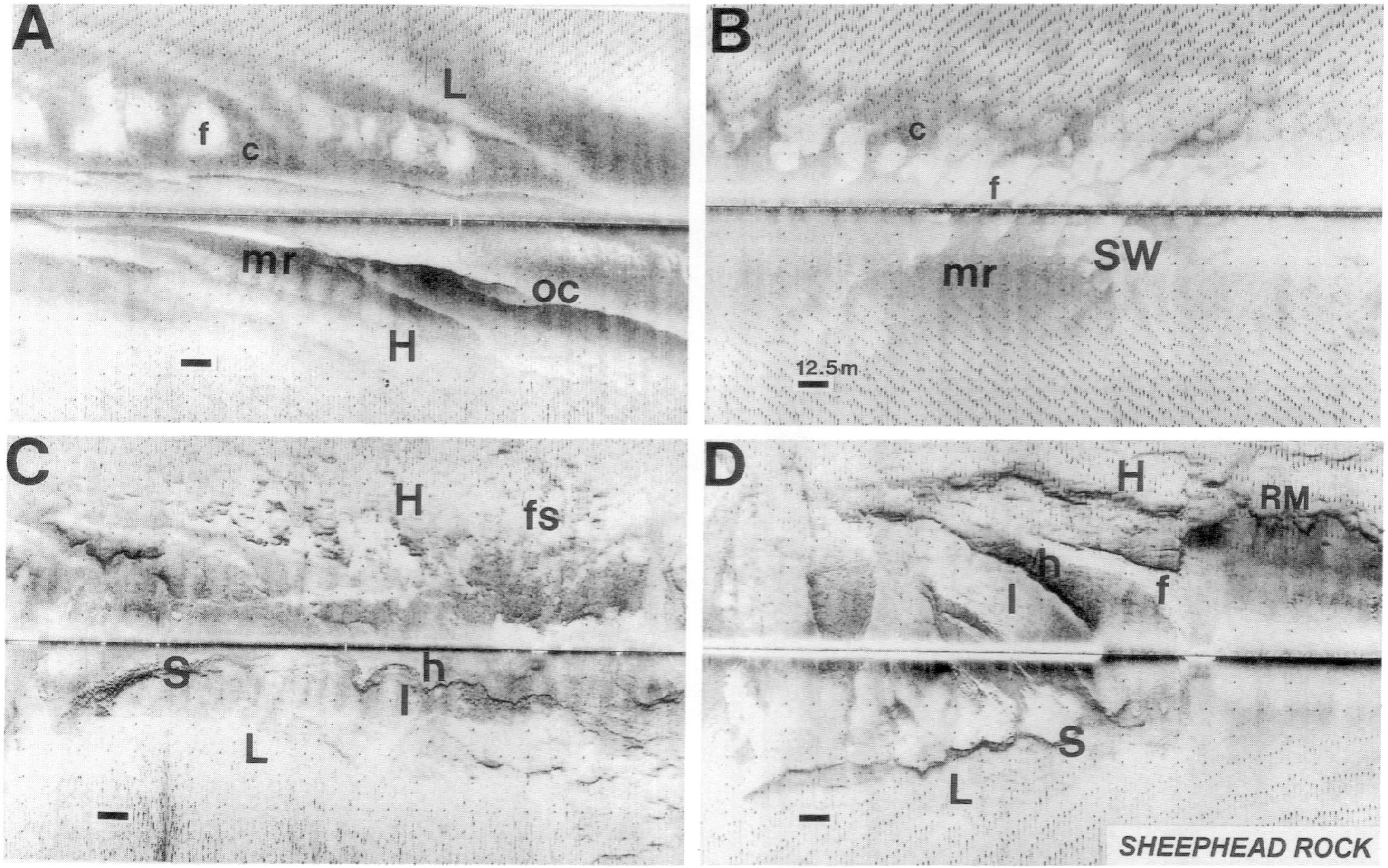

Fig. 5. Sonargraph images of shoreface. Exact relief cannot be determined from images. (**A**) Shore subparallel ridges of coquina, (oc; calcarenite) with sand waves abutting ridges. Light and dark patches represent alternating bottom types of fine sand (white) and coarse sands and gravels (dark). Lowest relief is at the top. (**B**) Large sand waves (>15 m) with megarippled (mr) coarse material on scarps with a small rubble mound area where overhangs have calved off. (**C**) The bottom part of the photo is an area of lower relief (L). A thin cover of fine sand over a flat hardbottom generally produces a fish-scale pattern (fs). Megarippled gravels occupy many of the depressions. (**D**) High relief on the flanks of a larger bathymetric feature known as Sheephead Rock. A series of ridges delineate well-indurated gravel-rich coquina units. Finer sand accumulates in intra-ridge areas. Bottom of image is topographically lower. A rubble mound fronts a scarp or vertical wall. Compare to bathymetric profile (Fig. 4). Scale bar on image represents 12.5 m. H, high; L, low; f, fine; c, coarse; mr, megaripples; RM, rubble mounds; oc, outcrops; SW, sandwaves; S, scarp: fs fish scale (thin veneer of sand on rock).

post-dates bedform development and may represent suspension-stage products of waning storm-driven currents.

Graded units are commonplace over much of this shoreface (Fig. 6). The basal portion of the upper graded unit typically contains whole 10 cm iron-stained pelecypod valves and other shell fragments often set in a fine sand to mud matrix. The shell-rich unit grades upward to a fine to medium iron-stained quartz sand. The composition of this unit, as well as many of the other graded units, suggests the material has been derived from remobilized sediments overlying the coquina sequences. Shells from the basal portions of the graded units are Pleistocene in age and have been amino-acid dated at 100–400 ka (Wehmiller *et al.* 1995). Similar age shells have been found on many of the washover fans that are common on the adjacent barrier beaches.

Core, side-scan data, and field maps compiled by divers indicate the sediment cover is both patchy and very thin across much of this region, and in many areas is totally lacking (Fig. 7). The modern sediment cover over the entire shoreface is highly mobile and composed of intercalated 5–70 cm thick graded fine quartz sand and shell gravel units. This thin veneer represents a mix of remobilized palimpsest and residual sediment. A recent ROV-video survey indicated that the bottom sediments of depressions and the talus at the base of scarps were mobile during moderate swell conditions.

The extensive series of coquina outcrops on the shoreface act as barriers that could significantly affect the refraction of incident waves, as well as affecting the movement of sediment across the shoreface. Reconnaissance mapping by divers seaward of the bluff shoreline shows a series of closed and partially connected irregular depressions that lie subparallel to the shoreline. These features act as sediment traps and ponds, some of which are perched on slightly elevated hardbottoms. Moorefield (1978) believed that sand derived from rapid shoreline erosion at Fort Fisher, as well as longshore sediment transport, is trapped seaward of the most landward outcrop during storm events. Trapped material is prevented from moving back onto the subaerial beach during subsequent low-energy periods (Fig. 2). The result of this process is a net historical sediment deficiency in which the rapidly retreating bluff shoreline is consuming the historic Fort.

The US Army Corp. of Engineers (USACE 1992) currently has several erosion mitigation and storm protection projects scheduled for this portion of the coast. An $8.6 \times 10^6\,m^3$ replenishment project is proposed for nearby Carolina Beach and portions of the study area. A seawall with associated beach-fill is currently under construction directly in front of the rapidly eroding sand bluff and the earthenworks of the nineteenth century fort. The distribution, morphology, and high relief of the shoreface hardbottom complex off Fort Fisher and Kure Beach

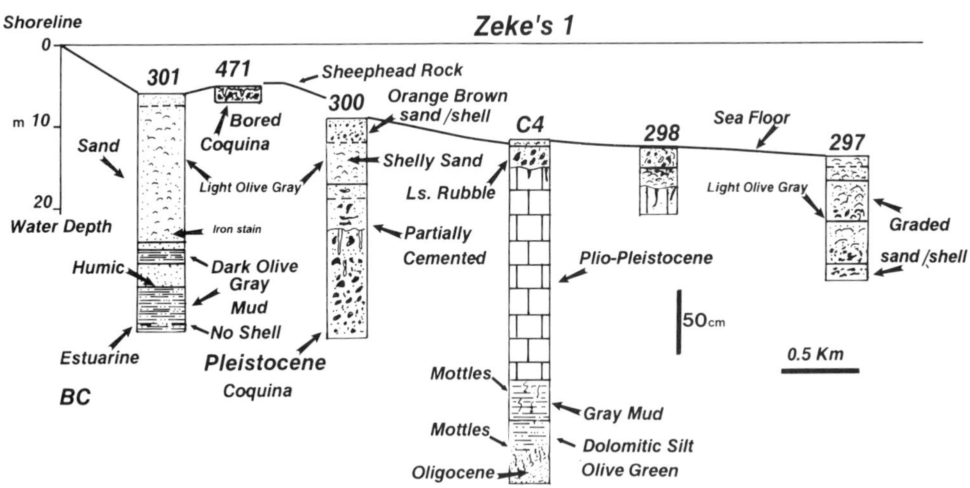

Fig. 6. Shore-normal vibracore transect (ZK). See Fig. 8 for location. A variety of hardbottom types dominate the shoreface. Diver-retrieved core 471 on flank of Sheephead Rock recovered a thin cover of residual sediment resting on biodegraded Pleistocene coquina. Calcarenite sequences are partially cemented and generally form local bathymetric highs. Sediment cover is highly variable in thickness and composition. Graded shell and lithoclast-rich units are characteristics of this area.

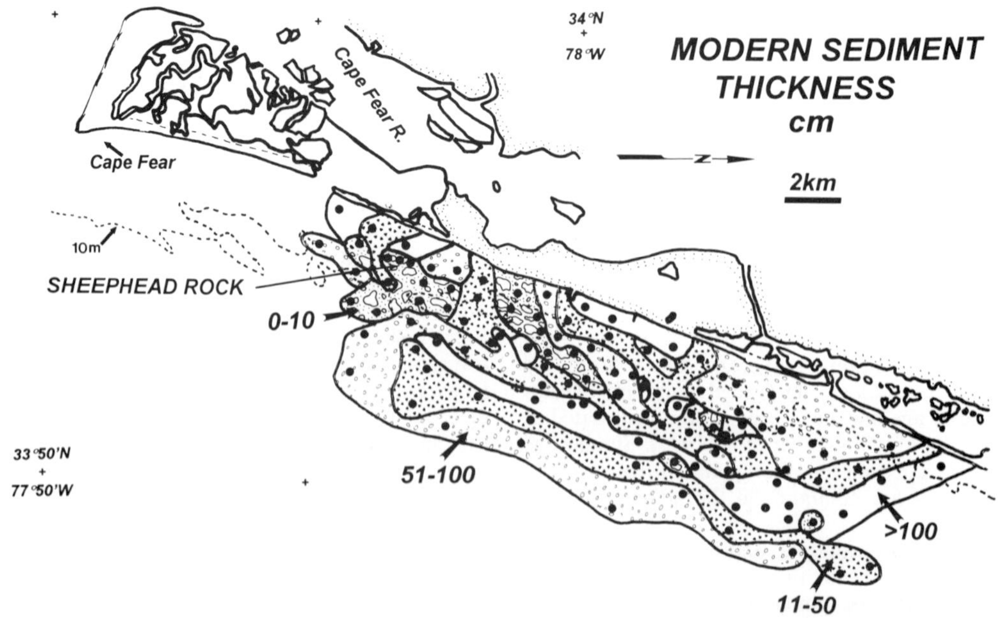

Fig. 7. The modern sediment cover is a complex mosaic and is highly variable in thickness (0 to >100 cm). Sediment cover is often absent on higher-relief calcarenite hardbottoms.

to the north are not only a major part of the ongoing erosion problems in this area, but they will probably have a significant, but as yet unknown impact upon the success and fate of the proposed engineering projects.

A number of concerns have been raised by environmental groups over the proposed dredging of several beach-fill borrow sites that have been targeted by the US Army Corps of Engineers in the environmentally sensitive hardbottom areas (Figs 8 and 9). The renourishment borrow sites are located in a complex of backfilled palaeochannels, 100–400 ka in age, of the ancestral Cape Fear River (Fig. 10). Portions of the channel complex are exposed at various localities across the shoreface. In the target area, the channels are infilled with clean, well-sorted graded quartzose sand and granule units. This fining upward sequence is often capped with a dense peat or woody material. The channel segments contain over 14×10^6 m^3 of quality compatible beach-fill sand in a 3.5 km^2 area. The areas identified are the only viable offshore localities where substantial amounts of compatible fill can be found on a shoreface dominated by hardbottoms.

New River Submarine Headland

The New River Inlet area, the second study site, protrudes seaward forming a small bulge in the coastline of central Onslow Bay (Fig. 1). The headland marks the boundary in Onslow Bay that separates the regressive barrier segment to the northeast from the transgressive barrier portion of the shoreline to the southwest. This shoreline bulge is the manifestation of a submarine headland composed of the Oligocene Silverdale Formation, an indurated unit composed of sandy, pelecypod mouldic limestone and calcareous-cemented quartz sandstone. The Silverdale Formation crops out at or slightly below sea level in the mouth of the New River estuary. It occurs extensively on dredge spoil islands of the Intracoastal Waterway behind Topsail Island and Onslow Beach, and forms a series of bathymetric ridges on the inner shelf on either side of New River Inlet (Crowson 1980). Crowson mapped these prominent submarine rock features as a series of ridges that occur just seaward of the lower shoreface with up to 5 m of relief above the surrounding ravinement surface. The ridges have steep landward facing scarps with smooth rock surfaces that dip gently away from the beach; they locally rise to about 5 m below sea level, which is higher than the elevation of the lower shoreface (Fig. 11).

The ridges are oriented at acute angles to the beach and intersect the shoreface off both Topsail Island and Onslow Beach (Fig. 12), subdividing each of these barriers into major

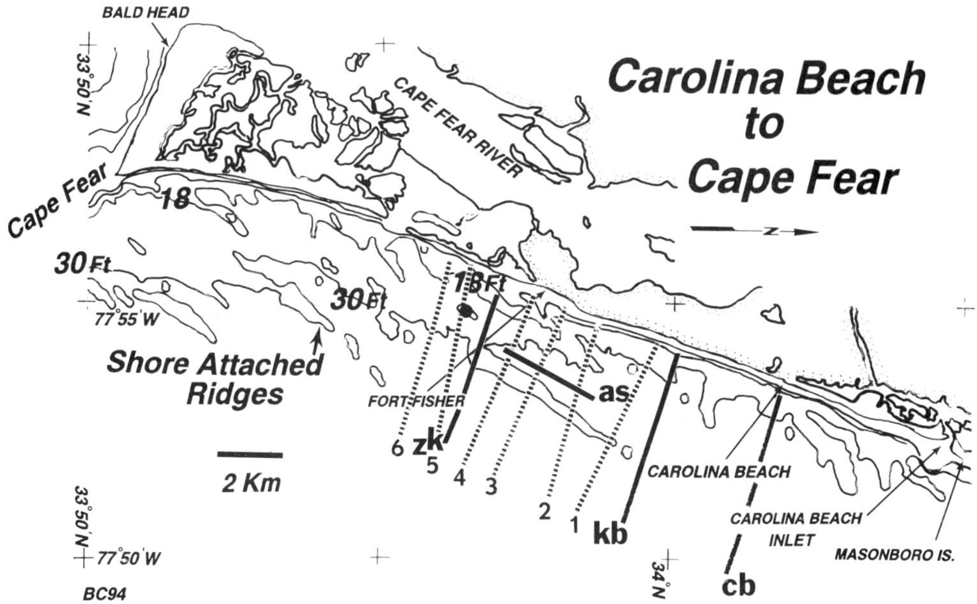

Fig. 8. Map of foreland region showing vibracore transect locations (lines as ZK, KB, CB) and shore-attached ridges. Black dot off Fort Fisher identifies Sheephead Rock. Stabilized shoreline extends from immediately south of Carolina Beach Inlet to Fort Fisher. Modified from NOAA Chart No. 11539 (contours in feet).

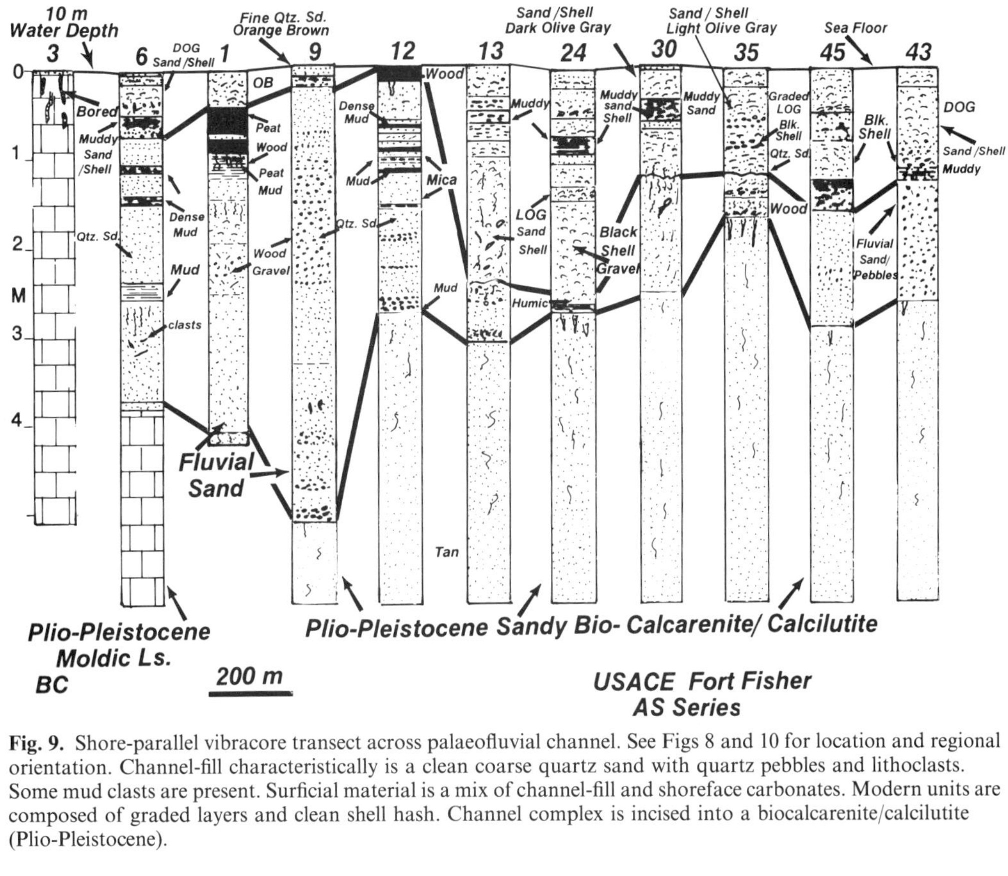

Fig. 9. Shore-parallel vibracore transect across palaeofluvial channel. See Figs 8 and 10 for location and regional orientation. Channel-fill characteristically is a clean coarse quartz sand with quartz pebbles and lithoclasts. Some mud clasts are present. Surficial material is a mix of channel-fill and shoreface carbonates. Modern units are composed of graded layers and clean shell hash. Channel complex is incised into a biocalcarenite/calcilutite (Plio-Pleistocene).

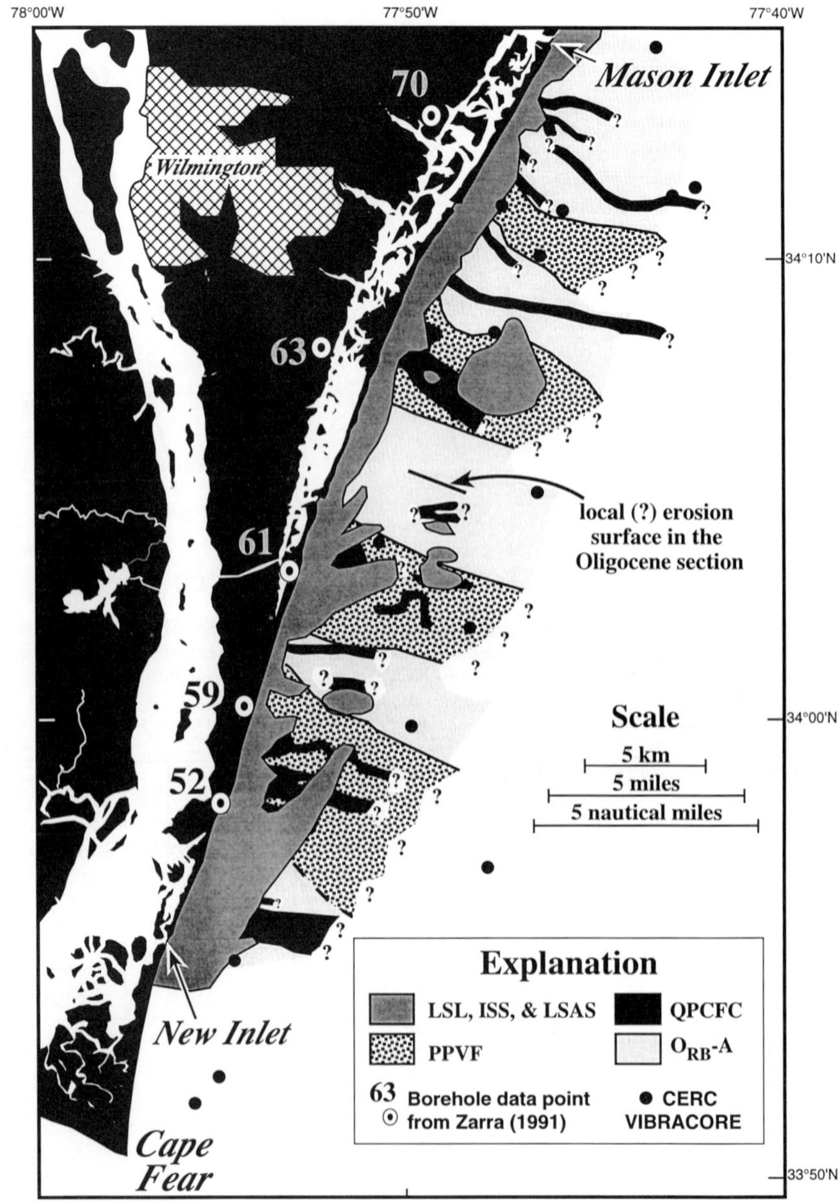

Fig. 10. Geological map of the shoreface (after Snyder *et al.* 1994).

coastal subcompartments that have different orientations and shoreface dynamics. The rock ridges continue under Onslow Beach and into the back-barrier estuarine system (Cleary & Hosier 1987), as demonstrated by the geologic cross-section depicted in Fig. 13 and based upon a series of core holes along the length of Onslow Beach (Fig. 14). Similar limestone ridges pass beneath Topsail Island and into the back-barrier estuarine system where the rock structures appear to be related to the geomorphic occurrence and orientation of Pleistocene bay barriers (Clark *et al.* 1986), such as Permuda Island (Fig. 12).

The presence of the Oligocene submarine headland has resulted in major changes in barrier

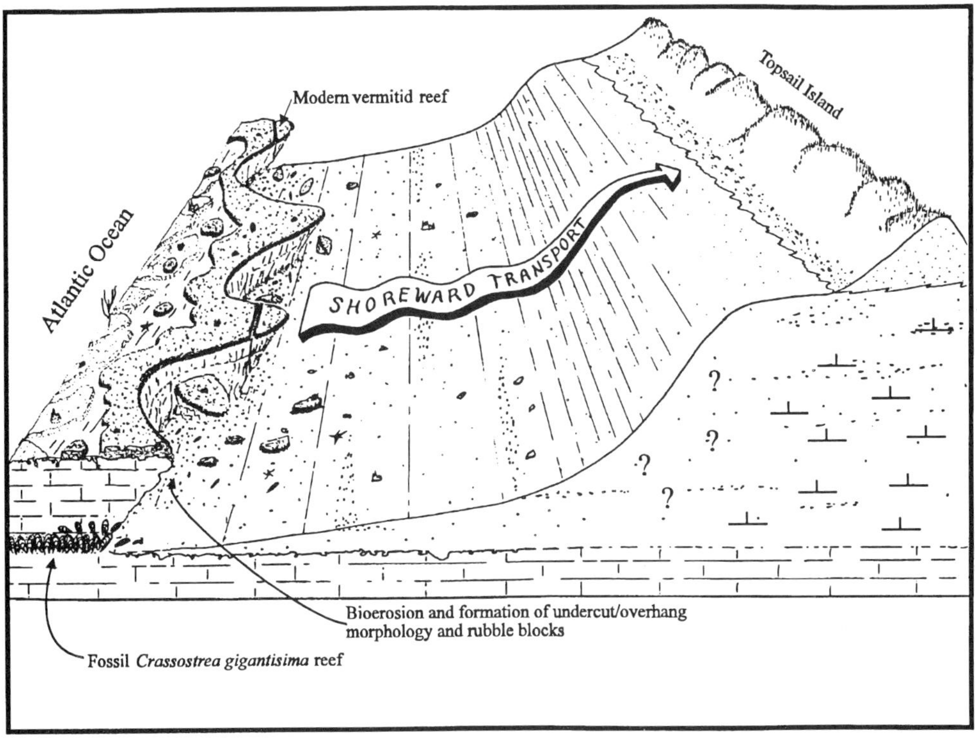

Fig. 11. Schematic geological diagram of the palaeotopographical scarps of the Oligocene Silverdale Formation cropping out on the shoreface off the West Onslow Beach and Topsail Island. Shore-parallel scarps, up to 5 m high, with major overhangs dip gently seaward. Extensive bioerosion and wave-related processes produce 'new' gravel sediment which is transported directly to the beach during storms. The sandy limestone crops out at sea level in the estuary behind the barrier; however, it is not known how the barrier is perched on top of this rock unit. Modified from Crowson (1980).

island orientation and distinct subcompartments with respect to coastal processes and shore-face dynamics (Fig. 14). Aerial photographs of Onslow Beach and Topsail Island (Figs 15A and 15B) show the major change in shoreline geometry that occurs coincidentally with the intersection of the Oligocene rock ridges. For example, Cleary & Hosier (1986) demonstrated that the northern segment of Onslow Beach occurs on the seaward side of the Oligocene ridges. It is characterized by a cuspate shoreline geometry with a wide intertidal beach, a recurved accretionary beach ridge segment, a nearly continuous high primary dune ridge, and a shoreline accretion rate that averages 2 m/year. By contrast, the southern segment occurs on the landward side of the Oligocene ridges and is characterized by a narrow shoreface with abundant rock gravel on the beach, a single discontinuous scarped foredune ridge, presence of major washover terraces, and current erosion rate up to 7 m/year.

In addition, the submarine rock ridges rise high enough in the shallow water column to cause storm-wave refraction and possibly affect the patterns of erosion and deposition on the adjacent beaches. Active bioerosion of the rock scarps represents a major source and supply of 'new sediment' to the adjacent beaches. Abundant gravel, up to boulder-size clasts, is derived from the rock scarps and lower shoreface and delivered to the beach during storms where it is rapidly broken down to sand-sized grains in the surf zone (Crowson 1980; Cleary & Hosier 1987).

Numerous bent core pipes and failures to retrieve sediment during reconnaissance sampling of the Topsail Island shoreface provide clues to the paucity of sediment. Preliminary mapping by divers substantiate the core data. Sediment is lacking over most of this area. The morphology and composition of the ridges dictate the characteristics of the surrounding sediment cover when present. In the intervening

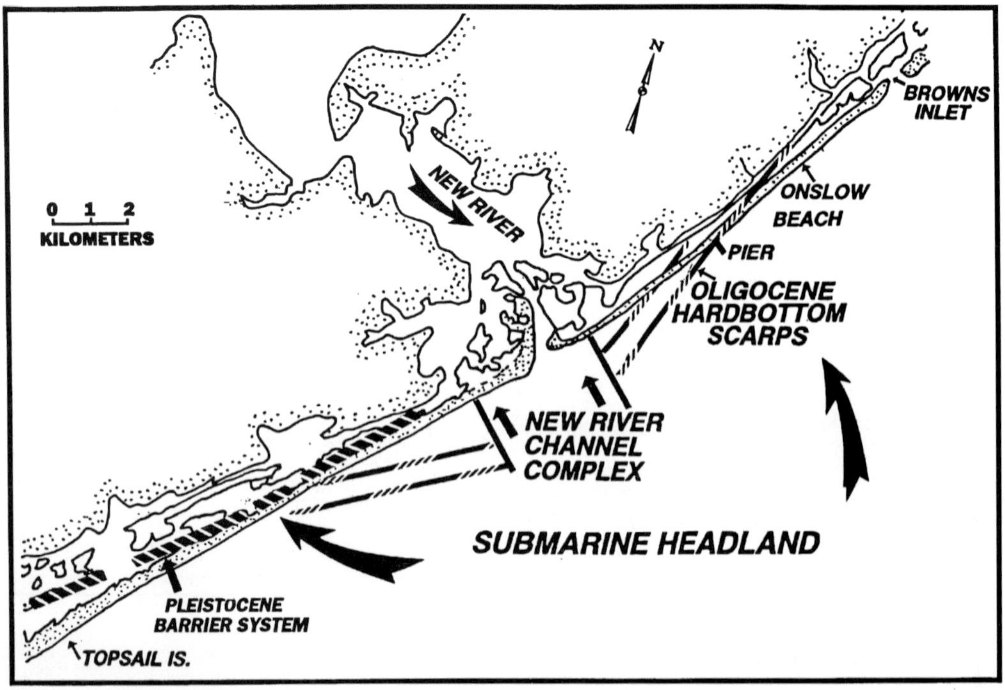

Fig. 12. Map showing the outcrop pattern of the Oligocene rock scarps on the shoreface formed by the Onslow submarine headland. The map shows the lowstand channel cut by the New River and the area of intersection where the topographically high rock features intersect the associated barrier islands. Notice the similarity of orientation patterns of both the estuaries and older Pleistocene beach ridges.

areas, between the ridges, the sediment cover is a mosaic of graded shell-rich sands and gravels. Remobilized residual sediments form ramps that front many of the scarps and overhangs. A fine sand veneer covers the elevated rock surfaces and rubble-filled elongate depressions on the platforms. Although it is conjectural, it is likely that this massive system of hardbottoms plays a key role in cross-shore transport of sediment onto and off the beaches during storm events.

Conclusion

Onslow Bay has a limited sand supply and, as such, the shoreface is not an infinitely thick prism of sand. Rather, it is a thin, variable, and temporal sequence of palimpsest and residual sediment, perched upon a pre-existing and highly dissected geologic framework. Holocene sea-level rise has produced a transgressive sequence of coastal sediments that has been deposited unconformably over irregular remnants of pre-existing sequences consisting of many units of variable ages, origins, and compositions. It is the complex variability in this underlying geologic framework, in conjunction with the dynamics of the coastal system, that ultimately determines the coastal morphology, composition, and texture of shoreface and beach sediments.

The basic structural, stratigraphic, and geomorphic characteristics of the pre-barrier topographic surface interact in a complex way with modern processes to determine beach morphology and shoreface dynamics. Each barrier and headland beach and associated shoreface are by-products of their heritage; the signature of their geologic history controls and influences the present morphology, shoreface dynamics, and rates of shoreline recession. Consequently, the concept of a common equilibrium profile for all shorefaces is neither realistic nor adequate when considering detailed processes along any given coastal segment. Management strategies can only be enacted if we understand the detailed geological framework underlying the shoreface, as well as the system. Only then can we realistically begin to model the decadal behaviour of these systems.

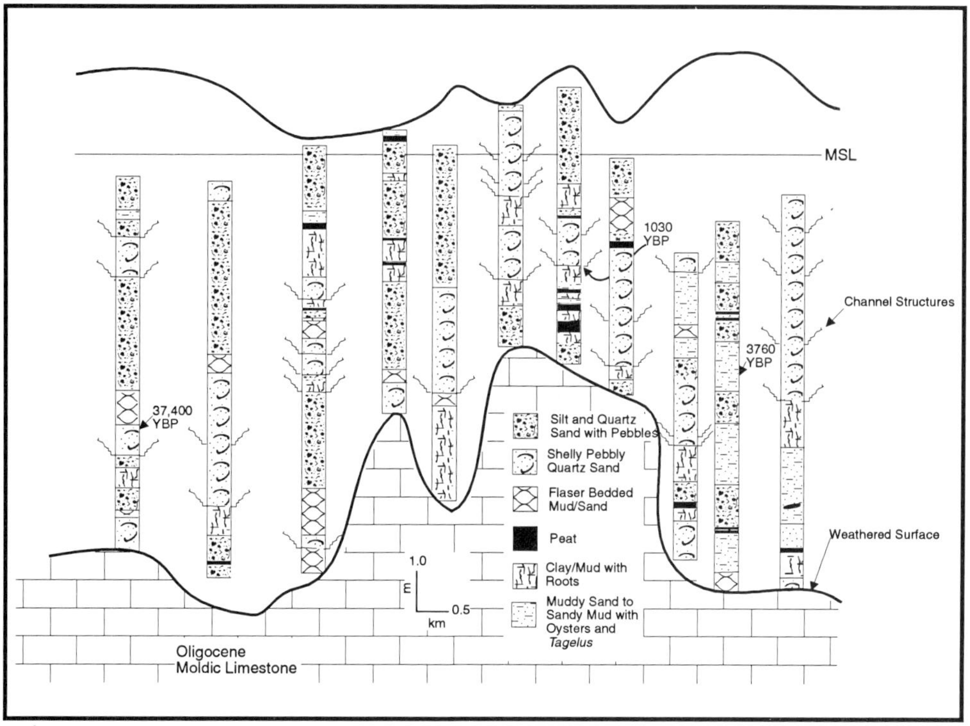

Fig. 13. Geological cross-section along Onslow Beach based upon continuous split-spoon drill holes located on Fig. 14. This section shows the location of two Oligocene rock ridges that rise to approximately 5 m below sea level and pass under Onslow Beach (Cleary, unpublished data).

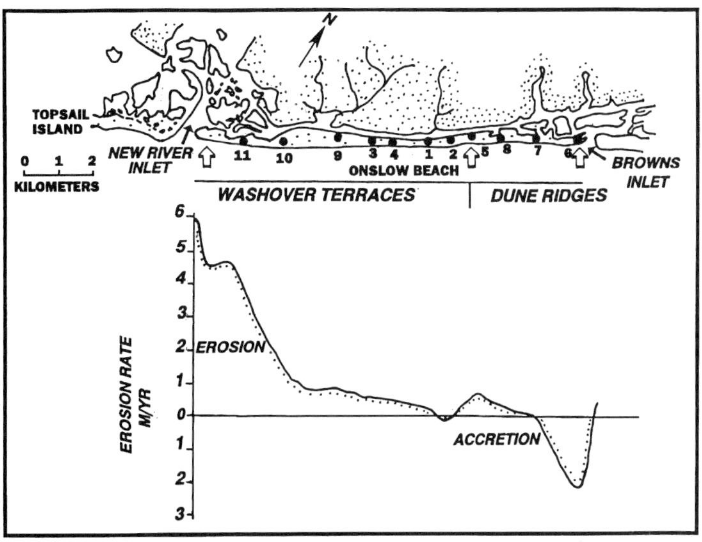

Fig. 14. Map of Onslow Beach showing the location of drill holes used in Fig. 13, location (middle arrow) of coastal inversion that takes place where the Oligocene rock ridges intersect the island, and a plot of the average annual rates of shoreline recession from the NC DEHNR (1993) along Onslow Beach. This map summarizes the effects of the underlying rocks in determining the shoreface dynamics: the southern portion of the island has steep shoreface profiles and is experiencing severe shoreline recession whereas the northern portion is able to develop more normal shoreface profiles and is actually experiencing accretion and dune ridge development. The area of most severe erosion adjacent to New River Inlet is a direct response to inlet modification.

A

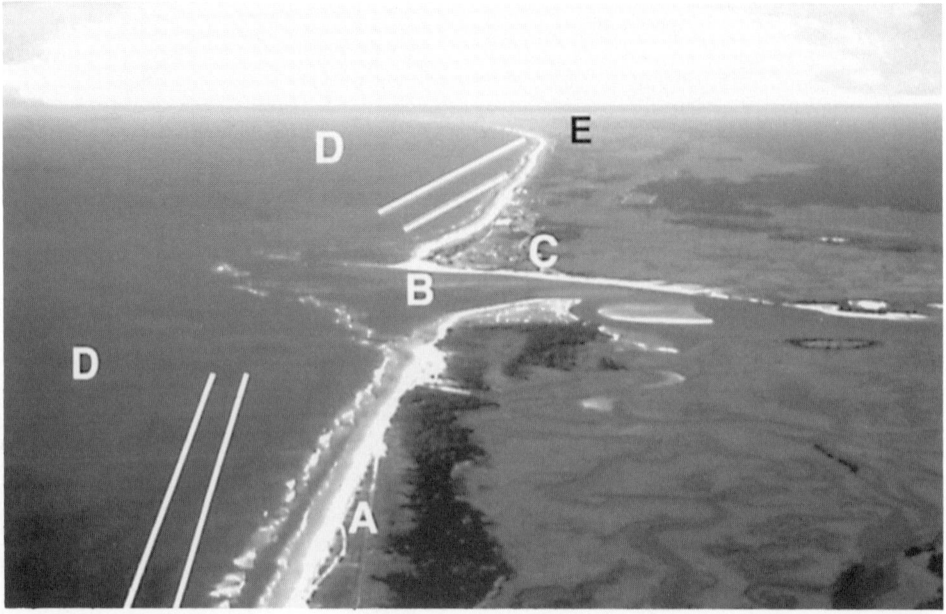

B

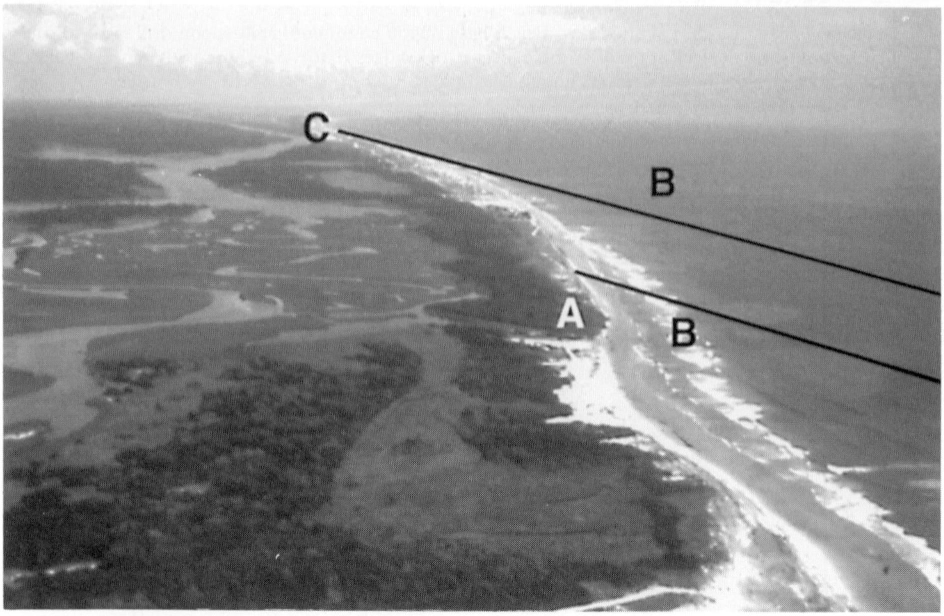

Fig. 15. (**A**) Oblique aerial photograph looking southwest from the southern end of Onslow Beach (A), across New River Inlet (B), and to Topsail Island (C). The approximate location and orientation of the submerged Oligocene rock ridges are indicated on the photo (D); notice the major change in orientation of Topsail Island southwest of the intersection of these ridges with the barrier island (E). (**B**) Oblique aerial photograph looking northeast from New River Inlet to Onslow Beach (A). The approximate location and orientation of the submerged Oligocene rock ridges are indicated on the photo (B); notice the major change in orientation of Onslow Beach northeast of the intersection of these ridges with the barrier island (C).

This paper is in part a product of research supported by the following grants: US Marine Corps at Camp Lejeune, NC (1985–1987) to W.J.C.; NOAA–National Undersea Research Center/University of North Carolina at Wilmington (1992–1994) to W.J.C. and S.R.R. Appreciation is expressed to the many students at the University of North Carolina at Wilmington and East Carolina University who have helped develop the level of understanding of North Carolina coastal systems that led to this article. A special debt of gratitude is extended to Sharon Kissling. This paper represents contribution no. 130 from the University of North Carolina at Wilmington Center for Marine Science Research.

References

BARTH, M. C. & Titus, J. G. (eds) 1984. *Greenhouse Effect and Sea-level Rise: A Challenge for This Generation.* Van Nostrand Reinhold, New York.

BLACKWELDER, B. W., MACINTYRE, I. G. & PILKEY, O. H. 1982. Geology of the continental shelf, Onslow Bay, North Carolina, as revealed by submarine outcrops. *AAPG Bulletin*, **66**, 44–56.

BRUUN, P. 1962. Sea-level rise as a cause of storm erosion. *Proceedings of the American Society of Civil Engineers. Journal of the Waterways and Harbors Division*, **88**(WWI), 117–130.

CLARK, P., CLEARY, W. J. & LAWS, R. A. 1986. A Late Pleistocene bay-barrier system: Topsail Sound, North Carolina. *Geological Society of America, Abstracts with Programs*, **18**(3), 215.

CLEARY, W. J. & HOSIER, P. E. 1979. Coastal geomorphology, washover history, and inlet zonation: Cape Lookout to Bird Island, North Carolina. *In*: LEATHERMAN, S. D. (ed.) *Barrier Islands from the Gulf of St. Lawrence to the Gulf of Mexico.* Academic, New York, 237–262.

—— & ——1986. *Barrier Island Physiography and Lagoon Infilling: Cape Fear Foreland.* Society of Economic Paleontologists and Mineralogists Annual Midyear Meeting, **3**, 23.

—— & ——1987. Onslow Beach, North Carolina: Morphology Stratigraphy, Proceedings, Coastal Sediments 1987. *American Society Civil Engineers*, New Orleans, 1745–1759.

—— & PILKEY, O. H. 1968. Sedimentation in Onslow Bay. *In*: *Guidebook for Field Excursion.* Geological Society of America, Southeastern Section, Durham, NC. Southeastern Geology, Special Publication, **1**, 1–17.

—— & THAYER, P. A. 1973. Petrography of Carbonate Sands on the Carolina Continental Shelf. *Gulf Coast Association of Geological Societies, Transactions*, **23**, 288–304.

——, THEILER, E. R. & RIGGS, S. R. 1992. A reconnaissance survey of shoreface sedimentation off a replenished barrier, Wrightsville Beach, N.C. IGCP 274 International Symposium on *Diversity in Coastal Evolution in the Quaternary*, Wellington. Geological Society of New Zealand, Miscellaneous Publication, **65A**, 13.

CROWSON, R. A. 1980. *Nearshore Rock Exposures and their Relationship to Modern Shelf Sedimentation, Onslow Bay, North Carolina.* MSc Thesis, East Carolina University, Greenville, North Carolina.

DUBAR, J. R., JOHNSON, H. S., THOM, B. G. & HATCHELL, W. O. 1974. Neogene stratigraphy and morphology, south flank of the Cape Fear Arch, North and South Carolina. *In*: OAKS, R. Q. & DUBAR, J. R. (eds) *Post Miocene Stratigraphy, Central and Southern Atlantic Coastal Plain.* Utah State University Press, Logan, Utah, 139–173.

FISHER, J. J. 1967. *Development Patterns of Relict Beach Ridges, Outer Banks Barrier Chain.* PhD Dissertation., University of North Carolina, Chapel Hill, North Carolina.

HINE, A. C. & SNYDER, S. W. 1985. Coastal lithosome preservation: evidence from the shoreface and inner continental shelf off Bogue Banks, North Carolina. *Marine Geology*, **63**, 307–330.

KOMAR, P. D. (Chairman) *et al.* 1991. The response of beaches to sea-level changes: A review of predictive models. *Journal of Coastal Research*, **7**(3), 895–921.

KRAFT, J. C. & JOHN, C. J. 1979. Lateral and vertical facies relations of transgressive barriers. *AAPG Bulletin*, **63**, 2145–2163.

LUTERNAUER, J. L. & PILKEY, O. H. 1967. Phosphorite grains: Their application to the interpretation of North Carolina shelf sedimentation. *Marine Geology*, **5**, 315–320.

MACINTYRE, I. G. & PILKEY, O. H. 1969. Tropical reef corals: Tolerance of low temperature on the North Carolina continental shelf. *Science*, **166**, 374–375.

MEARNS, D. L., HINE, A. C. & RIGGS, S. R. 1988. Comparison of sonographs taken before and after Hurricane Diana; Onslow Bay, North Carolina. *Geology*, **16**, 267–270.

MEISBURGER, E. P. 1979. *Reconnaissance Geology of the Inner Continental Shelf, Cape Fear Region, North Carolina.* US Army Corps of Engineers, Coastal Engineering Research Center, Technical Report, **TP79–3**.

MIXON, R. B. & PILKEY, O. H. 1976. *Reconnaissance Geology of the Submerged and Emerged Coastal Plain Province, Cape Lookout Area, North Carolina.* US Geological Survey, Professional Paper, **859**.

MOOREFIELD, T. P. 1978. *Geological Processes and History of the Fort Fisher Coastal Area, North Carolina.* MSc Thesis, East Carolina University, Greenville, North Carolina.

NC DEHNR. 1993. *Long-term Average Annual Erosion Rates for the Coast of North Carolina.* North Carolina Department of Environmental Health and Natural Resources, Division of Coastal Management.

NIEDORODO, A. W., SWIFT, D. J. P., FIGUEIREDO, A. G. & FREELAND, G. L. 1985. Barrier island evolution, middle Atlantic shelf, USA. Part II: evidence from the shelf floor. *Marine Geology*, **63**, 363–396.

PILKEY, O. H. & DAVIS, T. W. 1987. An analysis of coastal recession models: North Carolina coast. *In*: NUMMEDAL, D., PILKEY, O. H. & HOWARD, J. D. (eds) *Sea-level Fluctuation and Coastal Evolution*. Society of Economic Paleontologists and Mineralogists, Special Publication, **41**, 59–68.

——, BLACKWELDER, B. W., DOYLE, L. J., ESTES, E. & TERLECKY, T. M. 1969. Aspects of carbonate sedimentation on the Atlantic Continental Shelf off the southeastern United States. *Journal of Sedimentary Petrology*, **39**, 744–768.

——, YOUNG, R. S., RIGGS, S. R., SMITH, A. W. S., WU, H. & PILKEY, W. D. 1993. The concept of shoreface profile of equilibrium: A critical review. *Journal of Coastal Research*, **9**, 255–278.

RAMPINO, M. R. & SANDERS, J. E. 1981. Evolution of the barrier islands of southern Long Island, New York. *Sedimentology*, **28**, 37–48.

RIGGS, S. R., SNYDER, S. W., HINE, A. C., SNYDER, SC. W., ELLINGTON, M. D. & MALLETTE, P. M. 1985. Geologic framework of phosphorite resources in Onslow Bay, North Carolina continental shelf. *Economic Geology*, **80**, 716–738.

——, ——, SNYDER, SC. W. & HINE, A. C. 1990. Stratigraphic framework for cyclical deposition of Miocene sediments in the Carolina Phosphogenic Province. *In*: BURNETT, W. C. & RIGGS, S. R. (eds) *Neogene to Modern Phosphorites*. Phosphate Deposits of the World, **3**(29), Cambridge University Press, Cambridge, 381–395.

SNYDER, S. W., HINE, A. C. & RIGGS, S. R. 1982. Miocene seismic stratigraphy, structural framework, and sea-level cyclicity, North Carolina continental shelf. *Southeastern Geology*, **23**(4), 247–266.

——, HOFFMAN, C. W. & RIGGS, S. R. 1994. Seismic stratigraphic framework of the inner continental shelf: Mason Inlet to New Inlet, North Carolina. *North Carolina Geological Survey, Bulletin*, **97**.

SWIFT, D. J. P. 1969. Inner shelf sedimentation: processes and products. *In*: STANLEY, D. J. (ed.) *The New Concepts of Continental Margin Sedimentation*. American Geological Institute, Washington DC, 4–1 to 4–46.

——1976. Continental shelf sedimentation. *In*: STANLEY, D. J. & SWIFT, D. J. P. (eds) *Marine Sediment Transport and Environmental Managemen*. Wiley, New York, 311–350.

USACE. 1992. *Beach Erosion Control and Hurricane Wave Protection, Carolina Beach and Vicinity Area South Project, New Hanover County, North Carolina, Draft Environmental Impact Statement*. US Army Corps of Engineers, Wilmington District.

WEHMILLER, J. F., YORK, L. L. & BART, M. L. 1995. Amino acid racemization geochronology of reworked Quaternary mollusks on U.S. Atlantic coast beaches: implications for chronostratigraphy, taphonomy, and coastal sediment transport. *Marine Geology*, **124**, 303–337.

Nearshore sediment transport processes due to moderate hydrodynamic conditions

V. EITNER[1], R. KAISER & H. D. NIEMEYER

Niedersächsisches Landesamt für Ökologie – Forschungsstelle Küste, An der Mühle 5, D-26548 Norderney, Germany

[1]*Present address: DIN Deutsches Institut für Normung e.V., 10772 Berlin, Germany*

Abstract: On the beach of the island of Norderney sedimentological, morphological and hydrodynamic measurements have been carried out in order to investigate sediment transport processes in the course of a tidal cycle which was characterized by moderate hydrodynamic conditions. Wave heights varied between 0.3 and 0.5 m. An area of 90 m^2 within a groyne field was surveyed. Sediments samples were taken on a 5 m grid at low tide. Sediment flux was determined at each sampling site by means of depth-of-activity rods. The grain size distributions of the sediment samples were determined by dry sieving. Under the prevailing hydrodynamic conditions a landward sediment transport was observed. The mainly negative sediment budget in the seaward section and the landward directed distribution of fluorescent sediment tracers suggest a dominant onshore transport during the investigated tidal cycle. This is also reflected in the changes observed in sediment textures.

Wind, tidal and wave-induced currents act jointly in the nearshore zone, thereby shaping the morphology and distribution of sediment textures to produce a flexible dynamic equilibrium. Transport processes in the nearshore area of the island of Norderney (southern North Sea) have been described in a general way from both a sedimentological and a hydrodynamic point of view (Eitner *et al.* 1992; Eitner & Ragutzki 1994; Niemeyer 1991, 1992). On 19 October 1993 an interdisciplinary investigation dealing with sedimentological and morphological responses to specific hydrodynamic conditions was carried out in order to intensify knowledge about sediment transport processes in the nearshore zone. Sediment sampling, determination of sediment flux and beach levellings were carried out at low tide in the morning (M) and in the evening (A). Hydrodynamic parameters were measured continuously over the whole tidal cycle.

Study area

The study area is located in the groyne field D1–E1 on the northwestern beach of the East Frisian island of Norderney (Figs 1–3). The West and East Frisian Islands form a chain of siliciclastic barrier islands along the southern North Sea coast of The Netherlands and Germany. Norderney has a length of some 14 km and a mean width of 2 km. The island is separated from the adjacent islands of Juist (downdrift) and Baltrum (updrift) by the tidal inlets Norderneyer Seegat and Wichter Ee. The city of Norderney, situated at the western island head, is protected by solid coastal protection structures (groynes, seawalls and revetments). The wave climate is characterized by an annual offshore mean significant wave height of 0.7–1.0 m, the waves mainly approaching from western to northeastern directions (Niemeyer 1992). The mean tidal range is about 2.4 m with a variation of ±0.7 m due to neap and spring tides. Following Hayes (1979) this area is classified as a mixed energy, tide-dominated coast.

Methods

Hydrodynamic field measurements, data processing and analysis

Tidal water levels have been recorded continuously since the end of the last century by a gauge located at the western edge of the tidal inlet Norderneyer Seegat. On the basis of these data it was possible to relate the measurements made in the course of this study with long-term water level statistics of tidal peaks.

The measurements of waves and currents were carried out on the beach itself by pressure transducers and electromagnetic current meters (Fig. 4). Water level fluctuations were recorded at a sampling frequency of 5.9 Hz and current measurements at 11.8 Hz. The higher current measurement rate was chosen in order to allow investigations on the turbulence structure of currents due to breaking waves. Due to low-pass filter effects, higher sampling frequencies of pressure transducers will deliver insufficient benefits. Wave analysis was carried out both in the frequency and in the time domain. The bases for interpretation were characteristic wave heights and periods,

From De Batist, M. & Jacobs, P. (eds), 1996, *Geology of Siliciclastic Shelf Seas*, Geological Society Special Publication No. 117, pp. 267–288.

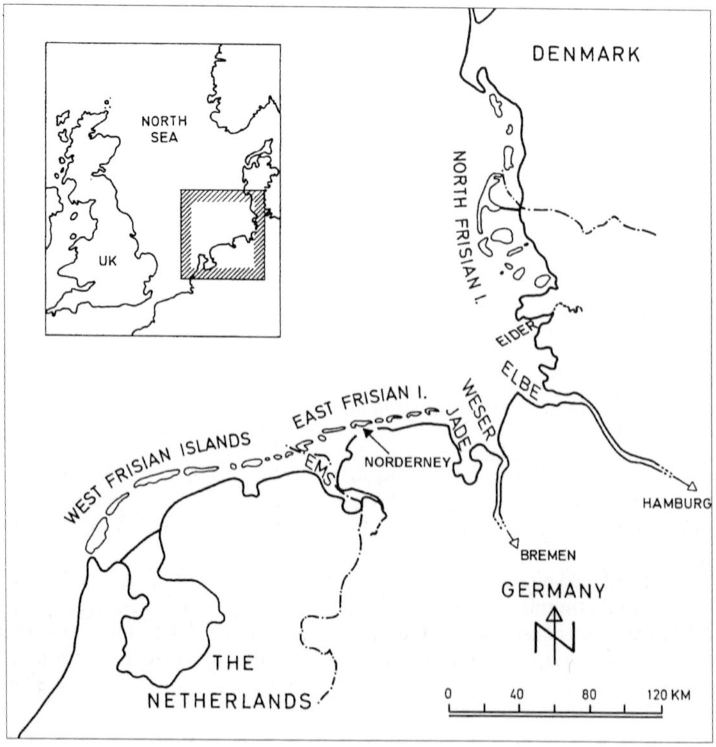

Fig. 1. Location map of the southern North Sea.

power spectra and – if providing additional benefits – spectral parameters. The interpretation of wave-induced currents on sediment transport was done with the aid of current roses following Niemeyer (1991, 1992).

Sediment sampling and analysis

Sediment samples were taken over the depth interval corresponding to the depositional height occurring within the tidal cycle (Fig. 5). The

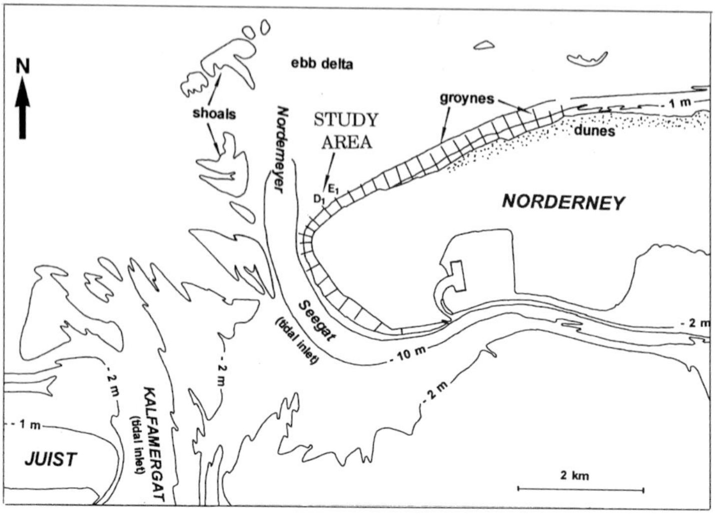

Fig. 2. Location map of the western spit of the island of Norderney.

Fig. 3. Aerial photograph of the study area (groyne field D1–E1).

determination of the depositional height was carried out by two different methods explained in the section dealing with sediment fluxes. Grain size distributions were determined by dry sieving using 1/4-phi sieve intervals. Carbonate contents were determined by the Scheibler method (Schultze & Muhs 1967), in this case reflecting the shell content of the sediment.

Heavy minerals were separated by a Frantz Magnetic Separator which separates heavy minerals from the sediment by an adjustable magnetic field (McAndrew 1957). The power of the magnetic field was increased in three steps (0.4 A, 0.8 A and 1.2 A). Following Veenstra & Winkelmolen (1976), the 0.106–0.125 mm sieve fractions were examined.

Tracer experiments

Tracers have been extensively used for the determination of sediment transport pathways. Fluorescent tracers are very effective for this purpose and have been used in a number of experiments on the beach of Norderney (e.g. Eitner *et al.* 1992; Eitner 1993). Natural sand was taken from the northwestern beach of Norderney and marked with water glass as binder and fluorescent colours following the method of Ruck (1967, 1972). Approximately 20 kg of fluorescent sand was spread over an area of 0.25 m^2 on the beach at mean sea level in a 5 cm thick layer.

Subsequent sampling was carried out over the depth corresponding to local depositional rates, because tracers will only occur in those sands which were transported during the tidal cycle. Local depositional rates were determined by means of 'depth-of-activity' rods or coloured sand columns, a method explained in the next section.

Counting of fluorescent sand particles was done visually using a chute installed on a bumping table which generates a grain layer (Eitner 1993). The fluorescent grains become visible when exposed to UV light. A constant bumping intensity establishes uniform counting conditions. Standardization guarantees the comparability of the counting procedure, e.g. by defining a constant sampling quantity (100 g).

Determination of sediment flux

Hydrodynamic forces lead to an intensive sediment flux in the nearshore. This was determined by so-called depth-of-activity rods at the sample

Fig. 4. Hydrographic measuring station with electromagnetic current meter and pressure gauge.

sites in the centre of the groyne field (Fig. 6) (Greenwood *et al.* 1980). In this experiment, a round steel rod (0.3 cm diameter, 30 cm long) was sunk 20 cm into the beach (Eitner 1993). A loose-fitting washer placed over the rod provided the control for determining scour or aggradation. Rod measurements were made at low tide and therefore do not reflect the erosional and depositional processes continuously over a tidal cycle. It is thus only possible to infer erosion and/or deposition between measurements from these data. The resulting net sediment flux can be calculated by the erosional and depositional rate. The depth-of-activity rods themselves cause small scours amounting to 10 mm at most as shown by control measurements with a column of coloured sand (Eitner *et al.* 1992; Eitner 1993).

Results and discussion

Hydrodynamical conditions

The consecutive tidal low, high and low water were at levels of 335 cm (M), 612 cm and 354 cm (A) above German gauge datum which is approximately 500 cm below mean sea level. The high-tide peak was nearly equal to the 10-year average of 616 cm (1984/93). In contrast, the low-water peaks were significantly lower

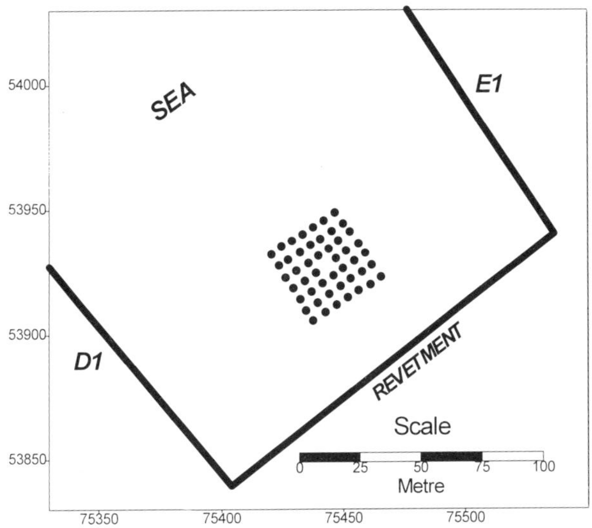

Fig. 5. Sample sites within the groyne field D1–E1 (co-ordinates from UTM reference grid).

than the corresponding mid-term average: 41 cm (M) and 22 cm (A). The tidal rise was thus 37 cm larger and the tidal fall 18 cm larger than the 10-year average tidal range. Tidal currents in the study area are relatively weak. Over 21 consecutive tides the mean ebb and flood velocities were less than 5 cm/s, ranging between values of 0.3 cm/s and 11.2 cm/s (Niemeyer 1987).

At the time of the investigation the wave climate was moderate. At high tide a maximum wave height of 62 cm and a significant one of 43 cm were measured at station M1 (Niemeyer *et al.* 1994). Time series of water levels and wave heights at mean sea level highlight the fact that there was no substantial increase in wave height during the tidal rise from about two hours before high tide, as would have been the case for stronger wave action (Fig. 7) (Niemeyer *et al.* 1994). Measurements of near-bottom velocities by means of hot film anemometers (Fig. 8) (Müller

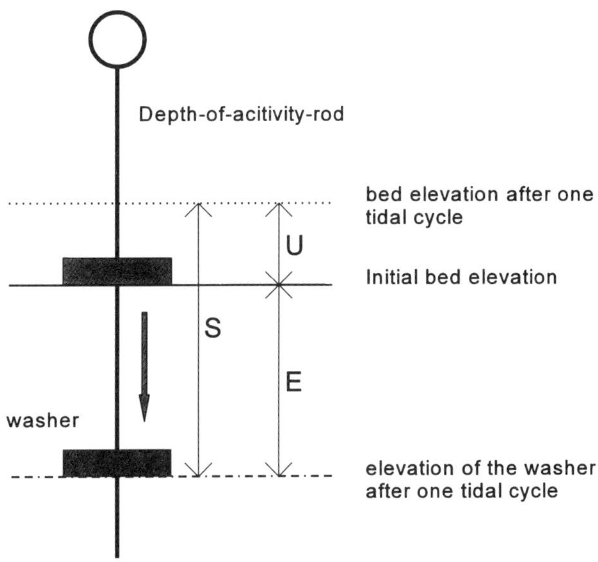

Fig. 6. Determination of the sediment flux by depth-of-activity rods (E, Erosion; S, deposition; U, net sediment flux).

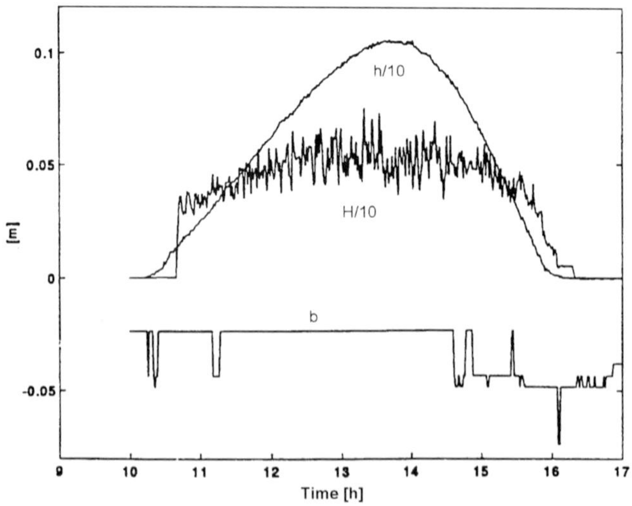

Fig. 7. Time series of water level and wave height at station M2, 19 October 1993 (h, water level; H, significant wave height; b, bottom level with reference 0 at the beginning of the measurement) (from Müller 1993).

1993) also show no significant correlation with tidal water levels and accompanying wave action. More remarkable are higher velocities for rising and falling tide during the transition phases between dry and wet beach states which can be explained as typical swash conditions (Yu *et al.* 1991). Unfortunately these measuring devices do not deliver directional information.

During rising tide the currents between the shoreface and the intertidal beach (M1, O1) are well balanced with no remarkable overshoot in any direction (Fig. 9A). Close to mean sea level on the intertidal beach the current roses reflect typical swash conditions (Yu *et al.* 1991), with high velocities in the offshore direction. At high tide again the current roses between shoreface and intertidal beach are again fairly well balanced, whereas those on the upper part of the intertidal beach are characterized by a more or less significant dominance of offshore directed components (Fig. 9B). It is possible that the current meters did not detect a very small near-bottom

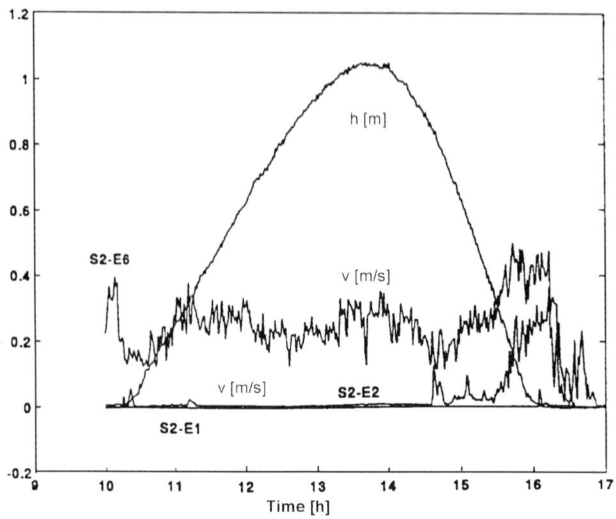

Fig. 8. Mean velocities over the tidal cycle at different levels, 19 October 1993 (v, velocity; S2-E6, S2-E2, S2-E1, recording device no. 2 with probes E6, E2, E1) (from Müller 1993).

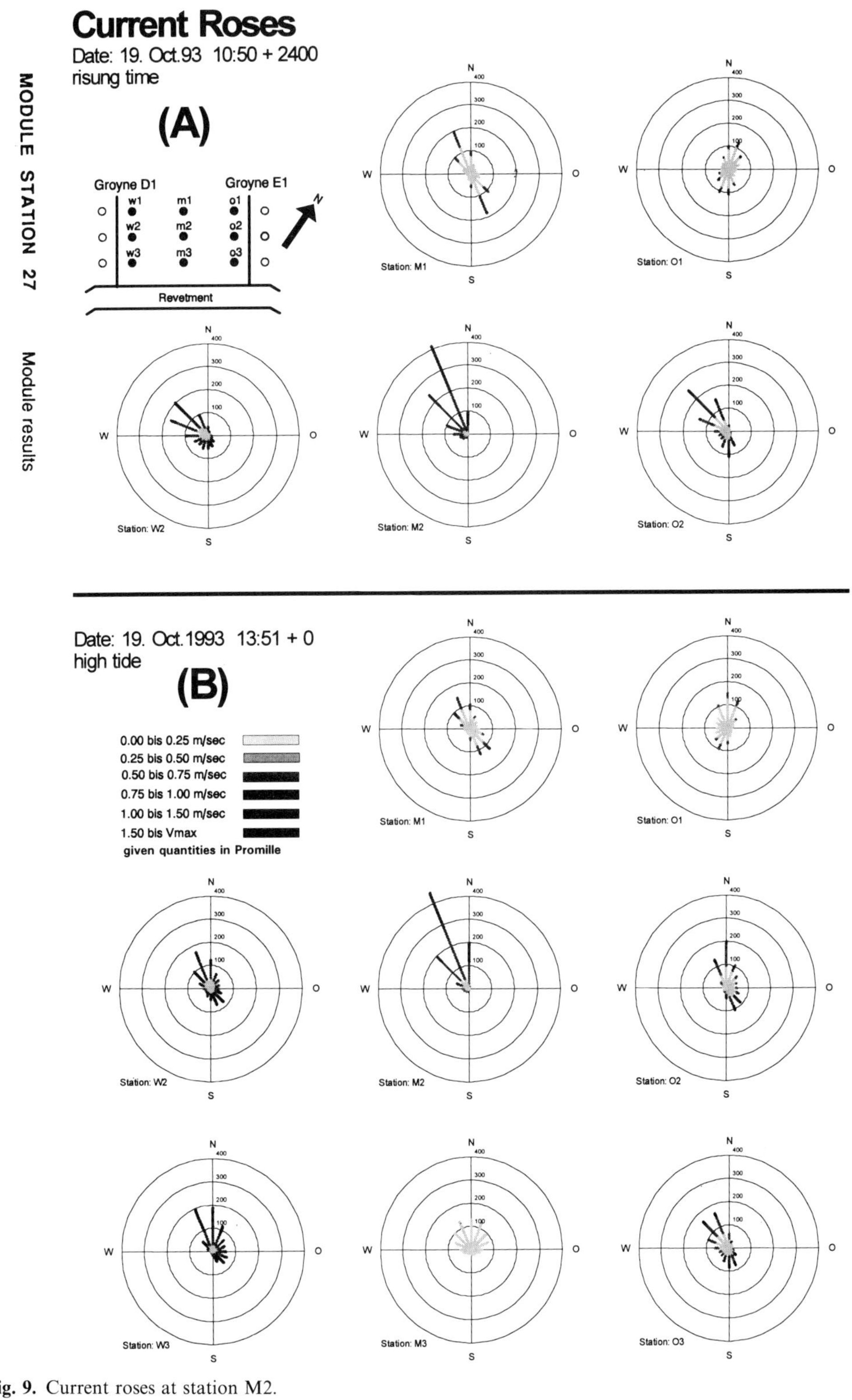

Fig. 9. Current roses at station M2.

onshore drift due to shoaling waves becase of the small layer thickness associated with moderate wave action. This onshore drift could explain the onshore directed sediment transport in the control box. It was found on previous occasions that such non-linear drift effects (Longuett-Higgins 1953) produce onshore sediment transport (Niemeyer 1991, 1992).

Morphological changes and sediment flux

On the basis of levellings at low tide (Fig. 10) changes of beach levels in the groyne field D1–E1 were found to be small, although an overall increase in beach elevation was observed. Furthermore, a channel formed parallel to groyne E1. Although the survey data do not reflect significant morphological changes, the erosional and depositional rates are remarkable. The depth-of-activity rods have proved a suitable tool for the determination of sediment fluxes by comparing the elevation measurements obtained at consecutive low tides. After the first tide (M) a maximum erosion depth of 10 cm was measured (Fig. 11). In particular, the area northeast and southwest of measuring station M2, situated approximately at mean sea level, was substantially eroded. In the seaward direction the erosion rate decreased to 3 cm and in the onshore direction to zero. By and large the data gained after the second tide were very similar but with a lower maximum erosion depth (6 cm). Erosion rates decreased in both offshore and onshore directions from the area of strongest relative erosion which itself had shifted onshore. In the area located southeast of the measuring station M2, which was strongly eroded in the course of the first tide, the highest depositional rates (8 cm) occurred after the second tide (Fig. 12). In general, the depositional rates decreased to heights of 2 cm in both offshore and onshore directions towards the high- and low-water lines respectively. After the second tidal cycle the area with maximum deposition moved landwards, corresponding to the depth of erosion, the values decreasing to a maximum of about 5 cm.

Descriptions of the resulting sediment flux elucidate an onshore directed sediment transport for both of the investigated tidal cycles (M and A) (Fig. 13), producing a negative sediment budget offshore and a positive one onshore of measuring station M2, positioned close to mean sea level during the first tidal cycles. The observed tendency was similar after the second tidal cycle, but the area with a negative sediment budget extended further onshore. The landward extension of the area with a positive sediment budget could only be detected within the limits of the test field but not further onshore. Nevertheless, the results document a pronounced landward directed sand transport.

Evolution of sediment distribution patterns

As the depositional height at all sampling sites was measured, it was possible to restrict sampling to those sediments which had been deposited in the course of the tidal cycle. Due to the installation of sensitive small scale sensors close to measuring station M2, no samples could be taken there. In those cases where no deposition had occurred, probes with a depth of 2 cm were taken. In the laboratory, grain size distribution, carbonate content and heavy mineral content of the sediment samples were determined and fluorescent sand grains were counted. The proportion of fine sand in the first sample series varied from 10 to 50% reflecting the fact that the onshore transport in the sample area is accompanied by a low proportion of fine sands (Fig. 14). The medium sand component decreased both from northeast to southwest and in the seaward direction from >40% to <20% (Fig. 15). The coarse sand fraction (Fig. 16) showed a very similar pattern to that of the medium fraction. The sediments are relatively coarse due to the high proportion of shell fragments (>30%) (Fig. 17). As a result, the median diameter is greater than 0.5 mm, classifying the sediment as coarse sand on the basis of the Wentworth scale. The heavy mineral content of about 1% is relatively small. Recent investigations have shown that a relatively high shell content generally excludes a high portion of heavy minerals because of the different hydrodynamic behaviour of the two sediment components (Eitner 1993). Under moderate hydrodynamic conditions local concentrations of shell fragments have been observed. Moreover, if the hydrodynamic conditions remain constant for a longer period of time, enormous shell accumulations can occur at the high-water line at the foot of the seawall (Eitner 1993).

The fine sand content in the second sample series (A) is markedly different to that of the first series: the values are much higher (up to 80%) and increase significantly in the offshore direction. The distribution of medium sand also changed. In the seaward part of the sample area all values decreased to 10% whereas in its landward part the values increased to more than 40%. The coarse sand content decreased substantially, particularly offshore and onshore of the measuring station M2. The increase in the fine sand content is reflected in the distribution

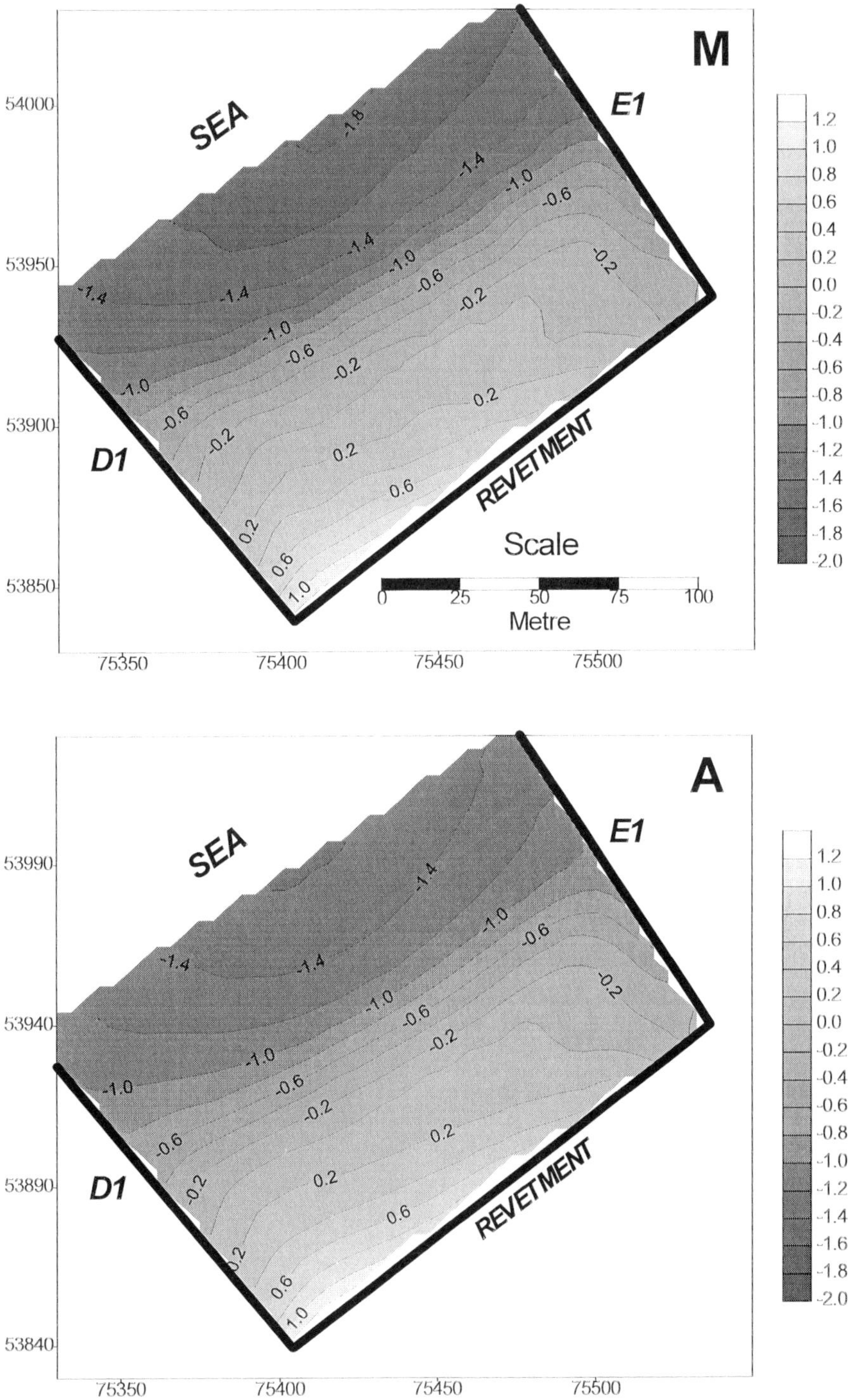

Fig. 10. Topography (in m NN (German datum)) of the groyne field D1–E1 at low tide in the morning (M) and the evening (A).

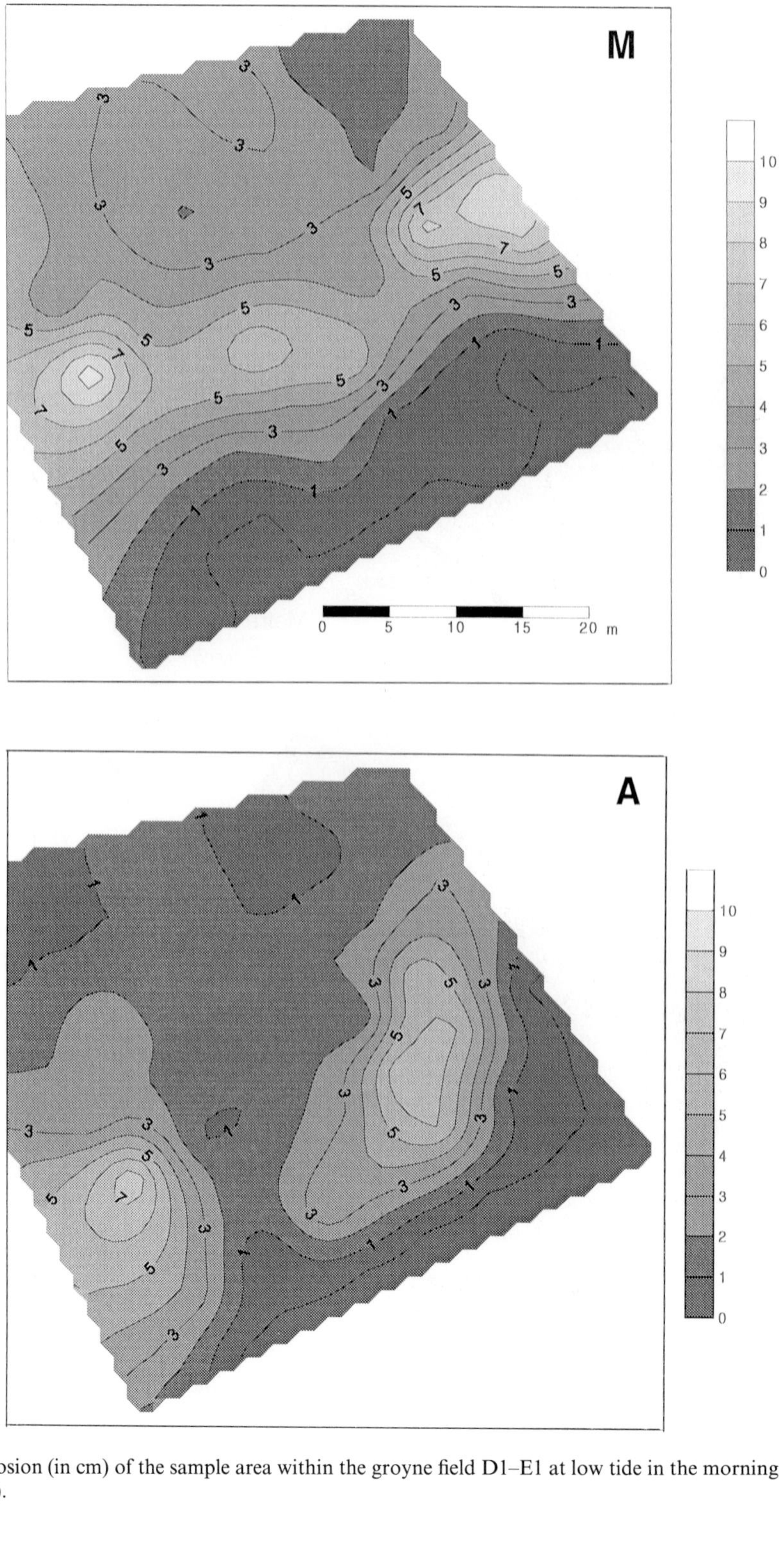

Fig. 11. Erosion (in cm) of the sample area within the groyne field D1–E1 at low tide in the morning (M) and the evening (A).

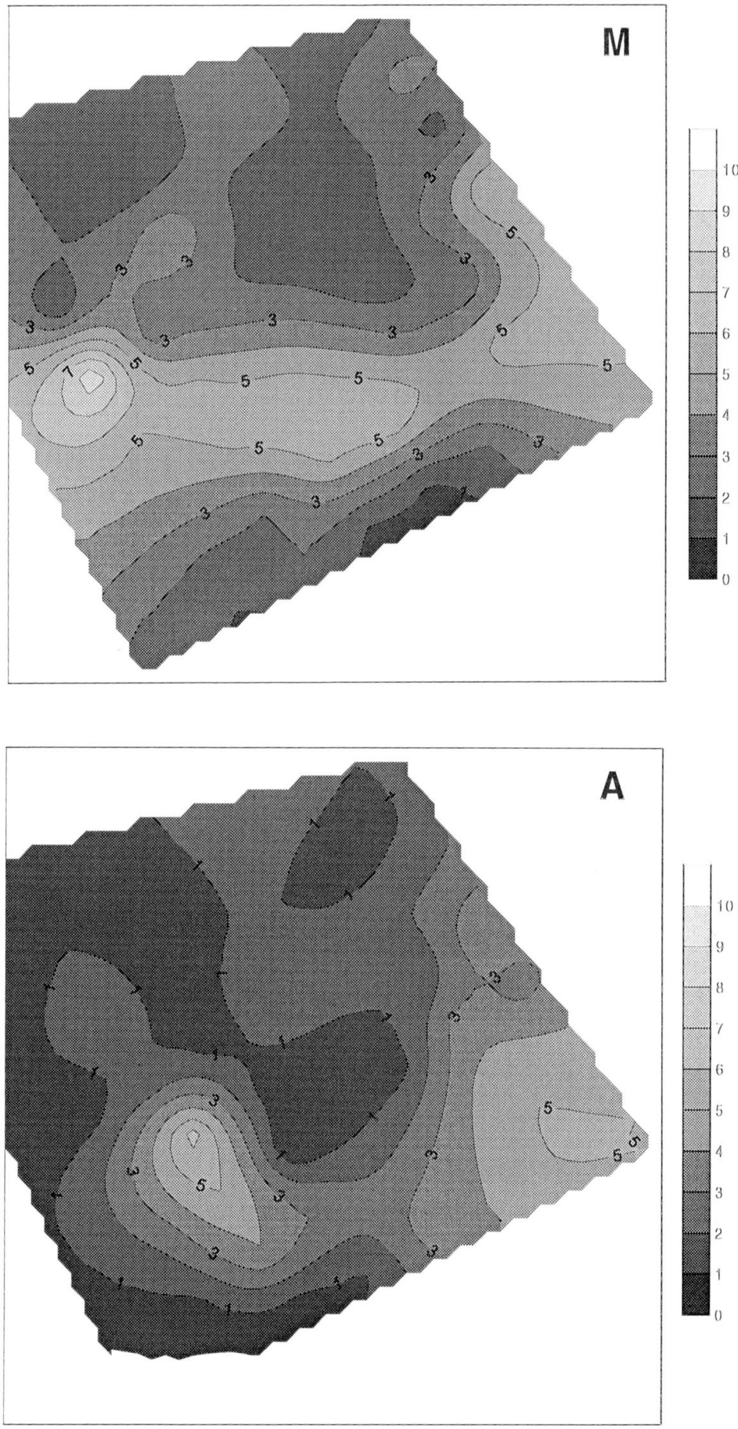

Fig. 12. Deposition (in cm) of the sample area within the groyne field D1-E1 at low tide in the morning (M) and the evening (A).

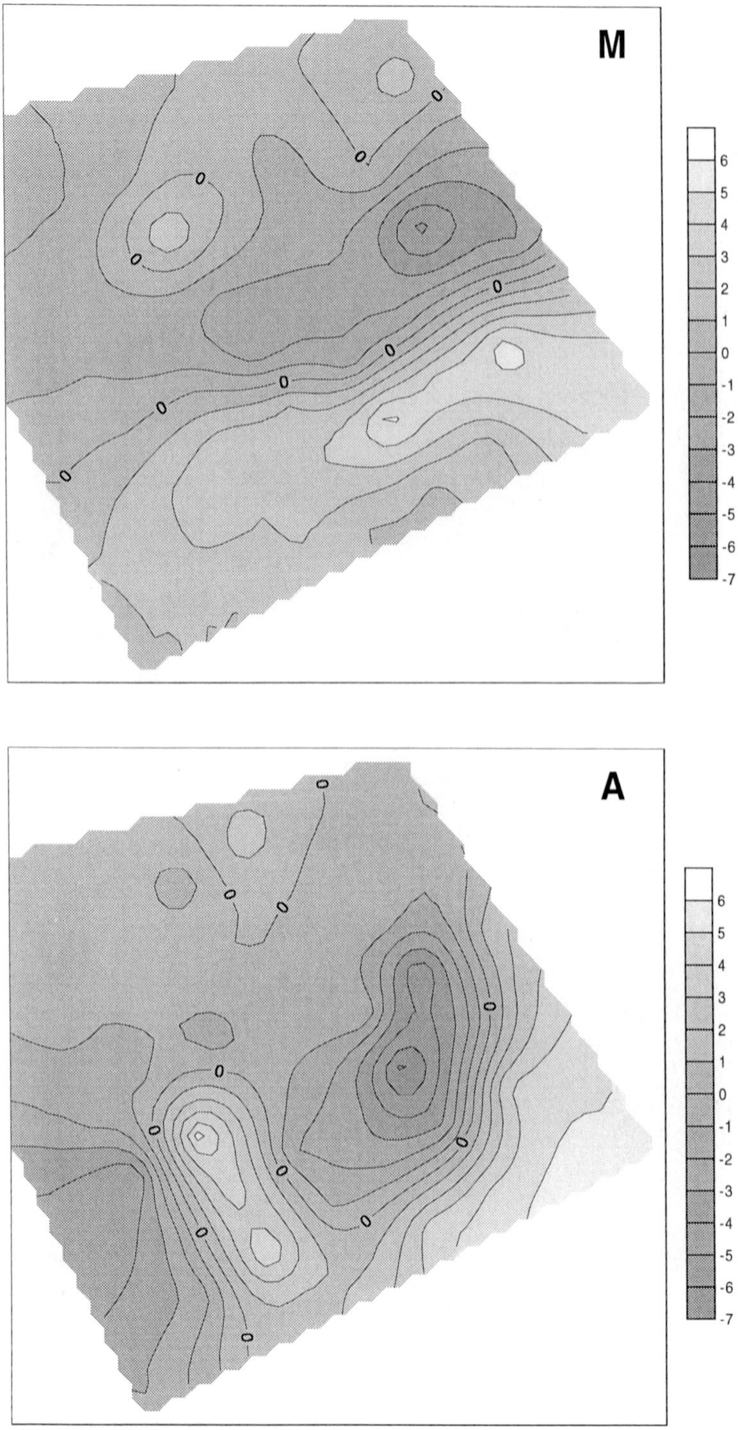

Fig. 13. Sediment flux (in cm) of the sample area within the groyne field 1–E1 at low tide in the morning (M) and the evening (A).

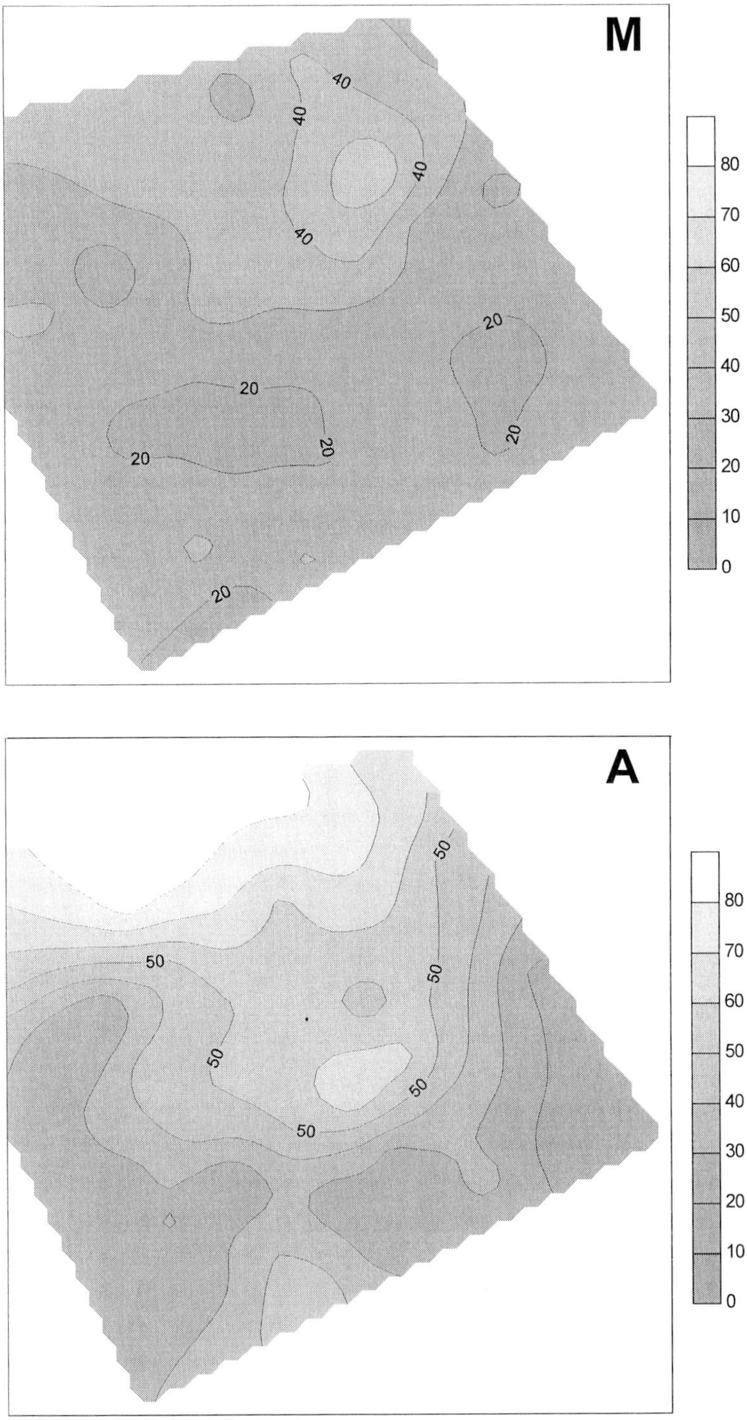

Fig. 14. Distribution of fine sand (in %) in the sample area within the groyne field D1–E1 at low tide in the morning (M) and the evening (A).

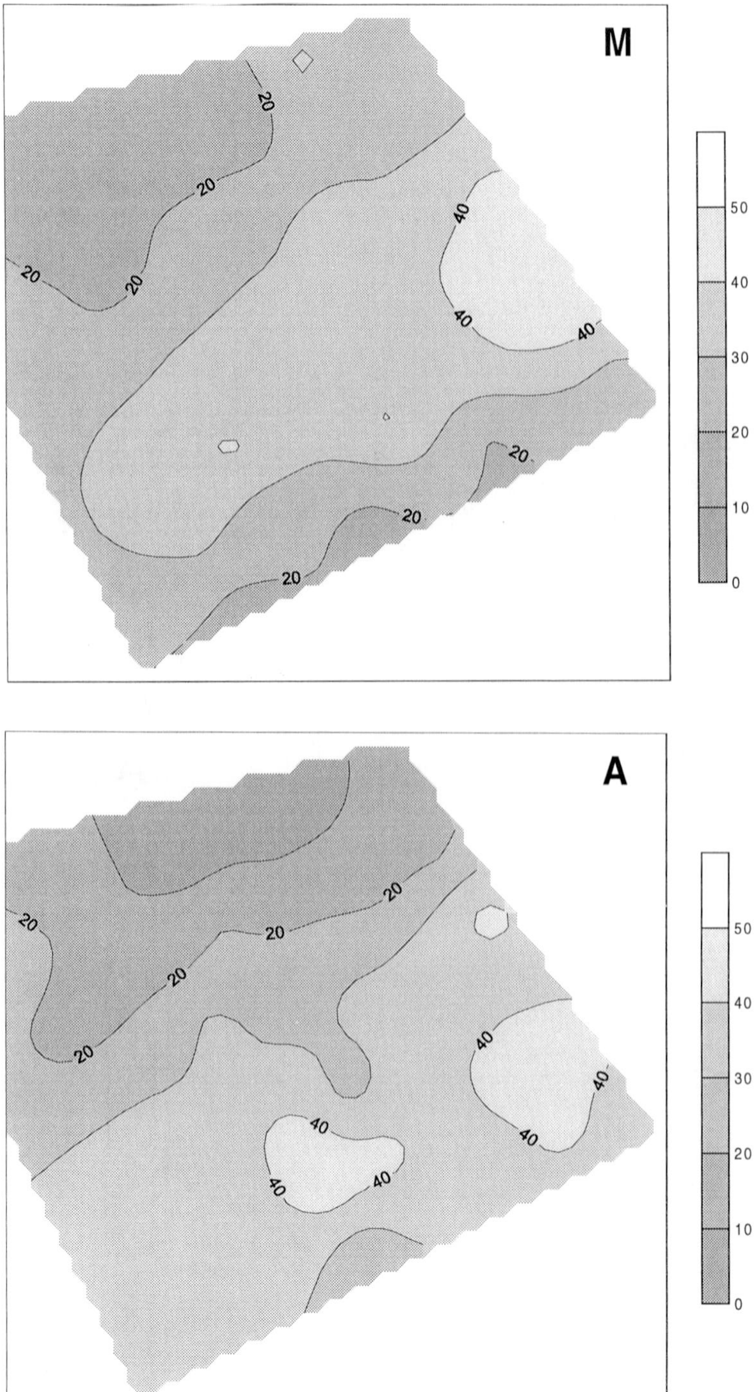

Fig. 15. Distribution of medium sand (in %) in the sample area within the groyne field D1–E1 at low tide in the morning (M) and the evening (A).

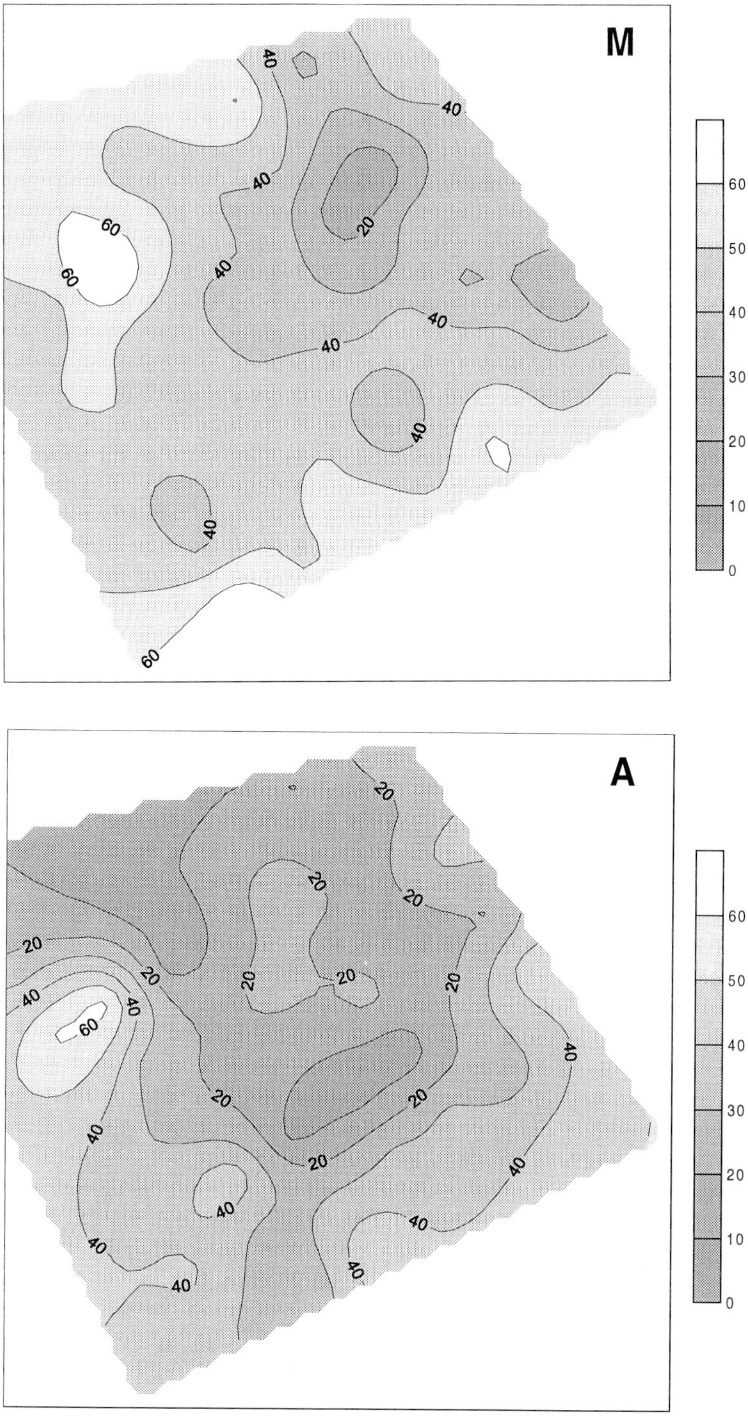

Fig. 16. Distribution of coarse sand (in %) in the sample area within the groyne field D1–E1 at low tide in the morning (M) and the evening (A).

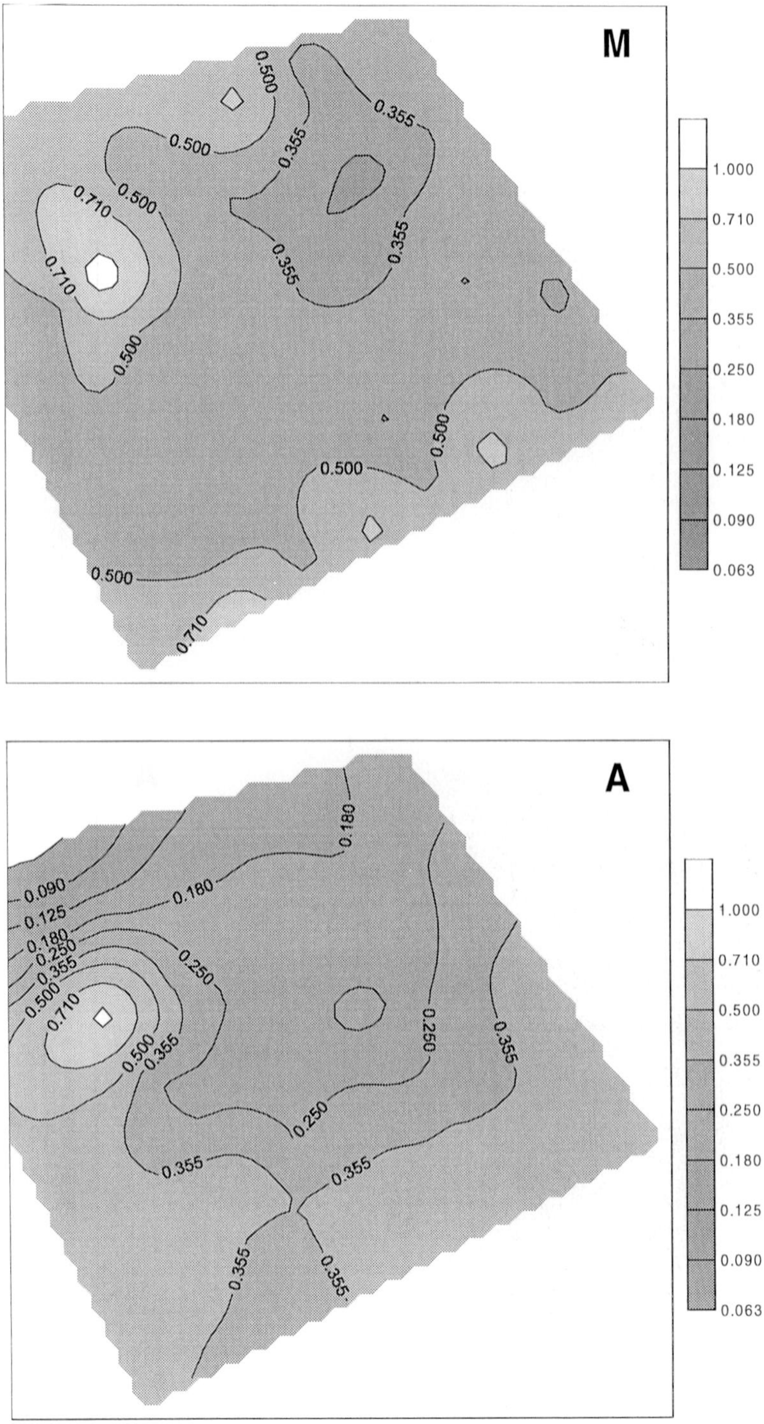

Fig. 17. Distribution of median diameter (in mm) in the sample area within the groyne field D1–E1 at low tide in the morning (M) and the evening (A).

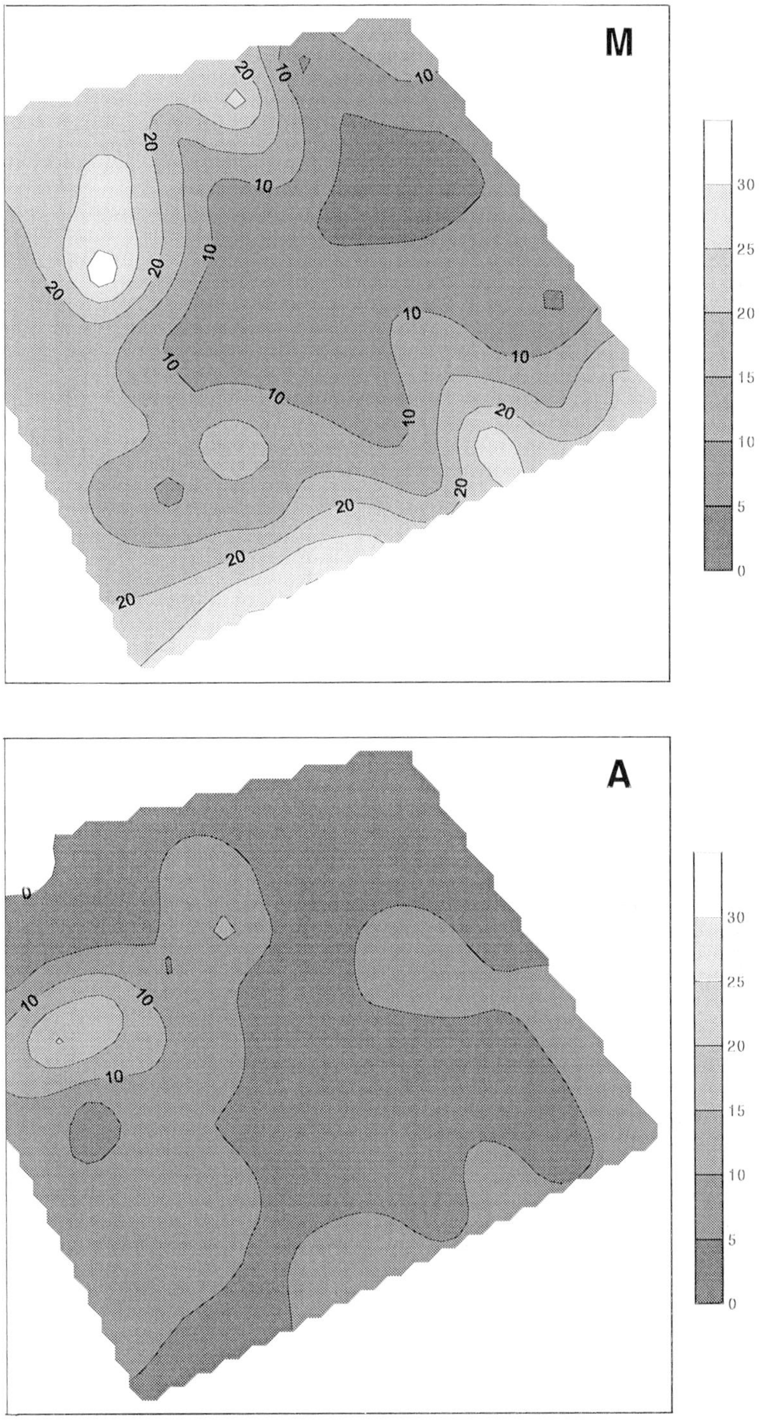

Fig. 18. Distribution of carbonate content (in %) in the sample area within the groyne field D1–E1 at low tide in the morning (M) and the evening (A).

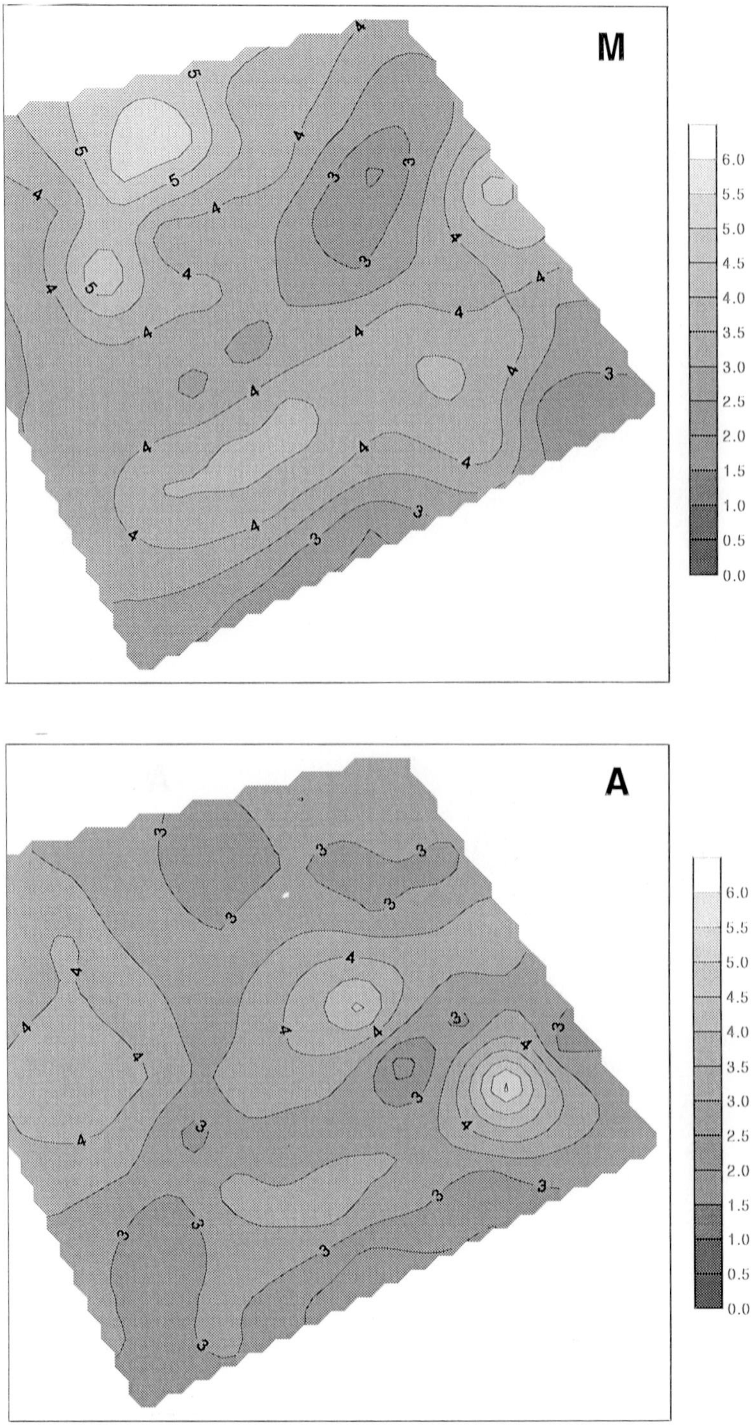

Fig. 19. Distribution of heavy mineral content (in %) in the sample area within the groyne field D1–E1 at low tide in the morning (M) and the evening (A).

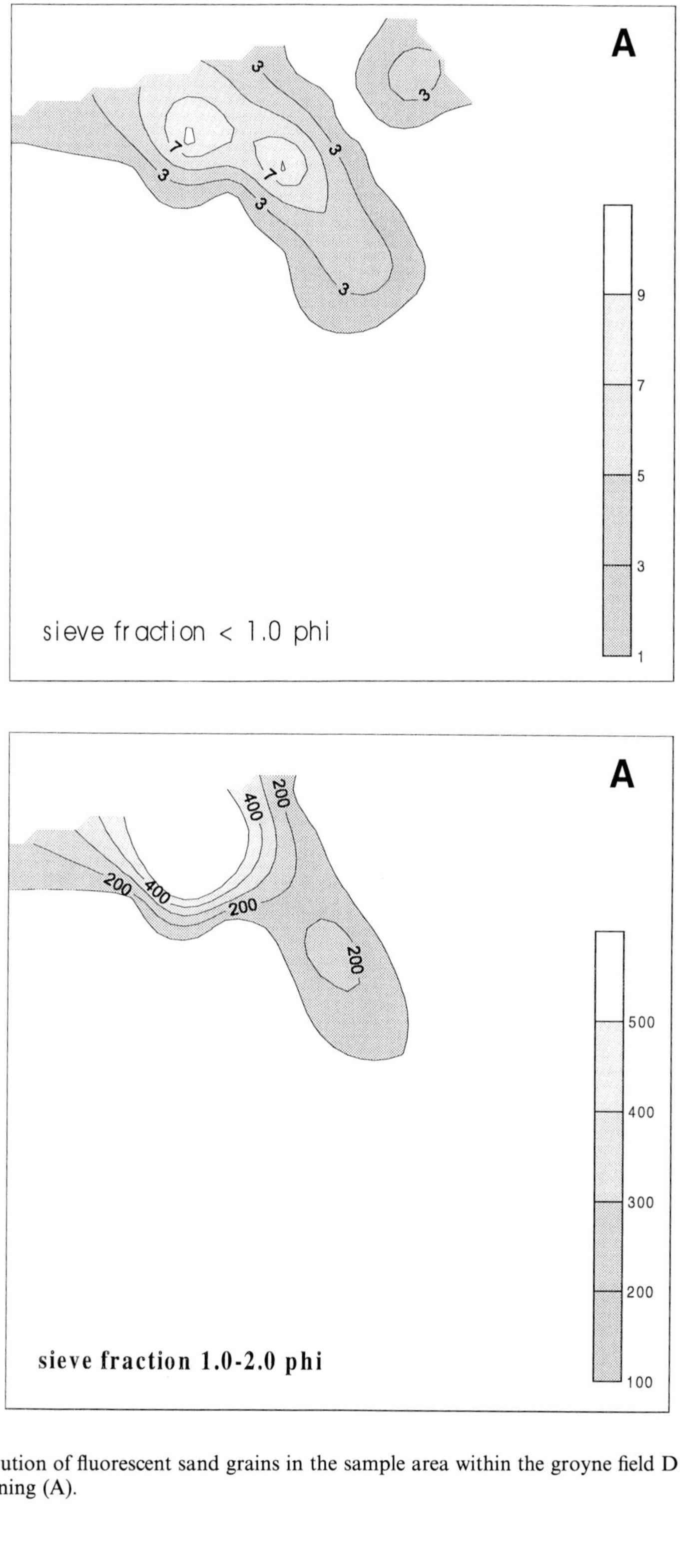

Fig. 20. Distribution of fluorescent sand grains in the sample area within the groyne field D1–E1 after one tidal cycle in the evening (A).

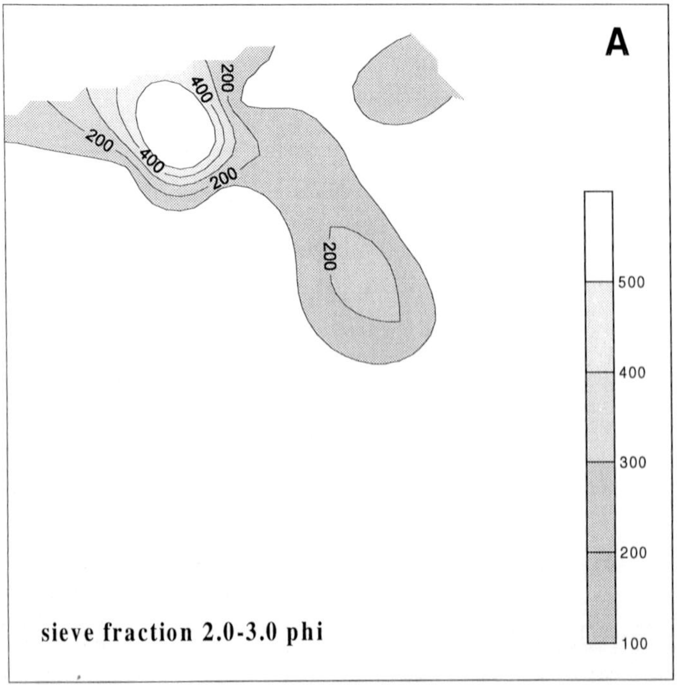

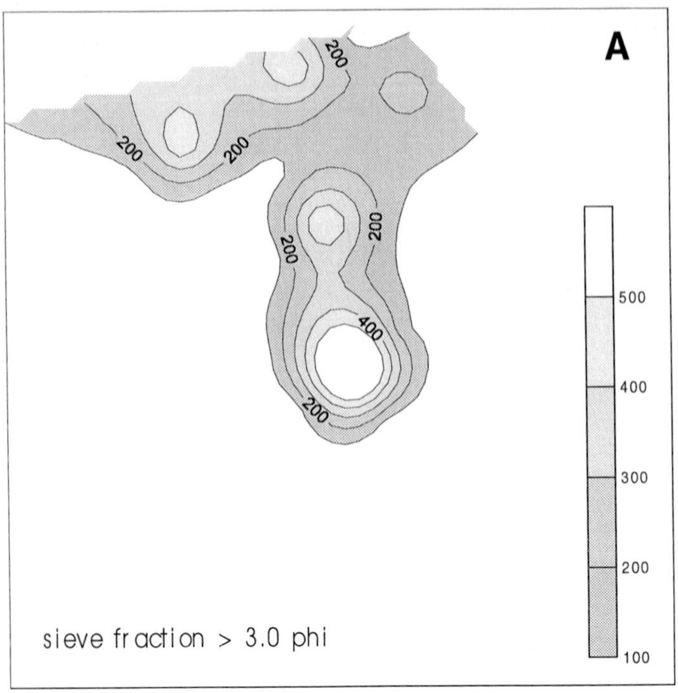

Fig. 20. *Continued.*

pattern of the medium grain size. In contrast to the samples taken in the morning, large quantities of fine sand (<0.25 mm) were found in the area around measuring station M2. Correspondingly, a decrease of shell fragments occurred which is documented in the distribution map of carbonate content (Fig. 18). The carbonate content never exceeded 10% in the study area. The heavy mineral content of the second sample series fluctuated around 3% (Fig. 19). A larger change was not detected. The small hydrodynamic energy input is insufficient to change the distribution of heavy minerals. Individual tides with high inputs, on the other hand, lead to a significant increase in heavy minerals (Eitner 1993, 1995).

Distribution of fluorescent sand grains

The distribution maps of fluorescent sand grains reflect a dominant onshore transport (Fig. 20). Furthermore, a second transport direction towards the northeast, i.e. alongshore, is evident. A transport route in a westerly direction is characterized by a small capacity. Fluorescent sand grains of different sieve fractions were additionally counted in order to assess the transport behaviour of different grain size components (Fig. 20). A significant effect of grain size on travel distance of transported grains is revealed. The distribution maps of fluorescent sand grains of individual grain size fractions show that the finer the sediment the further its onshore displacement.

Conclusions

For moderate hydrodynamic conditions an onshore directed sediment transport was observed in the study area which is particularly evident seaward of the mean sea level. The distribution of fluorescent sand grains reflects a dominant onshore sediment transport on the beach in the course of the investigated tidal cycle (Fig. 21). Variations in the distribution of sediments also confirm this dominant transport direction. These results agree well with earlier morphodynamical investigations considering onshore directed drift effects occurring with moderate wave conditions for beaches previously eroded during storm surges (Niemeyer 1991, 1992).

The sediments get finer with increasing water depth, in the offshore direction. The increasing energy dissipation with decreasing water depth has resulted in a characteristic size sorting pattern from fine to coarse grain sizes. This result agrees with earlier observations dealing with the distribution of sediments at the beginning and the end of a tidal cycle carried out in the same groyne field (Westhoff 1990; Eitner *et al.* 1992). Furthermore, a variation of grain

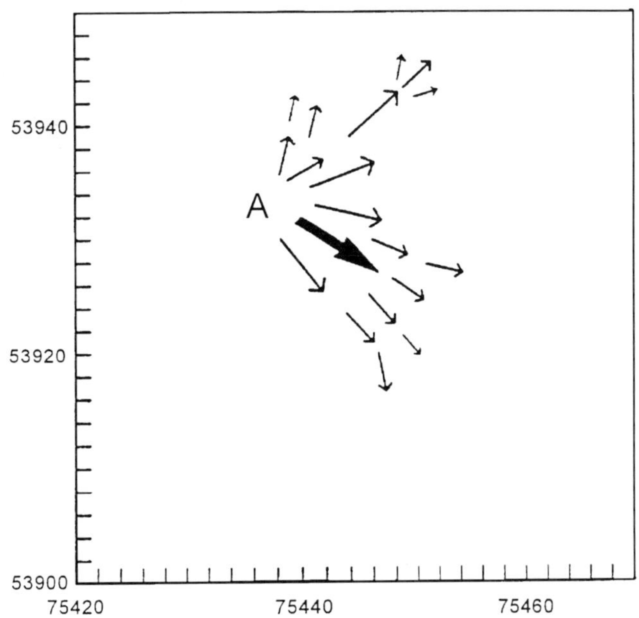

Fig. 21. Sediment transport model (A, point of distribution).

size distribution occurs in coincidence with water level variation. Thus with rising water levels during the flood tide grain sizes tend to become finer. In the course of the ebb tide grain sizes tend to increase again. The adaptation of common descriptive morphodynamical models to the beach of Norderney, e.g. from Wright *et al.* (1982) or Sunamura & Takeda (1984), made the changing beach geometry in interaction with hydrodynamic boundary conditions evident. For moderate hydrodynamic boundary conditions the beach is in accordance with the classification of Wright *et al.* (1982), with a beach state of $B_s = 2$ intermediate with a tendency to reflective (Eitner 1993). But with higher waves, which occur independently of the season but are more frequent in winter, the beach changes to a dissipative state. Sunamura & Takeda (1984) established a model for the transition from a dissipative to a reflective response of the beach due to wave climate. For a mean wave period of 6 s and a breaker height between 0.7 and 1.1 m it predicts an onshore transport. This prediction is in accordance with the observations made in studies on Norderney.

This study was supported by funds of the German Federal Ministry of Science and Technology (MTK 0545) in the framework of the research programme of the German Committee on Coastal Engineering Research (KFKI). We are grateful to our colleagues for valuable assistance during the measuring campaign and data evaluation phase, particularly Mr G. Brand, D. Glaser, H. Karow, G. Münkewarf and R. Taddigs, and to Ms H. Meyer for assistance by the preparation of figures. We thank all the colleagues engaged in the project who supported us during fruitful discussions.

References

EITNER, V. 1993. *Sedimentdynamik im Strandbereich einer brandungsbeeinflußten, mesotidalen Barriere-Insel unter Berücksichtigung der Auswirkungen künstlicher Strandauffüllungen (Norderney, südliche Nordsee).* PhD Thesis, Universität Münster.

—— 1995. Magnetic heavy mineral associations as sediment transport indicators on a beach of Norderney Island, southern North Sea. *Senckenbergiana Maritima*, **25**(4/6), 173–185.

—— & RAGUTZKI, G. 1994. Effects of artificial beach nourishment on nearshore sediment distribution (Island of Norderney, southern North Sea). *Journal of Coastal Research*, **10**, 637–650.

——, —— & WESTHOFF, T. 1992. Sedimentologische Untersuchungen zu den Auswirkungen einer künstlichen Auffüllung des Norderneyer Strandes auf die Transport-und Umlagerungsvorgänge. *Die Küste*, **54**, 93–110.

GREENWOOD, B., HALE, P. B. & MITTLER, P. R. 1980. Sediment flux determination in the nearshore zone, prototype measurements. Workshop on *Instrumentation for Currents and Sediments in the Nearshore Zone*. National Research Council of Canada, 99–119.

HAYES, M. O. 1979. Barrier island morphology as a function of tidal and wave regime. *In*: Leatherman, S. P. (ed.) *Barrier Islands*. Academic, New York, 1–29.

LONGUETT-HIGGINS, M. S. 1953. On the statistical distribution of the heights of sea waves. *Journal of Marine Research*, **XI**(3), 245–266.

MCANDREW, J. 1957. Calibration of a Frantz Isodynamic Separator and its application to mineral separation. *Proceedings, Australian Institute of Mineralogy and Metallurgy*, **181**, 59–73.

MÜLLER, V. 1993. *Meßbericht zum Feldversuch am 18.–20.10.1993, Norderney – Buhnenfeld D1/E1.* Internal report TU Hamburg-Harburg.

NIEMEYER, H. D. 1987. Tidestrommessungen in Buhnenfeldern. *Jahresbericht 1986 der Forschungsstelle Küste*, **38**, 127–150.

——1991. Field measurements and analysis of wave induced nearshore currents. *Proceedings of the 22nd International Conference on Coastal Engineering*, Delft, 783–797.

——1992. Die ursächliche Deutung von Transportphänomenen als Gestaltungsgrundlage für Strandauffüllungen. *Die Küste*, **54**, 53–92.

——, KAISER, R. & EITNER, V. 1994. *Combined Shoreface and Beach Nourishment – Island of Norderney/East Frisia.* NOURTEC Scientific and Technical Progress Report 6/93–6/94, MAS2-CT93-0049.

RUCK, K. 1967. Erfahrungen mit Sandwanderunguntersuchungen mittels Luminophoren. *Die Wasserwirtschaft*, **10**, 363–367.

——1972. Erfahrungen beim Präparieren von Sand für Leitstoffuntersuchungen. *Mitteilungsblatt der Bundesanstalt für Wasserbau*, **20**, 15–33.

SCHULTZE, E. & MUHS, H. 1967. *Bodenuntersuchungen für Ingenieurbauten.* Springer, Berlin.

SUNAMURA, T. & TAKEDA, I. 1984. Landward migration of inner bars. *Marine Geology*, **60**, 63–78.

VEENSTRA, H. J. & WINKELMOLEN, A. M. 1976. Size, shape and density around two barrier islands along the north coast of Holland. *Geologie en Mijnbouw*, **55**, 87–104.

WESTHOFF, T. 1990. *Sedimentgeologische Untersuchungen zur Klärung von Transportvorgängen im Bereich sandiger Küsten am Beispiel Norderney.* PhD Thesis, Universität Münster.

WRIGHT, L. D., GUZA, R. T. & SHORD, A. D. 1982. Dynamics of a high energy dissipative surf zone. *Marine Geology*, **45**, 41–62.

YU, Z., NIEMEYER, H. D. & BAKKER, W. T. 1991. Site investigations on sand concentration in the sheetflow layer. *Proceedings of the 22nd International Conference on Coastal Engineering*, 2360–2371.

Radiometry as a technique for use in coastal research

R. J. DE MEIJER, I. C. TÁNCZOS & C. STAPEL

Kernfysisch Versneller Instituut, Rijksuniversiteit Groningen, Zernikelaan 25, NL-9747 AA Groningen, The Netherlands

Abstract: Natural radioactivity in certain heavy minerals makes it possible to locate and follow, in time and space, such components in coastal sands. In this paper, the sensitivity of radiometric techniques is demonstrated in measurements carried out in the laboratory and in the field. In the laboratory, on the beach and on the sea floor, indications were found that transport processes acting on light and heavy minerals often result in net transport modes in opposite directions. A simplified transport model is discussed which describes this selective transport semi-quantitatively. This model calculates trajectories of individual grains, using physical expressions for the forces acting on them. The model incorporates turbulence and vortex motion in an effective viscosity coefficient.

Inspired by Bonka (1982), who discovered patches of sand with enhanced natural radioactivity on the beach of the German Frisian Island of Norderney, a search for similar areas was initiated along the Dutch coastline. Enhancements in natural radioactivity were found at the Dutch Frisian Island of Ameland; these were found to be due to concentrations of radiogenic heavy minerals.

Various investigations have followed on the dependence of the concentrations of radionuclides (^{40}K and the decay series of ^{238}U and ^{232}Th) on grain size and magnetic susceptibility (Schuiling *et al.* 1985; de Meijer *et al.* 1985, 1988, 1989). These investigations showed that the concentrations of radionuclides in the decay series of U and Th increased with decreasing (sand) grain size and that these activities were present in high-density minerals of small grain size. Calculations and measurements showed that these grains have a settling velocity in water similar to that of larger quartz grains.

Radionuclides such as ^{40}K, and ^{235}U, ^{238}U and ^{232}Th have been present in the crust of our planet since its origin and, since their half-life is similar to or longer than the age of our planet, their total concentrations have diminished only slightly. Of interest to the present contribution are mainly ^{40}K and γ-ray-emitting nuclei in the decay series of ^{238}U and ^{232}Th. For both, the half-life of their decay products is considerably shorter than that of the parent nuclei. In closed systems, where no nuclei disappear other than by nuclear decay, the activity concentrations of nuclei in the chain reach so-called secular equilibrium (Evans 1969). In secular equilibrium, the activity concentrations of all nuclei in a decay chain are the same. A closed system implies that measuring the activity concentration of one member of the decay chain provides information on the presence of all the members. For the ^{238}U decay series, measurements of the activity concentrations of ^{214}Bi and ^{214}Pb, both γ-ray emitting decay products, yield information on ^{238}U (which emits no γ-rays) provided no elements dissolve or escape, such as the gaseous member ^{222}Rn.

The relationship between the number of atoms (N) of a certain species and its activity (A) is defined as

$$N = T_{1/2}A/\ln 2 \qquad (1)$$

where $T_{1/2}$ is the half-life of the radionucleus. Since activities are expressed in becquerel (1 Bq corresponds to one decaying atom per second), $T_{1/2}$ has to be expressed in seconds. Using equation (1) and Avogadro's number, one can calculate that 1 ppm U and Th correspond to 12.3 and 4.0 Bq/kg, respectively. Similarly, one may calculate that 1% K_2O corresponds to 257 Bq/kg ^{40}K. In this paper, concentrations will be presented in Bq/kg .

Using the above relationship we have investigated to what extent heavy-mineral grains may be considered as closed systems. For this purpose, activity concentrations of ^{214}Pb and ^{214}Bi were compared with U concentrations based on X-ray fluorescence (XRF) measurements. Within the uncertainties in the two techniques, no evidence was found that the systems were not closed. For quartz, however, it is known that up to 20–30% of the radon, formed in the grains, escapes. Therefore, we will report our activity concentrations in the decay chain of ^{238}U as ^{214}Bi.

Coastal sands are the resistant products of mechanical weathering of rock. Sand mineralogy reflects, therefore, the mineralogy of the parent rock. Although dependent upon the amount of

From De Batist, M. & Jacobs, P. (eds), 1996, *Geology of Siliciclastic Shelf Seas*, Geological Society Special Publication No. 117, pp. 289–297.

sediment mixing during transportation from the source area to the site of deposition, sand mineralogy generally provides a reliable reflection of the mineral composition of the source rocks.

Sand may be divided into light- and heavy-mineral fractions. This division is based upon the traditional density-separation method in which sand is settled in organic liquids of high specific density, such as bromoform (specific density 2.82 g/cm^3). Light minerals consist mainly of quartz and feldspar. Quartz is almost pure SiO_2 and contains little U, Th or K. Feldspar is a mineral that often has a potassium-containing component.

Heavy minerals present in Dutch beach sands include garnet, zircon, epidote, ilmenite and magnetite. On average, the heavy-mineral fraction of these beach sands has a specific activity concentration 100 to 200 times higher than that of light minerals (de Meijer *et al.* 1990). This large difference permits the detection of heavy-mineral concentrations in sediments and in principle provides a method to determine radiometrically the approximate composition of the sediment by deriving the relative mass of a number of groups of minerals.

In a previous publication (de Meijer *et al.* 1990) the activity concentrations of groups of minerals with similar values of magnetic susceptibility were presented; for various groups an enhancement by two to four orders of magnitude is found in the Bi and Th activity concentrations with respect to the light minerals. Subsequently, de Meijer & Donoghue (1995) have shown, for the beaches of the Dutch, German and Danish coasts, that the total heavy-mineral activity concentrations may vary depending upon the origin of the sands. These concentrations may be used, therefore, to identify regions with the same source area for the heavy minerals in coastal sands (i.e. radiometric fingerprinting). Such patterns thereby reflect the effects of sediment transport over geological time scales.

Traditionally, heavy minerals have been used as indicators of transport processes. For example, Stapor (1973) measured a difference in the overall grain size on sheltered and open beaches. Coarser grains are removed from open beaches, resulting in a finer grain size. The reason for this difference is not yet clear. In addition, Stapor has pointed out that there are indications that, on these beaches, heavy minerals are deposited which were concentrated in deeper offshore waters. It is again not yet clear if there is a relation between the two observations.

The interpretation by Stapor is contrary to the more generally held opinion that heavy-mineral lag deposits result from erosion of beaches and dunes.

Sorting processes were studied by Slingerland (1984), who distinguished four sorting types:

(i) entrainment sorting in response to selective removal from the sediment bed due to size, density and shape;
(ii) suspension sorting, due to differences in settling velocities;
(iii) shear sorting, due to grain to grain interactions in a moving sediment layer; and
(iv) transport sorting due to differences in transport velocities of light and heavy minerals (in which sorting caused by entrainment and suspension plays a role).

Although the above investigations indicate the potential importance of heavy minerals in the interpretation of sediment transport processes, in practice such studies require extensive sampling in the field followed by a quite elaborate analysis procedure. Using radiometric techniques could change this situation. A feasibility study was carried out by Greenfield *et al.* (1989) on the beach of the Dutch island of Texel. At a number of test sites, changes in beach elevation and γ-radiation intensity were monitored, over time, for sites located in different erosion-sensitive areas. For sites dominated by aeolian transport, elevation changes showed a good inverse correlation with changes in the total γ-ray count rate, indicating that mainly light minerals were removed or deposited. At sites where erosion was caused by wave action, an opposite tendency was sometimes observed.

Monitoring elevations and γ-radiation intensity at two beach sites on the Dutch Frisian Island of Ameland has indicated that accretion of heavy minerals occurred at steeper slopes of the beach front under both eroding and accreting conditions (de Meijer *et al.* 1994*a*).

During the above investigation some major storm events took place, after which heavy-mineral concentrations were observed on the beach in volumes which seemed to exceed the heavy-mineral content of the dune and beach sands which were removed. These observations, which support the concepts of Stapor (1973), triggered a research programme to investigate the processes that selectively remove and deposit light and heavy minerals, respectively, during the same event. For this purpose, a towed seabed detector has been developed to map the sea floor radiometrically. Parallel to this development, laboratory experiments in a wave flume were undertaken.

Methodology and results

Sea-floor mapping

In collaboration with the British Geological Survey (BGS), a project was initiated to design and construct a towed sea-bed detector. This detector was based upon a prototype detector developed by BGS more than a decade ago. In comparison, the present KVI instrument is equipped with a detector with efficiency an order of magnitude larger for high-energy γ-rays. Such a detector is more sensitive, therefore, for monitoring heavy-mineral concentrations near the surface of the sea bed.

In May–June 1993 a trial tow of the system was undertaken in which both detectors, the prototype and the new one, were to be compared on the sea floor to the north of the island of Ameland, over a $20 \times 10\,km^2$ area. A grid pattern was set out with a spacing of 1 km, at angles of about 45° with respect to the coast. The detectors were towed over the sea floor behind a 25 m long Rijkswaterstaat vessel 'Blauwe Slenk'. For this purpose, the detectors were placed in watertight casings which were connected to the ship by an armoured coaxial cable. The cable serves as the physical as well as the electrical connection with the ship. To prevent the detector from being caught by underwater obstacles, like wrecks, the detector was placed at the end of a 30 m long PVC hose. The front of this eel was held up from the bottom by providing just the correct length of cable to maintain the detector in contact with the sea floor (see Fig. 1). The cable was mounted on a winch located at the stern of the ship. Signals from the detector were transmitted through the cable to an onboard computer, for data handling and storage. The vessel was equipped with a position and depth logging system (Syledis). Additionally, side-scan sonar was installed; its output was recorded on paper, for possible off-line analysis of the data with

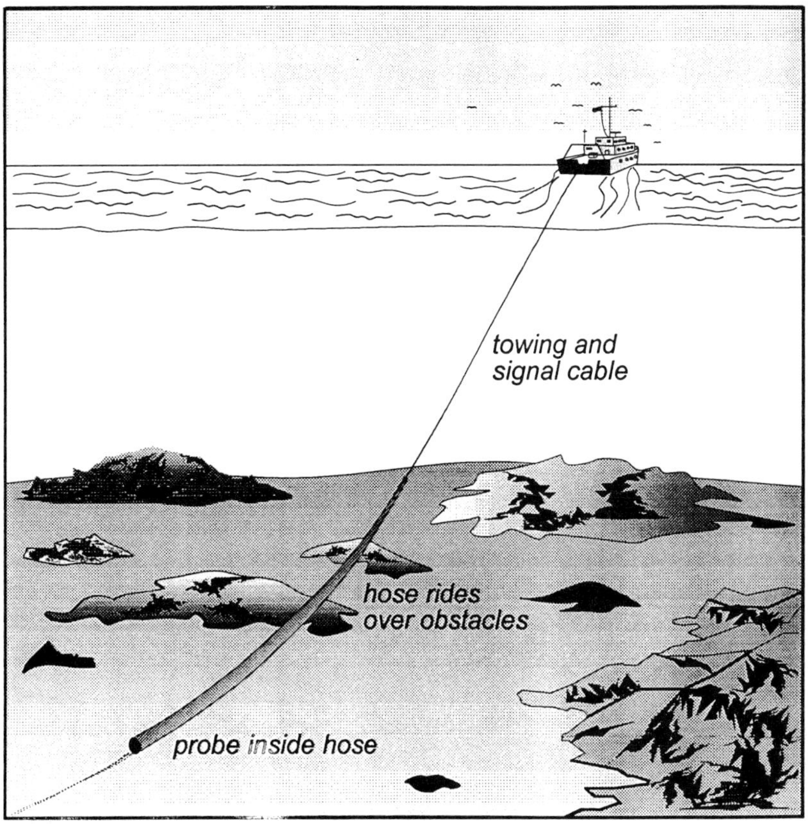

Fig. 1. Schematic presentation of the sea-bed detector system towed by a vessel over the sea floor (reproduced by permission of the Director, British Geological Survey: NERC copyright reserved).

Fig. 2. Map of The Netherlands and the adjacent part of the North Sea.

respect to bottom topography. The positioning accuracy for the ship is estimated to be about 10 m; the positioning accuracy of the detector system in the trial run was of the order of 100 m. Presently the ships are equipped with DGPS (accuracy <5 m) and the probe position is believed to be known to about 15 m accuracy.

Data from the ship's logging system were stored every 0.1 s, total counts of the γ-ray detector every 15 s and γ-ray spectra every 5 min. The towing speed was about 10 km/h. Due to changing weather conditions only the BGS prototype could be used on the North Sea (seaward of the island). Test comparisons of both systems were made during bad weather conditions in the sheltered Wadden Sea (see Fig. 2). During the tests the improvements in the KVI detector, previously measured on test slabs at BGS and on the beach of Ameland, were confirmed.

Figure 3 shows part of the coastline of Ameland and the area surveyed. The figure presents the survey lines towed with the BGS detector and the associated smoothed radiation data indicated by contour lines of 46 and 55 cps. Moreover the area on the beach, where concentrated heavy minerals were deposited during storms in January/February 1992, and sample sites are marked in the figure. The radiometric data are 'raw' data, i.e. not corrected for (varying) background levels due to cosmic radiation and K in sea water (these corrections are known to be less than 10%).

In addition to the measurements, grab samples were taken of the upper 20 cm of the sea-bed sediment using a Van Veen sampler. Sample locations were selected to cover the full range of radiation-intensity variations. Radiometric analysis of the samples yielded total heavy-mineral mass concentrations from <1 to about 10% (see Table 1). Based on the results, and assuming a layer thickness of 20 cm and a dry bulk density of sand of 1.7 kg/l, the area comprising the contour with the highest intensity in Fig. 3 corresponds to an estimated mass of 0.3 Mt of heavy minerals.

The smoothed radiometric data in Fig. 3 show that heavy minerals are concentrated in an area elongated parallel to the coast. The shape of the area of concentration, combined with the fact

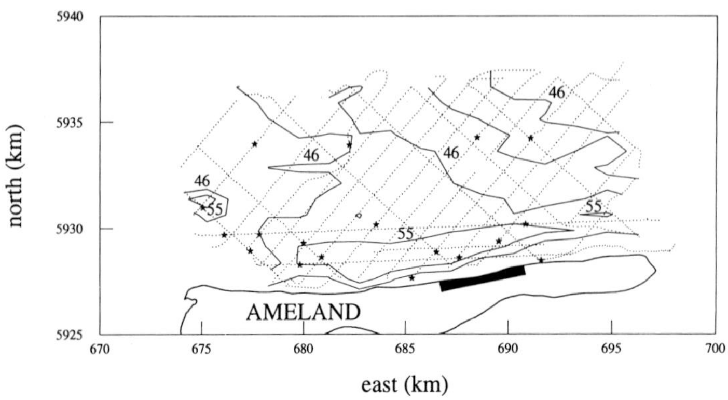

Fig. 3. Map (UTM coordinates) of the area investigated in the survey near Ameland. The dotted lines represent the survey lines, the contour lines total count rates in cps, the stars sampling sites, and the dark area on the beach the location of heavy-mineral deposition in 1992.

Table 1. *Total heavy-mineral concentrations (THM) expressed in mass percentages, in grab samples calculated from Bi and Th activity concentrations. The values represent weighted averages. The sample locations are presented in UTM coordinates (see also Fig. 3)*

East	North	THM (%)
677 562	5 933 963	1.18 (0.05)
679 972	5 929 306	1.41 (0.07)
680 878	5 928 650	1.93 (0.12)
679 797	5 928 306	4.94 (0.14)
677 814	5 929 716	4.45 (0.15)
677 380	5 928 928	0.27 (0.03)
676 094	5 929 688	5.4 (0.8)
675 056	5 930 973	6.8 (0.5)
683 565	5 930 176	2.10 (0.15)
686 515	5 928 886	2.54 (0.16)
689 537	5 929 385	2.25 (0.14)
690 835	5 930 196	3.50 (0.16)
691 058	5 934 237	2.23 (0.08)
688 484	5 934 273	1.37 (0.06)
682 224	5 933 925	1.02 (0.05)
685 325	5 927 655	0.09 (0.02)
687 637	5 928 631	8.5 (0.3)
691 577	5 928 501	1.02 (0.04)

that the highest concentrations occur in the middle of the enhanced area, are consistent with a cross-shore transport process. The location of the area relative to the beach deposits formed in 1992, supports the ideas of Stapor (1973) that heavy minerals are deposited on the beach from concentrations offshore.

Figure 4 shows depth and total count rate data for the western part of the most northern east–west survey line of Fig. 3. Here, the bathymetry reveals sand bars oriented northwest–southeast. From the figure one notices that the radiation is not evenly distributed but seems to be concentrated on the seaward slopes of some sand bars. This phenomenon is the same as may be seen on a beach where heavy minerals concentrate on the swash-backwash slope of the shoreface.

In June 1994 a second cruise was undertaken on which the KVI detection system was used. Besides γ-radiation, water depth at the probe and bottom roughness were also measured. During this cruise, the measurements of 1993 near Ameland were repeated; likewise, measurements north of the island of Terschelling were carried out. A preliminary analysis of the data near Ameland showed no significant changes in the heavy-mineral deposit, although smaller changes in bathymetry and γ-radiation intensity seem to be present.

Based on the absence of elevated heavy-mineral concentrations on the beach of Terschelling, no significant heavy-mineral concentrations in the coastal zone of this island were anticipated. To our surprise a $20 \times 4\,\mathrm{km}^2$ area with enhanced concentration of heavy minerals was located; this area contained some spot concentrations which exceeded those in the coastal zone of Ameland. An interesting feature is the presence of enhanced concentrations on top of part of an underwater renourishment. This

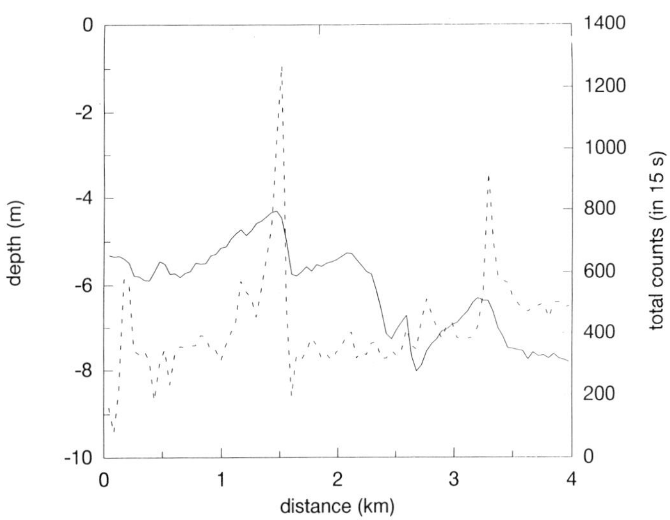

Fig. 4. Bathymetry (——) and total γ-ray count rate (----) measured along part of one of the survey lines. The distance along the horizontal axis refers to the distance to the point on the line nearest to the shore.

renourishment was built in 1993 with sand that originates from 20 km offshore and is supposed to be low-activity sand.

Laboratory experiments

A study was carried out in the $14 \times 0.8 \times 0.5\,m^3$ wave flume in the Laboratory of Fluid Mechanics at the University of Technology Delft, The Netherlands. The wave generator can be adjusted in tilt angle and frequency to allow for ranges in wave height, amplitude, frequency and asymmetry. A damping slope was present at the end of the flume to avoid wave reflection. In addition, by using a pumping system, a current could be induced in the opposite direction to the waves. In the experiments, a layer of beach sand ($30 \times 50\,cm^2$ and with a thickness of about 5 cm), naturally enhanced in heavy minerals (~30%), was placed near the middle of the flume.

For a number of wave and current conditions (see Table 2) sand transport was monitored for periods between 1 and 17 h. Prior to and after the experiments, samples were taken for measurements of density, grain size and radioactivity concentration. Combined flow velocities were measured at about 3 cm above the sediment bed and were recorded on paper.

During this study, it was observed that after some time 'ripple' areas were formed at both the upstream and downstream sides of the initial sand-deposit position. Usually the extent of the ripples in the direction of the largest peak velocity (wave plus current) was considerably larger than in the opposite direction. Moreover, the colour of the ripples became darker with increasing distance in that direction. The analysis of the sand samples revealed that, under all conditions, heavy minerals were transported in the direction of the greatest velocity near the bottom; light minerals either remained at the same location or were transported in the opposite direction. Characterization of the minerals using radionuclide concentrations was found to be a hundred times more sensitive than the use of either grain size or density. This observation may form the basis for a nondestructive method for use in sediment transport investigations (Tánczos & de Meijer 1996).

Table 2. *Range of parameters used in the experiments with the wave flume. The water depth was 25–30 cm; the wave heights in the crest and the trough were +17 and −9 cm, respectively, relative to the main water level*

	U_{cres} (m/s)	U_{trough} (m/s)	T (s)
No flow	0.3, 0.6	−0.4, −0.2	1.0, 2.4
Flow (0.7–0.09 m/s)	0.2, 0.65	−0.6, −0.2	1.3

A series of experiments in the sheet-flow regime was carried out at the wave tunnel of Delft Hydraulics at 'De Voorst' in May 1994. Preliminary results indicate clearly that even at these velocities (crest velocity, $U_c = 0.5$ to 1.8 m/s; through velocity, $U_t = -1.0$ to -0.3 m/s; $T = 6.5$ s), selectivity in the transport processes associated with light and heavy minerals was occurring.

Discussion: selective transport

The sensitivity of radiometric techniques for measuring heavy minerals in sands was demonstrated; this sensitivity is caused by the higher concentration, by two orders of magnitude, of radionuclides of the decay series of ^{238}U and ^{232}Th in the group of heavy minerals relative to the group of light minerals. However, the following limitations should be noted: (i) that absorption of γ-radiation by sand restricts the layer of sediment bed then can be monitored to about 50 cm, and (ii) this application is still in an exploratory phase and needs further investigation in environments of much less or much more (wave) energy.

In all the wave flume experiments it was hardly possible to avoid segregation, i.e. the phenomenon of segregation of light and heavy minerals is the rule rather than an exception. Thus, apparently, selective transport is an 'eigenmode'.

These observations raise the question of whether a dualism is present in the description of sediment movement. On the one hand, on a microscopic scale one is confronted with an extremely complicated description. All particles interact with each other and as a result of these interactions the boundary conditions (i.e. the sediment bed) change continuously. Turbulence or vortex motions generate complications that can only be handled in detail by the use of enormous computer capacity. On the other hand, nature confronts us with rather simple features. So on a macroscopic scale the challenge becomes to reduce the complexity of the description to a few quantities. In this section we would like to present a possible step in that direction. This step is strongly influenced by the observations through the glass wall of the wave flume, where we visually followed the motion of grains due to undulation.

A model for selective transport is presented in terms of gravity, a lift and a friction force. The description is semi-quantitative, utilizes almost exclusively physical relationships and incorporates an overall effective quantity, to cover chaotic motions like turbulence and vortices, and an empirical description of velocity within the boundary layer. Using this model the difference in the motion of heavy and light sand minerals can be accounted for.

The model calculates grain trajectories under oscillatory flow within a boundary layer. It is assumed that the fluid motion is only horizontal (in the x-direction) and varies logarithmically as a function of the distance, z from the sediment–water interface. Boundary conditions are given by:

$$U(z) = 0; \text{ at } z = z_0 \qquad (2)$$

$$U(z) = U(t)_{max}; \text{ at } z = z_0 + d$$

where z_0 is zero velocity level, d is the boundary layer thickness and $U(t)_{max}$ is the velocity at the edge of the boundary layer. The velocity is a function of time, following the wave motion.

In the first order, chaotic motions (like turbulence and vortices) are circular motions which lead to no net horizontal displacement and to effectively slower settling of the grains. This effect on the vertical motion is mimicked in the model by introducing an 'effective viscosity coefficient' in the friction force according to Stokes law:

$$F_{fric} = -6k\pi\eta r v_z \qquad (3)$$

with k being the effective viscosity coefficient, η the viscosity of the medium, r the radius of the grain and v_z the vertical velocity of the grain with respect to the flow. Since the coefficient k is greater than unity the friction force is enhanced. According to the general principle of separation of variables, and by including all viscosity effects into the friction force, the two other forces are acting on grains in a nonviscous fluid. Under these conditions the lift force can be described by Bernoulli's law:

$$F_{lift} = \tfrac{1}{2}A\rho_m[u^2(z + r, t) - u^2(z - r, t)] \qquad (4)$$

with A being the effective surface of the grain, ρ_m the density of the medium and z the height above the sediment bed.

Figure 5 presents a number of trajectories calculated with the model for a light (quartz), an intermediate and a heavy grain with density (grain sizes) of 2.6 kg/l (0.20 mm), 3.0 kg/l (0.25 mm) and 4.0 kg/l (0.15 mm), respectively. The results are presented for fixed values of the effective viscosity coefficient (20) and the thickness of the boundary layer and for various combinations of crest and through velocities, U_c and U_t. As can be seen in the figure, the light grains are moved forward and back by the wave motion, leading to an almost zero net displacement; the heavy minerals move only forward. Details on the experiments and the model may be found in Tánczos & de Meijer (1996).

The results obtained can be understood qualitatively by realizing that:

(i) the threshold velocity for moving light grains is lower than for heavy grains; and
(ii) under asymmetric waves the bottom velocities for the larger peak exceed the threshold velocity for both light and heavy grains; for the smaller peak this is so only for the light grains.

Based upon this model, it is to be expected that on the sea floor, segregation of light and heavy

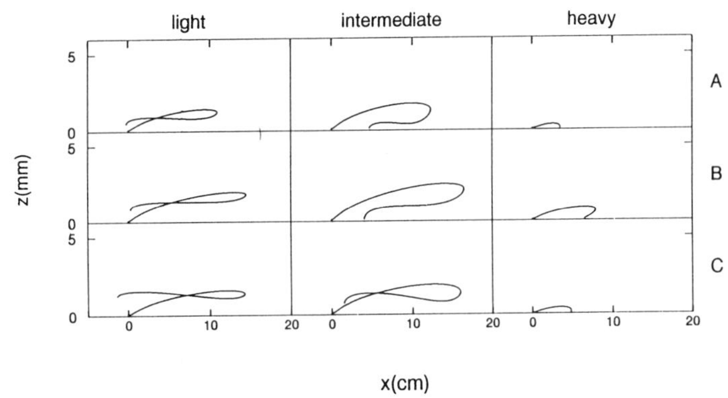

Fig. 5. Trajectories of a light (2.65 kg/l), intermediate (3.0 kg/l) and heavy (4.0 kg/l) grain under wave conditions with different peak velocities under crest and trough, respectively. The values for U_c and U_t are A, 0.5 and 0.3 m/s; B, 0.6 and 0.3 m/s; C, 0.5 and 0.4 m/s.

minerals will take place predominantly within a domain where waves and swell, in combination with current, will lead to asymmetric velocity patterns in the near-bottom boundary layer.

Experiments in an oscillating wave tunnel were recently carried out to further test and develop the model. In those experiments, the displacement of the minerals was monitored radiometrically (rather than sampling the sediment itself). This procedure allowed intermediate results to be obtained without too much interference with the experiment due to sediment removed by sampling. The results are being analysed.

Conclusions

The results of our field studies and the application of our simplified model have demonstrated that radiometric methods may contribute to the understanding of certain aspects of coastal dynamics. Restricting ourselves to selective transport features, we have become convinced that the combined differences in grain size and density play a much more important role than is usually acknowledged in coastal studies.

Measuring techniques, such as the towed sea-bed γ-ray detector system, enable the result of transport processes to be measured over time and space. These measurements may be carried out in the laboratory, under controlled conditions or at the beach and on the sea floor. They may provide supplementary information to stationary field experiments carried out, for example, rigs. The advantage of the technique is that information may be obtained over large areas and shortly after storm events, provided shiptime is available.

The advantage of large-scale mapping over spot measurements is that macroscopic information is obtained which may help to understand and identify eigenmodes in coastal dynamics. A clear example is that results from spot measurements near the Dutch coast indicate the importance of longshore transport (Hoekstra 1994), whereas large-scale mapping appears to indicate the importance of cross-shore transport.

On the basis of the present results it is intended to extend our activities in this field. In particular the radiometric fingerprinting model will be validated with detailed geological and sedimentological analyses. One possibility is to exploit further the data obtained from Bi, Th and K concentrations. As pointed out by de Meijer & Donoghue (1995), and de Meijer *et al.* (1994*a*), these three concentrations together with the (trivial) relation that relative masses sum to unity, allow in principle a derivation of the mass distribution of a sediment sample into four mineral groups. This method needs extensive calibration in the laboratory.

Combining the method of radiometrically deriving a mineral composition from Bi, Th and K activity concentrations from laboratory analysis or from measurements in the field with the towable sea-bed detector system may help to interpret the data in terms of changes of mineral composition. Thus it is hoped to unravel the segregation of quartz and feldspar and, for example, of garnet, zircons and monazite. Such data may shed light on the intricate transport processes occurring near the sea-bed.

On a longer time scale, it is hoped to answer the question of whether the concentration of heavy minerals is due to a local winnowing process or if there is a source from which the minerals are transported. In the latter case, the source may be an ancient deposit (pocket) which is presently eroding or, in the former case, the source may be more diffuse, with concentration being the result of selective transport and subsequent removal of light minerals.

Summarizing, radiogenic heavy minerals provide a new approach to coastal dynamics research. This approach seems worth exploiting on a large scale and further developing.

This work is part of the research programme 'Environmental Radioactivity Research' of the KVI, Groningen University. This work is also supported financially by the programme 'Kustgenese' (Coastal Genesis) of Rijkswaterstaat and by the Ministry of Economic Affairs (under the programme 'ISP-IV', administered by the Noordelijke Ontwikkelings Maatschappij).

The collaboration with BGS (Dr D. G. Jones and P. R. Roberts) and the assistance of D. Jellema in data collection and analysis during the 1993 cruise, together with the support of the personnel of Rijkswaterstaat on Ameland and on 'Blauwe Slenk', are gratefully acknowledged. The authors would also like to thank Dr ir. J. S. Ribberink, Prof. Dr ir. H. J. de Vriend, Prof. Dr ir. G. J. F. van Heijst, Dr J. Wiersma and Dr ir. J. van de Graaff for their critical support in the development of the transport model.

References

BONKA, H. 1982. Enhanced natural radiation exposure due to enriched heavy minerals at the coast of Northern Germany. *In*: VOHRA, K. G. *et al.* (eds) *Natural Radiation Environment*. Wiley, New Delhi, 58–66 (and references therein).

DE MEIJER, R. J. & DONOGHUE, J. F. 1995. Radiometric fingerprinting of sediments on the Dutch, German and Danish coasts. *Quaternary International*, **26**, 43–47.

——, LESSCHER, H. M. E., SCHUILING, R. D. & ELBURG, M. E. 1990. Estimate of the heavy-mineral content in sand and its provenance by radiometric means. *Nuclear Geophysics*, **4**, 455–460.

——, PUT, L. W., BERGMAN, R. *et al.* 1985. Local variations of outdoor radon in The Netherlands and physical properties of sand with enhanced natural radioactivity. *The Science of the Total Environment*, **45**, 101–109.

——, ——, SCHUILING, R. D., DE REUS, J. H. & WIERSMA, J. 1988. Provenance of coastal sediments using natural radioactivity of heavy-mineral sands. *Radiation Protection Dosimetry*, **24**, 55–58.

——, ——, ——, —— & ——1989. Natural radioactive heavy minerals in sediments along the Dutch coast. Proceedings of the 1987 KNGMG Symposium on *Coastal Lowlands, Geology and Geotechnology*. Kluwer, Dordrecht, 355–361.

——, TÁNCZOS, I. C. & STAPEL, C. 1994. Radiometric techniques in heavy-mineral exploration and exploitation. *Exploration and Mining Geology*, **3**, 389–398.

EVANS, R. D. 1969. *The Atomic Nucleus*. McGraw-Hill, New York.

GREENFIELD, M. B., DE MEIJER, R. J., PUT, L. W., WIERSMA, J. & DONOGHUE, J. F. 1989. Monitoring beach sand transport by use of radiogenic heavy minerals. *Nuclear Geophysics*, **3**, 231–244.

HOEKSTRA, P., HOUWMAN, K. T., KROON, A., VAN VESSEM, P. & RUESSINK, B. G. 1994. The NOURTEC experiment of Terschelling: Process-oriented monitoring of a shoreface nourishment (1993–1996). *In*: ARCILLA, A. S., STIVE, M. J. F. & KRAUS, N. C. (eds) *Coastal Dynamics '94: Proceedings of an International Conference on the Role of Large-scale Experiments in Coastal Research*. American Society of Civil Engineers, 402–431.

SCHUILING, R. D., DE MEIJER, R. J., RIEZEBOS, H. J. & SCHOLTEN, M. J. 1985. Grain size distribution of different minerals in a sediment as function of their specific density. *Geologie en Mijnbouw*, **64**, 199–203.

SLINGERLAND, R. 1984. Role of hydraulic sorting in the origin of fluvial placers. *Journal of Sedimentary Petrology*, **54**, 137–150.

STAPOR, F. W. 1973. Heavy mineral concentrating processes and density/shape/size equilibria in the marine and coastal dune sands of the Apalachicola Region, Florida. *Journal of Sedimentary Petrology*, **43**, 396–407.

TÁNCZOS, I. C. & DE MEIJER R. J. 1996. Selective transport of heavy mineral sands in a small-scale flame experiment. *Marine Geology*.

An interdisciplinary approach to the evaluation of physical parameters of shallow marine sediments

T. MISSIAEN[1], T. M. McGEE[2], D. PEARKS[3], G. OLLIER[4] & F. THEILEN[5]

[1] *Renard Centre of Marine Geology, Universiteit Gent, Krijgslaan 281 S8, B-9000 Gent, Belgium (currently at Département Géosciences Marines, IFREMER, BP 70, F-29280 Plouzané, France)*

[2] *Faculteit Aardwetenschappen, Rijksuniversiteit Utrecht, Budapestlaan 4, NL-3508 TA Utrecht, The Netherlands (currently at Thalassic Data Limited, PO Box 58097, Station 'L', Vancouver, British Columbia, V6P 6C5, Canada)*

[3] *School of Ocean Sciences, University of Wales Bangor, Menai Bridge, LL59 5EY, Gwynedd, UK*

[4] *Département Géosciences Marines, IFREMER, BP 70, F-29280 Plouzané, France (currently at DG XII-G1, Commission of the European Communities, Wetstraat 200, B-1049 Brussel, Belgium)*

[5] *Institut für Geophysik, Christian-Albrechts-Universitat zu Kiel, Olshausenstrasse 40–60, D-24098 Kiel, Germany*

Abstract: A large number of high-resolution seismic investigations and *in situ* geotechnical measurements were carried out in the southern Baltic, offshore Bornholm. The geotechnical measurements, carried out with Ifremer's Geotechnical Module and Kiel's GISP system, provided information on cone resistance, induced pore pressure and density of the surficial sediments. The seismic data were subjected to different processing techniques, including detrending and multiple quotient division, which made it possible to deduce the reflection coefficients and impedance contrasts at subbottom interfaces. The different geotechnical and seismic data sets were carefully integrated. There is a good correlation between the *in situ* results and the interpreted seismic units. Strong variations in geotechnical plots correspond with the reflectors identified on processed seismic sections. The general trend of calculated impedances agrees well with the lithological interpretation. However, it would appear that the inversion method does not always result in quantitatively correct values of the sediment parameters. Further development and refining of this method will undoubtedly provide better estimates of the *in situ* bulk properties of the shallow marine sediments.

The knowledge of the physical properties of shallow marine sediments is not only of interest in understanding the basic geoscientific processes controlling marine sedimentation, but also plays an important role in many geotechnical and environmental applications. Physical property measurements are frequently performed in the laboratory on sediment cores, but unfortunately the results are not always reliable due to sampling disturbance (Crusius & Anderson 1991). Recent studies, including *in situ* measurements with sensors directly penetrating into the surficial sedimentary layers, have proved to provide more reliable and often more precise information on the physical and geotechnical parameters of sea-floor sediments (e.g. Ollier *et al.* 1992; Fang *et al.* 1993; Baltzer *et al.* 1994).

Significant information for shallow marine sediments can also be obtained from the seismic waveform of reflection seismic records, provided the data are of good enough quality and high enough resolution. A well-established and efficient tool in this approach is the recovery of acoustic impedance through very careful, and often complicated, data processing (e.g. Oldenburg *et al.* 1983; Berteussen & Ursin 1983; Arntsen & Ursin 1993). Because of the theoretical and empirical relationships that exist between the seismo-acoustic and geotechnical parameters, it is then possible to estimate the sediment's bulk geotechnical properties (Buchan *et al.* 1972; Haynes *et al.* 1993). However, as yet no system is available that can extract meaningful results consistently. The degree of correlation of the relationships can vary widely, and there still remains a large amount of research to be undertaken on determining the precise nature of the geoacoustic relationships.

From De Batist, M. & Jacobs, P. (eds), 1996, *Geology of Siliciclastic Shelf Seas*, Geological Society Special Publication No. 117, pp. 299–322.

To date, few studies have tried to combine the results of *in situ* geotechnical measurements and seismic processing for physical properties. In the framework of the EEC MAST project 'GISP – Geophysical In Situ Probe' a large number of high-resolution seismic and *in situ* probing investigations were carried out in the southern Baltic Sea. Seismic interpretation, combined with some core data, resulted in a Late Quaternary stratigraphical model of the area. Estimates of the reflection coefficients and acoustic impedance were deduced from the digital seismic data using a relatively simple inversion scheme. The *in situ* geotechnical measurements, carried out with two different penetration systems, resulted in cone resistance, excess pore pressure and density data. An attempt was made to integrate these different results from seismic stratigraphy, seismo-acoustic processing and *in situ* probing. Such an interdisciplinary approach could help to provide a better definition of the surficial marine sediments, and an improved understanding of the geological processes involved.

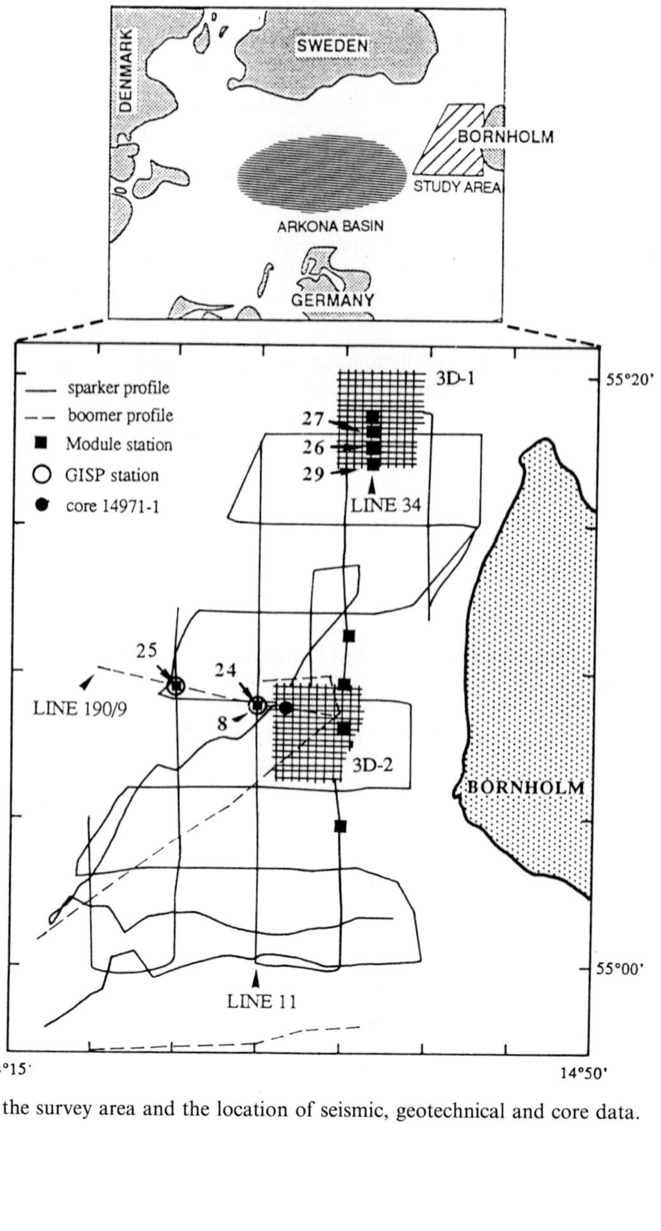

Fig. 1. Overview of the survey area and the location of seismic, geotechnical and core data.

Data acquisition

The seismic and geotechnical investigations of the GISP project were carried out in two adjacent areas of the southern Baltic Sea, namely the Arkona Basin and the Danish sector west of Bornholm (Fig. 1). In the scope of this paper only the survey area offshore Bornholm is considered. This area largely coincides with the Rønne Graben, and borders the eastern margin of the Arkona Basin.

Seismic investigations

Over 600 km of high-resolution reflection seismic profiles were collected offshore Bornholm in May 1991 with the 'RV Belgica' (Fig. 1). The data include a large-scale network for reconnaissance of the area (average line spacing ±5 km), and two small-scale dense networks (3D-1 and 3D-2) for detailed studies (average line spacing 300 and 450 m). The resolution and location of the latter were determined by the scale of the geological structures and the suitability of the site for *in situ* measurements. All profiles were shot using a 300 J comb-type sparker array and single-channel streamer. The data were recorded on paper and on DAT cassettes, and later digitized with the MDAS acquisition system. Positioning during this survey was done using the Decca system.

A few additional high-resolution seismic lines were acquired offshore Bornholm during a geotechnical survey with the 'RV Poseidon' in June–July 1992. The profiles were shot using a 300 J boomer source and single-channel streamer. The data were recorded digitally with the MDAS acquisition system, and also on paper. Positioning was done with GPS (Global Positioning System).

In-situ geotechnical investigations

Two different sea-floor penetration systems were involved in the GISP project: (1) the Geotechnical Module, developed by Ifremer, and (2) the GISP penetration system, developed by the University of Kiel. A large number of *in situ* geotechnical measurements were carried out offshore Bornholm with both probing systems during the 1992 'Poseidon' survey (Fig. 1). The

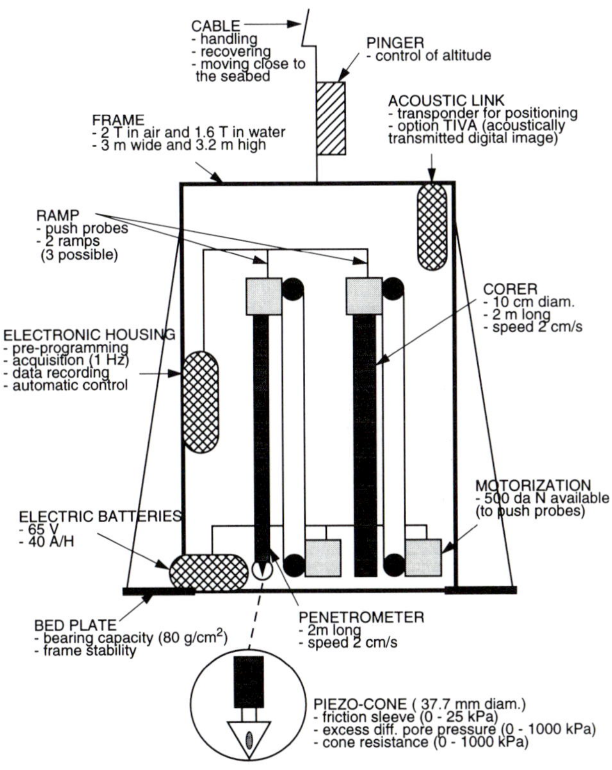

Fig. 2. Configuration of the Geotechnical Module.

different approaches of the two systems (shallow penetration with very high resolution versus deeper penetration with lower resolution) will allow a complementary study of the upper marine sediments.

Geotechnical Module. The Geotechnical Module is a sea-floor supported rig equipped with up to three different ramps for measurements of the sea-floor properties, or for coring (Fig. 2). The ramps penetrate the sea floor to a depth of 2 m in fine-grained sediments at a speed of 2 cm/s. The system is designed to operate in a maximum water depth of 6000 m. The coring and measurement operations are entirely automatic, and they are triggered by the acquisition system according to a pre-programmed timing (max. 20 operations per dive). The operations are automatically interrupted in case of tilting of the Module or when the encountered resistance is too big. The digital data are stored on an 8-bit acquisition microcomputer which is mounted on the Module frame. On board the data are then transferred to a standard PC.

During the 1992 survey two ramps were used, respectively equipped with a penetrometer probe and a coring device. The probe consists of a high-resolution piezo-cone equipped with a strength sensor (0–1000 kPa) and a pressure sensor (0–1000 kPa). The penetration speed is 2 cm/s (5% precision); cone resistance and excess pore pressure data are recorded every second. Several penetrometer runs were carried out per station, at a few metres spacing. The coring device operates like a conventional piston corer; it provides 2 m long push cores in unconsolidated sediments. The cores were subsequently analysed for grain-size distribution and gamma-ray density (using a ^{137}Cs source of 0.1 mCi).

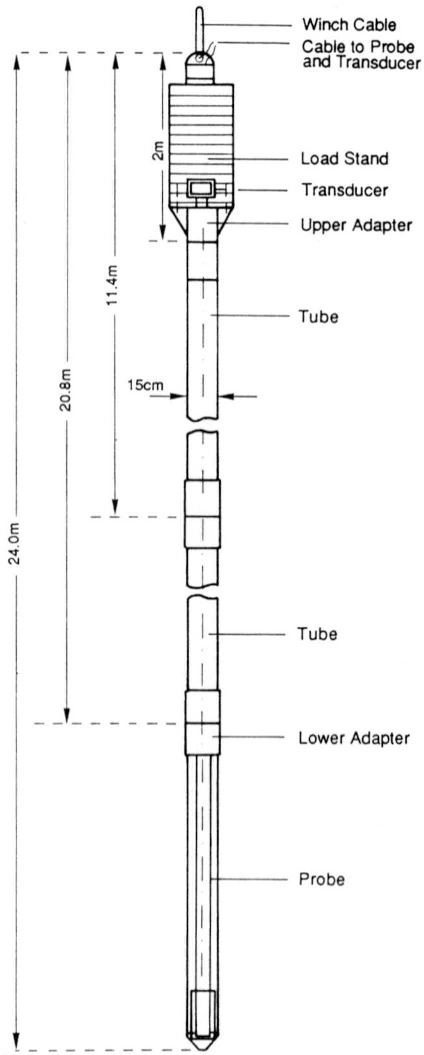

Fig. 3. Configuration of the GISP penetration system.

GISP penetration system. The GISP penetration system is based on conventional gravity coring techniques. It consists of a 9 m heavy tube of 15 cm diameter loaded with a lead weight stand of 1.5 tons (Fig. 3). The probe itself is mounted at the lower end of the tube, and is connected to a central data acquisition system and power supply on board. The system is developed for use in soft sediments, such as clays and mud, and penetrates the sea floor under its own weight with a speed of ±10 cm/s. The penetration depth is controlled by a pressure sensor with an accuracy of 1 cm, mounted just above the probe sensors. In order to guarantee safe handling even under rough sea conditions the system is deployed in a steel frame, similar to one normally used for coring devices.

During the survey in the southern Baltic different sensors were used to measure the natural gamma radiation and the wet bulk density. The sensor for natural gamma radiation is a conventional system as used in borehole logging techniques. The density data were acquired with a gamma–gamma sensor equipped with a protected ^{60}Co source of 3.5 mCi intensity, suitable for operations in a protection tube. This sensor was calibrated in sea water. Sampling intervals for both probes were about 10 cm. If possible, different runs were carried out per station, using the same sensors.

Lithological data

A 9 m long gravity core (14971-1) was acquired near Bornholm during a survey with 'RV Poseidon' in January 1990 (Fig. 1). The core contained three different lithological units (Fig. 4). The oldest sediments consisted of brown-red varved clay deposits marked by moderate carbonate content and extremely low total organic content (TOC). They are thought to represent the Baltic Ice Lake phase (Schubert 1991). A thin sandy layer, dated ±10 200 years BP, marked the top of these deposits. Further up, the core consisted of grey-brown silty clay deposits interrupted by thin sandy layers and showing an increase in silt content towards the top. Their high carbonate content and the occurrence of various plant fragments suggested a correlation with the Ancyclus Lake phase (Schubert 1991). The youngest sediments consisted of an olive-green marine mud rich in organic matter, and marked by a sandy level and various plant fragments in the upper part. The sharp drop in carbonate content and rise in TOC at the base of these sediments, which was dated ±7500 years BP, indicates the onset of the Litorina Sea transgression (Schubert 1991).

Resolution and positioning

In order to expect a realistic correlation between the different data, it is necessary to know the position of the geotechnical measurement sites with respect to the seismic lines with sufficient accuracy. This is especially important as the data were collected during different cruises, and various positioning systems were used. As all cruises were carried out within a period of some two years (1990–1992), any possible temporal variations in sediment properties were considered to be negligible.

The positioning accuracy of the seismic sparker lines was about 50 m whereas the accuracy of the boomer lines as well as the geotechnical sampling points was about 10 m. Although this can by no means be considered optimum, a correlation between the data is still thought quite feasible. Nevertheless, this correlation must be attempted carefully, as positioning errors may cause discrepancies in the depth of particular interfaces.

An examination of the vertical resolution of the data is also needed to ensure that different types of data are compatible, in terms of scale, in order to allow any correlation. The quality of

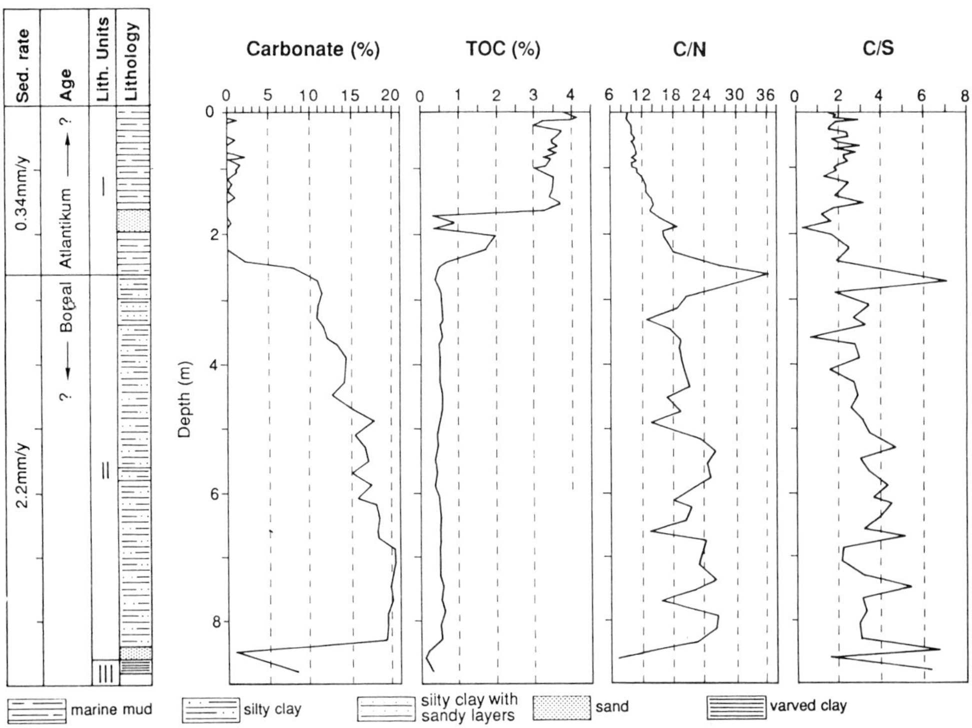

Fig. 4. Lithological description of core 14971-1 and its carbonate content, total organic content (TOC), C/N and C/S ratios (after Schubert 1991).

the recorded seismic data was very good, and resulted after processing in a vertical resolution of some 40 cm. The resolution of the *in situ* measurements with the Module and the GISP system was respectively 1–5 cm and 20–40 cm. The latter was linked to the distance between the source and detector in the probe. It was therefore reasonable to expect some correlations to be possible between the seismic and geotechnical data.

Stratigraphical interpretation

A sequence-stratigraphical interpretation of the high-resolution seismic data, in combination with the lithological information from core 14971-1, has led to the identification of two Quaternary depositional sequences containing up to nine different seismic-stratigraphical units (Fig. 5) (Perini *et al.* 1996). These units reflect the complex series of glacial, lacustrine and marine phases that have marked the recent history of the southern Baltic.

The oldest depositional sequence contains five units (A–E) of glacial and late-glacial origin. The glacial units A and B are relatively thin, with local thickenings inside channel depressions in the bedrock and where large mound-like structures are formed. They consist of till and moraine deposits, probably of Middle or early Late Weichselian age. The overlying units C, D and E contain glacial lacustrine clay and clayey-silty sediments, deposited during different Baltic Ice Lake phases of the Late Weichselian. The units are characterized by an aggradational style, and are affected by numerous gas patches and deformations related to mud diapirism.

The youngest depositional sequence contains four units (F–I) of postglacial (Holocene) lacustrine and marine origin. Units F and G are probably related to the Ancyclus Lake transgression. Unit F only occurs in the northeast, and consists of a complex body of backstepping beach or delta deposits, indicating a step-wise transgression. The onlapping unit G covers almost the entire survey area, and contains lacustrine silty clay deposits marked by vertical accretion. The uppermost marine mud deposits of units H and I can be correlated with the Littorina Sea transgression. Both units are rather thin, locally ≤ 1 m and therefore difficult to resolve on the seismic profiles.

Geotechnical interpretation

Geotechnical Module

Because the cone resistance and pore pressure both depend upon the type of sediment, the penetrometer measurements allow for a qualitative assessment of the lithology from the shape of the curves. Clayey sediments show a low cone resistance and relatively high pore pressure; sandy or silty layers, on the other hand, are indicated by a sudden increase in cone resistance and a drop in pore pressure (Couture 1990). The latter was clearly observed on several sites (e.g. Module station 24 – Fig. 16), often supported

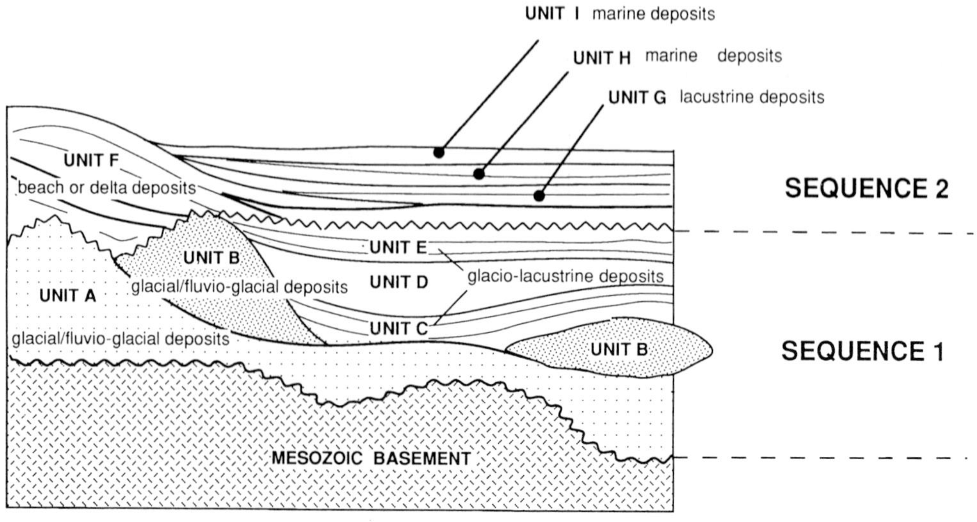

Fig. 5. Schematic diagram of the depositional units and sequences offshore Bornholm (after Perini *et al.* 1996).

by cores located nearby. Furthermore, the penetrometer results also illustrate the soil deposit stress history according to the fact that if the soil is normally consolidated, the tip resistance would increase linearly with depth (Cochonat *et al.* 1988; Baltzer *et al.* 1994). The measured cone resistance (Q_c) is related to the measured excess pore pressure (U_t) by $Q_t = Q_c + U_t(1 - a)$, with Q_t = true cone resistance which accounts for the water pressure

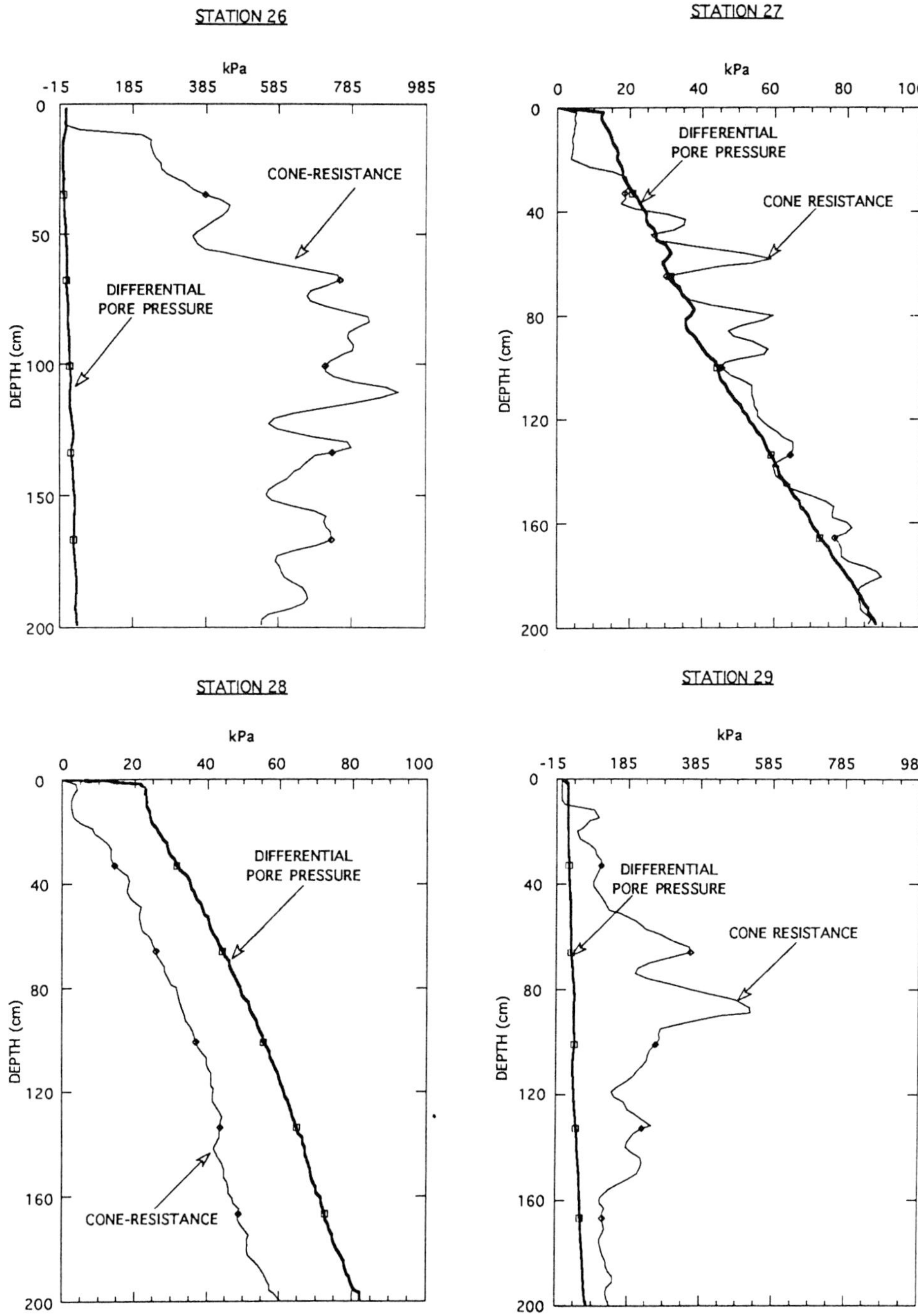

Fig. 6. Penetrometer results for different *in situ* Module stations offshore Bornholm.

acting on the back of the cone tip, and $a = 0.72$ (cone parameter). This correction of the cone resistance, due to the influence of the pore pressure, will become important for clayey sediments.

The cone resistance plots offshore Bornholm show a large variety of signature shapes, ranging in value from less than 10 kPa to more than 900 kPa (Fig. 6). This suggests a considerable variation in sedimentary facies going from very loose sediments with minimal cohesion, to more consolidated, resistant sediments. The latter, when reaching a resistance of over 1000 kPa, may cause the penetrometer to stop before reaching the maximum penetration depth of 2 m (Hevin 1993). The density values range from 1.1 g/cm^3 (soft sediments) up to a maximum of 2.2 g/cm^3 (dense sediments).

The excess pore pressure plots are almost perfectly linear, without large variations, and do not seem to support these sedimentary variations. This is probably due to an incomplete saturation of the piezo-cone caused by the low hydrostatic pressure of the shallow-water environment, which may have allowed small air bubbles to get caught in the cone (Couture 1990). Such compressible air bubbles will act like a filter, resulting in a smooth curve of average pore pressure values which are clearly not entirely representative.

GISP penetration system

The *in situ* density plots were calculated after conversion of the pressure values into depth data

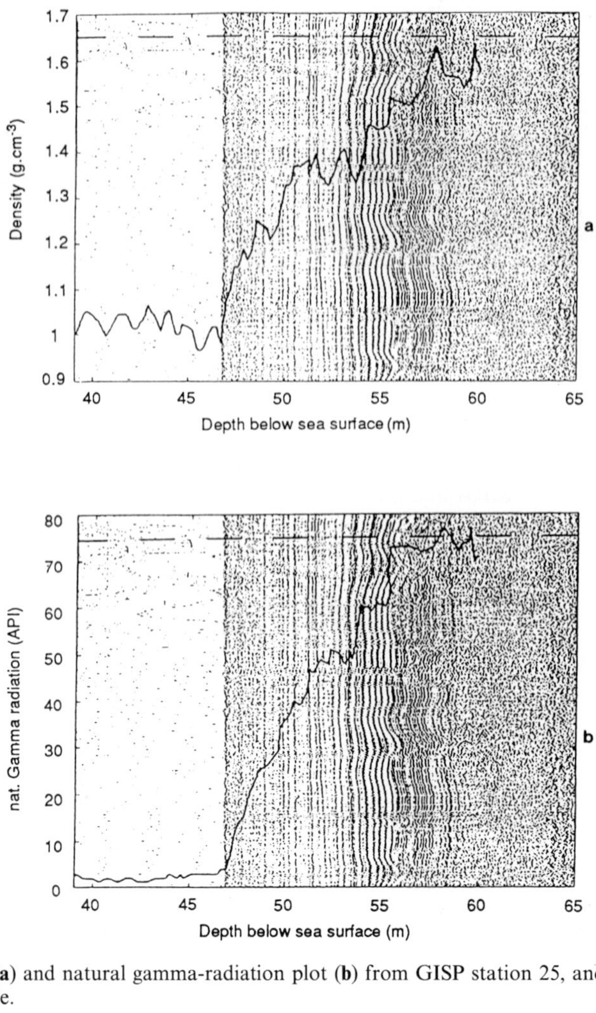

Fig. 7. Density plot (**a**) and natural gamma-radiation plot (**b**) from GISP station 25, and the corresponding seismic boomer profile.

and after calibration. In order to minimize the high-frequency fluctuation of the signals a Butterworth filter was applied, and the signals from different runs were stacked. Correlation with box core measurements indicated a deviation of 10% between the *in situ* density values and the box core densities. This deviation is most likely due to the calibration in water, and was taken into account by multiplying by a factor $(Z/A)_{sediment}/(Z/A)_{water}$.

The density–depth curves are often marked by a strong gradient obviously related to the increase in density due to consolidation with age and burial depth. In most cases, marked changes in the density gradient can be related to the subsurface structure observed on the corresponding seismic sections (Fig. 7a). This is not surprising, since layer boundaries in surficial marine sediments are more strongly characterized by a change in density than by a change in velocity. Some deviations can be observed in the deeper part of the density plot. The reason for this is not clear, although it could be related to the calibration.

The natural gamma-radiation curve is also characterized by a strong depth gradient, but the gradient variations are less pronounced (Fig. 7b). However, an interpretation cannot be given in a simple way; as the radiating particles are bound to clay particles, the measured radiation may reflect the clay content. Furthermore the density of the clay, the origin of the clay particles and the decay elements of the origin elements have also to be taken into account.

Seismic data processing

Several processing steps were applied to the high-resolution seismic data in order to recover the reflection coefficients at the reflecting interfaces, and calculate the acoustic impedance of subbottom layers. It is generally difficult to measure reflection coefficients directly from reflection seismograms because certain important system characteristics, i.e. total instrumental response and source signature, are usually not well known. It was, however, possible to circumvent some such difficulties by developing a relatively simple, indirect inversion procedure that utilizes a process of division by which features of the signal that occur in both primaries and multiples tend to cancel out (multiple quotient theory).

In order for this method to be successful, two conditions must be satisfied: (1) the offset between source and receiver must be very small so that reflections occur at nearly normal incidence, thereby minimizing mode conversion at reflecting interfaces, and (2) the water-bottom must comprise a system of nearly horizontal layers, i.e. dips less than 5 degrees, so that primaries and multiples propagate along virtually the same path. Also, interpretation of the result becomes easier if primary and multiple reflections do not overlap in time.

Pre-inversion processing

Any adequate description of the water-layer multiple sequence relies on relatively low-frequency information (McGee 1991*a*). If low-cut filtering is avoided, however, the seismic events will often be superimposed on slowly varying background signals (e.g. due to electrical induction and water wave action), the so-called trends. Since these vary in time, they do not cancel out and thus constitute a source of error for the inversion.

The procedure for trend removal consists of four principal operations: (1) numerically integrate the digital seismogram to obtain its 'primitive'; (2) smooth the primitive to suppress rapidly varying oscillations; (3) differentiate the smoothed primitive to obtain an estimate of the trend; and (4) subtract the estimated trend from the original seismogram (Brussaard & McGee 1992). The smoothing was done using a 'running-average' operator approximately as long as the source pulse, and the differentiation was done using a five-point operator based on a Taylor series expansion. The procedure proved to be quite fast in computer time, and gave good results.

Inversion–deconvolution of first water-layer multiple to primary

The primary seismic sequence produced by a noise-free source signal reflected from the sea floor and underlying interfaces can be represented in the frequency domain as a multiplication of the Fourier transforms of the system and earth response functions, i.e. $S(f)G(f)$. Except for the constructive interference of internal multiples (McGee 1991*a*), the first water-layer multiple will be $S(f)G(f)G(f)$. The Fourier transform of the earth response can therefore be represented as the spectral quotient $S(f)G(f)G(f)/S(f)G(f) = G(f)$. In practice, this quotient may not be stable due to zeros in the denominator. An estimate of its value may be calculated by using a well-known method that involves adding a small 'white noise level' to the denominator (Claerbout 1992). An estimate of the time function $g(t)$ can then be obtained by inverse Fourier transformation.

The estimate of $g(t)$ is a sequence of unit spikes, each convolved with an approximately symmetrical wavelet, plus noise. The spikes are located at the onset times of seismic reflections, both primaries and internal multiples. The maximum amplitude of each nearly symmetrical wavelet corresponds to the amplitude of the seismic reflection. As a result of the deconvolution, these wavelets generally have broader bandwidths than do the original seismic signals. This allows the processed results to exhibit a significant increase in resolution over the original data, as shown in Figs 8 and 9.

Correction for wavefront divergence

The estimate of $g(t)$ must be corrected for geometrical spreading before reflection coefficients may be recovered from it. If the wavefront is spherical, the amplitude loss due to geometrical spreading can be corrected by applying a predetermined linear gain function. Both the sparker and the boomer source have a non-spherical radiation pattern that is a function of frequency (Verbeek & McGee 1995; Verbeek 1995). In the vicinity of their main lobe, however, the wavefronts have been observed to spread more or less spherically at moderate frequencies (Verbeek, pers. comm.). Since the

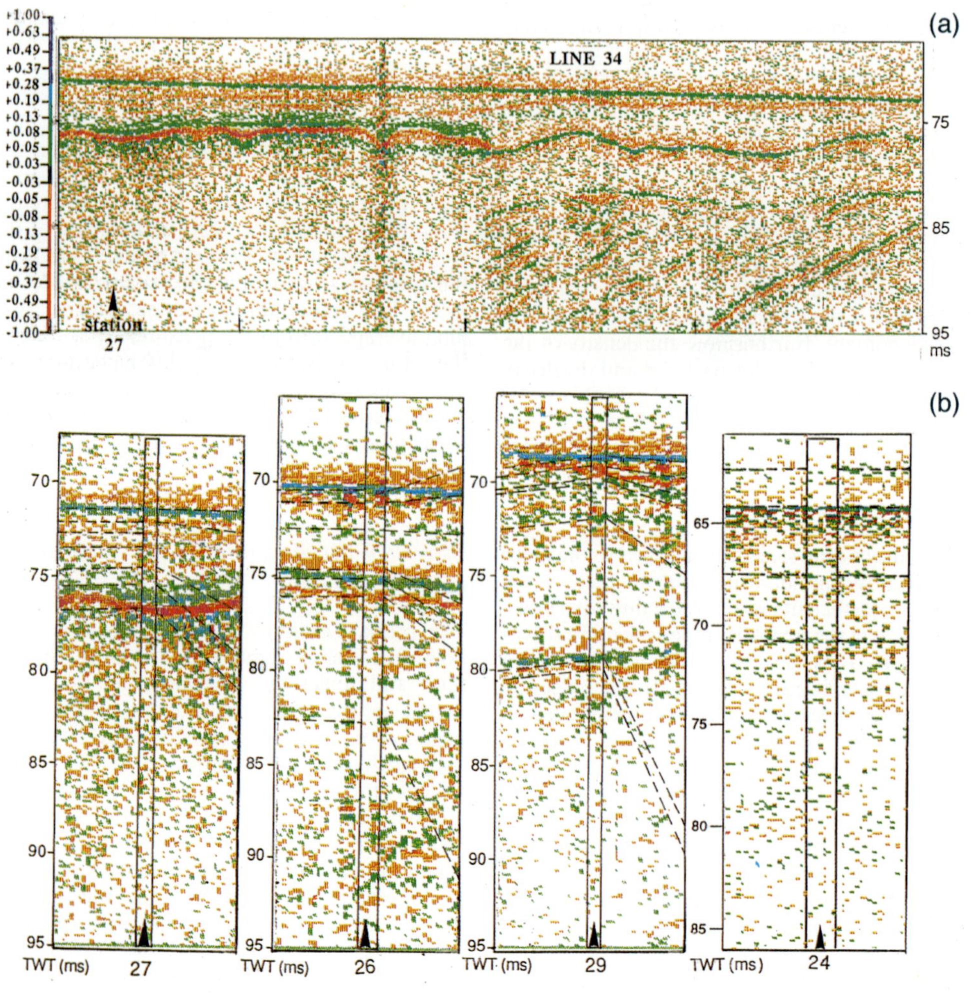

Fig. 8. (**a**) Seismic sparker section of line 34 after trend removal and multiple quotient division. The amplitude colour code is marked on the left; (**b**) processed sparker sections across Module stations 27, 26, 29 and 24.

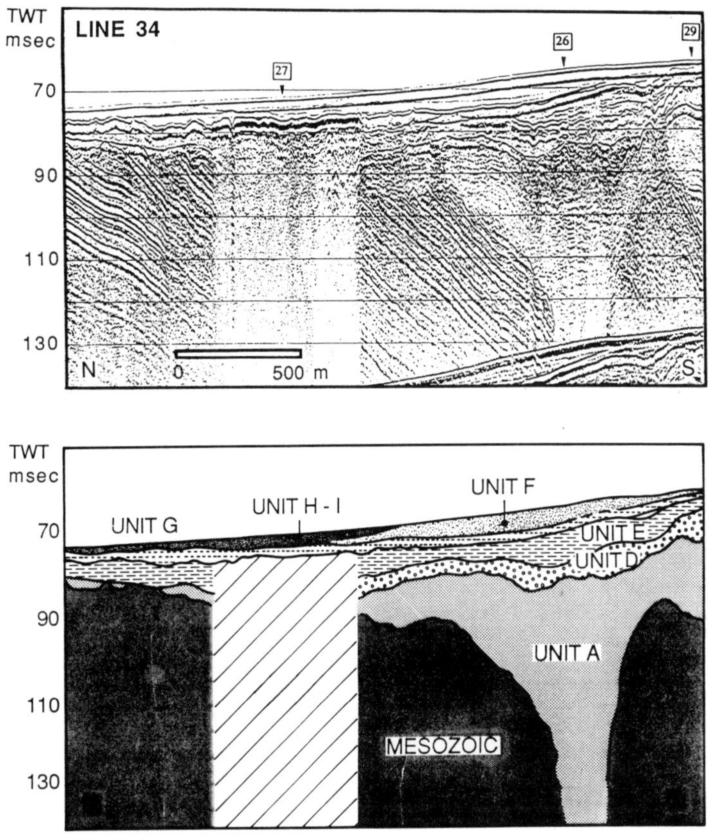

Fig. 9. Analogue profile and interpreted line-drawing of sparker profile 34. The 'striped' area indicates acoustic blanking due to the presence of gas.

distance between source and receiver was small and the sea floor was relatively flat, the data would have been recorded near the main lobe of the radiation pattern. It was therefore considered to be adequate, in the context of this study, to correct both sparker and boomer data for spherical divergence. These corrections were applied after the deconvolution procedure because they tend to distort waveforms and therefore can degrade the results.

Recovery of reflection coefficients

The obtained amplitude values may be used without further modification to recover the reflection coefficients if (1) the effects of absorption are negligible, and (2) the reflectors are smooth enough not to generate significant backscatter. Both of these conditions were assumed to be valid here. The recovery of reflection coefficients is done recursively, the amplitude of each subbottom primary reflection being an estimate of that reflection coefficient multiplied by the transmission coefficients of all primary reflections above it. The interpretation of primaries and multiples is quite important. Multiple reflections tend to occur in sequences that correspond to multiple passes between two particular interfaces (McGee 1991*b*). If the latter bound a region of locally higher (respectively lower) acoustic impedance, this will produce a sequence of negative (respectively positive) reflections. If the two interfaces reflect with the same polarity, they will produce multiple reflections of alternating polarity. In this case, the only clues indicating a multiple sequence are the regularity of occurrence in time and the decay rate.

Two different approaches were used to obtain values of reflection amplitudes. The first method involves the selection of a representative trace. The reflection amplitudes were obtained by referring to the values of the digital samples. The second approach was to print a processed section on which amplitudes are colour coded to represent normalized reflection amplitude values ranging from −1 to +1 (see Fig. 8a). This is a more subjective measurement but has the advantage of some visual averaging. This method turned out to be especially useful with noisy data showing a great trace variability, for which it was almost impossible to pick one representative trace. Both approaches were tried on various data sets, and the values obtained were mostly very similar. This demonstrated that there is a considerable degree of conformity between the two techniques, as would be expected since they are based on the same data.

Calculation of acoustic impedance

For normal incidence, the reflection coefficient R across an interface is related to the acoustic impedance of the material below (Ib) and above (Ia) it by the relationship $Ib = Ia(1 - R)/(1 + R)$. Since the acoustic impedance of water is well known, the impedance values of the layers below the sea floor can then be calculated recursively from the recovered reflection coefficients. It is, however, important to realize that the latter are only estimates and that an error in any one of them affects the results of all subsequent acoustic impedance calculations. This can have particular significance if the nature of a reflection, primary or multiple, is misinterpreted. Errors in deeper acoustic impedance calculations are then compounded because all deeper reflection coefficients are erroneous.

Geotechnical relationships

Within the literature, a large number of empirical relations exist between the seismo-acoustic properties of marine sediments (impedance, velocity, attenuation) and the most important geotechnical properties (density, shear strength, grain size, porosity, moisture content) (e.g. Hamilton 1970, 1972; Buchan *et al.* 1972). The degree of correlation of these relationships is often very poor, due to the large number of interrelated chemical and physical processes involved, and the large variety of sediments encountered (Haynes *et al.* 1993). The relationships are therefore particularly site-dependent and often not very well defined.

Nevertheless, good correlations have been obtained between values of acoustic impedance and porosity and density respectively, both from sea-floor investigations (Hamilton 1972) as well as from laboratory experiments on cores (Richardson & Briggs 1993). It is known that for shallow, unconsolidated marine sediments the acoustic impedance is controlled much more by variations in bulk density than by variations in acoustic velocity. Thus, for such sediments the reflectors visible on the seismic profiles should correspond well with changes in the density curves measured by geotechnical probes (Declerck 1990).

Acoustic impedance can also be an indicator of grain size and composition, although these relationships are more subject to scatter (Richardson & Briggs 1993). No empirical correlations between acoustic impedance and cone penetrometer resistance can be found in the literature. Cone penetration resistance, however, is dependent on density and porosity, both of which are strongly related to impedance, and therefore some amount of correlation would be expected between cone resistance and impedance.

Integration and interpretation of the data sets

Selected sites

Due to the time-consuming seismic processing procedures, it was decided to focus attention on two specific target areas where different data sets were available. The first area is located on the southern edge of the 3D-1 zone, near sparker profile 34 (Fig. 1). The Quaternary cover in this area is relatively thin, with the exception of local thickenings inside channel depressions in the bedrock, and is marked by several gas patches (Fig. 9). The uppermost marine mud layer is extremely thin or locally absent, and makes it impossible to distinguish between the individual units H and I. Four different *in situ* stations were investigated with the Geotechnical Module along profile 34. The stations are located at the foot of the prograding body of unit F (Fig. 10), which is partially interfingered with the onlapping unit G. Three of the stations, 26, 27 and 29, were selected for the discussion presented here.

A second target area is located in the central part of the basin area, at the intersection of sparker profile 11 and boomer profile 190/9, and close to core 14971-1 (Fig. 1). At this location *in situ* measurements were carried out both with the Geotechnical Module (Module station 24), and with the GISP penetration system (GISP station 8), which made it possible to compare both data sets. The Quaternary cover in this area consists largely of a thick sequence of glacio-lacustrine clays. The top of this sequence is marked by a sandy layer. The overlying postglacial deposits (units G, H and I) are also relatively thick in

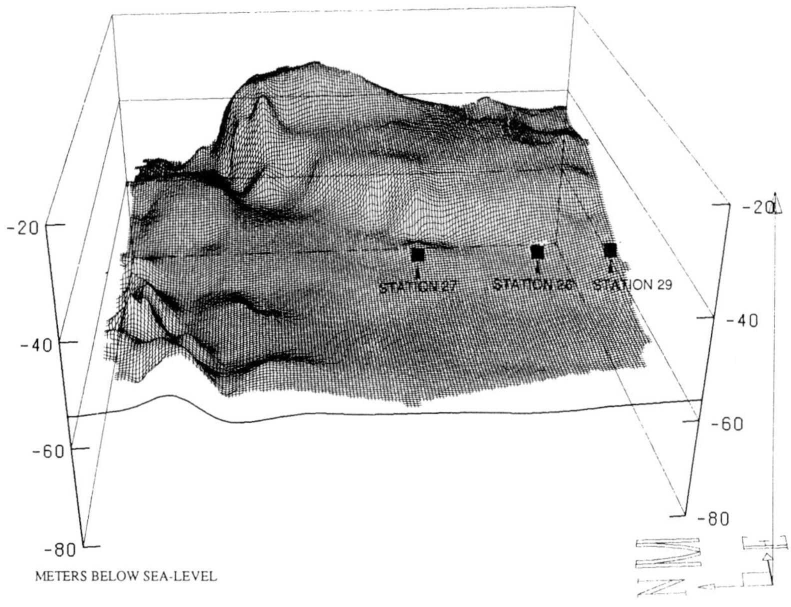

Fig. 10. Three-dimensional view of the prograding body of unit F in the 3D-1 area.

comparison with those further north (Fig. 11). Some internal stratification and onlapping can be observed in units G and H; the latter is separated from the uppermost unit I by a relatively strong unconformity.

The results from seismic stratigraphy, seismic processing and *in situ* geotechnical measurements were combined for the different Module–GISP stations. The reflection amplitudes were all obtained from the seismic records by using the colour scale method. The colour-coded seismic sections across the different stations are shown in Fig. 8b. Density estimates were obtained from the calculated acoustic impedance values by using a constant compressional wave velocity of 1500 m/s in the surficial sedimentary layers. As these layers mainly consist of cohesive soils that are fully saturated with water, i.e. clays and muds, this approximation is assumed to be valid.

Module station 27

The penetrometer results at Module station 27 (Figs 12 and 13) indicate the presence of a very thin layer (±20 cm) of surficial muds that are completely unconsolidated. Unfortunately, this layer is too thin to be resolved on the seismic section, but analogy with the observations at Module station 24 suggests that it could well represent unit I. Below this surficial muddy layer the sediments still show a low but slowly increasing consolidation (<100 kPa), indicating very soft sediments. The cone resistance plot is marked by several sharp excursions. The latter are often accompanied by a drop in pore pressure, indicating the presence of thin sandy or silty intercalations. This stratification agrees well with the seismic facies interpretation of the outcropping units G and H.

The processed seismic section is marked by a strong gas-enhanced reflection, below which acoustic blanking completely masks the record. Most likely the gas is contained in the upper part of the glacial lacustrine clay deposits (unit E). The small fissures observed at the bottom of the Module core can probably be linked to this presence of gas. A series of negative reflections is observed immediately below the sea floor. These could well be related to the thin sandy intercalations; indeed for such layers, almost beyond the limit of resolution, it is possible that only one negative reflection would be observed instead of a positive and a negative reflection at the upper and lower boundaries of the increased acoustic impedance. The variations observed in the density plot seem to support this stratification (Fig. 13). However, the reduced length of the core makes is difficult to correlate the individual variations with the different excursions in the cone resistance

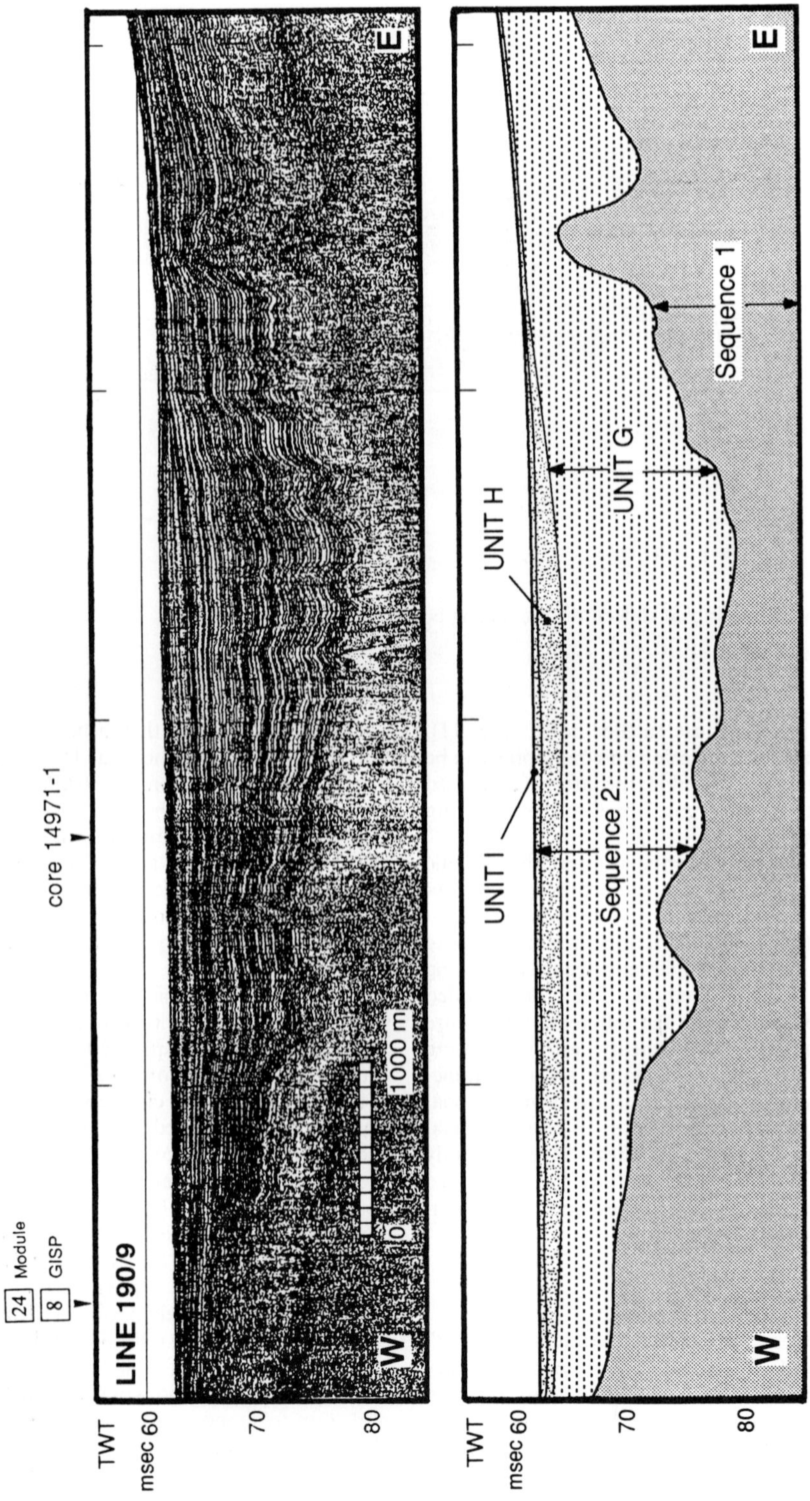

Fig. 11. Analogue profile and interpreted line-drawing of boomer profile 190/9.

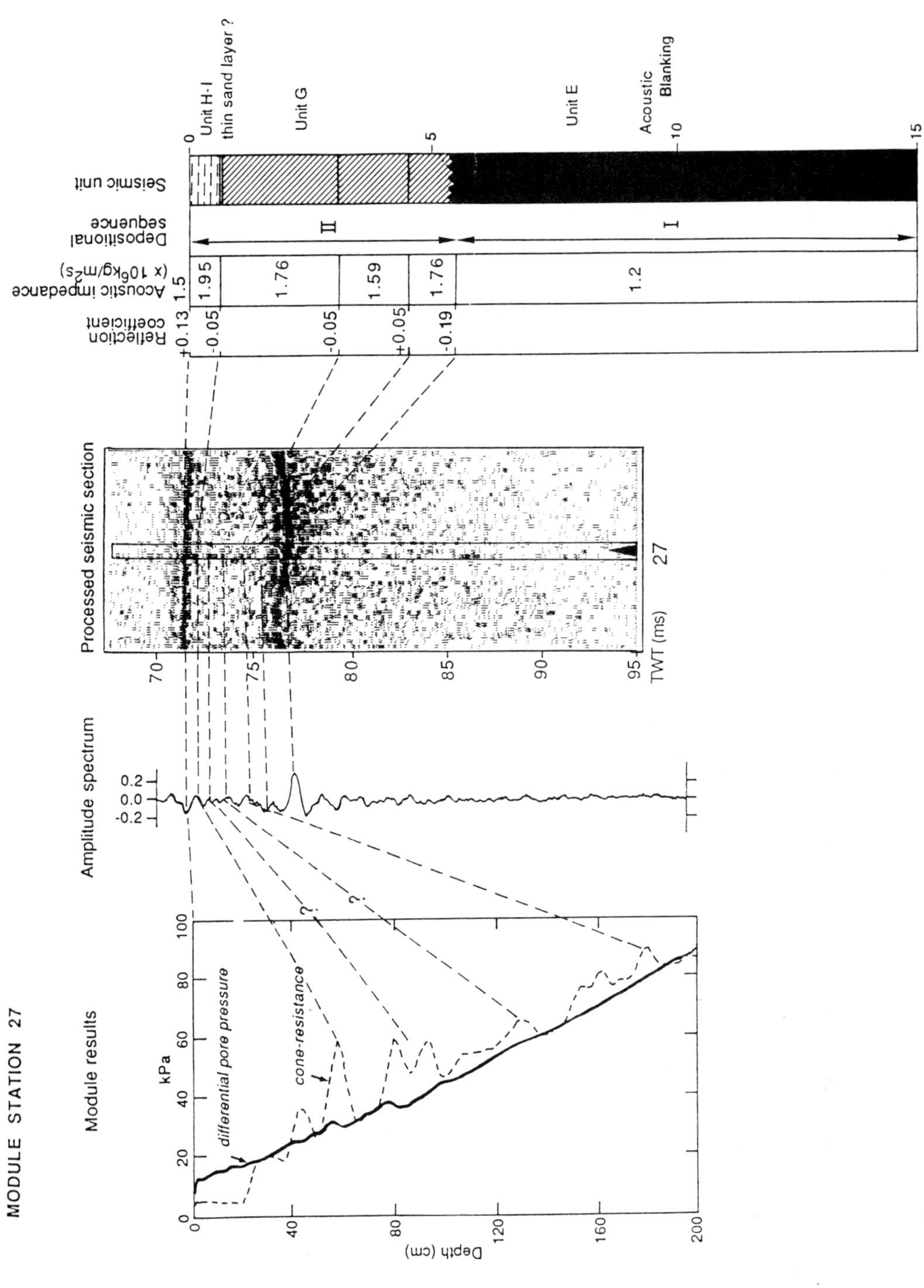

Fig. 12. Integration of different geotechnical, seismic and acoustic data from Module station 27. The colour-coded seismic section is shown in Fig. 8b.

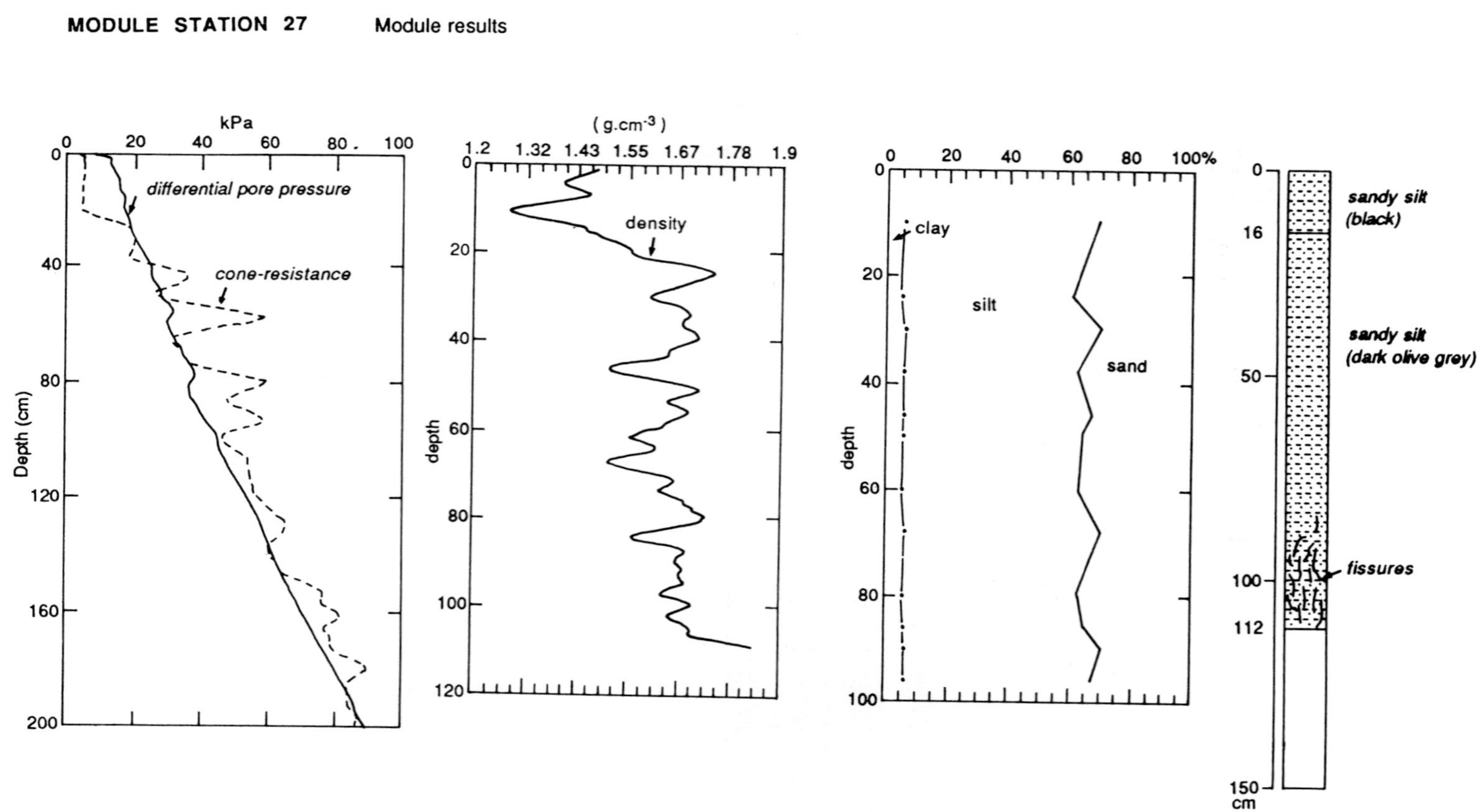

Fig. 13. Geotechnical data from Module station 27. From left to right penetrometer, density, granulometry and core results.

plot. As there is no evidence of a resistant layer that may have halted the coring operation, it is most likely that the bottom of the core corresponds with a depth of 2 m. The core reduction is probably due to a combination of sediment loss and consolidation of the soft sediments. The granulometry plot is marked by small variations (up to 10%) in sand content. Although these are not visible on the core itself, which has a rather uniform appearance, they may well be large enough to explain the density variations. This is quite important, as it seems to indicate that even slight modifications in physical properties can induce visible changes in geotechnical properties and seismic imaging.

It is a tempting idea to try to correlate the shallow reflections with the significant excursions in the *in situ* plots. The regularity of the reflection series, however, seems to suggest that not all observed reflectors are actually primaries. Indeed some of them could well represent multiples that have reverberated between the sea floor and the first negative reflector. The latter most likely marks the boundary between marine (unit H) and lacustrine deposits (unit G), and can probably be correlated with the sharp cone resistance peak at ±45 cm depth. The two negative reflections showing up below this primary reflector are believed to be internal multiples. The deeper reflections, observed above the negative gas-enhanced reflection, probably indicate some internal stratification of unit G.

The acoustic impedance values, calculated using the above interpretation, result in density estimates for the surficial sediments that are very reasonable for muddy or clayey deposits (±1.3 g/cm^3) (Hamilton 1972). They also agree well with the results obtained from the Module core. The density estimates obtained for the underlying sediments (unit G) are, however, lower than would normally be expected for sandy silt deposits; this is most likely related to the negative polarity of the reflections which tends to lower the calculated impedance values. The extremely low density of unit E clearly reflects the presence of gas-bearing sediments. In the case that all the negative reflections represented primaries, this would result in a series of impedance and density values that are far lower than can normally be expected, due to the negative polarity of these reflections.

Module station 26

The cone resistance values at Module station 26 (Fig. 14) are within an order of magnitude higher compared to station 27. The uppermost part of the plot is marked by a rather sharp increase, which is interrupted by a short stagnation at ±50 cm depth. Further down, the curve more or less oscillates around a constant, high cone resistance value of about 750 kPa. The change from low to high cone resistance suggests the presence of two different units, respectively made up of relatively soft and more resistant sediments. This seems to be supported by the seismic-stratigraphical interpretation; indeed station 26 is located at the foot of the prograding body of delta/beach deposits of unit F (Fig. 11), where it is interfingered with the onlapping lacustrine deposits of unit G. No core could be retrieved at this site.

The processed seismic section reveals a strong sea-floor reflection surrounded by a halo of negative noise, which could be due to a density stratification of suspended sediment near the sea floor, causing reverberations and reflections of the seismic waves. The strong negative reflection showing up below the sea floor may well correspond with the transition from unit G to unit F suggesting the presence of a thin layer of reduced impedance at this boundary. The excursions in the lower part of the cone resistance plot, which seem to indicate some sort of stratification of unit F could not be linked to the seismic data. This is quite surprising as the observed variations are relatively large (up to 250 kPa). A possible explanation could be the absorption of the signal in the more sandy material, or perhaps the strong negative reflector disguised possible weak reflections.

The deeper reflections on the seismic section can probably be related to the interfingering of units F and G. The negative reflection at 76 ms TWT (two-way travel time) marks the top of unit E. As the latter is known to contain very soft clayey sediments, this decrease in impedance is expected. However, the strong intensity of this reflection may also indicate the presence of a thin sand layer, which often marks the top of the glacio-lacustrine sequence.

The obtained impedance and density estimates for units F and G are acceptable for clays and silts (Hamilton 1972). Further down the values decrease, and the obtained estimate for unit E is therefore lower than might be expected for clayey sediments. This is most likely due to the occurrence of several negative reflections.

Module station 29

In general the cone resistance values at Module station 29 (Fig. 15) are considerably lower than at station 26, but still remain much higher than

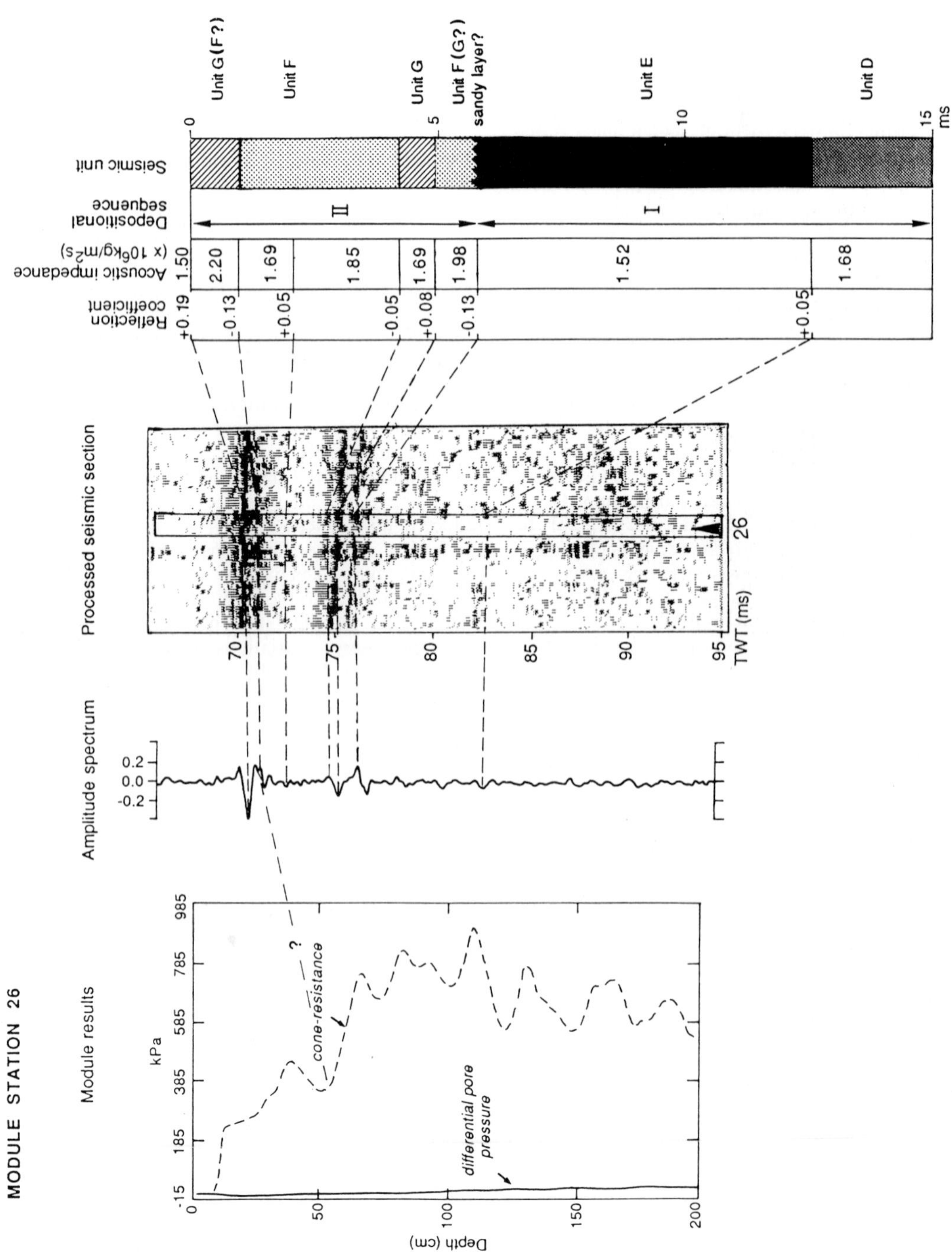

Fig. 14. Integration of different geotechnical, seismic and acoustic data from Module station 26. The colour-coded seismic section is shown in Fig. 8b.

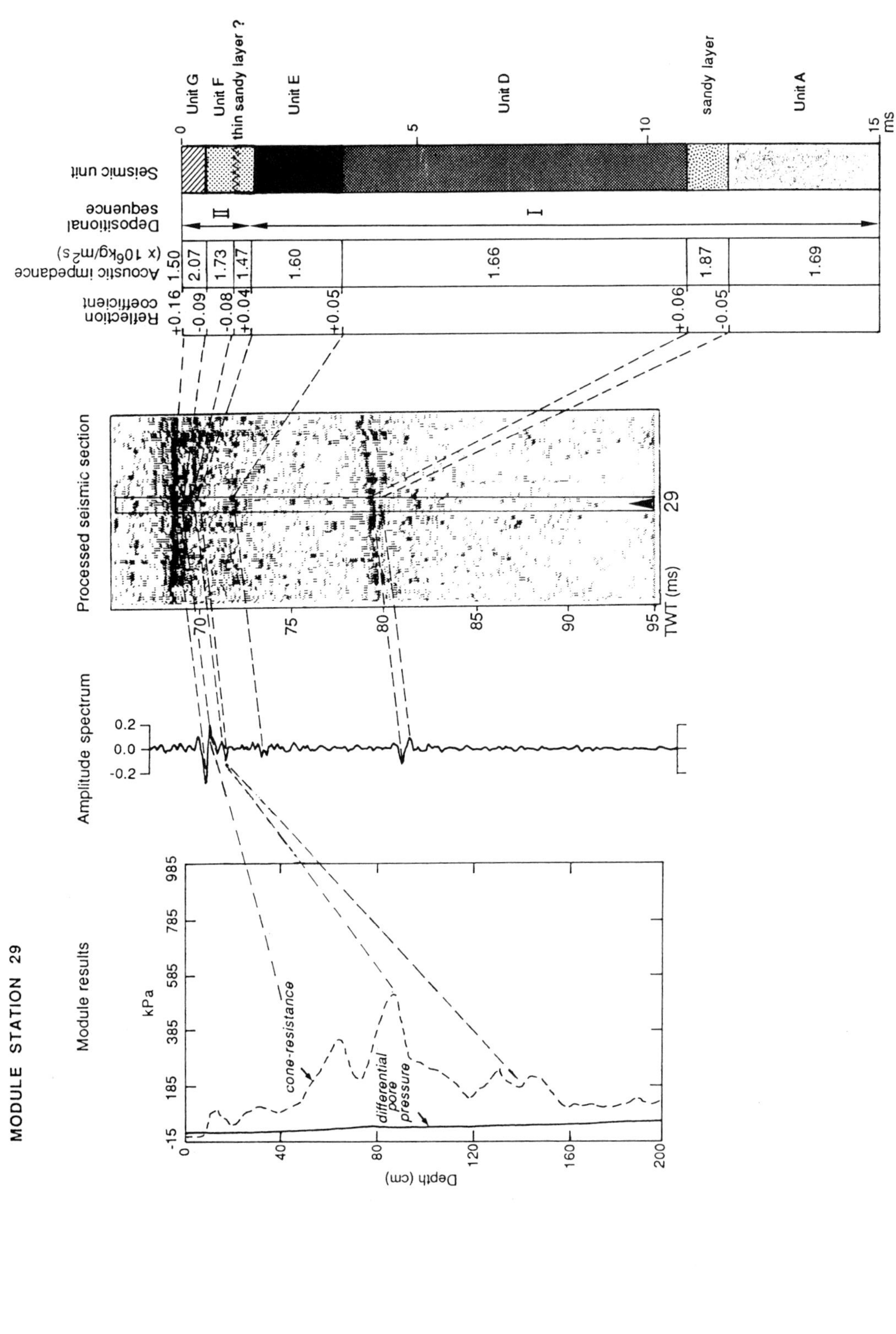

Fig. 15. Integration of different geotechnical, seismic and acoustic data from Module station 29. The colour-coded seismic section is shown in Fig. 8b.

at station 27. The penetrometer plot can be subdivided into three different parts. The upper part (up to 40 cm) has low values, indicating the presence of very soft sediments, which agrees well with the seismic interpretation of a thin layer of outcropping soft lacustrine sediments (unit G). The central part of the plot (between 40 and 160 cm) has markedly higher values, and contains several sharp peaks (up to 500 kPa), indicating the presence of more stiff sediments. This is also supported by the seismic interpretation of the beach deposits of unit F. The lower part shows relatively low cone resistance values, which can most likely be linked to the soft clays of unit E.

The processed seismic section is marked by a strong sea-floor reflection surrounded by a halo of negative noise, although the noise is less than that at station 26. Two prominent negative reflections can be observed immediately below the sea floor, which can most likely be correlated with the bases of units G and F. They may well correspond with the sharp peaks in the cone resistance plot at about 65 and 90 cm depth. The underlying positive reflections probably mark some internal stratification in unit E and the base of this unit, respectively. The pair of positive/negative reflections observed at ± 80 ms TWT clearly marks the sandy layer at the base of unit D. Again, no core could be recovered at this site.

When comparing the penetrometer results of Module stations 26 and 29, we can see that there is a marked difference in the cone resistance values for unit F. Indeed the values at station 26 (average 600 kPa) are much higher than those obtained at station 29 (average 300 kPa). This could be related to the fact that unit F represents a prograding body, and a decrease in sand content may occur towards the more distal parts of the body. Although both stations are located at the foot of the prograding body, the direction of progradation indicates that station 26 is more proximal than station 29.

The observed reflection coefficients result in reasonable impedance values for units F and G typical of clays or silts. The underlying unit E and to a lesser extent unit D show values that are lower than might normally be expected for clay deposits. Again, this is due to the negative polarity of the uppermost reflections, which seems to suggest that not all of them are primaries. However, this is most likely not the case here, as the reflections are clearly seen to divert from each other when going towards station 26. Nevertheless, the sandy layer at the base of unit D exhibits a markedly higher impedance and density than the surrounding layers.

Module station 24/GISP station 8

The cone resistance curve for Module station 24 (Fig. 16) indicates very soft sediments, and agrees well with the low density values from the core taken at this station. The sharp peak observed at 120 cm, associated with a slight drop in pore pressure, indicates the presence of a sandy layer. The latter is supported by the observation of sand at the base of the Module core, which suggests that the corer stopped upon reaching this layer. Apparently the resistance of the sand layer was too large for the relatively thick corer (10 cm diameter), whereas the much thinner penetrometer probe (3.6 cm diameter) could penetrate further. The discrepancy in depth between the two sand layers is probably due to compaction effects and an eventual loss of sediment in the upper part of the core. The sandy layer also probably accounts for the presence of the strong reflector below the sea floor, which marks the boundary between units H and I. The thin sandy level identified in the uppermost part of core 14971-1 supports these observations.

The results of the GISP penetrometer system (GISP station 8; Fig. 16) indicate the presence of three main sedimentary layers. The upper part of the density plot (<3.5 m) shows a sharp rise, modulated by a few small stagnations. The first stagnation, at ± 1.5 m probably marks the transition between units H and I although the sand layer associated with this boundary is not clearly evident (this is most likely due to the resolution of the GISP system). The deeper stagnations indicate some amount of internal stratification, which agrees well with the seismic facies of unit H. The middle part of the plot (3.5–7 m) is marked by relatively high-density values, which slowly increase with depth; it can probably be correlated with unit G. The sharp peak at 7 m may well correspond with the sandy layer identified at the base of unit G in core 14971-1. The lower part of the plot (>7 m) shows a marked drop in density, and most likely represents unit E. The natural gamma-ray curve, which is not shown in the figure, has a similar shape as the density curve and supports the above interpretation.

The density values from GISP and Module measurements are in good agreement with each other, and correlate well with the interpreted seismic units. The acoustic impedance values calculated from the processed seismic section show a general trend of increase with depth. However, the resulting density estimates are still lower than would normally be expected for the identified units. The latter is confirmed by the density values obtained from the *in situ* penetration measurements.

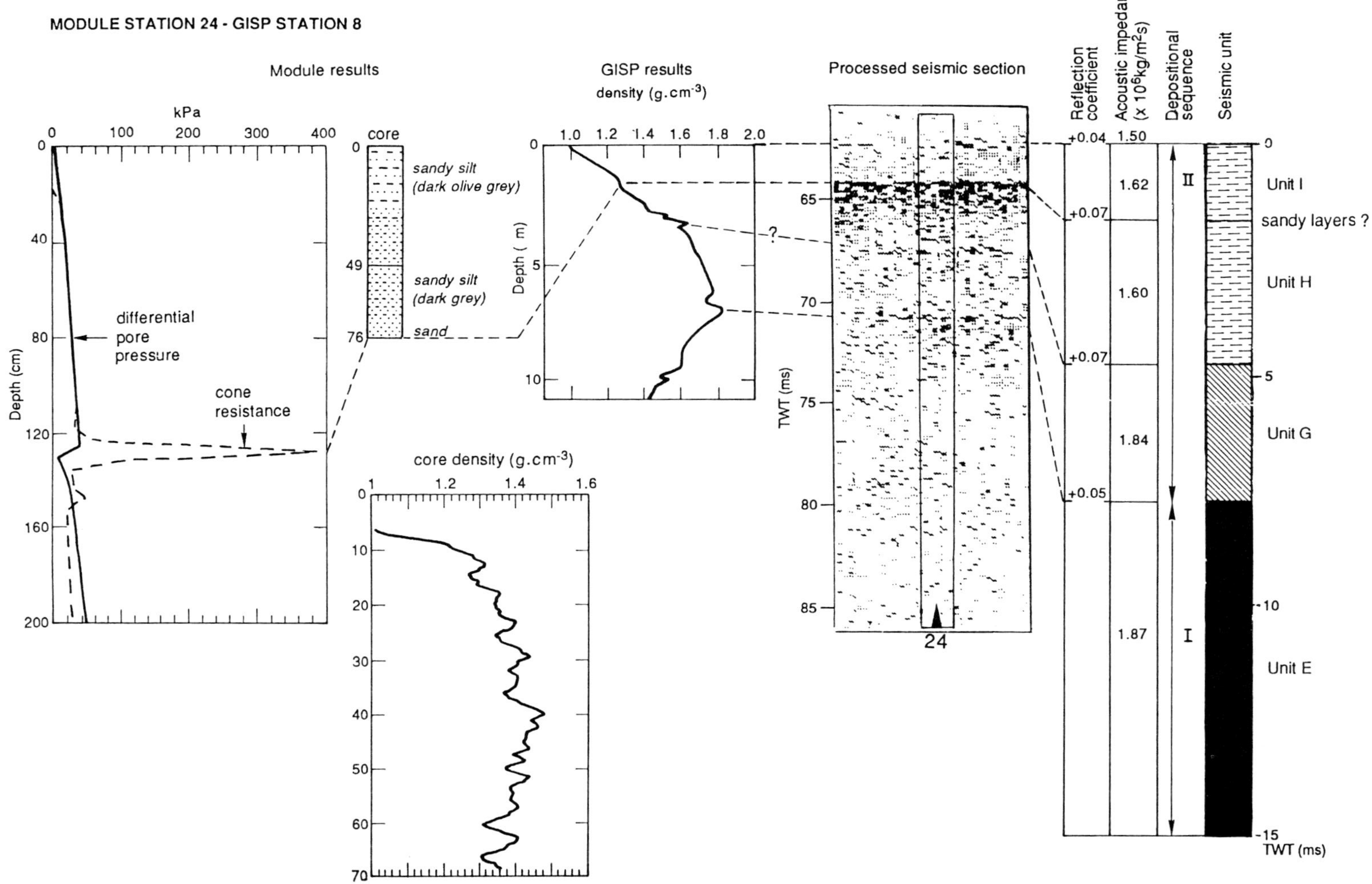

Fig. 16. Integration of different geotechnical, seismic and acoustic data from Module station 24/GISP station 8. The colour-coded seismic section is shown in Fig. 8b.

Discussion

In general, the data obtained from geotechnical measurements, seismic stratigraphy and seismic processing could be integrated satisfactorily. The geotechnical plots acquired with the GISP penetration system and the Geotechnical Module reflect the complex structure of the shallow marine deposits in the Bornholm area. Both data sets were in good agreement with each other, and could be correlated quite well with the seismic interpretation and lithology. Prominent excursions in the geotechnical plots could be related to the reflectors identified on the processed seismic sections. This was, however, not always the case for the plots from the Module, where the short penetration depth clearly limited the correlation possibilities.

The general trend of calculated impedances confirmed the seismic interpretation and, in most cases, the calculated impedance values agreed well with the lithological interpretation. In some situations, however, the calculated impedance values were lower than was considered reasonable for the soil types expected. This was usually associated with a strong prevalence of negative reflections in the considered seismic section. If all of them were interpreted to be primary reflections, the succession of negative reflection coefficients would serve to depress calculated impedance values to unacceptable levels. This seems to indicate that at least some of them represent multiple reflections which would not enter into impedance calculations.

Other explanations for the observed negative reflections were considered. Perhaps they are due to well-defined sharp decreases in impedance occurring in a deposit generally marked by a gradual increase of impedance with depth. In such a medium, there would be no positive reflectors, only negative ones. Also, if a layer of high impedance was too thin to be identified on the seismic section, the impedance values calculated for all deeper layers would be lower than they should be. The negative reflections might also be related to very thin layers, almost beyond the limit of resolution. For such layers only one (negative) reflection is observed instead of a positive and a negative reflection at the upper and lower boundaries of the increased acoustic impedance.

Some error may also be due to absorption effects, or to the assumption of spherical divergence. The latter implies the same propagation speed in marine soils as in water. That would be approximately true for low-impedance soils, but becomes less valid for denser sediments. Absorption and increased propagation speed would both decrease the measured reflection magnitude but leave polarity unchanged, producing impedance values either too large or too small (McGee 1991*b*). In a case history reported by Swart & McGee (1993), reflection coefficients recovered from boomer data in stiff clays needed to be multiplied by 1.2–1.3 to account for the combined effect of these sources of error.

The fact that a constant compressional wave velocity of 1500 m/s was assumed in the surficial marine sediments may also have caused some error in the calculation of the density values. The *in situ* measurements with the GISP penetration system, however, indicate that in some cases a strong density gradient can be observed in the uppermost layers, in which case the constant velocity model would no longer be valid.

Some parts of the processed sections were rather noisy, producing inconsistent results. Most likely this is due to the fact that these data were digitized at too low a rate (20 kHz instead of 40 kHz). This is supported by recent source signature studies, which indicate that digitization rates of respectively 36 and 50 kHz are required for the unprocessed sparker and boomer data (Verbeek & McGee 1995). At station 24 some reflections identified as primaries on the original data appear to have lost continuity or disappeared after processing. This suggests that they were so weak that they did not exceed the noise level of the processed plots, or that they were not digitized fast enough to be consistent between primary and first water-layer multiple.

Due to all these influences, the variation in impedance values obtained for the various sedimentary units at different positions was quite large. However, on the whole the mean values for all units overlap significantly when considering the standard deviation.

Conclusions

The aim of this study was to provide an evaluation of the physical parameters of the shallow sea floor by the integration of high-resolution seismic and geotechnical data. Different *in situ* geotechnical data sets were acquired at precise target stations located along seismic profiles offshore Bornholm. The cone resistance, excess pore pressure and density information obtained with the Geotechnical Module and the GISP system support the seismic interpretation, and correlate well with the subsurface structure observed on the seismic profiles. In some cases even slight modifications in physical properties were able to induce visible changes in geotechnical properties and in seismic imaging. This proves that the sensors are able to detect variations in lithology, and can therefore be used as a clear signal for the description of the sediment body.

The seismic data were processed in order to obtain estimates of the reflection coefficients and impedance. The processing procedure included, among others, a detrending routine and a seismic inversion technique based on multiple quotient division. The excursions in the geotechnical plots can be quite well related to the individual reflections on the processed seismic sections. The general trend of calculated impedances and densities agrees well with the lithological interpretation; however, they did not always result in quantitatively correct values of the sediment parameters.

As described here, the seismic inversion technique is at an early stage of development and is applicable only in situations where the sea floor is relatively smooth and flat and the source–hydrophone offset is small relative to the water depth. Also, it is best applied in situations where primary and water-bottom multiple sequences are clearly separated. Its precedence by detrending did, however, achieve significant improvements in resolution, thereby revealing useful information which is lost to conventional methods of processing.

Although the seismic processing procedures do not account for absorption and there are other sources of error which contribute to inaccuracies, some of the calculated results are quite positive and promising. Furthermore, there are indications that they could at least be somewhat improved by simply digitizing the field data at greater rates. Perhaps the greatest inaccuracies arose from difficulties experienced in distinguishing between primary reflections and internal multiple reflections. Since primaries are solitary events and multiples occur in sequences that are somewhat predictable, the solution to this interpretational problem may lie in making greater use of interactive computer analysis of the seismic sections.

The presented interdisciplinary approach is not a common practice and the study is considered to be only a first step in this line of research. Further development and refining of the inversion technique will undoubtedly provide better estimates of the acoustic properties. The efficiency of the *in situ* measurements can be improved by acquiring more geotechnical parameters during one single operation. The recent development of a gamma-ray density probe and vane shear sensor for the Geotechnical Module offers this potential. This will enable a better definition of the physical properties of shallow marine sediments, which can be used in a wide number of industrial and academic applications.

The study presented here was funded by the European Communities within the MAST programme (GISP Project; contract number CT90-0057). The Management Unit of the Mathematical Model of the North Sea and Schelde Estuary in Belgium and the Institut für Meereskunde an der Christian-Albrechts-Universität zu Kiel in Germany are greatly acknowledged for the supply of shiptime. The authors wish to thank the captain and crew of the research vessels 'RV Belgica' and 'RV Poseidon' for the fruitful cooperation and good working atmosphere on board. We are also very grateful to Dr F.-C. Kögler for providing the information on core 14971-1.

References

Arnsten, B. & Ursin, B. 1993. Estimation of reflection coefficients from zero-offset field data. *Geophysics*, **58**, 1634–1645.

Baltzer, A., Cochonat, P. & Piper, D. J. W. 1994. In situ geotechnical characterization of sediments on the Nova Scotian Slope, eastern Canadian continental margin. *Marine Geology*, **120**, 291–308.

Berteussen, K. A. & Ursin, B. 1983. Approximate computation of the acoustic impedance from seismic data. *Geophysics*, **48**, 1351–1358.

Brussaard, P. & McGee, T. M. 1992. Problems and constraints encountered in the digital processing of high-resolution single-channel marine seismic data for "GISP". *In*: Weydert, M. (ed.) *European Conference on Underwater Acoustics*. Elsevier, Amsterdam, 473–476.

Buchan, S., McCann, D. M. & Taylor Smith, D. 1972. Relations between the acoustic and geotechnical properties of marine sediments. *Quarterly Journal of Engineering Geology*, **5**, 265–284.

Claerbout, J. F. 1992. *Earth Sounding Analysis – Processing Versus Inversion*. Blackwell, Oxford.

Cochonat, P., Schieb, T., Guillaume, J., Kerbrat, R., Tisot, J. P., Auffret, G. A. & Müller, C. 1988. Geotechnical and sedimentological properties of Nice Slope and submarine Var Canyon deposits. American Association of Petroleum Geologists, *Mediterranean Basins Conference*, Nice.

Couture, R. 1990. Mesures Géotechniques in situ au Large de Nice (Baie des Anges – Module Géotechnique). Rapport de stage IFREMER-DRO/GM.

Crusius, J. & Anderson, R. F. 1991. Core compression and surficial sediment loss of lake sediments of high porosity caused by gravity coring. *Limnology and Oceanography*, **36**(5), 1021–1031.

Declerck, H. 1990. *Stratigraphie Sismique Haute Résolution sur le Talus Continental*. IFREMER-IFP Report 38503.

Fang, W. W., Langseth, M. G. & Schultheiss, P. J. 1993. Analysis and application of *in situ* pore pressure measurements in marine sediments. *Journal of Geophysical Research*, **98**(B5), 7921–7938.

HAMILTON, E. L. 1970. Reflection coefficients and bottom losses at normal incidence computed from Pacific sediment properties. *Geophysics*, **35**, 995–1004.

——1972. Compressional-wave attenuation in marine sediments. *Geophysics*, **37**, 620–646.

HAYNES, R., DAVIS, A. M. & REYNOLDS, J. M. 1993. The extraction of geotechnical information from high-resolution seismic reflection data. *Offshore Station Investigation and Foundation Behaviour*, **28**, 215–228.

HEVIN, G. 1993. *Analyse Sédimentologique du Bassin d'Arkona et de sa Bordure Orientale (Mer Baltique) – Campagne Poseidon, Module Géotechnique*. Rapport de stage de IFREMER-DRO/GM.

MCGEE, T. M. 1991*a*. Seismic reverberations and the remote estimation of properties of underwater soils. *International Journal of Imaging Systems and Technology*, **3**, 40–57.

——1991*b*. Modelling 1D wave propagation in a system of absorbing layers. *Geophysical Prospecting*, **39**, 29–49.

OLDENBURG, D. W., SCHEUER, T. & LEVY, S. 1983. Recovery of the acoustic impedance from reflection seismograms. *Geophysics*, **48**, 1318–1337.

OLLIER, G., COCHONAT, P. & DUMAS, A. 1992. Development of a gammadensity probe for *in situ* invesitigations in shallow seabeds. *54th EAEG Meeting*, Paris, 370–371..

PERINI, L., MISSIAEN, T., ORI, G. G. & DE BATIST, M. 1996. Seismic evidence of glacial to marine sedimentation during the Late Quaternary offshore Bornholm, southern Baltic. *Sedimentary Geology*, **102**, 3–21.

RICHARDSON, M. D. & BRIGGS, K. B. 1993. On the use of acoustic impedance values to determine sediment properties. *Proceedings of the Institute of Acoustics*, **15**(2), 15–23.

SCHUBERT, C. J. 1991. *Organischen Kohlstoff und Schwermetallgehalte in Jungquartären Ostseesedimenten*. PhD Thesis, Universität Giessen.

SWART, P. D. & MCGEE, T. M. 1993. Marine seismic single-channel investigation techniques to improve dredging performance – a case study. *Proceedings of the CEDA Dredging Days*, World Dredging Congress, Amsterdam.

VERBEEK, N. H. 1995. *Aspects of High Resolution Marine Seismics*. PhD Thesis, Universiteit Utrecht.

—— & MCGEE, T. M. 1995. Characteristics of high-resolution marine profiling sources. *Journal of Applied Geophysics*, **33**, 251–269.

Geophysics in offshore site investigation: a review of the state of the art

A. M. DAVIS

School of Ocean Sciences, University of Wales Bangor, Menai Bridge, Gwynedd, LL59 5EY, UK

Abstract: Geophysical techniques for seabed and sub-seabed site investigation can be divided into two categories: those providing information on gross structural (geological) and stratigraphical relations, and those producing quantitative data which can be used to predict sediment behaviour under applied loading. The past few years have seen key developments in both fields which have come about through fundamental improvements in data acquisition (technological and procedural), and through an improved understanding of geophysical–geotechnical property relations. For example, high-resolution seismic reflection signatures are now routinely recorded in digital format, and, given the appropriate data quality and processing software, it is now possible to extract geotechnically significant information from the reflection response. Seismic refraction techniques are also beginning to be more widely used to provide quantitative information on the engineering properties of seafloor materials; the seismic shear wave velocity is of particular interest, since it is now being recognized as an indicator of sediment strength and stiffness.

From a review of available data and the results of validation exercises, it would seem that offshore engineering and environmental surveys could reap considerable benefit from these and a range of other related geophysical developments.

One of the major goals of the offshore engineer/geological oceanographer is to obtain a truly representative description of the seabed and subsurface to whatever the required depth of interest. Dependent on the purpose of the investigation, this might include structural and stratigraphical mapping, delineation of geological hazards, and/or quantification of seabed engineering properties, e.g. dredgeability, anchor-holding capacity, bearing capacity, liquefaction potential etc. For many years geophysics has routinely been used during the initial stages of any offshore engineering investigation to provide a basic qualitative description of the site geology and to produce hazard distribution maps (natural and anthropogenic). Information of a more quantitative nature has generally come from the analysis of borehole samples (through routine mechanical testing) and from *in situ* geotechnical testing. However, through a combination of technological advances and an improved understanding of geophysical/geotechnical relations, it is becoming increasingly feasible and economically advantageous to extend the role of geophysics in offshore site investigation.

This paper aims to:

1. provide an overview of the current state of the art;
2. present details of related developments and experimental work at the author's own laboratories at the University of Wales Bangor (UWB); and
3. highlight potential new fields of application.

It thus provides the geotechnical engineer with the basic information required to incorporate geophysical methodologies into the design of an offshore site investigation.

Background

A seafloor sediment can be referred to as a poroelastic medium consisting of two or three phases: solid, fluid and/or gaseous. The way in which a seismo-acoustic wave propagates through it is governed by the physical composition, distribution and interaction of the various components, and by a number of other factors.

Theoretical models describing seismo-acoustic propagation relate the seismic wave velocities to the elastic properties of the sediment body, and these relationships form the basis for deriving the dynamic elastic properties from practical observations:

$$V_P^2 = (B + \tfrac{4}{3}eG)/\rho \qquad (1)$$

$$V_S^2 = G/\rho \qquad (2)$$

where V_p is the compressional wave velocity, V_s is the shear wave velocity, G is the shear modulus, B is the bulk modulus and ρ is the density.

Empirical methods can also be used to derive the physical properties of a sediment from measured geophysical quantities. For instance, there are numerous publications citing empirical

From De Batist, M. & Jacobs, P. (eds), 1996, *Geology of Siliciclastic Shelf Seas*, Geological Society Special Publication No. 117, pp. 323–338.

relations between acoustic and geotechnical properties (e.g. Stoll 1989; Taylor Smith 1986; Richardson & Briggs 1993), with the interdependence of porosity (or void ratio) and compressional wave velocity being particularly well documented. According to Stoll (1992), other important properties controlling acoustic propagation within marine sediments are density, overburden stress, degree and type of lithification, grain size and distribution, dynamic strain amplitude, material property of the grains and sediment structure.

For shear waves, the vibrations are translated through the skeletal frame of the sediment and it has been shown from practical and theoretical studies that the dynamic shear modulus derived from the seismic wave velocity is dependent on almost every geotechnical variable. However, the extent to which the shear modulus is affected varies considerably from one parameter to another, and Hardin & Drnevich (1972*a*, *b*) define the most important variables affecting the shear modulus of clays as strain amplitude, effective mean principal stress, void ratio and number of cycles of loading. This holds true for sediments exposed to small-amplitude dynamic loading, similar to that encountered in geophysical measurements, in which there is no change in the structure or in the strength of the sediment due to the applied load. Significantly though, this small-strain seismically derived modulus is able to be corrected (through practical measurement) to cover a range of strain levels and variable strain rates (see later) thus providing the potential for a diversity of applications.

The fundamental theory relating to transmission and reflection of seismo-acoustic waves at physical discontinuities within the Earth is also well understood and provides the means for spatial mapping of geological structures. The interpretation is based on the measurement of travel-times of waves that return to the surface after refraction or reflection at velocity or acoustic impedance boundaries respectively. In addition, reflection seismograms can be further analysed for stratigraphical information such as depositional environment, lithological characteristics, and other physical attributes (e.g. seismic velocity, fluid saturation or density). While the former geophysical approach, i.e. spatial mapping of geological structures, is routinely used by the engineering industry, the latter approach (generally referred to as seismic stratigraphy), although routinely used by the hydrocarbons exploration industry, has yet to become an established technique in engineering site investigation.

Advances in high-resolution seismic reflection profiling

Undoubtedly, the most widely used geophysical technique in offshore site investigation is high-resolution seismic reflection profiling. The major advantage of the technique is that it is capable of producing continuous seismic sections which bear close resemblance to geological cross-sections, and that these sections can be obtained rapidly and in a relatively cost-effective manner.

In recent years, the seismic data quality has improved significantly, as a result of fundamental advancements in source–receiver design, improvements in analogue display and, probably most significantly, through the adoption of digital data acquisition and processing. The latter has resulted largely from the adaptation of existing technology used in hydrocarbon exploration, and the advent of relatively inexpensive, powerful computers.

High-resolution seismo-acoustic classification systems can be broadly subdivided into those that analyse the seabed surface, and those that attempt to resolve the physical nature of the subsurface layers. The former are generally based around a single-tone piezoelectric ceramic transducer, with digital processing of the reflection waveform providing information on seabed character, e.g. systems based around echo-sounder transducers (e.g. the Roxann System; see Chivers *et al.* 1990), side-scan systems etc. The latter group either use single or multiple piezoelectric transducers or some type of impulsive device as source. Lambert *et al.* (1993), of the US Naval Research Laboratory, have demonstrated that it is possible to derive information on subbottom sediment properties using single-tone piezoelectric transducers of an appropriate frequency, and advanced signal processing. In their system, the Acoustic Sediment Classification System (ASCS), the acoustic impedance of specific depth increments of the subsurface sediment is computed from the amplitude and pulse characteristics of the return signal; empirical relations are then used to convert the impedance profile to a physical property profile. Several swept-frequency sonar systems have also been designed and developed to allow the remote determination of sediment character and properties. Schock *et al.* (1989) describe a method based on the determination of acoustic attenuation, and claim improved resolution and penetration through the use of a broad-band 'chirp' source. The chirp sonars are microcomputer controlled, the computer also undertaking the processing of the acoustic data, both the correlation processing and the attenuation estimation. However, in terms of basic hardware, experience and experimentation has shown that the more conventional broad-band impulsive sources such as the Uniboom, operating at frequencies of between 0.4 and 10 kHz, are particularly effective at providing high-resolution signatures (McGee *et al.* 1992). With this particular source, i.e. the boomer, penetration depths in the order of 40 m can be expected (Trabant 1984) although, given suitable bottom conditions, e.g. soft water-saturated muds, penetration depths of up to 90 m below seabed surface have been achieved.

Traditionally, with the boomer system the source and receiver are towed as separate entities, the boomer being mounted on a catamaran and the receiver being a towed 'eel' consisting of a series of pressure-sensitive elements arranged in an oil-filled tube. Data quality is largely dependent on the prevailing weather conditions, although water depth becomes important when surveying either in very shallow (a few metres) or relatively deep water (greater than 100 m). The shallow-water problem is chiefly one of multiple suppression with sea-surface multiples masking primary subbottom reflectors. One commercially available system which has been specifically designed to help overcome this problem is the IKB Seistec (Simpkin & Lewis 1992). In this a line and cone receiver is mounted on the catamaran alongside the boomer plate, and for single-channel recording this replaces the towed eel. Experience has shown that this type of configuration not only suppresses multiple reflections but significantly reduces environmental noise. For surveys in deeper water, problems associated with signal attenuation and beam spreading are overcome by using a deep-tow subbottom profiler. Such systems often also incorporate high-frequency (around 100 kHz) side-scan transducers to provide images of the seabed surface simultaneously with subbottom structure.

Data recorded with the above can be displayed in analogue form or stored as a digital disc or tape. The advantages of recording digitally are numerous, the most obvious being that some form of post-processing can be applied to improve the data quality. A number of single- and multi-channel digital acquisition systems designed specifically for high-resolution subbottom profiler surveys are now commercially available (e.g. the GeoAcoustics Sonar Enhancement System, the Elics Delph 1-, 2- and 24-channel systems), and some offer a limited degree of real-time processing (generally in the form of fast Fourier transform and deconvolution for multiple suppression). The GeoAcoustics and Delph systems have already been adopted by a number of oil industry contractors operating in the North Sea (particularly for hazard evaluation surveys), and are beginning to find application in environmental and engineering studies within the coastal zone.

At the processing end, a number of relatively inexpensive commercial packages are now available to handle high-resolution seismic data, e.g. SierraSeis ISX, Micromax, and its more powerful relative, Promax. These allow the interpreter to follow basic processing routines leading, for instance, to the construction of stacked and migrated sections. More importantly though for engineering and environmental studies, it should also become possible to take the interpretation of high-resolution seismic reflection data a stage further and extract geotechnically significant information. Haynes *et al.* (1993*a*) recently reviewed the scientific literature and concluded that it should be possible to develop robust computer programs designed to extract a number of seismic parameters (velocity, acoustic impedance, attenuation) from digital reflection data and use these to produce depth profiles (to 20–30 m below seabed surface) of a marine sediment's bulk properties (e.g. density, void ratio, moisture content and grain size) via empirical relations. Given this information and a limited amount of borehole control, they argued that it should be theoretically possible to map the spatial variability of parameters such as shear strength and permeability. In order to support their argument Haynes *et al.* (1993*a*) produced a series of summary schematics illustrating the various interrelationships, shown here as Fig. 1.

Thus to date, while seismic reflection profiling is a well established and widely used technique in offshore site investigation, its potential has probably not yet been fully recognized. Clearly, the recent fundamental improvements in data acquisition have led to corresponding improvements in data quality and processing facilities, and it is the latter which will continue to enhance the scope of the technique for quantitative engineering studies.

Marine seismic refraction surveying

While seismic compressional wave refraction shooting is the preferred geophysical technique (over reflection methodologies) for mapping subsurface structures for engineering applications on land, it has to date found limited use in the marine environment, largely because of practical operational difficulties associated with the presence of the water-column. The technique utilizes the principle of critical refraction of

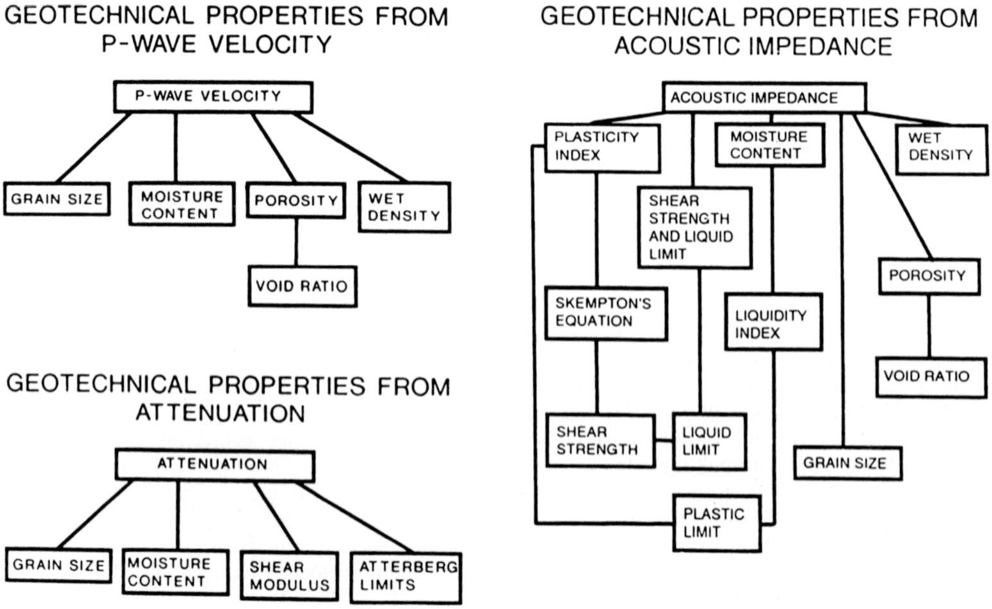

Fig. 1. Schematic diagrams showing empirical interrelationships between the geotechnical properties of marine soils and their seismo-acoustic properties (from Haynes *et al.* 1993*a*).

seismic waves impinging on boundaries between media of differing velocities (Telford *et al.* 1976). For subsurface investigations the essential criteria for successful refraction shooting are an increase in velocity with depth, and for the velocity contrast to be large enough to produce a recognizable head wave (the wave returning to the surface after critical refraction at the velocity interface). The head wave is used both to determine depth to the interface and to provide fundamental information on the properties of the lower medium (via the seismic wave velocity). In the marine environment, if utilizing a surface tow source and receiver, the energy generated will be compressional and the first physical boundary will be the seabed surface. Generally, however, there is very little difference between the compressional wave velocity in water and that in unconsolidated marine sediments, thus making differentiation of the seabed surface difficult if not impossible. Similarly, the velocity contrast between individual sediment layers may also be proportionately small and the head wave correspondingly indeterminate. There are cases, however, where marine refraction shooting using surface-towed instruments can provide useful additional information. For instance, if it is required to map rockhead or provide an indication of the rock quality designation, this can probably be achieved most cost-effectively using refraction techniques (under suitable operational conditions). Some examples of its use for the latter application are given by Gardener (1992).

The situation with regard to shear waves and marine applications is somewhat different. Firstly, as shear waves require a finite rigidity in order to propagate, the water-column has to be eliminated at the data acquisition stage and thus any seismic survey automatically has to commence at the seafloor, i.e. with source and receivers in contact with the seabed. Secondly, as shear waves propagate through the particle 'frame' rather than via the pore phase (as with compressional waves) then the shear wave senses different properties of the sediment and the velocity gradient or velocity contrast at physical discontinuities tends to be proportionately more significant (Davis & Bennell 1988). Thirdly, and most importantly, in engineering terms the shear wave velocity probably holds considerably greater significance over any other geophysical parameter, particularly for predictions of dynamic responses, and it is the well proven land-based surface refraction methodology which offers the most cost-effective means of providing these geotechnical quantities in a reconnaissance mode offshore.

In situ shear wave methodologies

A number of research groups have attempted to develop hardware and a range of methodologies (including seabed-surface seismic refraction shooting) for *in situ* shear wave velocity data acquisition. A brief review of some of these developments is given below, followed by more specific detail on related research efforts at the author's own laboratories.

Non-invasive techniques

The non-invasive techniques can be divided into two categories: (i) those utilizing the seismic body wave, i.e. the shear wave, and (ii) those based on analysis of surface/interface waves.

In 1974 Schwarz & Conwell reported the results of experiments carried out off the coast of the USA using an electromagnetic source device deployed on the seafloor and adopting the refraction approach. More recently, a number of other research groups, namely Kiel University, Germany, the Lamont-Doherty Observatory, USA, and the author's own laboratories at UWB have also reported shear wave experiments in which specific hardware has been constructed to allow the analysis of the shear wave velocity structure of the seabed using refracted wave arrivals. Whereas both Kiel and Lamont-Doherty have used a modified airgun to produce horizontally-polarized shear waves (Stumpel *et al.* 1984; Stoll *et al.* 1988), Stoll (1992) has more recently described a new type of focused explosive source which has been shown to be an efficient generator of both body waves and interface waves. Lamont-Doherty have investigated a number of marine sites in order to build-up a portfolio of data (including 'ground truth' information) to be used to test physical property predictions from theoretical models. For the interface waves, Stoll's group have developed a semi-automatic iterative model which predicts the shear wave velocity of the varying layers using dispersion methods. Although a powerful technique, there still remains the question of 'uniqueness' of the interpretation and Stoll recommends that the model be tested against synthetic seismograms wherever possible. However, a number of other groups working on inversion of the surface wave component of the wave-train appear to be making considerable progress towards solving the velocity inversion problem, e.g. Addo & Robertson (1992).

UWB's approach to shear wave generation and detection is similar to that first described by Schwarz & Conwell (1974). Shear waves are generated using an electromagnetic device deployed on a bottom-towed sledge, with signals detected by geophone units also coupled to the seafloor and towed behind the sledge. The UWB hardware has been developed for both near-surface investigation (uppermost 2 to 3 m of the sediment column) and deeper penetration (up to 50 m below seabed surface), with both systems aiming to provide information on the spatial variability of *in situ* sediment properties.

Invasive approaches

To obtain greater detail of the subsurface velocity structure and its depth variability, it is generally necessary to utilize some form of probe device. For the purpose of this paper reference will be made to one probing technique, that of seismic cone penetration testing (CPT), in order to provide the justification for using the UWB bottom-towed shear wave sledge as a reconnaissance survey tool (see later section on research developments at UWB).

Seismic cone penetrometers. On land, borehole seismic techniques, e.g. crosshole and downhole shooting, are generally used to obtain a more precise definition of the *in situ* dynamic elastic properties. Such techniques become a necessity when dealing with complex sequences, and particularly ones where velocity inversions are prevalent. Offshore, with the cost of drilling boreholes specifically for geophysical testing being largely prohibitive, engineering geophysicists have sought other means of acquiring site-specific *in situ* dynamic data. In this respect, the tool which is most widely used for offshore site investigation is the seismic cone penetrometer. Typically, the basic tool is modified to carry seismic sensing devices behind the cone head (Robertson *et al.* 1986; Hepton 1989), and seismic records are collected during penetration testing by activating a source at the surface (to preferentially generate either compressional or shear waves) and recording the vertically propagating vibrations arriving at the sensing devices, i.e. method according to the downhole procedure. In this way a seismic velocity depth profile can be drawn up which can be used either for the determination of dynamic elastic properties (if the density structure is known), or for inferring other geotechnical quantities (through correlations with data from the CPT or other tests).

Supporting laboratory studies. Few site investigations rely entirely on *in situ* measurements. In fact more typically, geotechnical engineers have tended to rely upon laboratory testing of recovered core samples, supplementing these with a limited amount of *in situ* data. Until relatively recently this meant high strain measurements under static load conditions. However, with the significance of dynamically derived variable strain measurements now being recognized, geophysical methodologies are increasingly finding a niche in the soil mechanics laboratory. In fact measurements in a carefully controlled laboratory environment probably offer the best means of evaluating the relationships between shear wave propagation and the physical and geotechnical properties of sediment samples under various load conditions (Bennell & Taylor Smith 1991), and furthermore, provide an invaluable aid to the interpretation of the more site-specific *in situ* geophysical data.

The resonant column apparatus. The resonant column device is probably the most versatile laboratory instrument for studying the dynamic properties of unconsolidated soils. While its main use is in defining the dynamic shear stiffness and damping characteristics of recovered core samples, it can also be used in a range of other modes to produce additional information, e.g. on the effects of varying the strain rate, confining pressure etc. (Bennell & Taylor Smith 1991).

Geophysically instrumented triaxial and consolidation cells. The resonant column device is a highly sophisticated apparatus which may well prove beyond the scope of many sediment-testing laboratories (both in terms of basic cost and availability of a specialist operator). However, if the requirement is solely for information on the low-strain dynamic properties, a more cost-effective solution can be provided by instrumenting 'off the shelf' laboratory soil mechanics cells for geophysical measurements. Piezoelectric transducers have for many years been used for laboratory measurements of the compressional wave velocity of rock samples (e.g. King 1966). In the 1970s, Shirley and co-workers at the Applied Research Laboratories in Texas developed the bender bimorph element and successfully made shear wave velocity measurements on samples of marine sediment (Shirley 1977). Following on from Shirley's work, Schultheiss (1981) instrumented soils test cells at UWB and began to make simultaneous geophysical and mechanical measurements in triaxial and consolidation cells.

As with the resonant column device, predictions of the depth variation in dynamic properties can be produced by varying the confining pressure. Concerning elastic moduli and static–dynamic relations, although in absolute terms the two data sets, i.e. static versus dynamic, generally differ by two to three orders of magnitude, this can adequately be explained in terms of differing strain levels and drainage conditions (Davis 1982). Given this basic understanding of the relation between the two data sets, it then becomes possible to derive useful working empirical relations which can be applied to solving specific engineering problems (Taylor Smith 1993).

Research developments at the University of Wales Bangor

Over the past two decades at UWB, a continuing research theme of the Marine Geophysical Group has been the development of geophysical remote sensing methodologies for quantitative site evaluation. The past few years have seen significant progress in this respect, largely coming about through involvement in major collaborative research programmes which have given the opportunity to validate the various new methodologies. This section aims to summarize related research developments at the author's laboratories, concentrating initially on methodologies, then illustrating their survey potential through presentation of a series of summarized case studies relating to continental shelf investigations.

Seismic reflection profiling

At UWB, research relating to the enhancement of the role of high-resolution seismic reflection profiling in marine investigation has progressed through a number of projects. With regard to surveying very shallow-water areas, the Natural Environment Research Council's LOIS (Land Ocean Interaction Studies) Special Topic programme provided UWB with both the impetus and funding for an appraisal of the state of the art regarding instrumentation and methodologies, whilst at the same time yielding a high quality data set able to be used for stratigraphical interpretation (see later).

The work of Haynes *et al.*, mentioned earlier formed part of a parallel and recently completed collaborative research project on high-resolution seismics between UWB and Applied Geology Ltd (funded under the SERC Teaching Company Scheme). This project initiated UWB's research into the development of methodologies for the extraction of geotechnical information from seismic reflection responses (see Haynes *et al.* 1993*b*). The current thrust is now towards

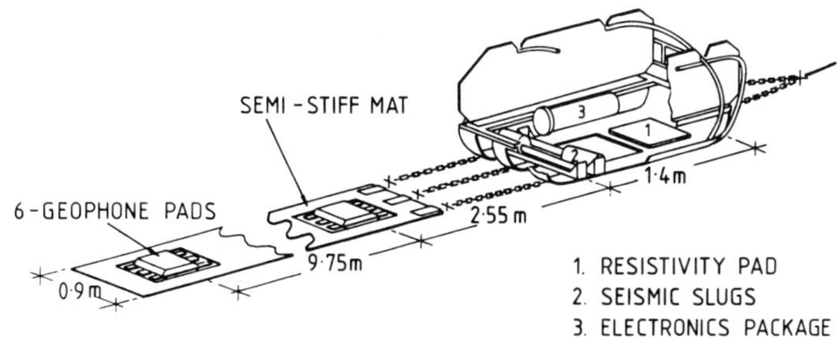

Fig. 2. The UWB pseudo-underway bottom-towed seafloor geophysical system 'magic carpet'.

validation of the developed methodology through co-ordinated measurements in areas offering the relevant control information (Davis *et al.* 1995). Of particular importance in this respect is the US Office of Naval Research's Coastal Benthic Boundary Layer (CBBL) programme (Richardson 1994) which has provided UWB with a unique opportunity to ground-truth the geophysical interpretation (see following section on Continental shelf investigations).

In situ shear wave studies

UWB's experience in shear wave research spans some 15 to 20 years, with initial programmes concentrating on land-based investigations and laboratory studies, these ultimately providing the required expertise to develop a marine capability. Currently the Geophysical Group's shear wave research thrust is towards sediment classification and behavioural prediction through the integration of *in situ* marine survey data with results from laboratory testing.

Non-invasive approach. The UWB bottom-towed geophysical sledge 'magic carpet', developed primarily for high-resolution investigations of near-surface sediment properties, comprises a sledge-mounted seismic source designed to preferentially generate horizontally polarized shear waves, and a towed geophone array (Fig. 2). The geophone elements, which are gimbal-mounted, are also housed in small sledges, and these are attached to a rubber mat. The mat essentially acts as an aid to coupling and also allows a constant receiver offset to be maintained. The mat forms an integral part of the towed array and measurements are made in a pseudo-underway fashion as the sledge is towed along a predetermined survey track. The data collected are in the form of individual seismograms for each measurement point along the track and these are analysed using a conventional time–distance approach (the seismic refraction approach). The 'magic carpet' has already been used to investigate a number of sites for a range of specific purposes. Example results and comments on potential applications are in the section on Continental shelf investigations.

A seismic refraction approach has also been used for deeper penetration measurements, providing information on the velocity gradient structure down to a few tens of metres below seabed surface. A twin geophone array is deployed some 100–200 m behind the stationary seabed source using a 'long offset' cable (see Fig. 3), and a multi-offset seismogram is built up by making a series of shots as the geophone array is progressively moved towards the source. A series of measurements made with the long-range system in the Belgian sector of the North Sea, as part of an EC contract awarded by the Hydrocarbons Sector of the Energy Directorate, have been published in a PhD thesis by Huws (1993). Figure 4 is an

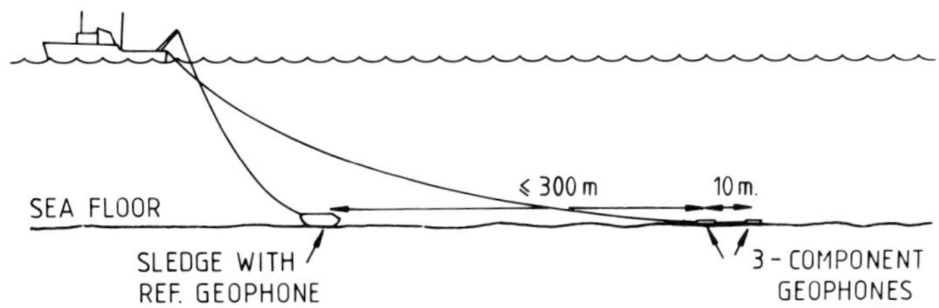

Fig. 3. Shear wave refraction configuration using a long offset geophone cable.

example data set collected at one specific location. The seismogram shown in the figure has been interpreted to provide information on the velocity gradient structure down to 50 m below seabed surface.

Invasive approach: seismic CPT. The seismic penetrometer device developed at UWB (with the aid of funding from the Science and Engineering Research Council (SERC) and Soil Mechanics Ltd) was designed to be capable of being set up with either a cone tip or a dilatometer device ahead of a pair of seismic sensors (spaced 1 m apart). In the proving stage, the tool was tested on a land-based test-bed site (the SERC Bothkennar site in Scotland) and the acquired seismic and geotechnical data compared with velocity depth profiles interpreted from conventional surface seismic refraction shooting. Figure 5 has been included to show a comparison between data sets obtained at this site using the non-invasive refraction method and the invasive seismic CPT approach. Clearly there is very good agreement between the two seismic velocity profiles for this particular formation (soft alluvial silts) and it is reasonable to conclude that surface refraction methodologies may be applicable in a quantitative appraisal of the subsurface for any site

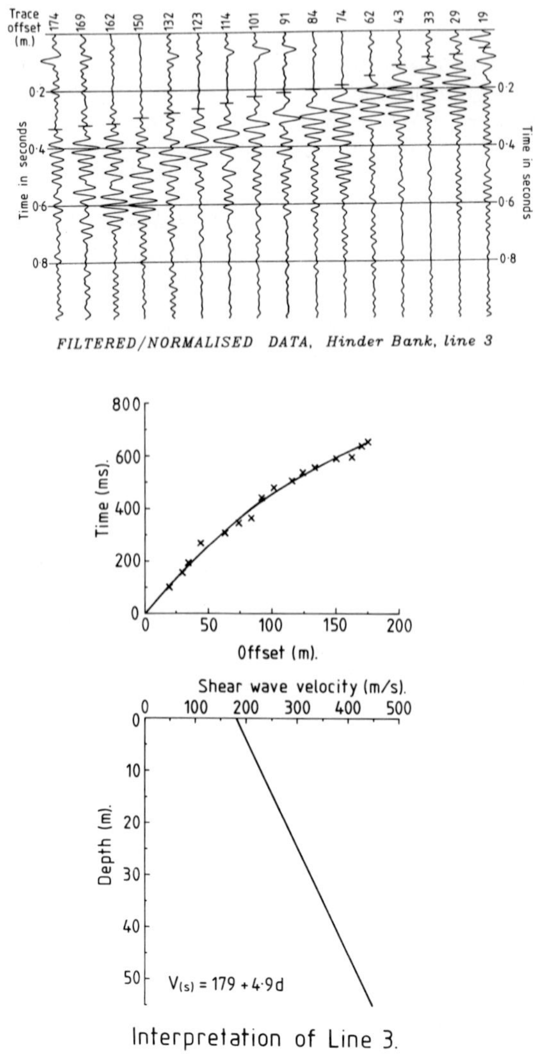

Fig. 4. Example seismogram (top) and interpreted velocity structure (bottom) for a shear wave refraction line shot in the southern North Sea (from Huws 1993).

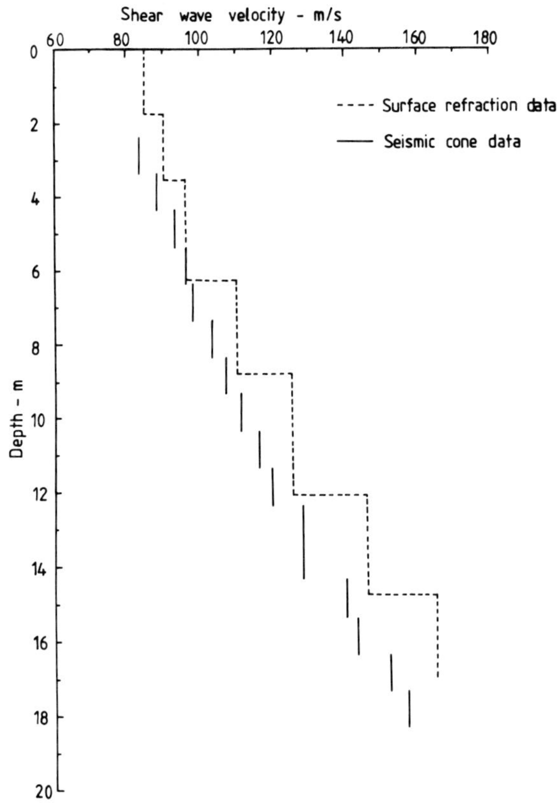

Fig. 5. Comparison of shear wave velocity depth profiles interpreted from surface seismic refraction shooting and measurements made with a seismic cone penetrometer.

where the velocity increases approximately linearly with depth. Furthermore, Hepton's (1989) plot of cone resistance against shear wave velocity for the same seismic CPT run (included here as Fig. 6) illustrates the potential for using the measured shear wave velocity for predicting other geotechnical properties and lends support to the argument for using seismic shear wave refraction shooting as a cost-effective technique for engineering site investigation. However, it should be pointed out that, for more complex geological sequences, while the correlation between the cone resistance and the seismic velocity obtained from shooting 'downhole' is expected to be maintained, the refraction versus 'downhole' velocity correlation will start to break down as the refraction interpretation cannot accommodate velocity inversions. Under these circumstances the surface refraction approach can at best be used as a reconnaissance technique to provide a general impression of the velocity gradient structure.

In a separate study, Finn *et al.* (1989) and Hunter *et al.* (1991) have reached similar conclusions to those cited above. Their work, which has concentrated in the area around the Fraser River, British Columbia, and has been essentially land-based, has provided the impetus for applying the UWB marine shear wave refraction methodology to the stability evaluation of the subaqueous portion of the Fraser River delta (see section on Liquefaction studies).

Laboratory studies

While accurate and appropriate *in situ* testing will clearly provide the more representative data on seafloor sediment properties, the marine environment is a particularly challenging one for the geotechnical specialist and there remains much to be gained from controlled laboratory testing.

Resonant column. At UWB two different resonant column devices are currently being used to provide fundamental sediment data in connection with a number of applications. Both devices

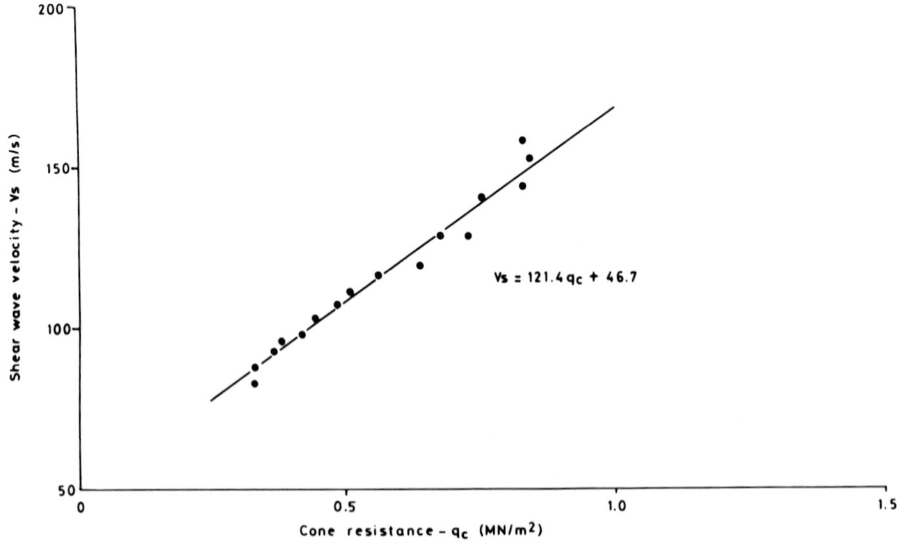

Fig.6. Comparison of shear wave velocity and cone resistance measured during penetration testing (from Hepton 1989).

can be used in conventional resonance mode to calculate shear stiffness and damping ratios under a range of stress–strain conditions, and both can also provide additional information on the very low-strain dynamic response via piezoelectric transducers which have been inserted into the end platens of the apparatus.

Over a period of several years a portfolio of information has been built up for a variety of sediments and a range of applications. In the future it is anticipated that resonant column data will be used to further enhance the interpretation of *in situ* shear wave velocity data, particularly in relation to their use for earthquake loading and mine burial predictions.

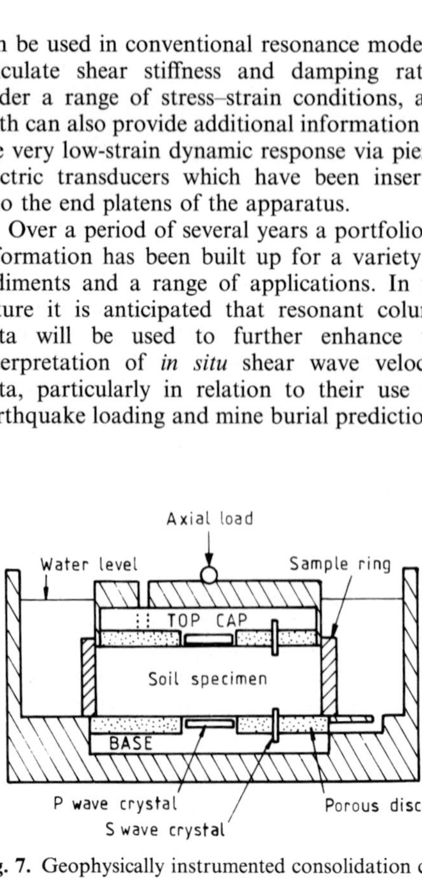

Fig. 7. Geophysically instrumented consolidation cell.

Geophysically instrumented triaxial and consolidation cells. Figures 7 and 8, which show schematic diagrams of the UWB geophysically instrumented test cells, are included primarily to illustrate the geophysical components (P- and S-wave transducers) of the UWB modified cells. Seismic velocities are obtained by measuring the elapsed time for waves propagating over known transducer separations, and these are converted to elastic moduli using information on sample density gained from the standard test. The moduli are typically referred to as G_{max} or E_{max}, representing strain levels of the order of 10%. Since its conception in the early 1980s (Schultheiss 1981), the geophysically instrumented triaxial apparatus has been installed in a number of specialist research and commercial laboratories, finding use in a range of applications. At UWB, its current role is in providing data to help define the likely liquefaction susceptibility of seafloor sediments (see later).

Continental shelf investigations

Land Ocean Interaction Studies

UWB's role within the NERC LOIS programme was primarily a reconnaissance task: the Geophysics Group was invited to undertake a series of high-resolution geophysical surveys in the Tees and Humber estuaries (east coast UK), the

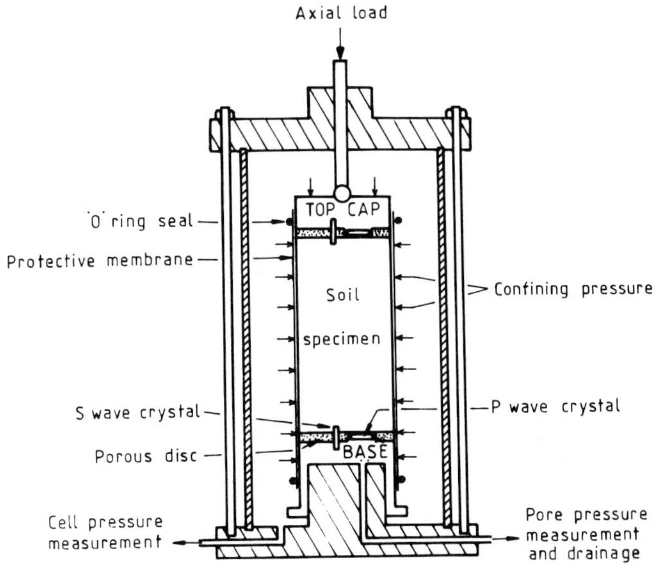

Fig. 8. Geophysically instrumented triaxial cell.

data to be used to help specify locations for geological sampling boreholes. Through these surveys, UWB were also able to advance shallow-water geophysical survey methodologies through demonstrating that, given appropriate instrumentation, it is possible to obtain high quality analogue and multi-channel digital seismic reflection data in less than 2 m of water (see Fig. 9; after Richardson *et al.* 1993). Furthermore, by applying land-based seismic reflection methodologies in the intertidal zone at low water, it was also demonstrated that given favourable conditions it is possible to extend marine surveys even further inshore to provide an interpretation of the 'problematic' interface zone between land and sea.

Gassy sediment studies

Geophysical techniques hold significant potential for investigating materials which are difficult to sample, physically, and probably most particularly for investigating those sediments which suffer extreme disturbance on sampling/removal from the seabed environment. Gas-charged sediments fall into this category. However, while analogue subbottom seismic reflection profiling is accepted as a reliable tool for identifying 'gas-charged' sediments, very little progress has been made beyond this towards physical quantification. In an attempt to improve on this situation, UWB are concentrating efforts on digital data processing, with seismic attribute analysis being investigated as a more quantitative approach to gas signature recognition. Examples of processed data from the Irish Sea, presented in Davis *et al.* (1994), illustrate advanced interpretations based on instantaneous frequency and amplitude envelope sections. Areas of acoustic turbidity, which on analogue sections are generally used in a purely qualitative sense to infer the presence of gas, on the digital record appear to be frequency significant, being predominantly high frequency; this is probably a function of bubble resonance effects. Amplitude envelope sections, which are essentially displays of reflectivity contrast, also appear suited to gas signature enhancement and recognition of gas-induced features; in several instances 'bright spots', similar to those seen on hydrocarbon exploration sections, were observed in association with shallow gas.

In a separate study centred around the southern Baltic (as part of the US Navy's CBBL programme), similar processing routines and interpretational approaches are being applied in an experimental study of environmental processes and physical property interactions for the soft gassy mud of Eckernfoerde Bay (Davis *et al.* 1994). Particularly interesting are the reflectivity contrasts observed around gas-associated features such as 'pockmarks', and those observed around 'acoustic windows', features which, in the case of Eckernfoerde Bay, interrupt the otherwise laterally continuous gas blanket (see Davis *et al.* 1995).

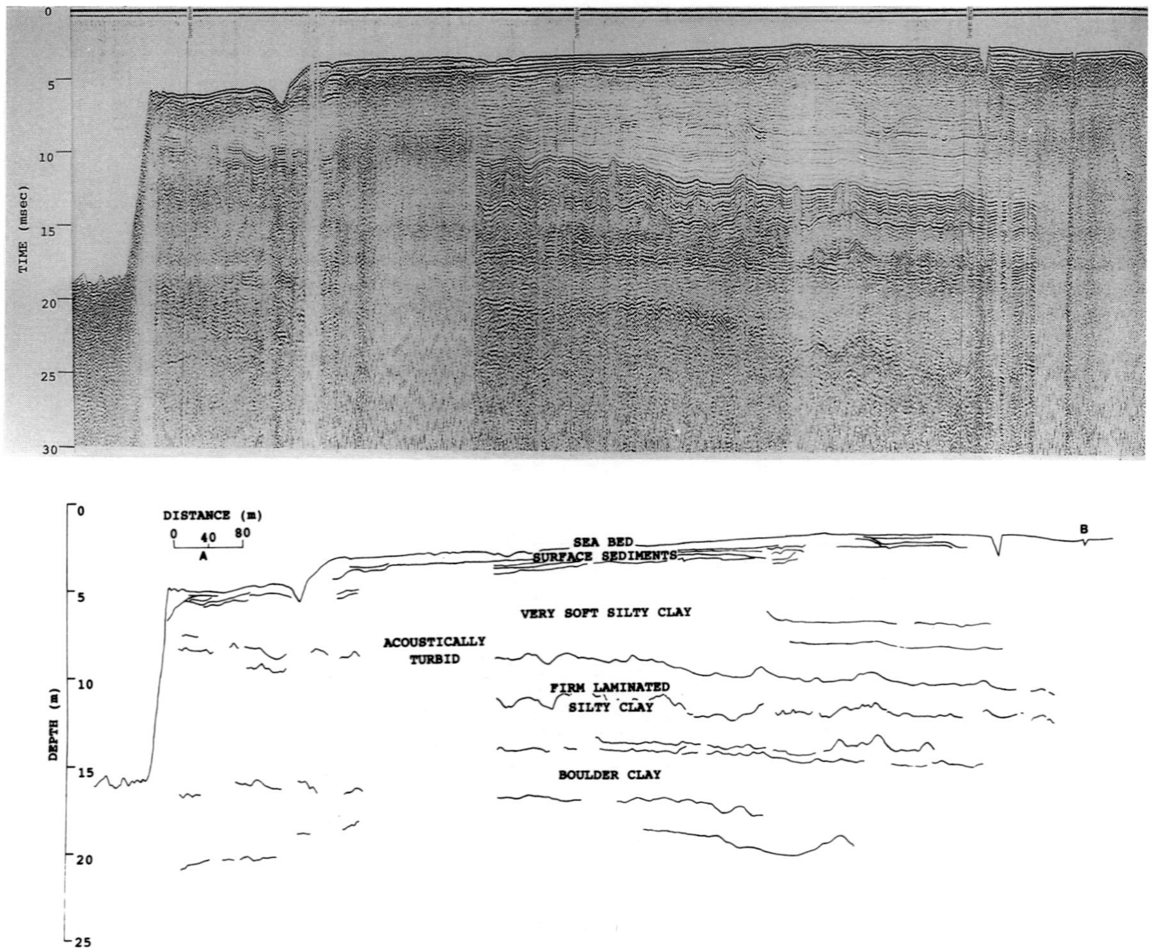

Fig. 9. Example analogue record and record interpretation for seismic subbottom profiler data collected with the IKB Seistec in the Tees estuary (from Richardson *et al.* 1993).

Sediment classification studies

The UWB 'magic carpet' has, since its development as a research tool, been used to investigate a number of sites for a range of specific purposes, one important purpose being physical characterization of seafloor sediments. The data presented in Fig. 10 (from Davis *et al.* 1991) were collected with the system during initial sea trials carried out off the North Wales coast. The geophysical sledge used for that particular survey was also fitted with an electrical resistivity device (see Fig. 2), thus allowing simultaneous measurement of seismic shear wave velocity and electrical formation factor. In the event, it also became possible to analyse the seismic records for compressional wave velocity (the electromagnetic hammer source generating a small quantity of compressional as well as shear energy). The velocity data shown in the figure relate to the upper 1–2 m of seabed sediment and indicate that the shear wave is sensitive to subtle variations in seabed properties, which are presumably controlled to a certain extent by the grain size distribution. These initial conclusions were subsequently reinforced by measurements made in the Firth of Clyde (Huws *et al.* 1991).

The past two years have seen increased activity and significant advances in UWB's sediment classification research, primarily as a result of involvement in major multi-disciplinary research programmes such as the US Navy's CBBL programme. As mentioned previously, the CBBL project has provided a unique opportunity both to compare UWB's site classification approaches with other classification methodologies, e.g. through comparison of UWB's digital boomer data with data from the Naval Research Laboratory's ASCS and the Chirp Sonar (operated by Florida Atlantic University), or comparison of 'magic carpet' shear wave velocity data with interpretations based on analysis of interface waves or *in situ* probe data (Davis *et al.* 1995), and to groundtruth the geophysical interpretation through access to the wealth of information provided by other researchers experimenting in the same areas. Within the CBBL programme a range of sedimentary environments are being investigated to provide a better understanding of sedimentary processes and their impact on physical properties. Using the UWB approach, it is proving possible to provide important information on sub-metre to kilometre sized features and especially on their spatial variability.

Liquefaction studies

There is growing evidence to support the use of seismic shear wave velocity as a predictor of sediment liquefaction potential. In collaboration with the Geological Survey of Canada (GSC), UWB are currently investigating an integrated

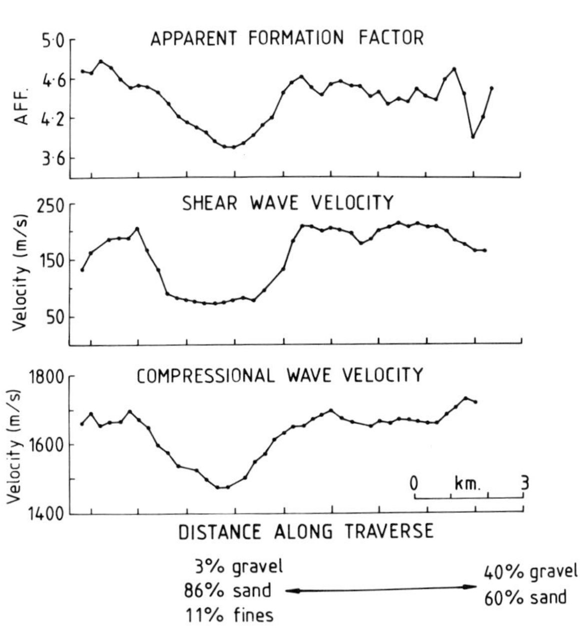

Fig. 10. Example data set for a traverse completed with the 'magic carpet' system.

shear wave approach (*in situ* and laboratory methods) to seabed stability evaluation for two sites where seismically induced liquefaction has been recognized as a possible sediment failure mechanism. On the Canadian west coast, measurements have been concentrated around the Fraser Delta, British Columbia (offshore of the densely populated Vancouver area); on the eastern shelf, the area of interest is the Hibernia site on the Grand Banks of Newfoundland, the proposed location for the installation of an offshore oil platform.

Based on *in situ* shear wave velocity data collected with the 'magic carpet' system, and using approaches similar to those of Finn *et al.* (1989) and Hunter *et al.* (1991), where predictions are based on empirical relations developed through analysis of known earthquake response, it would appear that significant volumes of sediment on the subaqueous portion of the Fraser Delta might be prone to liquefy under the influence of a large-magnitude earthquake, shear wave velocities of the uppermost few metres of the sediment column appearing to fall below the 'critical' value denoted as that separating liquefiable from non-liquefiable material for a magnitude 6.5 earthquake (Davis *et al.* 1993).

To further substantiate the geophysical prediction, a parallel laboratory study has been undertaken at UWB to determine the relationship between shear wave velocity, sediment state, and liquefaction potential for a range of non-cohesive sediments (Pyrah *et al.* 1995). The conventional laboratory sediment mechanics approach involves an investigation of void ratio/effective stress relationships, and provides an indication of the sediment's likely behaviour – contractive or dilative – when subjected to cyclic shear stresses such as those imposed by earthquake loading. In the tests carried out at UWB, the shear wave velocity is also monitored to establish velocity–void ratio–effective stress relations, leading to behavioural predictions based on shear wave velocity. In practical terms this provides the means to extrapolate the laboratory findings to the field situation: given a knowledge of the laboratory behaviour and a measure of the *in situ* shear wave velocity (provided by the 'magic carpet'), together with information on grain characteristics (from core/grab sampling) and static stress regime (calculated from water depth and depth within the sediment column), it becomes possible to provide an estimate of the sediment's stability under cyclic load conditions. For the specific case of the Fraser Delta, the initial findings, i.e. those based purely on field data collected with the 'magic carpet', were further substantiated by laboratory data, confirming the sediment's likely tendency to liquefy under earthquake loading. In practical terms, at present, shear wave velocity measurement probably offers the only realistic and cost-effective approach to reconnaissance surveying for marine seismic hazard (liquefaction) evaluation.

Summary and conclusions

In recent years significant advances have been made in the field of engineering geophysics, and particularly in the *in situ* measurement of seabed sediment properties. In this paper the author has attempted to highlight some of these developments and review the state of the art.

The paper has inevitably concentrated primarily on seismic techniques, as probably the most significant developments have come in this area. However, the reader should be aware that there are other geophysical parameters (e.g. electrical resistivity, thermal conductivity), methodologies and approaches (e.g. geoacoustic model predictions) which can provide additional useful information on sediment properties.

Thus, while the review is by no means exhaustive, it is hoped that the content will be sufficiently detailed to provide the engineer and geologist alike with an overall appreciation of the subject. In this way it may be possible to promote the use of geophysics as a quantitative tool for use in offshore site investigation alongside more traditional geotechnical methods.

References

ADDO, K. O. & ROBERTSON, P. K. 1992. Shear wave velocity measurements using Rayleigh surface waves. *Canadian Geotechnical Journal*, **29**(4), 558–568.

BENNELL, J. D. & TAYLOR SMITH, D. 1991. A review of laboratory shear wave techniques and attenuation measurements with particular reference to the resonant column. *In*: HOVEM, J. M., RICHARDSON, M. D. & STOLL, R. D. (eds) *Shear Waves in Marine Sediments*. Kluwer, Dordrecht, 83–94.

CHIVERS, R. C., EMERSON, N. & BURNS, D. 1990. New acoustic processing for underway surveying. *Hydrographic Journal*, **56**(April).

DAVIS, A. M. 1982. *Dynamic–static Properties of Sedimentary Materials*. PhD Thesis, University of Wales.

—— & BENNELL, J. D. 1988. Resolving overburden characteristics via shear wave propagation. *Bulletin of the Australian Society of Exploration Geophysicists*, **19**, 41–44.

——, CHRISTIAN, H., HUWS, D. G., PARROTT, R. & BARRIE, V. 1993. Seabed stability evaluation studies using shear wave propagation phenomena. *Proceedings of the Institute of Acoustics*, **15**(2), 25–32.

——, HAYNES, R. & HUWS, D. G. 1994. Geophysical approaches to determining the geotechnical charteristics of sea-floor sediments: Acquisition and analysis of shear wave, electrical resistivity, and digital seismic subbottom profiler data. *In*: WEVER, T. (ed.) *Proceedings of the Gassy Mud Workshop*. Forschungsanstaslt der Bundeswehr fur Wasserschall-und Geophysik, Kiel, Germany, Report **14**, 59–64.

——, HUWS, D. G. & BENNELL, J. D. 1991. Seafloor shear wave velocity data acquisition: procedures and pitfalls. *In*: HOVEM, J. M., RICHARDSON, M. D. & STOLL, R. D. (eds) *Shear Waves in Marine Sediments*. Kluwer, Dordrecht, 329–336.

——, —— & HAYNES, R. 1995. Geophysical groundtruthing experiments in Eckernfoerde Bay. *Geo-Marine Letters*, in press.

FINN, W. D., WOELLER, D .J., DAVIES, M. P., LUTERNAUER, J. L., HUNTER, J. A. & PULLEN, S. E. 1989. *New Approaches for Assessing Liquefaction Potential of the Fraser Delta. British Columbia.* Current Research, Paper **1989-1E**, Geological Survey of Canada, 221–231.

GARDENER, R. 1992. Seismic refraction as a tool in the evaluation of rock quality for dredging and engineering purposes: case studies. *In*: *Eurock*. Telford, London, 153–158.

HARDIN, B. O. & DRNEVICH, V. P. 1972*a*. Shear modulus and damping in soils: measurement and parameter effects. *Journal of the Soil Mechanics and Foundations Division. American Society of Civil Engineers*, **98**(SM6), 603–624.

—— & —— 1972*b*. Shear modulus and damping in soils: design equations and curves. *Journal of the Soil Mechanics and Foundations Division. American Society of Civil Engineers*, **98**(SM7), 667–692.

HAYNES, R., BENNELL, J. D., DAVIS, A. M. & REYNOLDS, J. M. 1993*a*. Advantages offered by the routine acquisition of digital high-resolution subbottom profiling data. *Proceedings of the Institute of Acoustics*, **15**(2), 165–172.

——, DAVIS, A. M., REYNOLDS, J. M. & TAYLOR, D. I. 1993*b*. The extraction of geotechnical data from high-resolution seismic reflection data. *In*: *Offshore Site Investigation and Foundation Behaviour*. Society for Underwater Technology, Kluwer, Dordrecht, **28**, 215–228.

HEPTON, P. 1989. *Shear Wave Velocity Measurements during Penetration Testing*. PhD Thesis, University of Wales.

HUNTER, J. A., WOELLER, D. J. & LUTERNAUER, J. L. 1991. *Comparison of Surface, Borehole and Seismic Cone Penetrometer Methods of Determining the Shallow Shear Wave Velocity Structure in the Fraser River Delta. British Columbia.* Current Research, Geological Survey of Canada, Paper **1991-lA**, 23–26.

HUWS, D. G. 1993. *Measuring and Modelling the in situ Physical Properties of Marine Sediments*. PhD Thesis, University of Wales.

——, DAVIS, A. M. & BENNELL, D. G. 1991. Mapping of the sea bed via in situ shear wave (SH) velocities. *In*: HOVEM, J. M., RICHARDSON, M. D. & STOLL, R. D. (eds) *Shear Waves in Marine Sediments*. Kluwer, Dordrecht, 3–12.

KING, M. S. 1966. Wave velocities in rocks as a function of changes in overburden pressure and pore fluid saturants. *Geophysics*, **31**, 50.

LAMBERT, D. N., CRANFORD, J. C. & WALTER, D. 1993. Development of a high resolution acoustic seafloor classification survey system. *Proceedings of the Institute of Acoustics*, **15**(2), 149–156.

MCGEE, T., DAVIS, A. M., ANDERSON, H. & VERBEEK, N. 1992. High-resolution marine seismic source signatures. *In*: WEYDERT, M. (ed.) *European Conference on Underwater Acoustics*. Elsevier, Amsterdam, 639–643.

PYRAH, J., DAVIS, A. & HUWS, D. 1995. A combined geotechnical geophysical method for the prediction of flow liquefaction with special reference to the Fraser River delta, British Columbia. *In*: MAUND, J. G. *et al.* (ed.) *Geohazards and Engineering Geology. International Hazard Assessment and Mitigation*. Special Issue of the Engineering Group of the Geological Society of London, September 1995, 69–78.

RICHARDSON, I. R., SIMPKIN, P. G., DAVIS, A. M., BENNELL, J. D. & BUTCHER, J. A. 1993. Some developments in high resolution seismic reflection profiling relevant to investigations of very shallow water areas. *Proceedings of the Institute of Acoustics*, **15**(2), 253–259.

RICHARDSON, M. D. 1994. Investigating the coastal benthic boundary layer. *EOS Transactions*, **75**(17), 201–206.

—— & BRIGGS, K. B. 1993. On the use of acoustic impedance values to determine sediment properties. *Proceedings of the Institute of Acoustics*, **15**(2), 15–24.

ROBERTSON, P. K., CAMPANELLA, R. G., GILLESPIE, D. & RICE, A. 1986. Seismic CPT to measure in situ shear wave velocity. *Journal of the Geotechnical Engineering Divisions American Society of Civil Engineers*, **112**, 791–803.

SCHOCK, S. G, LEBLANC, L. R & MEYER L. A. 1989. Chirp subbottom profiler for quantitative sediment analysis. *Geophysics*, **54**(4), 445–450.

SCHULTHEISS, P. J. 1981. Simultaneous measurement of P and S wave velocities during conventional laboratory soil testing procedures. *Marine Geotechnology*, **4**(4), 343–367.

SCHWARZ, S. D. & CONWELL, F. R. 1974. A technique for the in situ measurement of shear wave velocities (Vs) for deep marine foundations. *Proceedings of Offshore Technology Conference* **2014**, Houston, Texas, 755–762.

SHIRLEY, D. J. 1977. An improved shear wave transducer. *Journal of the Acoustics Society of America*, **62**, 1028–1032.

SIMPKIN, P. G. & LEWIS, J. F. 1992. Seiszmic profiling in shallow water using a large aperture. line and cone array. *Proceedings of the Marine Technology Society*, Washington DC, USA.

STOLL, R. D. 1989. *Sediment Acoustics*. Lecture Notes in Earth Sciences, **26**, Springer Berlin.
——1992. Shear waves in marine sediments – bridging the gap from theory to field applications. *In*: HOVEM, J. M., RICHARDSON, M. D. & STOLL, R. D. (eds) *Shear Waves in Marine Sediments*, Kluwer, Dordrecht, 3–12.
——, BRYAN, G. M., FLOOD, R., CHAYES, D. & MANLEY, P. 1988. Shallow seismic experiments using shear waves. *Journal of the Acoustics Society of America*, **83**(1), 93–102.
STUMPEL, H., KAHLER, S., MEISSNER, R. & MILKEREIT, B. 1984. The use of seismic shear waves and compressional waves for lithological problems of shallow sediments. *Geophysical Prospecting*, **32**, 662–675.
TAYLOR SMITH, D. 1986. Geotechnical characteristics of the sea bed related to seismo-acoustics. *In*: AKAL, T. & BERKSON, J. M. (eds) *Ocean Seismo-Acoustics*. Plenum, New York, 483–500.
——1993. Geotechnical–geophysical predictions. *Proceedings of the Institute of Acoustics*, **15**(2), 1–14.
TELFORD, W. M., GELDART, L. P., SHERIFF, R. E. & KEYS, D. A. 1976. *Applied Geophysics*. Cambridge University Press, Cambridge.
TRABANT, P. K. 1984. *Applied High-resolution Geophysical Methods: Offshore Geoengineering Hazards*. Reidel, Boston.

Index